NONTHERMAL LIGHT-BASED TECHNOLOGIES IN FOOD PROCESSING

Innovations in Agricultural and Biological Engineering

NONTHERMAL LIGHT-BASED TECHNOLOGIES IN FOOD PROCESSING

Edited by

C. K. Sunil, PhD

Preeti Birwal, PhD

Megh R. Goyal, PhD, PE

R. Mahendran, PhD

AAP | APPLE ACADEMIC PRESS

First edition published 2025

Apple Academic Press Inc.
1265 Goldenrod Circle, NE,
Palm Bay, FL 32905 USA

760 Laurentian Drive, Unit 19,
Burlington, ON L7N 0A4, CANADA

CRC Press
6000 Broken Sound Parkway NW,
Suite 300, Boca Raton, FL 33487-2742 USA

4 Park Square, Milton Park,
Abingdon, Oxon, OX14 4RN UK

Library and Archives Canada Cataloguing in Publication

CIP data on file with Canada Library and Archives

Library of Congress Cataloging-in-Publication Data

CIP data on file with US Library of Congress

ISBN: 978-1-77491-750-3 (hbk)
ISBN: 978-1-77491-751-0 (pbk)
ISBN: 978-1-00350-374-3 (ebk)

DEDICATION

To our esteemed mentor

Professor Megh R. Goyal, PhD, P.E.

Senior Technical Editor-in-Chief (Agricultural and Biological Engineering)
Apple Academic Press Inc., (Copartner CRC Press/Taylor and Francis Group)

For your leadership, motivation, and guidance have been instrumental in shaping this book volume.

For your invaluable contributions to the field of science and technology research.

For being a guiding light, providing us with profound impact both personally and academically, throughout our professional careers.

For your mentorship at every stage of our professional and personal development is truly priceless.

For your unwavering support and dedication.

ABOUT THE BOOK SERIES: INNOVATIONS IN AGRICULTURAL AND BIOLOGICAL ENGINEERING

Under this book series, Apple Academic Press Inc. is publishing book volumes over a span of 8–10 years in the specialty areas defined by the American Society of Agricultural and Biological Engineers (<asabe.org>). Apple Academic Press Inc. aims to be a principal source of books in agricultural and biological engineering. We welcome book proposals from readers in areas of their expertise.

The mission of this series is to provide knowledge and techniques for agricultural and biological engineers (ABEs). The book series offers high-quality reference and academic content on agricultural and biological engineering (ABE) that is accessible to academicians, researchers, scientists, university faculty and university-level students, and professionals around the world.

Agricultural and biological engineers ensure that the world has the necessities of life, including safe and plentiful food, clean air and water, renewable fuel and energy, safe working conditions, and a healthy environment by employing knowledge and expertise of the sciences, both pure and applied, and engineering principles. Biological engineering applies engineering practices to problems and opportunities presented by living things and the natural environment in agriculture.

ABE embraces a variety of the following specialty areas (<asabe.org>): aquaculture engineering, biological engineering, energy, farm machinery and power engineering, food and process engineering, forest engineering, information and electrical technologies, soil and water conservation engineering, natural resources engineering, nursery and greenhouse engineering, safety and health, and structures and environment.

For this book series, we welcome chapters on the following specialty areas (but not limited to):

1. Academia to industry to end-user loop in agricultural engineering
2. Agricultural mechanization
3. Aquaculture engineering
4. Biological engineering in agriculture

5. Biotechnology applications in agricultural engineering
6. Energy source engineering
7. Food and bioprocess engineering
8. Food and water security
9. Forest engineering
10. Hill land agriculture
11. Human factors in engineering
12. Impact of global warming and climatic change on agriculture economy
13. Information and electrical technologies
14. Irrigation and drainage engineering
15. Nanotechnology applications in agricultural engineering
16. Natural resources engineering
17. Nursery and greenhouse engineering
18. Potential of phytochemicals from agricultural and wild plants for human health
19. Power systems and machinery design
20. GPS and remote sensing potential in agricultural engineering
21. Robot engineering and drones in agriculture
22. Simulation and computer modeling
23. Smart engineering applications in agriculture
24. Soil and water engineering
25. Micro irrigation engineering
26. Structures and environment engineering
27. Waste management and recycling
28. Rural electrification
29. Sanitary engineering
30. Circular biosystems in agriculture
31. Farm-to-fork technologies in agriculture
32. Any other focus areas

For more information on this series, readers may contact:

Megh R. Goyal, PhD, PE
Senior Technical Editor-in-Chief:
Book Series: Innovations in Agricultural and Biological Engineering
E-mail: goyalmegh@gmail.com

OTHER BOOKS ON AGRICULTURAL AND BIOLOGICAL ENGINEERING BY APPLE ACADEMIC PRESS, INC.

Management of Drip/Trickle or Micro Irrigation
Megh R. Goyal, PhD, PE, Senior Editor-in-Chief

Evapotranspiration: Principles and Applications for Water Management
Megh R. Goyal, PhD, P.E. and Eric W. Harmsen, Editors

Book Series: Research Advances in Sustainable Micro Irrigation
Senior Editor-in-Chief: Megh R. Goyal, PhD, PE

- **Volume 1:** Sustainable Micro Irrigation: Principles and Practices
- **Volume 2:** Sustainable Practices in Surface and Subsurface Micro Irrigation
- **Volume 3:** Sustainable Micro Irrigation Management for Trees and Vines
- **Volume 4:** Management, Performance, and Applications of Micro Irrigation Systems
- **Volume 5:** Applications of Furrow and Micro Irrigation in Arid and Semi-Arid Regions
- **Volume 6:** Best Management Practices for Drip Irrigated Crops
- **Volume 7:** Closed Circuit Micro Irrigation Design: Theory and Applications
- **Volume 8:** Wastewater Management for Irrigation: Principles and Practices
- **Volume 9:** Water and Fertigation Management in Micro Irrigation
- **Volume 10:** Innovation in Micro Irrigation Technology

Book Series: Innovations and Challenges in Micro Irrigation
Senior Editor-in-Chief: Megh R. Goyal, PhD, PE

- Engineering Interventions in Sustainable Trickle Irrigation: Water Requirements, Uniformity, Fertigation, and Crop Performance
- Management Strategies for Water Use Efficiency and Micro Irrigated Crops: Principles, Practices, and Performance

- Micro Irrigation Engineering for Horticultural Crops: Policy Options, Scheduling, and Design
- Micro Irrigation Management: Technological Advances and Their Applications
- Micro Irrigation Scheduling and Practices
- Performance Evaluation of Micro Irrigation Management: Principles and Practices
- Potential of Solar Energy and Emerging Technologies in Sustainable Micro Irrigation
- Principles and Management of Clogging in Micro Irrigation
- Sustainable Micro Irrigation Design Systems for Agricultural Crops: Methods and Practices

Book Series: Innovations in Agricultural and Biological Engineering
Senior Editor-in-Chief: Megh R. Goyal, PhD, PE

- Advanced Research Methods in Food Processing Technologies
- Advances in Food Process Engineering: Novel Processing, Preservation and Decontamination of Foods
- Advances in Green and Sustainable Nanomaterials: Applications in Energy, Biomedicine, Agriculture, and Environmental Science
- Advances in Sustainable Food Packaging Technology
- Analytical Methods for Milk and Milk Products, 3-volume set:
    - Volume 1: Sampling Methods, Chemical and Compositional Analysis
    - Volume 2: Physicochemical Analysis of Concentrated, Coagulated and Fermented Products
    - Volume 3: Microbiological Analysis, Isolation, and Characterization
- Applications of Supercritical Fluid Extraction in Food Processing
- Biological and Chemical Hazards in Food and Food Products: Prevention, Practices, and Management
- Bioremediation and Phytoremediation Technologies in Sustainable Soil Management, 4-volume set:
    - Volume 1: Fundamental Aspects and Contaminated Sites
    - Volume 2: Microbial Approaches and Recent Trends
    - Volume 3: Inventive Techniques, Research Methods, and Case Studies
    - Volume 4: Degradation of Pesticides and Polychlorinated Biphenyls

- Dairy Engineering: Advanced Technologies and Their Applications
- Developing Technologies in Food Science: Status, Applications, and Challenges
- Emerging Technologies in Agricultural Engineering
- Engineering Interventions in Agricultural Processing
- Engineering Interventions in Foods and Plants
- Engineering Practices for Agricultural Production and Water Conservation: An Interdisciplinary Approach
- Engineering Practices for Management of Soil Salinity: Agricultural, Physiological, and Adaptive Approaches
- Engineering Practices for Milk Products: Dairyceuticals, Novel Technologies, and Quality
- Enzyme Inactivation in Food Processing: Technologies, Materials, and Applications
- Field Practices for Wastewater Use in Agriculture: Future Trends and Use of Biological Systems
- Flood Assessment: Modeling and Parameterization
- Food Engineering: Emerging Issues, Modeling, and Applications
- Food Process Engineering: Emerging Trends in Research and Their Applications
- Food Processing and Preservation Technology: Advances, Methods, and Applications
- Food Technology: Applied Research and Production Techniques
- Functional Dairy Ingredients and Nutraceuticals: Physicochemical, Technological, and Therapeutic Aspects
- Handbook of Research on Food Processing and Preservation Technologies, 5-volume set:
    - Volume 1: Nonthermal and Innovative Food Processing Methods
    - Volume 2: Nonthermal Food Preservation and Novel Processing Strategies
    - Volume 3: Computer-Aided Food Processing and Quality Evaluation Techniques
    - Volume 4: Design and Development of Specific Foods, Packaging Systems, and Food Safety
    - Volume 5: Emerging Techniques for Food Processing, Quality, and Safety Assurance
- Modeling Methods and Practices in Soil and Water Engineering
- Nanotechnology and Nanomaterial Applications in Food, Health, and Biomedical Sciences

- Nanotechnology Applications in Agricultural and Bioprocess Engineering: Farm to Table
- Nanotechnology Applications in Dairy Science: Packaging, Processing, and Preservation
- Nanotechnology Horizons in Food Process Engineering, 3-volume set:
    - o Volume 1: Food Preservation, Food Packaging and Sustainable Agriculture
    - o Volume 2: Scope, Biomaterials, and Human Health
    - o Volume 3: Trends, Nanomaterials, and Food Delivery
- Nonthermal Plasma Technology for Food and Food Products
- Nonthermal Pulsed Electric Field Technology for the Food Industry
- Novel and Alternative Methods in Food Processing: Biotechnological, Physicochemical, and Mathematical Approaches
- Novel Dairy Processing Technologies: Techniques, Management, and Energy Conservation
- Novel Processing Methods for Plant-Based Health Foods: Extraction, Encapsulation and Health Benefits of Bioactive Compounds
- Novel Strategies to Improve Shelf-Life and Quality of Foods: Quality, Safety, and Health Aspects
- Ohmic Heating Technology for Processing of Foods and Food Products: Design, Food Quality, and Applications
  Editors: Preeti Birwal, PhD, Megh R. Goyal, PhD, and
  Baldev Singh Kalsi, PhD
- Phytochemicals and Medicinal Plants in Food Design: Strategies and Technologies for Improved Healthcare
- Processing of Fruits and Vegetables: From Farm to Fork
- Processing Technologies for Milk and Milk Products: Methods, Applications, and Energy Usage
- Quality Control in Fruit and Vegetable Processing: Methods and Strategies
- Scientific and Technical Terms in Bioengineering and Biological Engineering
- Soil and Water Engineering: Principles and Applications of Modeling
- Soil Salinity Management in Agriculture: Technological Advances and Applications
- State-of-the-Art Technologies in Food Science: Human Health, Emerging Issues and Specialty Topics

- Sustainable and Functional Foods from Plants: Health Impact, Bioactive Compounds, and Production Technologies
- Sustainable Green Nanomaterials: Synthesis, Characterization, and Engineering Applications
- Sustainable Biological Systems for Agriculture: Emerging Issues in Nanotechnology, Biofertilizers, Wastewater, and Farm Machines
- Sustainable Nanomaterials for Biomedical Engineering: Impacts, Challenges, and Future Prospects
- Sustainable Nanomaterials for Biosystems Engineering: Trends in Renewable Energy, Environment, and Agriculture
- Sustainable Use of Wastewater in Agriculture: Applications, Methods, and Technology
- Technological Interventions in Dairy Science: Innovative Approaches in Processing, Preservation, and Analysis of Milk Products
- Technological Interventions in Management of Irrigated Agriculture
- Technological Interventions in the Processing of Fruits and Vegetables
- Technological Processes for Marine Foods, From Water to Fork: Bioactive Compounds, Industrial Applications, and Genomics
- The Chemistry of Milk and Milk Products: Physicochemical Properties, Therapeutic Characteristics, and Processing Methods

ABOUT THE EDITORS

C. K. Sunil, PhD
Associate Professor (Food Engineering), Department of Food Engineering and Centre of Excellence for Grain Sciences, National Institute of Food Technology, Entrepreneurship and Management-Thanjavur (NIFTEM-T), Tamil Nadu, India

C. K. Sunil, PhD, is currently working as Associate Professor in the Department of Food Engineering and Centre of Excellence for Grain Sciences at the National Institute of Food Technology, Entrepreneurship and Management-Thanjavur (NIFTEM-T), Tamil Nadu, India. He completed his BTech and master's in Agricultural Engineering from the University of Agricultural Sciences, Bangalore. He obtained his PhD in Food Process Engineering from IIFPT, Thanjavur. He was awarded the University Gold Medal for outstanding performance in the bachelor's degree program. With over 20 years of teaching, research, and industry experience, before joining NIFTEM-Thanjavur, he worked as Manager-QA at the SAFAL Market (Unit of MDFVPL, NDDB), Bangalore.

Sunil has published scientific papers in peer-reviewed journals and book chapters. Currently, he is working on processing of millet, design and development of processing machinery, application of novel thermal and nonthermal technologies, and valorization of processing waste. He teaches various areas of food processing/food engineering, including unit operations, introduction to food processing, advances in food processing, processing of cereals, pulses, and oilseeds, thermal and nonthermal processing of foods, etc. He serves as a review editor for *Frontiers in Sustainable Food Systems* and *Frontiers in Food Science and Technology-Food Packaging and Preservation*. Additionally, he acts as a peer reviewer for international journals and is a member of various professional bodies.

Preeti Birwal, PhD
Scientist (Processing and Food Eng.)
Department of Processing and Food Engineering, College of Agricultural Engineering and Technology, Punjab Agricultural University, Ludhiana, Punjab, India

Preeti Birwal, PhD, is working as a Scientist (Processing and Food Eng.) in the Department of Processing and Food Engineering, College of Agricultural Engineering and Technology, Punjab Agricultural University, Ludhiana, Punjab, India.

She holds a BSc (2012) in Dairy Technology from ICAR-National Dairy Research Institute (NDRI), Karnal, an MSc (2014) in Food Process Engineering and Management from NIFTEM, Haryana; and a PhD (Dairy Engineering) from ICAR-NDRI, Bangalore. She is the recipient of MHRD (2008), Nestle India (2009), GATE (2012–14), and UGC-RGN fellowships (2014–18).

She is currently working in the area of nonthermal food preservation, fermented beverages, food packaging, and technology of millet-based beer. She has served Jain Deemed to be University, Bangalore, as a member of the board of examiners and placements. She is advising several MTech. scholars in Food technology. She has participated in several national and international conferences and seminars. She has delivered lectures as a resource person on doubling the farmer's income through dairy technology in a training sponsored by the Directorate of Extension, Ministry of Agriculture and Farmers Welfare, Government of India. She has received an Outstanding Reviewer of the Month award. She has completed AUTOCAD 2D and 3D certification.

She has 18 research papers, one edited book, 5 book chapters, about 28 popular articles, and five conference papers and 56 abstracts, two editorial opinions to her credit. She has successfully guided five postgraduate students in their dissertation work. She is serving as an external examiner for various Indian state agricultural universities. She is also serving as editor and reviewer of several journals. She is a life member of IDEA. Readers may contact her at: preetibirwal@gmail.com.

Megh R. Goyal, PhD, PE

Retired Professor in Agricultural and Biomedical Engineering, University of Puerto Rico, Mayaguez Campus; Senior Acquisitions Editor, Biomedical Engineering and Agricultural Science, Apple Academic Press, Inc.

Megh R. Goyal, PhD, PE, is, currently a retired professor of agricultural and biomedical engineering from the General Engineering Department at the College of Engineering at the University of Puerto Rico–Mayaguez Campus (UPRM); and Senior Acquisitions Editor and Senior Technical Editor-in-Chief for Agricultural and Biomedical Engineering for Apple Academic Press Inc.

During his professional career, he has worked as a Soil Conservation Inspector; Research Assistant at Haryana Agricultural University and Ohio State University; Research Agricultural Engineer/Professor at the Department of Agricultural Engineering of UPRM; and Professor of Agricultural and Biomedical Engineering in the General Engineering Department of UPRM. He spent a one-year sabbatical leave in 2002–2003 at the Biomedical Engineering Department of Florida International University, Miami, USA.

Dr. Goyal was the first agricultural engineer to receive the professional license in agricultural engineering from the College of Engineers and Surveyors of Puerto Rico. In 2005, he was proclaimed the "Father of Irrigation Engineering in Puerto Rico for the Twentieth Century" by the American Society of Agricultural and Biological Engineers, Puerto Rico Section, for his pioneering work on micro irrigation, evapotranspiration, agroclimatology, and soil and water engineering.

During his professional career of 52 years, he has received many awards, including Scientist of the Year, Membership Grand Prize for the American Society of Agricultural Engineers Campaign, Felix Castro Rodriguez Academic Excellence Award, Man of Drip Irrigation by the Mayor of Municipalities of Mayaguez/Caguas/Ponce and Senate/Secretary of Agriculture of ELA, Puerto Rico, and many others. He has been recognized as one of the experts "who rendered meritorious service for the development of [the] irrigation sector in India" by the Water Technology Centre of Tamil Nadu Agricultural University in Coimbatore, India, and ASABE who bestowed on him the 2018 Netafim Microirrigation Award.

Dr. Goyal has authored more than 200 journal articles and edited more than 100 books. AAP has published many of his books, including *Management of Drip/Trickle or Micro Irrigation; Evapotranspiration: Principles and Applications for Water Management;* ten-volume set on *Research Advances in Sustainable Micro Irrigation.* He has also authored the textbooks *Elements of Agroclimatology* (Spanish) by UNISARC, Colombia, and two *Bibliographies on Drip Irrigation.* Dr. Goyal has also developed several book series with AAP, including Innovations in Agricultural & Biological Engineering (with over 60 titles in the series to date), Innovations and Challenges in Micro Irrigation; and Innovations in Plant Science for Better Health: From Soil to Fork.

Dr. Goyal received his BSc degree in Engineering from Punjab Agricultural University, Ludhiana, India, and his MSc and PhD degrees from the Ohio State University, Columbus, Ohio, USA. He also earned a Master of Divinity degree from the Puerto Rico Evangelical Seminary, Hato Rey, Puerto Rico, USA.

Readers may contact him at goyalmegh@gmail.com.

R. Mahendran, PhD
Professor and Head, Centre of Excellence in Nonthermal Processing (CENTP)
National Institute of Food Technology, Entrepreneurship and
Management -Thanjavur (NIFTEM-T), India

R. Mahendran, PhD, is working as a Professor and Head of the Center of Excellence in Nonthermal Processing (CENTP) at the National Institute of Food Technology, Entrepreneurship, and Management–Thanjavur (NIFTEM-T).

Dr. R. Mahendran completed his BTech in Chemical Engineering and MTech in Industrial Biotechnology from SASTRA University, Thanjavur, Tamil Nadu, and PhD in Food Process Engineering from Tamil Nadu Agricultural University, Coimbatore, India. Before his association with NIFTEM-Thanjavur, he worked as a Production Officer in the Fermentation Department of M/s. Krebs Biochemicals and Industries Ltd. Nellore, Andhra Pradesh, and also worked as a Scientist in TIFAC/Department of Science and Technology, Government of India at New Delhi from 2006 to 2010.

Dr. Mahendran's research focus includes cold plasma applications on agri-food, food structure shape transformation, and novel nonthermal processing on food systems. He has authored over 100 publications in reputed journals, popular articles, and book chapters. He has been awarded the Dr. V. Subrahmanyan Best Scientist Award for the year 2021 for his outstanding research contributions. He is a fellow of the Institute of Food Science and Technology (IFST) and a fellow of the Institution of Engineers (IEI). Dr. Mahendran is a life member of various professional bodies including The Institution of Engineers, AFST(I), ISTE, ISC, IIChE, ISAE, and ISFE. He also serves as a guest editor, editorial board member, and reviewer for reputed food processing journals. Readers may reach him at: mahendran@niftem-t.ac.in.

CONTENTS

CONTRIBUTORS

J. Apoorva
Research Scholar, Department of Food Technology, FLAHS, Ramaiah University of Applied Sciences, Bangalore, Karnataka, India

K. A. Athmaselvi
Associate Professor and Head, Center of Excellence for Grain Science, National Institute of Food Technology and Entrepreneurship Management, Thanjavur, Tamil Nadu, India

Surendra Babu Ayenampudi
Assistant Professor, Department of Food Science and Technology, School of Agricultural Sciences, Malla Reddy University, Hyderabad, Telangana, India

Smruthi Bhat
PhD Research Scholar, Department of Food Technology, FLAHS, Ramaiah University of Applied Sciences, Bangalore, Karnataka, India

Rewati Raman Bhattarai
Lecturer, School of Molecular and Life Sciences, Faculty of Science and Engineering, Curtin University, Bentley, Western Australia, Australia

Malini Buvaneswaran
PhD Research Scholar, Department of Food Engineering, National Institute of Food Technology and Entrepreneurship Management-Thanjavur, Thanjavur, 613005, India

Yadigar Seyfi Cankal
Research Assistant, Department of Food Engineering, Izmir Institute of Technology, Gülbahçe Campus, Urla, Izmir, Turkey

Ashok Kumar Chakka
Department of Post Graduate Studies and Research in Food Science, St. Aloysius College (Autonomous), Mangaluru, Karnataka, India

Monisha Chandran
Center of Excellence for Grain Science, National Institute of Food Technology and Entrepreneurship Management, Thanjavur, Tamil Nadu, India

V. Chandrasekar
Associate Professor, Department of Food Product Development, National Institute of Food Technology, Entrepreneurship, and Management (NIFTEM), Thanjavur, Tamil Nadu, India

Pintu Choudhary
Assistant Professor, Department of Food Technology and Nutrition, School of Agriculture, Lovely Professional University, Phagwara, Punjab, India

G. D. Arlene Christina
Food Technology Manager, Ahara Technology Group, Hyderabad, Telangana, India

Vidya D' Cruz
Assistant Professor, Center of Excellence for Grain Science, National Institute of Food Technology and Entrepreneurship Management, Thanjavur, Tamil Nadu, India

Ali Demirci
Professor, Department of Agricultural and Biological Engineering, The Pennsylvania State University, University Park, PA, USA

Sudha Devi Gollaru
Assistant Professor, Department of Processing and Food Engineering, College of Agricultural Engineering, University of Agricultural Science, 584104 Karnataka, India

N. A. Nanje Gowda
Department of Food Science, University of Arkansas, Fayetteville, Arkansas, USA

Megh R. Goyal
Senior Technical Editor-in-Chief (Agriculture and Biomedical Engineering) for AAP, and Retd. Professor in Agricultural and Biomedical Engineering, University of Puerto Rico, Mayagüez, Puerto Rico

Sadvatha Ramanna Haromuchadi
Scientist, ICAR–Central Institute of Agricultural Engineering, Regional Center, Coimbatore, Tamil Nadu, India

P. Himashree
BTech Student, Department of Food Engineering, National Institute of Food Technology, Entrepreneurship, and Management, Thanjavur, Tamil Nadu, India

Nene Meltem Keklik
Associate Professor, Department of Gastronomy and Culinary Arts, Faculty of Tourism, Sivas Cumhuriyet University, Main Campus, Sivas, Turkey

K. Kumara
Department of Post Graduate Studies and Research in Physics, St. Aloysius College (Autonomous), Mangaluru, Karnataka, India

M. N. Lavanya
Department of Nutrition and Dietetics, Faculty of Allied Health Sciences, Manav Rachna International Institute of Research and Studies, Faridabad, Haryana, India

R. Mahendran
Professor and Head, Center of Excellence in Non-Thermal Processing, National Institute of Food Technology, Entrepreneurship, and Management, Thanjavur, Tamil Nadu, India

Sivashankari Manickam
Scientist, ICAR–National Rice Research Institute, Cuttack, Odisha, India

Marwa Mayan
Research Scholar, Department of Food Technology, FLAHS, Ramaiah University of Applied Sciences, Bangalore, Karnataka, India

Nikitha Modupalli
PhD Research Scholar, Department of Food Engineering, National Institute of Food Technology, Entrepreneurship and Management-Thanjavur, Thanjavur, 613005, India

Vinass Jamali Pitchai Mydeen
PhD Research Scholar, Department of Academics and HRD, National Institute of Food Technology, Entrepreneurship, and Management, Thanjavur, Tamil Nadu, India

Ravindra Naik
Principal Scientist, ICAR–Central Institute of Agricultural Engineering, Regional Center, Coimbatore, Tamil Nadu, India

Venkatachalapathy Natarajan
Professor and Head, Department of Food Engineering, National Institute of Food Technology and Entrepreneurship Management-Thanjavur, Thanjavur, 613005, India

Udaykumar Nidoni
Professor and Head, Department of Processing and Food Engineering, College of Agricultural Engineering, University of Agricultural Science, Karnataka, India

P. Preetha
Teaching Assistant, Department of Food Process Engineering, Agricultural Engineering College and Research Institute, Tamil Nadu Agricultural University, Coimbatore, Tamil Nadu, India

Pavuluri Srinivasa Rao
Professor, Agricultural and Food Engineering Department, Indian Institute of Technology, Kharagpur, West Midnapur, India

K. G. Sanjay
Center of Excellence for Grain Science, National Institute of Food Technology and Entrepreneurship Management, Thanjavur, Tamil Nadu, India

S. Shahir
Associate Professor, Department of Agricultural Engineering, Kalasalingam Academy of Research and Education, Srivilliputhur, Tamil Nadu, India

A. Jamna Sherin
MTech Student, Department of Food Engineering, National Institute of Food Technology, Entrepreneurship, and Management, Thanjavur, Tamil Nadu, India

Eleonora Shylla
MTech Student, Department of Food Engineering, National Institute of Food Technology, Entrepreneurship, and Management, Thanjavur, Tamil Nadu, India

N. U. Sruthi
PhD Research Scholar, Agricultural and Food Engineering Department, Indian Institute of Technology, Kharagpur, West Midnapur, West Bengal, India

Jeyamkondan Subbiah
Department of Food Science, University of Arkansas, Fayetteville, Arkansas, USA

Chikkaballapur Krishnappa Sunil
Assistant Professor, Department of Food Engineering, National Institute of Food Technology and Entrepreneurship Management-Thanjavur, Thanjavur, 613005, India

Shweta Suri
Department of Food Engineering, National Institute of Food Technology, Entrepreneurship, and Management, Sonipat, Haryana, India

Humeera Tazeen
Assistant Professor, Department of Food Technology, Jain University, Bangalore, Karnataka, India

Radhika Theagarajan
Department of Home Science, Faculty of Food Processing and Management, Thassim Beevi Abdul Kader College for Women, Kilakarai, Ramanathapuram, Tamil Nadu, India

Sevcan Ünlütürk
Professor, Department of Food Engineering, Izmir Institute of Technology, Gülbahçe Campus, Urla, Izmir, Turkey

T. Veena
Scientist, KVK University of Agricultural Science, Karnataka, India

Eyarkai Nambi Vijayaram
Associate Professor, Department of Academics and HRD, National Institute of Food Technology, Entrepreneurship, and Management, Thanjavur, Tamil Nadu, India

ABBREVIATIONS

(6-4PPs)	pyrimidine 6–4 pyrimidone photoproducts
4CL- 4	coumarate-CoA ligase
6-4 PP	6-4 pyrimidone
6-4PPs	6-4 pyrimidone photoproducts
ACP	atmospheric plasma processing
AFB1	aflatoxin B1
AFT	aflatoxin
AJ	apple juice
ALA	aminolevulinic acid
ALA	5-aminolevulinic acid
AlGaAs	aluminum gallium arsenide
AlGaInN	aluminum gallium indium nitride
AlGaInP	aluminum gallium indium phosphide
AlGaN	aluminum gallium nitride
AlGaP	aluminum gallium phosphide
AlGaP	gallium arsenide phosphide
AlN	aluminum nitride
APC	total aerobic plate count (APC)
aPDT	antimicrobial photodynamic treatment
ATCC	American type culture collection
br	bromine
CFU	colony forming unit
Cl	chlorine
DE	delta E
DNA	deoxyribonucleic acid
E coli	*Escherichia coli*
EFSA	European food safety authority
EPA	environmental protection agency
FCV	feline calcivirus
GAE	gallic acid equivalent
HAV	hepatitis a virus
HHP	high hydrostatic pressure
I	iodine
IQF	individual quick freezing
LED	light emitting diodes
LPHO	low pressure high output UV lamp
LPM	low pressure mercury

MAP	modified atmosphere packaging
MCL	maximum contamination level
MNV-1	murine norovirus
NoV	norovirus
PAL	phenylalanine ammonia lyase
PE	polyethylene
PL	pulsed light
PP	polypropylene
RNA	ribonucleic acid
RTE	ready to eat
SASP	small acid-soluble proteins
TCID50	mean tissue culture infectious dose
TPC	total plate count
USDA	U.S. Department of Agriculture
UV	ultraviolet
UV-A	ultraviolet-A
UV-B	ultraviolet-B
UV-C	ultraviolet-C
UVL	ultraviolet light
ATCC	American type culture collection
CFU	colony forming units
dA	cross-sectional region of the imaginary sphere
dN	incident photons on an imaginary sphere
E	amount of energy LED light/area
E_c	energy of electron in the conduction band
Eg	energy in the bandgaps of diode
Ev	energy of the holes in the valence bond
h	planck's constant
hv	energy of a photon released
I	irradiance of the LED light
J	joule
LED	light emitting diodes
mA	milliampere
NADH	nicotinamide adenine dinucleotide
Ø	fluence of photon (cm^2)
t	treatment time for the LED light
UV	ultraviolet
v	frequency of photon of light
V	volt
Aw	water activity
BL	blue light
BRT	*Brochothrix thermosphacta*
BSPL	broad spectrum pulsed light

C	inductance-capacitance C (Farads)
C4H	cinnamate-4-hydroxylase
CAJ	cloudy apple juice
CB	carbon black
CE-LED	capillary electrophoresis-light emitting diode
CHI	chalcone isomerase
CHS	chalcone synthase
Cm	centimeter
COD	chemical oxygen demand
CPD	cyclobutane Pyrimidine Dimer
CS/TP	casein/total protein
DBD	double-strand breaks
DNA	deoxyribonucleic acid
DWPI	denatured whey protein isolate
E	energy
EDTA	ethylenediaminetetraacetic acid
EFSA	European Food Safety Authority
EGCG	epigallocatechin gallate
EM	electromagnetic
ENT	enterobacteriaceae
Eo	energy stored in capacitor
EU	european union
f	flow rate
F	fluence
FDA	Food and Drug Administration
FFA	free fatty acids
FL	fluorescent light
FR	fresh milk
FSSAI	Food Safety and Standards Authority of India
FTIR	fourier transform infrared region
GaAs	gallium arsenide
GaAsP	gallium arsenide phosphide
GaP	gallium phosphide
GRAS	generally recognized as safe
HACCP	hazard analysis critical control points
HDPE	high density polyethylene
HID	high-intensity discharge
HIPL	high-intensity pulsed light
H-PET	high polyethylene terephthalate
HPP	high hydrostatic pressure processing
HPP	high pressure processing
HPP	hydrostatic pressure processing
Hyp-	hypericin

Hz	hertz
I	lamp current
ILP	intense light pulsed
InGaN	indium gallium nitride
IR	infrared
IS	innoculum size
ISC	inter system crossing
J/cm^2	joule per square centimeter
KI	potassium Iodide
Ko	impedance constant of the lamp
LA	arc length of the lamp, mm.
LA	light A
LAB	lactic acid bacteria
LB	light B
LC	light C
LCR	ligase chain reaction
LDPE	low density polyethylene
LE	light exposed
LED	light-emitting diode
LEDF	light emitting diode + folic acid
LIS	*Listeria* spp.
LP	light protected
LP	low pressure
LPA	light protective agent
LPM	low pressure mercury lamp
MAA	mycosporine like amino acid
MAP	modified atmosphere packaging
MDA	malondialdehyde
MNV-1	murine norovirus 1
MP	medium pressure
M-PET	moderate polyethylene terephthalate
MPM	medium-pressure mercury lamp
MVC	mesophilic viable counts
Na-Chl	sodium chlorophyllin
NASA	national aeronautics and space administration
NC	near code milk
NFDM	non-fat dry milk
NIR	near infrared region
Nm	nanometre
NMB	new methylene blue
NPU	net protein unit
N_{Re}	Reynolds number
NUVVis	near ultraviolet visible

OCC	off-line chemical clean procedure
OSI	oxidative stability index
Pa	Pascal
PA/PE	polyamide/polyethylene
PAL	phenylalanine ammonia lyase
PBC	psychotropic bacteria count
PBS	phosphate buffered saline
PDI	photodynamic inactivation
PEF	pulsed electric field
PET	polyethylene terephthalate
PG	polygalacturonase
PHP	*Photobacterium phosphoreum*
PL	pulsed light
PLA	polylactic acid
PLT	pulsed light technology
PME	pectin methylesterase
POD	peroxidase
PP	polypropylene
PPO	polyphenol oxidase
PPS	pulse-per-second
PRR	pulse-repetition-rate
PSE	*Pseudomonas* spp.
PUV	pulsed ultra-violet
PUVL	pulsed UV light
PV	peroxide values
PVC	polyvinyl chloride
PVC	psychrophilic viable counts
PVP-C	curcumin bound to polyvinylpyrrolidone
PWL	pulsed white light
RAPD-PCR	random amplified polymorphic DNA-polymerase chain reaction
RH	relative humidity
RMWE	real municipal wastewater effluent
RNA	ribonucleic acid
ROS	reactive oxygen species
RP	roast pork
RPR	raw pork roast
RSME	root square mean error
RTE	ready to eat
SARS	severe acute respiratory syndrome
SDS-PAGE	sodium dodecyl-sulfate polyacrylamide gel electrophoresis
SiC	silicon carbide
SMWE	synthetic municipal wastewater effluent
SPs	spore photoproducts

SS	soluble solids
SS	stainless steel
SSB	single-strand breaks
SSE	sum of squared estimate of errors
t	processing time(s)
TBARS	thiobarbituric acid reactive substances
Td	duration of pulse
TEM	transmission electron microscope
TL lamp	luminescent tube lamp
TOC	total organic carbon
TSS	total soluble solids
TVBN	total volatile base nitrogen
U	ultraviolet
US FDA	United States Food and Drug Administration
US	ultrasound
US	United States
USc	ultrasounication
UV-TiO_2	ultraviolet titanium dioxide
UV	ultraviolet
UVA	ultraviolet A
UVB	ultraviolet B
UV-C	*ultraviolet-C*
UV-LED	ultraviolet light emitting diode
V	volume
VL	visible light
W/cm^2	watt per square centimetre
WHO	world health organization
WPI	whey protein isolate
α	damping coefficient
ΔE	color change
θ	residence time
μ	viscosity
ρ	density

PREFACE

Food preservation has existed since the beginning of human civilization. Food is not important only for relief from hunger, but food with nutrition is also the demand of today's millennium. Food with convenience, good sensorial attributes, and longer shelf-life is today's challenge. The preservation in the current scenario is done by heating, cooking, drying, concentration, freezing, and cooling. All these techniques have their own pros and cons. Scientists and researchers are continuing to work on new ways to address the challenge to save food from deteriorating and from losing its quality in terms of loss of nutrients, texture, and sensorial characteristics and to provide longer shelf-life.

This volume, Nonthermal Light-Based Technologies in Food Processing, is an exploration of various emerging techniques in the food processing sector and is quite relevant for the food industry and academic community. The technologies such as ozone, ohmic heading, high-pressure processing, cold plasma, pulsed electric field (PEF), microwave, irradiation, light-based technologies (ultraviolet, pulsed light, light-emitting diodes [LED]), supercritical fluid extraction, and cold plasma are investigated in this book. Starting with an introduction and going on to design and application to quality and regulatory aspects, the chapters discuss each technology. Literature on successful applications in different food products ranging from, milk, meat, fish, grains, fruits and vegetables, and other foods is included as well.

This book explores and provides in-depth knowledge and understanding of the ultraviolet, pulsed light, and light-emitting diode and also introduces the role of nonthermal technology in food processing applications.

This book is divided into four parts: Part 1 provides and introduction to light-based technologies. Part 2 covers ultraviolet (UV) light technology, providing an introduction and discussing UV light sources, equipment, design aspects, inactivation of microorganisms-mechanisms, the effect of UV light modeling, and evaluation of physicochemical and microbial changes. Part 3 looks at pulsed light (PL) technology and covers the design, microbial inactivation mechanisms, modeling and its application in liquid foods, solid, and semi-solid foods and meat, fish and poultry products. Part 4 discusses light-emitting diodes (LED) in food processing, introducing the principles, sources, design, microbial inactivation mechanisms and their application in liquid foods and solid and semi-solid foods.

Proving in-depth knowledge and understanding of all light-based technologies and their applications in food and highlighting the role of nonthermal technology for novel food product development and for ensuring the quality of food products, this book volume will serve as a rich source of information for scientists, researchers, growers, students, traders, industry, and processors for food processing and quality aspects of food products.

This book has taken its present shape due to the excellent contributions by the contributing authors, who have been souls to this compendium. We have mentioned their names in each chapter and also in the list of contributors. We are indeed indebted to them for their knowledge, dedication, and enthusiasm.

We appeal to our readers to share their productive suggestions that may form the basis for improving future editions.

We take this opportunity to thank: (1) our families for their motivation, moral support, and blessings in counteracting every obstacle coming in the way; (2) our spouses for their understanding, patience, and encouragement throughout this project; (3) our institutes: NIFTEM-T, Thanjavur, Tamil Nadu, India; PAU, Ludhiana, Punjab for their support during the compilation of this book.

—Editors

PART I

LIGHT-BASED TECHNOLOGIES: INTRODUCTION

CHAPTER 1

INTRODUCTION TO LIGHT-BASED TECHNOLOGIES FOR FOOD PROCESSING: CURRENT STATUS, FUTURE SCOPE, AND OPPORTUNITIES

SUDHA DEVI GOLLARU,
ARLENE CHRISTINA GANDHAM DEVASIKAMANI,
UDAYKUMAR NIDONI, and VEENA TELGOL

ABSTRACT

In this chapter, light-based technologies used for the processing or preservation of food products are emphasized, focusing on the history of the use of light in the food processing sector, principles and methods, basic concepts and characteristics, factors affecting various parameters used, economic importance, and commercial applications. It also deals with the current status and future scope of foods, opportunities, and challenges.

1.1 INTRODUCTION

These days, there is a huge demand for minimally processed foods along with awareness of food safety and quality. Due to this rapid increase, there is a need for the development of nonthermal technologies that aid in the preservation of nutrients and sensory attributes of products. Conventionally, heat processing of foods has its limitations, especially in terms of nutrient preservation and longer shelf life. Hence, non-thermal technologies like

Nonthermal Light-Based Technologies in Food Processing.
C. K. Sunil, Preeti Birwal, Megh R. Goyal, and R. Mahendran (Eds.)

light-based technologies, pulsed electric fields, ultrasound, high hydrostatic pressure, etc., are gaining popularity [40].

Light-based technologies are the most sophisticated processes in today's world, and in the food sector, they are mainly used for food preservation. Technological advancements in the generation of light ensure that these technologies would be promising in reducing microbial load. According to the World Health Organization (WHO) report on foodborne diseases, it is evident that a significant burden globally causes higher rates of mortality and morbidity [42]. These are caused by microbes that pose a disaster to public health and a significant economic issue for stakeholders in the food industry, leading to recalls, costs, drops in sales, and a scarcity of epidemiological data. Difficulty in the detection of emerging problems can only be rectified by ensuring microbial decontamination.

The usage of light for food processing is a non-thermal technique that involves the discharge of high-voltage pulses of electricity into food products that have been placed between electrodes for a short period of time. This technology also aims to reduce pests and pathogens in food without significantly affecting the quality. It can also be used to sterilize packaging materials, equipment, and food surfaces.

This chapter explores light-based technologies for processing or preserving food products and discusses the commercial applicability of this technology to specific food types and food contact surfaces.

1.2 HISTORY OF NON-THERMAL LIGHT-BASED FOOD PROCESSING TECHNOLOGIES

For the enhancement of shelf life, various treatments using thermal processes are being used, which result in nutritional deterioration and reduction in food's sensory qualities. Foods are today changed due to many lifestyle changes across the globe. Most consumers are willing to pay a premium for safe and clean food without compromising sensory and nutritional aspects. This has led to the development of non-thermal light-based technologies, which are environmentally friendly, safe, and preserve the nutritional quality of food.

In these technologies, food is processed at near-ambient temperatures, resulting in the intact preservation of heat-sensitive nutrients. These technologies have been developed in the last decades and are utilized for treating pulses, meat, fruits, fish, vegetables, [24], etc. Also, unlike traditional food preservation methods that expose food to high temperatures, non-thermal technologies reduce contamination by not using heat. This helps reduce the microbial load in food and also prevents the loss of temperature-sensitive nutritional components [18]. In thermal processing, food is exposed for a longer duration, resulting in low-grade food production [20]. The thermal techniques used for preservation also lead to the formation of toxicants in food that can be carcinogenic [33] and harmful to health. The formation of these toxicants depends on the cooking method and properties of the food.

Microwave technologies or fat frying cause the formation of aromatic heterocyclic amines that can lead to changes in the body, which can be mutagenic. Due to the change in fatty acid composition, there is a loss of water, and lipid oxidation takes place in thermal treatments. Other thermal treatments like barbequing cause the meat to lose its juices, thus decreasing the saturated fatty acid content and increasing the polyunsaturated content, which leads to lipid oxidation. The lipid oxidation imparts an off flavor to the meat and decreases its quality [30].

Today, consumers are more aware of food standards and safety, thus demanding food with high nutritional and quality aspects. This has led to the study of better alternatives such as non-thermal technologies, where food is exposed for a short amount of time at ambient temperature for decontamination, resulting in no change in texture, mouthfeel, and nutritional composition [21]. Additionally, less energy consumption is used, resulting in a higher quality food product.

These non-thermal technologies are radiation, light, and sound-based technologies, which are environmentally friendly, consumer-friendly, and economical for processing and preserving food. These technologies include cold plasma, microwave, pulsed electric field, supercritical technology, ultraviolet (UV) light, light-emitting diode (LED), ultrasonication, etc. Microbial load reduction occurs within seconds, increasing the shelf life and retaining sensory characteristics. There are no toxic residues or changes in food composition after the technology is used, resulting in an increased shelf life of the product. The non-thermal effects also inhibit enzymatic activity, reducing spoilage in perishables [32].

There is a decrease in the microbial load due to alterations in the membrane structure of the microbes during non-thermal treatment, leading to the unfolding of the DNA's helical structure, resulting in the death of the microbial cell in a short time. This chapter focuses on light-based non-thermal techniques such as ultraviolet light, pulsed light, and LED technologies. The brief concepts and introductions of these technologies are summarized in this chapter.

1.3 EXAMPLES OF LIGHT-BASED TECHNOLOGIES

Ultraviolet light (UV) is a light-based technology that is an alternative to thermal processing, which is the traditional method for processing and preservation. UV light is used mainly for water treatment, disinfection of air, and decontamination of surfaces [34]. This is promising, and for food, it is used for juice pasteurization, sterilization of meat, treatment of food contact surfaces, and aids in extending the shelf life of perishables and solid or liquid foods. Continuous UV (CW UV) is another technology [9] that is used for the higher penetrating and energetic efficiency of UV light. But nowadays, only pulsed light is mainly used instead of this technology. This is because the peak power during the pulses is superior to continuous UV light, and also the temperature that builds up is less [7].

The Light-emitting diode (LED) is an emerging technology where light energy is utilized with a wavelength of 200 to 780 nm. This results in the reduction of microbial load, bacterial inactivation, and decontamination, especially in water or liquid foods. This is also caused by the photosensitization and photodynamic inactivation of the compounds that absorb light and the damage of DNA, which is interdependent on the light's wavelength used. This technology is used commercially for air ventilation, medical applications, algaculture, water treatments, and liquid food decontamination.

Pulsed light is a case with light-based technology used for disinfecting. It generates flashes [26]. These are high-intensity and polychromatic light which range from infrared to UV. This is done by storing electrical energy in capacitors discharged to a xenon lamp. Pulsed light inactivates microbes faster than conventional light and is more efficient [1]. This technology helps in the pasteurization of liquid food (milk) where short pulses of high intensity for 100–200 microseconds are used. The PL has a pulse generator (which is of high voltage DC, pulse controlling system, and a capacitor) and a chamber for treatment (which has a lamp that flashes, reflectors, and a cooling system [2]. The light is usually a xenon lamp which emits UV light in pulses, and the cooling system reduces the treatment chamber's temperature.

1.4 PRINCIPLES AND APPLICATIONS

The principle of inactivation by the usage of light-based technologies is briefly explained below. The research on non-thermal light-based treatments is focused on huge consumer demands for nutritious and safe food. The major principle in these technologies is the exposure of the food product to the treatment for a minimum or very short period of time at the ambient temperature.

1.4.1 UV LIGHT FUNDAMENTALS

The inactivation principle of UV light is given as a function of energy impinged on the surface target. This is called fluence (F, J/cm^2), which is the total energy traversing an imaginary transparent spherical target at the point of consideration, divided by the area of the cross-section of the target. UV light suppresses microbial proliferation, resulting in decontamination. Although this technology has been discovered and used for ages, very little has been added to the principle, which is the structural modifications that happen in the microbes resulting in their inactivation. The applications of UV light are mainly for water disinfection and to inactivate bacteria in foods [9]. The treatment dosage is represented by the radiant energy which is received by the unit area which is given in J/cm^2.

1.4.2 LED FUNDAMENTALS

The inactivation is due to the semiconductor which emits light energy that works on the principle of electroluminescence. This means that the magnetic or electric field created will excite the electrons, resulting in the emission of electromagnetic radiation. Also, when the LED semiconductor material is doped [39] with impurities, it creates an interface or boundary (called a p-n junction) between the two types of semiconductors. One type (p-type, also known as positive) has excess holes, and the other type (n-type, also known as negative) has excess electrons, creating a depletion region. Therefore, the color or wavelength of the emitted light from the LED depends on the semiconductors and impurities used. LEDs are similar to conventional diodes, having a p-side (anode) and an n-side (cathode).

1.5 ULTRAVIOLET LIGHT IN FOOD PROCESSING

Ultraviolet (UV) light is the electromagnetic spectrum which falls in the wavelength range of 100–400 nm. This is further divided into UV-A from 315–400 nm, UV-B from 315–280 nm, UV-C from 280–200 nm, and vacuum UV from 200–100 nm. For the inactivation of microbes and surface decontamination, UV-B, and UV-C [27] are used up to 300 nm.

1.5.1 SOURCES OF UV LIGHT

Artificial UV light is provided by lamps that produce the electromagnetic spectrum. The low-pressure mercury lamp is the most common type, which emits a quasi-monochromatic output at 253.7 nm. Next are the medium-pressure mercury lamps, which have polychromatic emission. Also, light-emitting diodes that can generate narrow bandwidth emissions are used. The UV wavelength of about 254 nm is used for the destruction of all microorganisms in seconds [2]. Excimer lamps, which contain gas of halides such as XeBr* (282 nm) or KrCl* (222 nm), are also available sources of UV light.

1.5.2 APPLICATIONS OF UV LIGHT

UV light is used for a large variety of applications in the food industry. Majorly, UV light sources which are of high performance are used to disinfect food processing surfaces reliably. This helps in reducing chemical usage and cleaning economically and in an environmentally friendly manner. For fruit or vegetable juice processing, the performance criterion is defined as a 5-log10 reduction in target pathogen number as per FDA guidelines. The use of UV light for disinfecting surfaces is mainly for equipment cleaning (in meat plants, bakeries, and dairy plants), sanitizing premises [28], and for decontamination purposes (for packaging material and food contact surfaces). This technology is very efficient for the food industry, but the applications are limited as it requires a perfect surface structure; otherwise, shielding effects may occur. This technology is mostly used for decontaminating meat products and is a potential treatment post-lethality to control meat pathogens.

In Poultry facilities, many studies have shown a reduction in *Listeria monocytogenes* in meat chicken, and *Salmonella Typhimurium* on carcasses of poultry. Also, studies showed a reduction in *Escherichia coli* [6] and *Salmonella* levels on the skin and muscle of pork. This technology is also available commercially, called the UV tumbling process developed by C&S Equipment Company (www.csequipment.com). This process involves a rotating drum that lifts and creates a tumbling effect, exposing the product to a UV source for decontamination. This unit can also be used to treat all perishable items (fruits, vegetables, meats, seafood), bakery products, and frozen and refrigerated items. It is a patented technology using Steril-AireTM UV Emitters, which are plastic-sleeved to meet food safety guidelines in the food industry.

For liquid foods like coconut water and skim milk, an experimental set-up of two types using UV light has been used, employing T1UV and MS2 phages [5]. One is a stirred batch reactor with a collimated beam, and the second is a continuous reactor with a dean vortex. For MS2, the reduction in microbial load reached 4 log cycles, and for T1UV, there was a reduction of more than 5 log cycles in coconut water. Additionally, there was at least a 5-log reduction for both MS2 and T1UV in skim milk. The efficacy of UV for liquid foods decreases with suspended solids and turbidity, as suspended solids absorb or scatter light before reaching the microbial particles. This issue is addressed by reactor designs that promote turbulent flow, maximizing exposure to the microbial particles. In the aforementioned cases, the transmittances of UV were 0.57–0.89%/cm in skim milk and 9.7%/cm in coconut water, providing satisfactory results. It was also observed that there was no formation of potentially toxic compounds from the usage of UV light on the food components.

As per studies, the pulsed light has been tested in agitated water filled with berries to avoid the heating of food and facilitate rotation to allow uniform exposure. It also shows a reduction in allergens [8]. It has been reported to cause 1.8 to 3.6 Log reductions in the MNV-1 titer for raspberry and strawberry, respectively. This was significantly higher than an equivalent (1 min) time in water of 10 ppm chlorine for inactivation. Also, in the UV-C [38] light scenario, the inactivation of microbes on the smoother fruit (raspberry) was greater than that on strawberries. Furthermore, efficacy is higher during the treatment with water assistance versus chlorinated water. This suggests that this method could be a potential alternative to treatment with chlorine, which generates byproducts like trihalomethanes that are harmful.

1.5.3 UV LIGHT LIMITATIONS

The limitations of using UV for decontamination are that a sanitizing method or agent needs to be identified [17], which has low power of penetration, and has a shadowing effect. The agent should also reduce the efficacy in the presence of any extraneous matter. Another limitation is that on fruit skins and water, UV has a low penetration power [23]. The depth of penetration on the food surfaces ranges from 2.2 μm to 6 mm, which makes UV mostly unreliable for the disinfection of porous food matrices. The effect of shadowing effect on convoluted or rough surfaces can cause uneven treatment, but this is somewhat overcome by the usage of rotating lamps [22] for illumination at all angles. Hence, bivalve shellfish, berries, and pork liver are some examples of foods that are problematic. Also, the presence of any organic matter on the food reduces the effectiveness of the treatment as it acts as a hurdle for UV light to pass through to the food product [11]. So, proper pretreatment of food by washing delicately is recommended. Lastly, UV treatment is expected to induce mutations in the microbes present, causing genome reassortment which could also allow another infection by the generation of another strain; therefore, it is recommended that after UV treatment, the food is packed and stored to reduce this effect [38].

1.6 LED (LIGHT EMITTING DIODE) IN FOOD PROCESSING

The light-emitting diode is used with different wavelengths for the disinfection of water used for agriculture and also for human consumption [22]. The wavelengths used in LEDs range from 780 nm to 200 nm. The material used for making the light-emitting diodes (LEDs) is semiconductor in nature and these generate monochromatic light which helps in the destruction of microorganisms present in food materials. These LEDs are made of non-hazardous material without using mercury [3] and are of a compact size; due to which the usage is more in food processing decontamination. LED studies show that they offer high performance with robustness and a long lifetime (> 10,000 h) [3]. This is combined with low power use which is economical and cost-effective. The treatment is given using specific wavelengths which could be used in continuous or pulsed operation mode, making LEDs more effective [4].

The semiconductor material chip used in LEDs is doped with impurities which enables the emission of light with a particular wavelength and light. Thus, this works on the principle of electroluminescence, where light is produced when electricity is passed through it. The disinfection takes place when this conversion of light energy to heat occurs in a magnetic field forming electromagnetic radiation, which helps in the increase of the nutritional value of foods and an increase in shelf life.

The use of LED with the combination of Pulsed light, UV, and high-intensity pulsed light is called hurdle technology for cranberry and apple juice (90:10 (volume/volume)), which improves the flavor and odor of the juice. In the study, the ultraviolet light of 5.3 Joule/cm^2 and high-intensity pulsed light of 3.3 Joule/cm^2, LED and pulsed light of 34 kilovolt/cm [3], and 18 Hertz for 93 microseconds were used. The treated juice was stored at –20°C pre- and post-processing, and it was concluded that this combination helped in improving the quality of the juice. Sensory and quality analysis were done, and the optimum combination was obtained as mentioned above, which could also be used commercially for smaller capacities.

This technology is also used for controlled fruit ripening, synthesizing bioactive products, and reducing microbial contamination. The blue light LEDs, which are 410 nm to 470 nm, have been tested to destroy pathogenic bacteria in milk and orange juice. *Salmonella* inactivation up to 2 log cycles was observed when treated with orange juice. Also, when this technology is used with other conventional applications (like heating and drying), it will result in better microbial inactivation.

LEDs create low emissions of radiant heat and more monochromatic light and have a long life expectancy (the diode), making them more suitable for cold storage food processing applications. There are several challenges, such as difficulty in optimizing LED lighting for post-harvest storage and maintaining the sensory quality of foods treated by LEDs. Nevertheless, this LED technology provides an alternative for decontaminating and storing nutritious and safe food.

1.7 GENERAL INFORMATION ON REGULATIONS AND INTERNATIONAL STANDARDS

The Indian food industry requires the upgrading of food quality, development of new or improved products, venturing into new marketing opportunities, and/or reduction in production costs. Hence, the potential applications of the novel non-thermal, light-based technologies need to be marketed well and can be clustered into groups of three: (i) pasteurization or sterilization of food (solids or liquids); (ii) surface decontamination of packaging materials; and (iii) improvement of nutritional properties and functional properties of food. Thus, based on the group these technologies are used in, the regulations and international standards differ. A few standards are given below for UV light, which is standardized for commercial use; other technologies are in the pipeline.

The usage of UV light is subject to many legislations that regulate it, and these regulations vary based on the country. It is legally used in the USA but only for novel foods [10] in other countries. In these regulations, the use of UV light is considered not only for its antimicrobial efficacy but also for other intended uses, such as vitamin D photogeneration. Some countries also differentiate their regulations depending on the type of light source and if there is no intended use for virus inactivation [12].

The usage of mercury lamps in low pressure for decontamination of food has been limited under the regulation of the Food and Drug Administration (FDA) Code of Federal Regulations Title 21. Thus, this restricts the use of the lamp for any surface decontamination or microbial control in food products. This is only used for water sterilization, which is used for the production of food, aiding in reducing pathogenic microorganisms in liquid food products like fruit juices [13]. There are no specific values of fluence as per the FDA, but regulations state that the usage of light technology treatment should be consistent with the good practices of manufacturing [14]. For the usage of pulsed light technology for food treatment, the regulation followed is the regulation of the Food and Drug Administration (FDA) Code of Federal Regulations Title 21 of 179 and 41 [15]. This restricts the maximum fluence to 12 J/cm^2. For the use of UV light for treating or sterilizing food, the regulation of the Food and Drug Administration (FDA) Code of Federal Regulations Title 21 of 172 and 381 [16]. For the European Union, UV-treated foods are considered to be novel foods, especially used for mushrooms for 800–200 nm 315–240 nm for bread, and 310–200 nm for milk.

1.8 ECONOMIC IMPORTANCE OF LIGHT-BASED FOOD PROCESSING TECHNOLOGIES

The light-based food processing technologies are of economic importance, as they help in the decontamination of complex surfaces, especially of perishable foods like fish and meat, by illuminating the foods evenly for a sterilizing effect [29]. The equipment used for food preservation in all light-based technologies depends on the product's surface area and the required degree of decontamination. Additionally, for some processes, an additional cooling facility is required if the treated food is sensitive to temperature [31].

Microbial inactivation: The pulsed light treatment system results in the extension of shelf life for ready-to-cook or ready-to-eat meat products by inactivating the microbe *Listeria monocytogenes*, which causes foodborne disease [35]. A treatment of 1,000 pulses of high-power pulsed light for 200 seconds, along with an ultraviolet dose of 5.4 Joules/cm^2, also reduces *Listeria monocytogenes* and *Salmonella typhimurium microbes* [36].

Packaging Material Decontamination: Exposure to pulsed light with a fluence of 0.244 to 9.77 Joules/cm^2 inactivates spores of *Aspergillus niger, Cladosporium herbarum, Aspergillus cinnamomeus,* and *Aspergillus repens* [37] by a reduction of 2.7 logs. This helps reduce recall costs when outbreaks occur, making these technologies economically important.

Application of UV on decontamination equipment used for food processing: The light treatment studies for decontamination of equipment made with stainless-steel contacting surfaces, when treated with fluence of 0.255 to 0.877, caused inactivation of *Escherichia coli* O157:H7 and *Listeria monocytogenes* by 6.5 log of reduction [19]. This treatment also helped in keeping the quality of milk by reducing the fouling effect [25]. Thus, increasing the life of the equipment for the industry is also of great advantage to the whole food supply chain from the farm-to-fork value chain. These help in reducing food wastage, and food preservation for longer periods, and play a significant role in ensuring food safety.

1.9 COMMERCIAL APPLICATION OF LED-BASED FOOD PROCESSING TECHNOLOGIES

As of now, commercially light-based technologies are used in combination with other non-thermal processing technologies. The drawback of using these technologies is that it is very difficult to scale up the fluence of the light used for continuous processing. Additionally, there is superficial absorption where the light does not penetrate food, resulting in a lower level of microbial inactivation that is not commercially viable. Currently, there is a growing concern for nutrition and food safety, leading to a high demand for novel technologies to provide healthy, safe, high-quality foods with high nutritional value and no added preservatives. Therefore, more studies need to be conducted for these technologies to be used commercially [41].

1.10 SUMMARY

The use of light-based technologies can be a boon for food processing and preservation, knowing the complete physical and biochemical aspects of its system and operation, and its effects. Future trends show that for microbial inactivation, UV light and pulsed light are consistent with results from extensive research. UV light is highly effective for transparent or opaque liquids with consistent turbulent flow. For solid foods with surface irregularities, UV or pulsed light technologies have limitations in inactivation. There is limited information available and high variability in the response of UV light for solid foods. UV-C is less effective on berries, meat, and dry herbs. However, when combined with other technologies as a hurdle, it can yield better results. Agitated-water UV treatment is another effective method for rinsing with chlorinated water and disinfecting solid foods.

Further research is needed to explore the potential formation of unpalatable and toxic by-products after treatment, the resistance of common pathogens or surrogate organisms to pulsed light treatments, differences between this technology and the more conventional UV (254 nm) light treatment, mechanisms of microbial inactivation to determine if they are significantly different from those proposed for UV light, and understanding the mechanism and quantification of the benefits attributed to the pulse effect.

ACKNOWLEDGMENTS

The authors appreciate Raichur Agriculture University and NIFTEM–Thanjavur for their continuous support in this research work and also for the development of important technologies for future use.

KEYWORDS

- **contamination**
- **food processing**
- **light emitting diode**
- **light-based technologies**
- **preservation**
- **pulsed light**
- **UV light**

REFERENCES

1. Anderson, J. G., Rowan, N. J., MacGregor, S. J., Fouracre, R. A., & Farish, O. (2000). Inactivation of food-borne enteropathogenic bacteria and spoilage fungi using pulsed light. *IEEE Transactions on Plasma Science, 28*(1), 83–88.
2. Arboleya, J. C., Artiguez, M. L., Fernández, E., & Martínez-de-Marañón, I. (2011). Method for improving functional properties by means of pulsed light, samples with improved functional properties, and uses thereof. Patent number WO 2011/113968 A1.
3. Artiguez, M. L., & Martínez-de Marañón, I. (2015). Improved process for decontamination of whey by a continuous flow-through pulsed light system. *Food Control, 47*, 599–605.
4. Baka, M., Mercier, J., Corcruff, R., Castaigne, F., & Arul, J. (1999). Photochemical treatment to improve storability of fresh strawberries. *Journal of Food Science, 64*, 1068–1072. https://doi.org/10.1111/j.1365-2621.1999.tb12284.x.
5. Bhullar, M., Patras, A., Kilonzo-Nthenge, A., Pokharel, B., & Sasges, M. (2019). Ultraviolet inactivation of bacteria and model viruses in coconut water using a collimated beam system. *Food Science and Technology International, 25*, 562–572.
6. Bialka, K. I., Demirci, A., & Purl, V. M. (2008). Modeling the inactivation of *Escherichia coli* O157 and *Salmonella enterica* on raspberries and strawberries resulting from exposure to ozone or pulsed UV-light. *Journal of Food Engineering, 85*(3), 444–449. https://doi.org/10.1016/j.jfoodeng.2007.08.007.
7. Bohrerova, Z., Shemer, H., Lantis, R., Impellitteri, C. A., & Linden, K. G. (2008). Comparative disinfection efficiency of pulsed and continuous-wave UV irradiation technologies. *Water Research, 42*, 2975–2982.
8. Chung, S. Y., Yang, W., & Krishnamurthy, K. (2008). Effects of pulsed UV-light on peanut allergens in extracts and liquid peanut butter. *Journal of Food Science, 73*(5), 400–404. https://doi.org/10.1111/j.1750-3841.2008.00784.x.
9. Delorme, M. M., Guimarães, J. T., Coutinho, N. M., Balthazar, C. F., Rocha, R. S., Silva, R., Margalho, L. P., Pimentel, T. C., Silva, M. C., & Freitas, M. Q. (2020). Ultraviolet radiation: An interesting technology to preserve the quality and safety of milk and dairy foods. *Trends in Food Science & Technology, 102*, 146–154.
10. Demirci, A. (2002). Novel processing technologies for food safety. *Association of Food Drug Officials, 64*(4), 1–8.
11. Di Cola, G., Fantilli, A. C., Pisano, M. B., & Ré, V. E. (2021). Foodborne transmission of hepatitis A and hepatitis E viruses: A literature review. *International Journal of Food Microbiology, 338*, 108986.
12. Dunn, J. E., et al. (1989). *Methods for Preservation of Foodstuffs*. US Patent Number: 4871559 A.
13. Fernández, E., et al. (2012). Effect of pulsed-light processing on the surface and foaming properties of β-lactoglobulin. *Food Hydrocolloids, 27*, 154–160.
14. Gómez-López, V. M. (2007). Pulsed light for food decontamination: A review. *Trends in Food Science & Technology, 18*, 464–473.
15. Gómez-López, V. M., & Bolton, J. R. (2016). An approach to standardize methods for fluence determination in bench-scale pulsed light experiments. *Food and Bioprocess Technology, 9*, 1040–1048.

16. Gunter-Ward, D. M., Patras, A., Bhullar, M., Kilonzo-Nthenge, A., Pokharel, B., & Sasges, M. (2018). Efficacy of ultraviolet (UV-C) light in reducing foodborne pathogens and model viruses in skim milk. *Journal of Food Process and Preservation, 42*, e13485.
17. Gwynne, P. J., & Gallagher, M. P. (2018). Light as a broad-spectrum antimicrobial. *Frontiers in Microbiology, 9*, 119.
18. Hernández-Hernández, H. M., Moreno-Vilet, L., & Villanueva-Rodríguez, S. J. (2019). Current status of emerging food processing technologies in Latin America: novel non-thermal processing. *Innovative Food Science & Emerging Technology, 58*, 102233.
19. Huang, Y., & Chen, H. (2015). Inactivation of *Escherichia coli* O157, *Salmonella*, and human norovirus surrogate on artificially contaminated strawberries and raspberries by water-assisted pulsed light treatment. *Food Research International, 72*, 1–7.
20. Iqbal, A., Murtaza, A., Hu, W., Ahmad, I., Ahmed, A., & Xu, X. (2019). Activation and inactivation mechanisms of polyphenol oxidase during thermal and non-thermal methods of food processing. *Food and Bioproducts Processing, 117*, 170–182.
21. Jadhav, H. B., Annapure, U. S., & Deshmukh, R. R. (2021). Non-thermal technologies for food processing. *Frontiers in Nutrition, 8*, 1–14.
22. Jubinville, E., Girard, M., Trudel-Ferland, M., Fliss, I., & Jean, J. (2021). Inactivation of murine norovirus suspended in organic matter simulating actual conditions of viral contamination. *Food Environmental Virology, 13*, 544–552.
23. Zoschke, K., Börnick, H., & Worch, E. (2014). Vacuum-UV radiation at 185 nm in water treatment – A review. *Water Research, 52*, 131–145.
24. Kim, T., Silva, J., & Chen, T. (2002). Effects of UV irradiation on selected pathogens in peptone water and stainless steel and chicken meat. *Journal of Food Protection, 65*(7), 1142–1145.
25. Koutchma, T. (2008). UV light for processing foods. *Ozone: Science & Engineering, Journal of the International Ozone Association, 30*(1), 93–98.
26. McDonald, K. F., Curry, R. D., & Hancock, P. J. (2002). Comparison of pulsed and CW ultraviolet light sources to inactivate bacterial spores on surfaces. *IEEE Transactions on Plasma Science, 30*, 1986–1989.
27. Moon, Y., Han, S., Son, J. W., Park, S. H., & Ha, S. D. (2021). Impact of ultraviolet-C and peroxyacetic acid against murine norovirus on stainless steel and lettuce. *Food Control, 130*, 108378.
28. Moraru, C. I., & Uesugi, A. R. (2010). Pulsed-light treatment: Principles and applications. In T. N. Koutchma, L. J. Forney, & C. I. Moraru (Eds.), *Ultraviolet Light in Food Technology: Principles and Applications* (pp. 235–265). CRC Press.
29. Nasheri, N., Harlow, J., Chen, A., Corneau, N., & Bidawid, S. (2021). Survival and inactivation by the advanced oxidative process of foodborne viruses in model low-moisture foods. *Food Environmental Virology, 13*, 107–116.
30. Niu, D., Zeng, X. A., Ren, E. F., Xu, F. Y., Li, J., & Wang, M. S. (2020). Review of the application of pulsed electric fields (PEF) technology for food processing in China. *Food Research International, 137*, 109715.
31. Oms-Oliu, G., Martín-Belloso, O., & Soliva-Fortuny, R. (2010). Pulsed light treatments for food preservation. *Food and Bioprocess Technology, 3*, 13–23.
32. Ortiz-Solà, J., Viñas, I., Aguiló-Aguayo, I., Bobo, G., & Abadias, M. (2021). An innovative water-assisted UV-C disinfection system to improve the safety of strawberries frozen under cryogenic conditions. *Innovative Food Science & Emerging Technology, 73*, 102756.

33. Oz, E. (2020). Effects of smoke flavoring using different wood chips and barbecuing on the formation of polycyclic aromatic hydrocarbons and heterocyclic aromatic amines in salmon fillets. *PLoS ONE, 15*, e0227508.
34. Pexara, A., & Govaris, A. (2020). Foodborne viruses and innovative non-thermal food-processing technologies. *Food Microbiology, 9*, 1520.
35. Ravi, S. (2022). Intense pulse light in food technology: A review. *International Journal for Research in Applied Science & Engineering Technology (IJRASET)*, 10(2), 403–415.
36. Rice, J. K., & Ewell, M. (2001). Examination of peak power dependence in the UV inactivation of bacterial spores. *Applied Research and Environmental Microbiology, 67*, 5830–5832.
37. Rodov, V., Vinokur, Y., & Horev, B. (2012). Brief postharvest exposure to pulsed light stimulates coloration and anthocyanin accumulation in fig fruit (*Ficus carica* L.). *Postharvest Biology and Technology, 68*, 43–46.
38. Sanchez, A. G., & Smart, W. D. (2021). Surface disinfection using ultraviolet light with a mobile manipulation robot. *Food Microbiology, arXiv*, arXiv:2104.10739.
39. Sauer, A., & Moraru, C. I. (2009). Inactivation of *Escherichia coli* ATCC 25922 and *Escherichia coli* O157 in apple juice and apple cider, using pulsed light treatment. *Journal of Food Protection, 72*, 937–944.
40. Schalk, S., Adam, V., Arnold, E., Brieden, K., Voronov, A., & Witzke, H. D. (2006). UV-lamps for disinfection and advanced oxidation: Lamp types, technologies, and applications. *IUVA, 8*, 32–37.
41. Uysal, P. Ç., & Kirca, T. A. (2011). Effect of ultraviolet-C light on anthocyanin content and other quality parameters of pomegranate juice. *Journal of Food Composition and Analysis, 24*(6), 790–795.
42. World Health Organization (WHO). (2015). WHO estimates of the global burden of foodborne diseases: Foodborne Disease Burden Epidemiology Reference Group 2007–2015. https://apps.who.int/iris/handle/10665/199350 (accessed on 3 June 2024).

PART II

ULTRAVIOLET LIGHT TECHNOLOGY: ADVANCES IN FOOD PROCESSING

CHAPTER 2

INTRODUCTION TO ULTRAVIOLET LIGHT: PRINCIPLES, ADVANTAGES, LIMITATIONS, AND APPLICATIONS

VINASS JAMALI PITCHAI MYDEEN and EYARKAI NAMBI VIJAYARAM

ABSTRACT

The deployment of Ultraviolet (UV) light in processing sectors has been a potential source of microbial inactivation in recent years that works on improving the safety of foods with minimal influences on characteristic features of the food product. UV-light irradiation has a favorable public perception as a physical preservation technology and has a higher prominence in the processing area as an alternative to thermal inactivation of microbes. This chapter focuses on basics of Ultraviolet light, principle of microbial inactivation, advantages and limitations over using the UV light on the food materials.

2.1 INTRODUCTION

Food security includes practices for processing, preserving, and maintaining food that are intended to prevent chronic conditions from infectious pathogens. Various enhanced concerns potentially come into interaction with food items as they move from the farmland to the industry to the plate. At every stage of the food production chain, effective food handling procedures and techniques are used to minimize these risks and safeguard consumers. In order to increase storability while retaining functional and dietary qualities, non-thermal methods like pulsed light, HPP, and UV light treatment for food products have been intensively investigated for preserving dietary and functional features in response to shifting consumer requirements. The

Nonthermal Light-Based Technologies in Food Processing.
C. K. Sunil, Preeti Birwal, Megh R. Goyal, and R. Mahendran (Eds.)

utilization of UV light on food safety by surface treatment of food products and sterilization of food packaging has been growing to a large extent in non-thermal processing methods by reducing the microbial load [68].

UV light sterilization is becoming more popular as an alternative to thermal pasteurization techniques because it utilizes minimal power compared to conventional pasteurization approaches and marginally alters the nutritional constituents and quality of produce. However, the selection of the type of UV source, dose level, and application in the food sector necessitates a far greater understanding of the resistance level of food-borne pathogens and microorganisms, as well as their behavior under varied environmental circumstances and processing parameters [25]. Despite its benefits, UV light treatment has significant limits in that it only penetrates to a shallow depth of the food product's surface. In general, the level of treatment is determined by intrinsic qualities such as absorptivity, range of wavelength, and intensity of light with the presence of suspended solids, and total soluble solids of the foodstuffs [24].

The ultraviolet light process is identified to be a promising non-thermal method which is simple in operation and deployment, and has a low toxicity level, making it environmentally benign [29]. Between the wavelength ranges of 100 nm and 400 nm, it is electromagnetic radiation that exists between X-rays and visible lights. UV light disinfection has several applications in the food sector, although it is most commonly used for surface disinfection of packing materials, as well as air and water treatment. It is classified as non-ionizing radiation because the shorter the wavelength, the shorter the ionization [39]. The use of UV light disinfection in food processing areas will eradicate residues from chemical procedures while also lowering customer exposure to food-borne illness [35]. UV light treatment is found to be efficient for inactivating a diverse array of microbes like bacteria, yeast, and molds due to its germicidal effect [64].

Research on ultraviolet light treatment and its efficacy in controlling airborne tuberculosis was proven in 1961 by exposing ventilation air from tuberculosis-affected patients' wards to pig warehouses with UV light. In comparison to the untreated air, ultraviolet-treated air caused reduced infection in guinea pigs [47].

However, considering bioterrorism and the airborne spread of emerging and re-emerging infections like SARS (COVID virus) and other diseases, interest in the use and research of ultraviolet light as a microbial inactivator and disinfectant has been increasing in recent times [54].

In the published studies, the biocidal activities of UV light and the microbiological resistance of various pathogens with their intraspecific diversity have received little attention. However, comparing experimental results becomes more challenging due to UV equipment structure and geometry, flow sequence, and optical characteristics of liquid samples that influence UV efficacy. There have been several studies on the influence of UV light on pathogen physiological features such as development stages and chronic conditions response before and after treatment, which is still limited, and the results are inconclusive [45].

This chapter discusses the application of UV light on food products, along with design considerations, as well as the parameters involved in improving the efficacy of UV light treatment in depth.

2.2 FUNDAMENTALS OF ULTRAVIOLET LIGHT TECHNOLOGY

The electromagnetic spectrum is a classification of energies based on electrical and magnetic properties, as well as photo ionic interactions, of various molecules. Wavelengths range from 1×10^{-7} nm to exceptionally extended frequencies of high vibrations. The spectrum includes ranges from gamma rays (3×10^{-3} nm) to MW radiation and radio waves of low-energy levels (3×10^{13} nm). Sunlight is the plant's natural supply of UV radiation, which may be separated into three wavelength zones.

- UVC (short-wave ultraviolet) – Wavelength 200 to 280 nm;
- UVB has wavelengths ranging from 280–320 nm; and
- UVA (ultraviolet) wavelength – 320 to 400 nm.

The shorter the wavelength, the detrimental effect of the UV radiation will be higher [5]. Johann Wilhelm Ritter, a German physicist, was the one to discover UV radiation in 1801 while investigating the possibility of energy beyond ultraviolet and visible light. Niels Ryberg Finsen, a Danish physician, observed microbicidal activity in the early 1900s [13]. When compared to gamma radiation, which leaves residual radioactivity in the final product with major repercussions, UV light treatment is considered an interventional technique because the product does not acquire any radiation as a residue [52].

UV radiation is emitted by the sun, and the amount of light that reaches the atmosphere is determined by the scattering and absorption patterns. The ozone layer entirely absorbs UV-C light, so it never reaches the Earth's atmosphere. UV-B is equally blocked, while some UV-B does traverse the Earth's stratosphere. The longest wavelength, UV-A, barely impacts the atmosphere, and the environment is exposed to wavelengths ranging from 290 to 400 nm[2].

UV-C has the shortest wavelength of the three UV radiations and has been proven to have a germicidal impact, causing a photochemical reaction on bacterial DNA that is deleterious to the pathogen and causes inactivation [39]. It is critical to choose an appropriate source that produces the radiation that causes the photochemical reaction on the reaction sites of the absorbent (food material). The behavior of the reactants, when exposed to the absorption spectrum, must be completely understood. Emission consistency, ease of handling, extended life, appropriate physical parameters, and high efficiency are all significant aspects to be considered when selecting radiation sources.

UV radiation affects the microorganism's genetic traits, which is the principal site of inactivation, while it also affects other cell components, most notably the protein structure. The microbial cell absorbs UVC light, which causes the development of photoactive components such as cyclobutane pyrimidine dimers and other pyrimidone photoproducts, which hinders transcription of DNA and mitosis where UV light treatment effectiveness eventually leads to microbial cell death and mutagenesis. The effective resistance of the microbes to the exposure of the DNA repair mechanism and also the extent to which the treatment causes damage to the genetic material. In addition to all these, some other factors are also implied in the effectiveness of the treatment such as processing, environmental and microbial factors that should be taken into consideration before, during, and after the treatment [2, 43].

Bacteria, for example, have enhanced resilience to various stress conditions and have changed their physiological behavior to respond to such changes (temperature, growth rate, and media under incubation). The use of stress treatments like heat, osmotic, and starvation stress to endorse defensive adaption reflexes may enhance the lethality of UV treatment [69]. UV triggers photolysis in aflatoxin compounds by causing the production of OH radicals [15, 53].

2.2.1 PRINCIPLES OF FOOD DECONTAMINATION BY UV LIGHT

The mechanism of microbial annihilation varies depending on the intensity of the wavelength imparted on the food material. The impact of UV light radiation on the Deoxyribonucleic acid of the bacterial cell leads to the formation of lesions on the DNA structure which interferes with the replication and growth cycle of the microbes [24]. The incident light will release the photons at the effective wavelength between 250 and 260 nm (Figure 2.1). The photons thus reach will concentrate on the thymine–cytosine, and purine nitrogenous base region of the DNA [7]. The nitrogenous base purine and pyrimidines are more likely to absorb the UV radiation where the absorption level of the pyrimidines is more than 10 times that compared to the purine nitrogenous bases. The major lesions that are induced by the UVC treatment are cyclobutene pyrimidine dimers (CPDs) and pyrimidine 6–4 pyrimidone photoproducts (6-4PPs) [21]. The basic principle involved in photoproducts CPD formation is the reaction of absorption of a photon by one with two adjacent pyrimidines. This results in the formation of a ring structure when C5 and C6 carbon get saturated where the pyrimidines become covalently bonded. Usually, dimers containing thymine are the key site for the incidence of UV light. CPD formation takes place between the thymine-cytosine or cytosine-thymine bases [21].

The formation of pyrimidine 6-4 pyrimidone photoproducts (6-4PPs) also follows a similar process, but the reaction happens between C5–C6 carbon to its C4 carbonyl or amino groups for its 3′ neighbor. As a result, unstable oxetane and azetidine products formed will rearrange carbonyl or amino group to the 5′ position of DNA from the 3′ position of the thymine-cytosine bases. The 6-4PPs component when treated with UV-A or UV-B source gets converted into Dewar valence isomers [57].

Another set of microorganisms taken into consideration is the bacterial spores. The treatment of bacterial spores with UV light forms the spore products due to the reaction of the methyl constituent of neighboring thymine linked to the carbon placed at the 5th position [27]. The impacts caused by spore products are comparable to those of photoproducts [8]. The isolation of the vegetative cells should be undertaken at relatively low humidity conditions to improve the treatment's potency [48].

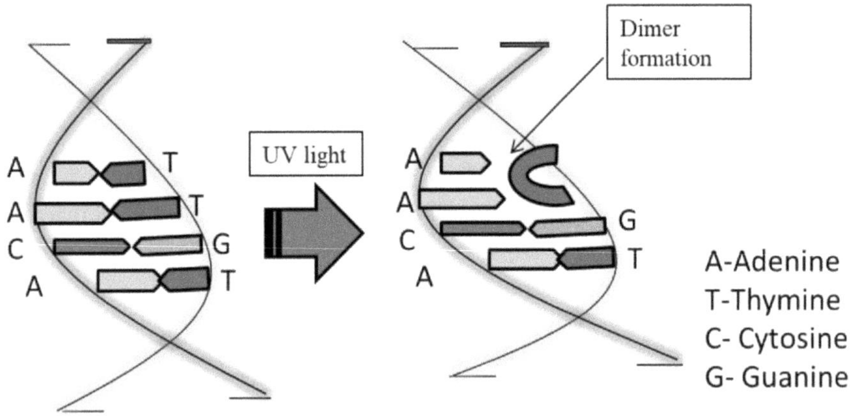

FIGURE 2.1 Action of ultraviolet light on the genetic material of microorganism.

The inactivation process using the UV technique will be effective while considering the two factors, i.e., one is the wavelength, and the other is the UV dosage provided. The dose level varies per species, and no one dose is effective against all of the pathogens in question. UV radiation causes phyto-chemical alterations. UV-A radiation is less prominent due to the DNA's limited absorbability at that wavelength [26]. UV-A causes photo-oxidation of the guanine base, involved with the production of n 8-oxo-7,8-dihydro-2′-deoxyguanosine [17].

According to earlier research, gram-positive species are significantly susceptible to UV radiation than gram-negative microbes. It is more difficult for photons to efficiently permeate through the bacterial cell membrane owing to the prevalence of the robust peptidoglycan layer on the cell wall's exterior [17]. Pigmentation is the primary hindrance to UV light passing through pigmented spores, which are readily absorbed. Yeast, on the other hand, has large cells and a lower amount of the pyrimidine base target site, making it less vulnerable to inactivation. The absorption of photons by other components in the cell before targeting the DNA would also contribute to the bacteria's resistance [23]. Viruses having a complex capsid structure are more resistant to UV-induced denaturation [22].

Photo reactivation and dark repair are the subsequent reactions that are formed due to dimer formation. Photo reactivation was also known as "light repair," mainly due to the photolyase enzyme carried out by irradiated bacteria to repair themselves by rearranging the bonds of (6–4) PP to their basic state [61]. Photo reactivation can be reduced by properly storing the UV-treated products. To make the products safer and clearer, exposure of the treated products to light energy from blue to near-UV light energy for the reactivation process can be avoided [36].

Another mechanism is dark repair, also known as Nucleotide excision repair, which involves the excision of DNA lesions [60]. This mechanism does not involve the absorption of light, but some proteins are responsible for this repair mechanism. It is a multistep process that involves cutting the damaged 5′ and 3′ strands with the help of helicase. The complementary strands will be developed by the DNA polymerase enzyme, and the phos-phodiester backbone will be linked with ligase. The gene responsible for this repair is the UvrABC gene [36].

Photooxidation is another indirect method through which inactivation takes place and is referred to as a photosensitizing reaction. Photosensitizers that are in the ground state will be excited to a state that produces short-span protonated species to the long-span triplet state, which mediates this whole process. The produced triplet, in turn, reacts in two pathways: one by the hydrogen or electron transfer process and the other by energy transfer. The photodegradation of organic components is formed due to the formation of hydrogen peroxide ion or superoxide anion. These components, in further reactions, undergo hydrogen or electron transfer reactions. Some components in the food [12].

When a surface area is exposed to a certain amount of photons, the inactivation of bacteria and other microbes begins at the quantum level. The energy that every photon provides is given by:

$$E = h\,\nu (1)$$

where; 'E' represents energy value of single-photon; 'h' represents Planck's constant (6.626×10^{-34} joules); and 'v' represents frequency of photon with the unit of Hertz (Hz).

From previous research works, it was estimated that for the inactivation of microbes having a cell with a diameter of 0.1 μm and a cross-section of 3.14×10^{12} cm^2, 4,01,000 photons per second will be sufficient and it passes through the cell [11].

At that particular wavelength, the UV light targets the double bond between the adjacent carbons and affects its stability. The sigma orbitals of the two connected atoms are adjacent to the more energetic nucleus. π orbitals are highly stable whereas conjugated bonded components like pyrimidine and purine have large π orbitals. Purine is responsible for the basic structure of the adenine and guanine nucleobases. Pyrimidines provide the basic structure of Uracil and thymine nucleobases. When a photon strikes the molecular level, the nucleobases in the Pi orbital are targeted, where the high energy of the photons will be converted into vibrational energy because of the excitation of the electrons [40]. The π orbital is moved to an unsteady level, where it lasts for 10 to 15 seconds, attributable to the vibrational energy. This orbital from the unsteady level returns to the ground level by dissipation of some energy or bond change brought on by rotation [13].

2.2.2 ADVANTAGES AND LIMITATIONS

UV light inactivation is gaining popularity as a non-thermal processing approach that can be utilized for a variety of purposes. Several factors, such as cost and performance improvements over existing processing methods, influence the adoption of new technology at the industrial level [56]. Ultraviolet treatment of foods has proven to be an intriguing method in the food sector, which has both merits along with limitations. It outperforms traditional processing procedures because it inactivates a wider range of spoilage and harmful bacteria in less time. The FDA (Food and Drug Administration) and the European Food Safety Authority (EFSA) have both identified this method as a simplistic, affordable technology that enables integration with different technologies. With the decreased formation of residuals and hazardous compounds and zero waste generation, the sensory, nutritional, and textural attributes of foods are unaffected [23]. In comparison to other technological breakthroughs such as High-Pressure Processing (HPP), membrane filtration, and Pulsed electric field (PEF), Ultraviolet radiation sterilizes with reduced energy consumption, and the procedure is cost-effective when compared to advanced techniques [59]. It was reported that the commercial pasteurization procedure requires 1,000–10,000 times greater energy than the Ultraviolet light treatment. UV light efficiency is influenced by a variety of parameters, including the flow pattern of the product, the attributes of the product (nutritional composition, viscosity, density, and so on), and the type of UV light source used. UV light irradiation has been established as an appealing option for reducing and eliminating post-process contamination [42].

There is no heat generated throughout the process, which means lower maintenance costs as well as installation and operational costs. The UV treatment technique has the feature of being versatile, which can be coupled with other processing technologies [13].

The primary drawback of the UV light treatment approach is the modest depth to which it can permeate the surface of food products. The time duration of the effective inactivation of several detrimental microorganisms was more. From the perspective of industrial operations, the long duration of processing is not highly recommendable as it may involve the deterioration of the fresh produce [13]. The main problem with UV light is the photo degradation process which leads to the raise of public apprehensiveness, but many studies have proved that the treatment by-products do not produce any cytotoxicity. Another limitation is the exposure of radiation to the entire food surface is still troublesome [19]. Treatment efficiency for opaque liquids is extremely poor. For example, milk. With long-term radiation exposure, this technique is harmful to humans, affecting the skin and eyes [30]. UV treatment disinfection rates on food surfaces are challenging to anticipate and model. Foods with a high amount of suspended particles have a lower UV light penetration level [13].

2.2.3 APPLICATIONS OF UV LIGHT

2.2.3.1 SURFACE TREATMENT OF FRESH PRODUCE

UV light treatment is widely used in the processing of fresh produce like fruits, vegetables, and meat products. After UV treatment, substantial disinfection levels in the target microbes found on the surface of the foods were observed. The type of microbes, frequency, and strength of UV lamps, sample height from the lamps, exposure period, and the quantity of applied UV dosage to the solid substrates were all shown as critical in achieving the necessary level of disinfection. For example, the target microbes in apricots and fresh-cut broccoli, namely *E. coli*, aerobic organisms, and *Salmonella*, were drastically reduced by more than 3 log reduction cycle with the supply of UV dosage ranging from about 1.5 J/cm^2 to 3.1 J/cm^2 with a processing time of 0–25 minutes [32].

To analyze the efficiency of ultraviolet light on the processing of food products, changes in the number of physicochemical characteristics like color, pH, TSS, and TA must be investigated. When it comes to formulating the optimum processing conditions, the pH and acidity of the food product are critical [45]. For example, pH and acidity are two crucial characteristics to consider while selecting process conditions. The texture and color of fresh produce and liquid foods are significantly more connected to turbidity and absorption coefficient [49]. According to the research, there are no notable changes in pH, color, or turbidity immediately after UV treatment [6]. TSS content in fruit juices has been discovered to vary as a result of the photolysis of fructose using UV light treatment, as reported in previous works [68]. When exposed to 3.1 to 5.6 kJ m^2, discoloration and dark surface development were identified in the apple and pear, with substantial alterations in the L*, a*, and b*. The color shift was linked to color pigment breakdown or the browning reactions (Enzymatic and Maillard reactions) that occurred after UV treatment of fresh food [66]. Based on the requirements for the food, an adequate UV apparatus having a low ultraviolet intensity and a long exposition duration must be developed to prevent food color degradation [32].

The changes in the different physicochemical parameters during the storage period were investigated, and it was found that the change in these parameters was associated with the spoilage of microbes [14]. The color consistency of foods can be maintained by more effectively applying UV radiation to products, depolarizing enzymes that cause browning processes, and adopting appropriate storage circumstances, according to the findings [32].

Because of its health benefits, the combination of processed foods with high levels of micronutrients (vitamins and phytochemicals) is highly valued, and it also contributes to the importation and exportation of the produce. The effects of UV light irradiation on fresh food should be explored, as it is important to maximize processing and storage conditions for active ingredients that impact human life and product commercial viability [29].

Plant defense systems contain phenolic chemicals, which serve a vital importance in strengthening vegetative tolerance to abiotic and biotic stress factors. They contribute to the fresh produce's enticing color and flavor. The biosynthesis of chemicals in plants changes based on the developmental stage, heredity, and external circumstances [3]. Phenylalanine ammonia-lyase (PAL) is the crucial enzyme released when the product is subjected to any wounding, stress, hormonal, biotic stress, chilling injury, and exposure to radiation. On whole or freshly cut fresh vegetables, it is possible to use UV irradiation as a stress factor. It activates the enzymes associated with the formation of bioactive molecules [3]. Ultraviolet radiation was adopted as a processing technique to accelerate the synthesis of specific flavonoids and phenolic acids in tomato fruit. After 21 days of storage periods, chlorogenic acid was found to be the most predominant compound from the phenolic group, with caffeic acid, p-coumaric acid, gallic acid, and protocatechuic acid in the list [1]. The considerably elevated activity of Phenylalanine ammonia lyase on tomatoes subjected to Ultraviolet-C light has been attributed to this revelation. Moreover, the amplification of genes that are engaged in the phenylpropanoid pathway has been linked to the augmentation of several enzymes (C4H, Phenylalanine ammonia-lyase, CHS, 4-coumarate—CoA ligase, and chalcone isomerase) that are attributable to the phenol compounds present in Ultraviolet C-irradiated tomatoes in preservation [44, 55].

In contrast to the findings of the previous study, shredded carrots do not show any alterations when exposed to UV light, but there were significant alterations that occurred in storage periods [55]. The impact of Ultraviolet light on phenolic compounds was split into three phases, with an increase in PLA activity when storage duration and temperature are reduced (24 hours at 15°C). The PLA activity increases in the second and third stages as the duration extends from 24 to 72 hours. On the other hand, PAL activity in shredded carrots was inhibited by UV-C irradiation, which may imply substantial conformational changes of PAL due to the increased absorption coefficient of photons than UV-B [3].

The formation of ROS is a key disadvantage in using Ultraviolet radiation-B in the treatment of perishable fruits and veggies [34]. Under the impact of UV-B radiation, different phenolic components react differently. On the surface of apples, phenolic components were found to be reduced. For example, exposure to UV-B irradiation has been shown to enhance hydroxycinnamic acid while decreasing flavonols in apple skin. However, during the 21-day storage period, the concentrations of hydroxycinnamic acid, flavonol, and anthocyanin, as well as antioxidants, increased in all UV-treated apples [3].

Lycopene is a carotenoid pigment that is in charge of determining the color of tomato fruit, melon, pink grapefruit, and other fruits. The content of lycopene generally depends on the maturity index, harvesting and cultivation methods, and the processing conditions. The treatment of the tomato fruit with Ultraviolet radiation-C showed a substantial decrease in the lycopene content under the prolonged storage period of about 25 days compared with untreated samples [4].

The enzyme activities like Polyphenol Oxidase, Peroxidase, and Pectin Methyl Ester in fresh produce play a significant role in enhancing the keeping quality of the produce. The enzyme responsible for oxidative browning is copper-containing polyphenol oxidase, in which o-dihydroxy phenols oxidize to form o-quinones. PPO enzyme activity serves as an indicator of enzymatic browning in fresh produce. Ding and Ling experimented and studied the impact of Ultraviolet light on PPO enzyme activity in Berangan banana fruit. They found that the fruit subjected to radiation intensities of 0.003 and 0.004 J/cm^2 concluded that the radiation of 0.003 J/cm^2 is sufficient for the formation of discoloration that can be best correlated with PPO activity [16]. Conversely, exposing cells to UV-C for a substantial amount of time enhances enzyme activity because the longer the treatment time, the more enzyme and substrate come into contact, causing neurodegeneration. In fresh-cut apples, it has been noted that there will be an increase in PPO enzyme impact on produce with storage time [13].

Mycotoxin contamination in the food produced is a severe issue worldwide. Various processing techniques (thermal and non-thermal approaches) are adopted in order to reduce the level of mycotoxins. The presence of mycotoxins poses a major threat to human health. These mycotoxins are formed by filamentous fungi that produce secondary metabolites, which are carcinogenic, mutagenic, and toxic [51]. There were over 400 toxins found, with *Aspergillus* toxins being the most harmful, which are produced by *Aspergillus parasiticus* and *Aspergillus flavus*. G1, G2, B1, and B2 are the four major chemicals found in Afs [46]. AFB1 metabolism via cellular enzymes produces a variety of breakdown products, including Aflatoxin M1 and Aflatoxin Q1 with oxidative derivatives and DNA acetylated with guanine [46]. Ultraviolet radiation can be utilized to inhibit the AFB1 components as these components readily absorb the wavelengths of 222 nm, 265 nm, and 360 nm, and the absorption happens in the range of 362 nm [46]. The target site of the UV light is the double-bonded furan ring at the terminal of Aflatoxin B1 where the carcinogenic and poisonous properties of the toxins are coded. The furan ring undergoes changes, and consequently, the lactose ring gets broken. Hassan reduced the level of AFB1 in several canned goods by using UV lamps generating light at 365 nm [33]. After 30 minutes of exposure, UV irradiation was shown to reduce AFB1 levels from 75 ppb to 8 ppb in tuna flakes and from 975 ppb to 111 ppb in luncheon made of beef [33]. From previous research works, it was determined that Ultraviolet treatment proves to be useful as a non-thermal approach to inhibit fungal activity.

2.2.3.2 TREATMENT OF LIQUID FOODS

According to the US FDA, the permissible limit for the pasteurization of liquid foods is a 5-log reduction [65]. The UV light radiation at a 253.7 nm wavelength was adequate to surpass the FDA's minimal standards. In processed fruit juices, *E. coli* and LAB bacteria are prevalent, and UV light can be used to attain a decrease in the cycle of 6 logs in processed apple juice. Moreover, carrot juice when subjected to the equivalent UV dosage, where the UV inactivation of psychotropic bacteria was found to be substantially greater in comparison to yeast and mold according to Riganakos [58]. Furthermore, coliforms in watermelon juice were shown to be more UV sensitive than yeast and molds [20]. The lemon-melon juice mixture processed using the annular flow UV batch reactor, subjected to a 2.46 J/mL dose, was found to have a decreased population of the K12 strain of *E. coli* to 6 logs [20]. However, several researchers analyzed that molds and yeasts are sturdier to UV radiation compared to bacteria [37]. The variations in pH and TSS content are influenced by changes in the sensory characteristics of liquid diets. The type of radiation sources also implies more about the organoleptic properties of the product. UV treatment has a significant impact on two essential optical parameters: absorption coefficient and turbidity. The turbidity variations are mostly attributable to suspended particles generated as a result of pectin breakdown and pectin complex formation with other proteins and phenolic components, according to Kaya [38]. The elimination of yeast and molds correlates to the reduction in turbidity observed in Ultraviolet light treated lemon-melon juice [63].

Due to its success in producing improved quality beverages from the fruits and vegetables without considerably lowering the content of bioactive components, it has been stated that an alternative processing method to heat treatment is Ultraviolet-C irradiation. Interestingly, UV irradiation is being used more frequently in liquid foods like skim milk, mango juice, apple juice, and cranberry flavored water [28]. Anthocyanin is the important water-soluble pigment responsible for the pigmentation of fresh produce. The anthocyanin was degraded easily in various processing like clarification, filtration, and UV treatment is the better alternative for the retention of the anthocyanin [62]. In particular, in pomegranate juice irradiated at 62.35 J/mL with Ultraviolet-C dosage resulted in degradation of anthocyanin ranged from 8.1% to 16.3%. However, when compared to untreated pomegranate juice, anthocyanin percent was determined to be minor in relation with treated one [50].

The losses in the ascorbic acid content in the grapefruit were significantly lower when compared with the thermal treatment (60% losses) where the other study indicated that the reductions in the ascorbic acid concentration increased to 2%–17%, 20%–29%, and 25%–35%, for the treatment dosage level of 1.83, 2.84, 3.94 J/cm^2 [9].

The UV light treatment has been employed in the treatment of the allergens in the allergenic milk protein sample where it alters the binding capacity of the proteins of allergens with IgE antibodies. This is mainly due to the alterations that happen in the structure of protein like agglomeration, and crosslinking with the absorption of the photons from the UV light by the chromophore of the protein. The allergenicity of α-casein and α-Lactalbumin was reduced at the dosage level ranging from 1.00 to 3.01 J/cm^2 [67].

2.2.3.3 DISINFECTION OF PACKAGING MATERIALS

One of the most essential processes to carry out to maintain the quality, the commercial value of processed food goods and their storage stability is the packaging. To significantly decrease the prevalence of microorganisms in perishable foods, disinfection of the packing materials that store the food product is vital. To disinfect cartons, bottles, packs, foils, seals, wrappings, lids, and caps, a combination of thermal and non-thermal technologies is being used [18]. At an industrial level, UV light treatment is the best alternative for packing disinfection compared with the chemical and thermal methods because these are low in cost, effective at lower dosage levels, and are more successful [41]. The minimum lethal dose required to degrade the is a crucial factor in the adoption of radiation sources. Alongside, geometry of the packaging system, and the speed at which the system is operated (cup and seal system) are some other criteria to be considered for an effective disinfection process of packaging materials [10].

The packaging materials with a polymer base will readily absorb the photons and create free radicals from the UV light. The transmission range of polyvinyl chloride (PVC) and polyethylene terephthalate (PET) was minimal, which may be due to the presence of residuals and other impurities that act as a receptor. The permeability of polyethylene and polypropylene was high compared to PET and PVC [10].

The thickness of the packaging material also influences the disinfection process. Film materials with a higher thickness of 0.10 to 0.13 mm PE or PP-packed inoculated cheese slices showed a lesser effect on lowering *S. Typhimurium, L. monocytogenes*, and *E. coli* O157:H7, compared to samples packed under 0.07 mm PE or PP packed films. At the commercial level, cheese slices are packed under PP or PE films under 0.04 mm. This concludes that thicker films interfere with the transmission and reduce the efficacy of disinfection by UV light [31].

2.3 SUMMARY

The mechanism by which ultraviolet light treatment inhibits the growth of pathogens and its positive effects on food products are gaining attention in the food processing industries as viable solutions for the processing and preservation of foods. The mechanism is quite variable and depends on how UV radiation affects the genetic structure of the microorganisms. Every aspect of the food processing market has made use of UV light applications, from product processing through packaging. UV light technology has been discovered to be a promising approach in the processing of fresh produce, packing materials disinfection, and surface treatment of food. In addition to its benefits, UV light treatment has some downsides that should be taken into account before utilizing it for preservation. In order to make this technology a plausible one in the near future, more exploration on modest penetration depth is required.

KEYWORDS

- **dark repair**
- **dimer formation**
- **light repair**
- **phenylalanine ammonia-lyase**
- **photolysis**
- **pyrimidines**

REFERENCES

1. Alegria, C., Pinheiro, J., Duthoit, M., Gonçalves, E. M., Moldão-Martins, M., & Abreu, M. (2012). Fresh-cut carrot (cv. Nantes) quality as affected by abiotic stress (heat shock and UV-C irradiation) pre-treatments. *LWT – Food Science and Technology, 48*(2), 197–203.
2. Antonio-Gutiérrez, O. T., López-Díaz, A. S., López-Malo, A., Palou, E., & Ramírez-Corona, N. (2019). UV-C light for processing beverages: Principles, applications, and future trends. In *Processing and Sustainability of Beverages* (pp. 205–234). Woodhead Publishing.
3. Assumpção, C. F., Hermes, V. S., Pagno, C., Castagna, A., Mannucci, A., Sgherri, C., Pinzino, C., Ranieri, A., Flôres, S. H., & Rios, A. de O. (2018). Phenolic enrichment in apple skin following post-harvest fruit UV-B treatment. *Postharvest Biology and Technology, 138*, 37–45.
4. Bhat, R. (2016). Impact of ultraviolet radiation treatments on the quality of freshly prepared tomato (*Solanum lycopersicum*) juice. *Food Chemistry, 213*, 635–640.
5. Bhattacharjee, C., Saxena, V. K., & Dutta, S. (2019). Novel thermal and non-thermal processing of watermelon juice. *Trends in Food Science and Technology, 93*, 234–243.
6. Biancaniello, M., Popović, V., Fernandez-Avila, C., Ros-Polski, V., & Koutchma, T. (2018). Feasibility of a novel industrial-scale treatment of green cold-pressed juices by UV-C light exposure. *Beverages, 4*(2), 1–15.
7. Brem, R., Guven, M., & Karran, P. (2017). Oxidatively generated damage to DNA and proteins mediated by photosensitized UVA. *Free Radical Biology and Medicine, 107*, 101–109.
8. Cadet, J., Sage, E., & Douki, T. (2005). Ultraviolet radiation-mediated damage to cellular DNA. *Mutation Research/Fundamental and Molecular Mechanisms of Mutagenesis, 571*(1–2), 3–17.
9. La Cava, E. L. M., & Sgroppo, S. C. (2019). Combined effect of UV-C light and mild heat on microbial quality and antioxidant capacity of grapefruit juice by flow continuous reactor. *Food and Bioprocess Technology, 12*(4), 645–653.

10. Chawla, A., Lobacz, A., Tarapata, J., & Zulewska, J. (2021). UV light application as a means for disinfection applied in the dairy industry. *Applied Sciences (Switzerland), 11*(16), 7285.
11. Cutler, T. D., & Zimmerman, J. J. (2011). Ultraviolet irradiation and the mechanisms underlying its inactivation of infectious agents. *Animal Health Research Reviews, 12*(1), 15–23.
12. Davies, M. J., & Truscott, R. J. W. (2001). Photo-oxidation of proteins and its role in cataractogenesis. *Journal of Photochemistry and Photobiology B: Biology, 63*(1–3), 114–125.
13. Delorme, M. M., Guimarães, J. T., Coutinho, N. M., Balthazar, C. F., Rocha, R. S., Silva, R., Margalho, L. P., Pimentel, T. C., Silva, M. C., & Freitas, M. Q. (2020). Ultraviolet radiation: An interesting technology to preserve the quality and safety of milk and dairy foods. *Trends in Food Science and Technology, 102*, 146–154.
14. Demir, H., & Oral, M. O. (2018). Nonthermal pasteurization of onion juice by continuous UV-C reactor. *Journal of Food Safety, 38*(6), 1–9.
15. Deng, L. Z., Tao, Y., Mujumdar, A. S., Pan, Z., Chen, C., Yang, X. H., Liu, Z. L., Wang, H., & Xiao, H. W. (2020). Recent advances in non-thermal decontamination technologies for microorganisms and mycotoxins in low-moisture foods. *Trends in Food Science and Technology, 106*, 104–112.
16. Ding, P., & Ling, Y. S. (2014). Browning assessment methods and polyphenol oxidase in UV-C irradiated Berangan banana fruit. *International Food Research Journal, 21*(4), 1667–1674.
17. Douki, T., Reynaud-Angelin, A., Cadet, J., & Sage, E. (2003). Bipyrimidine photoproducts rather than oxidative lesions are the main type of DNA damage involved in the genotoxic effect of solar UVA radiation. *Biochemistry, 42*(30), 9221–9226.
18. Evrendilek, G. A., Tok, F. M., Soylu, E. M., & Soylu, S. (2008). Inactivation of Penicillium expansum in sour cherry juice, peach, and apricot nectars by pulsed electric fields. *Food Microbiology, 25*(5), 662–667.
19. Fan, X., Huang, R., & Chen, H. (2017). Application of ultraviolet C technology for surface decontamination of fresh produce. *Trends in Food Science and Technology, 70*, 9–19.
20. Feng, Z. G., Yu, J. Y., Zhang, H. L., Kuang, D. L., & Xue, L. H. (2013). Effect of ultraviolet aging on rheology, chemistry, and morphology of ultraviolet absorber modified bitumen. *Materials and Structures, 46*(7), 1123–1132.
21. Friedberg, E. C., Aguilera, A., Gellert, M., Hanawalt, P. C., Hays, J. B., Lehmann, A. R., Lindahl, T., Lowndes, N., Sarasin, A., & Wood, R. D. (2006). DNA repair: From molecular mechanism to human disease. *DNA Repair, 5*(8), 986–996.
22. Garcia, L. A. T., & Barardi, C. R. M. (2019). Performance of a storage tank coupled with UV light on enteric virus inactivation in drinking water. *Water Science and Technology: Water Supply, 19*(4), 1103–1109.
23. Gayán, E., Condón, S., & Álvarez, I. (2014). Biological aspects in food preservation by ultraviolet light: A review. *Food and Bioprocess Technology, 7*(1), 1–20.
24. Gayán, E., Serrano, M. J., Álvarez, I., & Condón, S. (2016). Modeling optimal process conditions for UV-heat inactivation of foodborne pathogens in liquid foods. *Food Microbiology, 60*, 13–20.
25. Gayán, E., Serrano, M. J., Pagán, R., Álvarez, I., & Condón, S. (2015). Environmental and biological factors influencing the UV-C resistance of *Listeria monocytogenes*. *Food Microbiology, 46*, 246–253.

26. Girard, S., Ouerdane, Y., Marcandella, C., Boukenter, A., Quenard, S., & Authier, N. (2011). Feasibility of radiation dosimetry with phosphorus-doped optical fibers in the ultraviolet and visible domain. *Journal of Non-Crystalline Solids, 357*(8–9), 1871–1874.
27. Gollasch, S., & David, M. (2010). Testing sample representativeness of a ballast water discharge and developing methods for indicative analysis. *European Maritime Safety Agency, Report No. 4*, 124.
28. Gopisetty, V. V. S., Patras, A., Kilonzo-Nthenge, A., Yannam, S., Bansode, R. R., Sasges, M., Burns, S. M., Vergne, M. J., Pan, C., & Xiao, H. (2018). Impact of UV-C irradiation on the quality, safety, and cytotoxicity of cranberry-flavored water using a novel continuous flow UV system. *LWT, 95*, 230–239.
29. Guerrero-Beltrán, J. A., & Barbosa-Cánovas, G. V. (2004). Review: Advantages and limitations of processing foods by UV light. *Food Science and Technology International, 10*(3), 137–147.
30. Guneser, O., & Karagul Yuceer, Y. (2012). Effect of ultraviolet light on water- and fat-soluble vitamins in cow and goat milk. *Journal of Dairy Science, 95*(11), 6230–6241.
31. Ha, J. W., Back, K. H., Kim, Y. H., & Kang, D. H. (2016). Efficacy of UV-C irradiation for inactivation of food-borne pathogens on sliced cheese packaged with different types and thicknesses of plastic films. *Food Microbiology, 57*, 172–177.
32. Hakguder Taze, B., & Unluturk, S. (2018). Effect of postharvest UV-C treatment on the microbial quality of 'Şalak' apricot. *Scientia Horticulturae, 233*, 370–377.
33. Hassan, F. F. (2017). Detection of Aflatoxin B1 in some canned foods and reduction of toxin by ultraviolet radiation. *Iraqi Journal of Science, 58*, 2349–2343.
34. Hideg, É., Jansen, M. A. K., & Strid, A. (2013). UV-B exposure, ROS, and stress: Inseparable companions or loosely linked associates? *Trends in Plant Science, 18*(2), 107–115.
35. Hinds, L. M., O'Donnell, C. P., Akhter, M., & Tiwari, B. K. (2019). Principles and mechanisms of ultraviolet light-emitting diode technology for food industry applications. *Innovative Food Science and Emerging Technologies, 56*.
36. Jones, D. L., & Baxter, B. K. (2017). DNA repair and photoprotection: Mechanisms of overcoming environmental ultraviolet radiation exposure in halophilic archaea. *Frontiers in Microbiology, 8*, 1–33.
37. Kaya, Z., & Unluturk, S. (2016). Processing of clear and turbid grape juice by a continuous flow UV system. *Innovative Food Science and Emerging Technologies, 33*, 282–288.
38. Kaya, Z., Yildiz, S., & Ünlütürk, S. (2015). Effect of UV-C irradiation and heat treatment on the shelf life stability of a lemon-melon juice blend: Multivariate statistical approach. *Innovative Food Science and Emerging Technologies, 29*, 230–239.
39. Keklik, N. M., Krishnamurthy, K., & Demirci, A. (2012). Microbial decontamination of food by ultraviolet (UV) and pulsed UV light. In *Microbial Decontamination in the Food Industry* (pp. 344–369). Woodhead Publishing.
40. Koutchma, T. (2008). UV light for processing foods. *Ozone: Science and Engineering, 30*(1), 93–98.
41. Koutchma, T., & Barnes, G. (2013). Shelf life enhancement of milk products. *Food Technology, 67*(10), 68–70.
42. Koutchma, T., & Popović, V. (2019). UV Light-Emitting Diodes (LEDs) and food safety. *Ultraviolet LED Technology for Food Applications: From Farms to Kitchens*, 91–117.

43. Kowalski, W. (2009). *Ultraviolet germicidal irradiation handbook: UVGI for air and surface disinfection. Ultraviolet Germicidal Irradiation Handbook: UVGI for Air and Surface Disinfection*, 1–501.
44. Liu, Y., Wang, Y., Wang, Q., Pan, J., & Zhang, J. (2018). Simultaneous removal of NO and SO_2 using vacuum ultraviolet light (VUV)/heat/peroxymonosulfate (PMS). *Chemosphere, 190*, 431–441.
45. Lu, G., Li, C., & Liu, P. (2011). UV inactivation of milk-related microorganisms with a novel electrodeless lamp apparatus. *European Food Research and Technology, 233*(1), 79–87.
46. Magzoub, R. A. M., Yassin, A. A. A., Abdel-Rahim, A. M., Gubartallah, E. A., Miskam, M., Saad, B., & Sabar, S. (2019). Photocatalytic detoxification of aflatoxins in Sudanese peanut oil using immobilized titanium dioxide. *Food Control, 95*, 206–214.
47. Manzocco, L., & Nicoli, M. C. (2015). Surface processing: Existing and potential applications of ultraviolet light. *Critical Reviews in Food Science and Nutrition, 55*(4), 469–484.
48. Moeller, R., Douki, T., Cadet, J., Stackebrandt, E., Nicholson, W. L., Rettberg, P., Reitz, G., & Horneck, G. (2007). UV-radiation-induced formation of DNA bipyrimidine photoproducts in Bacillus subtilis endospores and their repair during germination. *International Microbiology, 10*(1), 39–46.
49. Mohd-Hanif, H., Shamsudin, R., & Adzahan, N. M. (2016). UVC dosage effects on the physico-chemical properties of lime (*Citrus aurantifolia*) juice. *Food Science and Biotechnology, 25*, 63–67.
50. Pala, Ç. U., & Toklucu, A. K. (2011). Effect of UV-C light on anthocyanin content and other quality parameters of pomegranate juice. *Journal of Food Composition and Analysis, 24*(6), 790–795.
51. Pankaj, S. K., Shi, H., & Keener, K. M. (2018). A review of novel physical and chemical decontamination technologies for aflatoxin in food. *Trends in Food Science and Technology, 71*, 73–83.
52. Park, S. H., Kang, J. W., & Kang, D. H. (2018). Inactivation of foodborne pathogens on fresh produce by combined treatment with UV-C radiation and chlorine dioxide gas, and mechanisms of synergistic inactivation. *Food Control, 92*, 331–340.
53. Patras, A., Julakanti, S., Yannam, S., Bansode, R. R., Burns, M., & Vergne, M. J. (2017). Effect of UV irradiation on aflatoxin reduction: A cytotoxicity evaluation study using human hepatoma cell line. *Mycotoxin Research, 33*(4), 343–350.
54. Pexara, A., & Govaris, A. (2020). Foodborne viruses and innovative non-thermal food-processing technologies. *Foods, 9*(11), 1520.
55. Pinto, E. P., Perin, E. C., Schott, I. B., da Silva Rodrigues, R., Lucchetta, L., Manfroi, V., & Rombaldi, C. V. (2016). The effect of postharvest application of UV-C radiation on the phenolic compounds of conventional and organic grapes (*Vitis labrusca* cv. 'Concord'). *Postharvest Biology and Technology, 120*, 84–91.
56. Priyadarshini, A., Rajauria, G., O'Donnell, C. P., & Tiwari, B. K. (2019). Emerging food processing technologies and factors impacting their industrial adoption. *Critical Reviews in Food Science and Nutrition, 59*(19), 3082–3101.
57. Rastogi, R. P., Kumar, A., Tyagi, M. B., & Sinha, R. P. (2010). Molecular mechanisms of ultraviolet radiation-induced DNA damage and repair. *Journal of Nucleic Acids, 2010*.

58. Riganakos, K. A., Karabagias, I. K., Gertzou, I., & Stahl, M. (2017). Comparison of UV-C and thermal treatments for the preservation of carrot juice. *Innovative Food Science and Emerging Technologies, 42*, 165–172.
59. Rodriguez-Gonzalez, O., Buckow, R., Koutchma, T., & Balasubramaniam, V. M. (2015). Energy requirements for alternative food processing technologies-principles, assumptions, and evaluation of efficiency. *Comprehensive Reviews in Food Science and Food Safety, 14*(5), 536–554. https://doi.org/10.1111/1541-4337.12142.
60. Sancar, A. (1996). DNA excision repair. *Annual Review of Biochemistry, 65*, 43–81.
61. Sancar, G. B. (2000). Enzymatic photoreactivation: 50 years and counting. *Mutation Research–Fundamental and Molecular Mechanisms of Mutagenesis, 451*(1–2), 25–37.
62. Santhirasegaram, V., Razali, Z., George, D. S., & Somasundram, C. (2015). Comparison of UV-C treatment and thermal pasteurization on quality of Chokanan mango (*Mangifera indica* L.) juice. *Food and Bioproducts Processing, 94*, 313–321.
63. Shamsudin, R., Mohd Adzahan, N., Pui Yee, Y., & Mansor, A. (2014). Effect of repetitive ultraviolet irradiation on the physico-chemical properties and microbial stability of pineapple juice. *Innovative Food Science and Emerging Technologies, 23*, 114–120.
64. Shin, J. Y., Kim, S. J., Kim, D. K., & Kang, D. H. (2016). Fundamental characteristics of deep-UV Light-Emitting Diodes and their application to control foodborne pathogens. *Applied and Environmental Microbiology, 82*(1), 2–10.
65. Sommers, C. H., Sites, J. E., & Musgrove, M. (2010). Ultraviolet light (254 nm) inactivation of pathogens on foods and stainless steel surfaces. *Journal of Food Safety, 30*(2), 470–479.
66. Syamaladevi, R. M., Lupien, S. L., Bhunia, K., Sablani, S. S., Dugan, F., Rasco, B., Killinger, K., Dhingra, A., & Ross, C. (2014). UV-C light inactivation kinetics of Penicillium expansum on pear surfaces: Influence on physicochemical and sensory quality during storage. *Postharvest Biology and Technology, 87*, 27–32.
67. Tammineedi, C. V. R. K., Choudhary, R., Perez-Alvarado, G. C., & Watson, D. G. (2013). Determining the effect of UV-C, high-intensity ultrasound, and nonthermal atmospheric plasma treatments on reducing the allergenicity of α-casein and whey proteins. *LWT – Food Science and Technology, 54*(1), 35–41.
68. Taze, B. H., Akgun, M. P., Yildiz, S., Kaya, Z., & Unluturk, S. (2020). UV processing and storage of liquid and solid foods: Quality, microbial, enzymatic, nutritional, organoleptic, composition and properties effects. *Innovative Food Processing Technologies: A Comprehensive Review, 2*, 277–305.
69. Van der Veen, S., & Abee, T. (2011). Bacterial SOS response: A food safety perspective. *Current Opinion in Biotechnology, 22*(2), 136–142.

CHAPTER 3

ULTRAVIOLET LIGHT: SOURCES, EQUIPMENT, DESIGN CONSIDERATIONS, AND PARAMETERS AFFECTING ITS EFFICACY

VINASS JAMALI PITCHAI MYDEEN and EYARKAI NAMBI VIJAYARAM

ABSTRACT

The employment of ultraviolet (UV) light treatment in the food industry is expanding as a validated technique for the microbial deterioration in the food system. Due to its germicidal effect on the targeted bacteria, the UV-C spectrum, with a wavelength range of 200–280 nm, is commonly employed in the food industry. Different sources can generate ultraviolet light, which was covered in-depth in this chapter along with the special technologies involved in the generation of the UV light. This chapter emphasizes UV reactor equipment and its various forms, criteria that should be considered when selecting a UV light source, and the impact of various parameters on the caliber of food products.

3.1 INTRODUCTION

The choice of appropriate sources according to the type of food products being processed is an important component in increasing the effectiveness of ultraviolet light treatment. Mercury vapor and xenon sources are just a few examples of sources that are generated economically and emit radiation in the UV spectrum. The mercury vapor discharge type sources are among the most preferred ones for germicidal treatments. The dose and exposure duration have an inverse relationship between them. Using UV dose rather than time as a function is the most efficient measure to provide kinetic data

Nonthermal Light-Based Technologies in Food Processing.
C. K. Sunil, Preeti Birwal, Megh R. Goyal, and R. Mahendran (Eds.)

for UV-induced processes [2]. To determine the concentration of by-products of specific photochemical processes, the dose of ultraviolet light imparted to food products must be quantified [23]. Prior to actually processing products, it is essential to consider the use of the appropriate photo reactor for the inactivation of bacteria. Two aspects, including the target's characteristics and the inherent attributes of the food products under consideration, have a significant impact on treatment productivity to achieve maximal decontamination [22].

The UV dose intensity will not have the same effects on all microorganisms, and when it is applied, it will generate a variety of photoproducts. In rare circumstances, the spores may revive during germination and repair DNA damage caused by UV light [23]. However, it becomes increasingly difficult to compare experimental results since UV equipment geometry, flow order, and liquid sample optical properties will affect the UV lethal efficiency. To render ultraviolet treatment as an efficient technology in the near future, extensive studies and investigations must be accomplished in categories such as reactor design, dosage level, and the effects of various constraints on the competence of the UV treatment.

3.2 SOURCES OF ULTRAVIOLET LIGHT

The UV source must be cost-effective and robust for the treatment to be effective. UV light can be obtained from different sources either naturally or artificially. There are many different types and varieties of UV lamps that emit monochromatic or polychromatic light, and not every lamp is ideal for every use. The wavelength spectra, productivity, working temperature, and lamp longevity are only a few of the attributes that affect the UV light treatment [13]. Some of the natural and artificial UV sources have been discussed in brief in this section.

3.2.1 NATURAL SOURCES

3.2.1.1 ENERGY FROM SOLAR RADIATION

The natural sources for the UV lamp are the sunlight with a spectrum wavelength distributed from 250 nm to 1,200 nm. The sunlight that reaches the Earth's atmosphere contains varied spectra where some spectra will be absorbed, transmitted, and reflected. The ozone layer will block the Ultraviolet-C radiation by absorbing it, preventing it from penetrating the atmosphere and oxygen species which form the upper layer of the atmosphere. The UV-B radiation will be absorbed partially by the upper layer, but it causes some medium to severe health effects in humans. The UV-A radiation is hardly affected by the ozone layer [8].

3.2.2 ARTIFICIAL SOURCES

The artificial sources are man-made, and they contain a range of radiation filters that involve the absorption of a certain wavelength of interest and eliminate the rest. Mercury lamps are widely used for the production of UV spectra and based on the emission spectrum and the wavelengths, the lamps can be categorized into Short Ultraviolet lamps, Medium Ultraviolet lamps, and Long Ultraviolet lamps (Table 3.1) [6].

TABLE 3.1 Different Types of Artificial UV Light Sources

Types of Lamps	Source of Excitation	Experimental Setup	Intensity and Efficiency	Lifetime	References
		Pulse UV Lamp			
Flash lamp	Inert gas (argon, krypton, xenon, and its mixture.	Two confined electrodes.	600 W/cm^2	Less	[13]
Surface discharge lamp SD) lamps.	Plasma production	Xenon-confined fused silica tube.	30,000 W/cm^2	Longer lifetime	
Low-pressure mercury lamp	Monochromatic operating conditions – 30 to 50°C at 0.1 to 10Pa.	The germicidal effect is about 30 to 40%.	0.01 W/cm^2	18–24 months	[13]
Xenon lamp	Point sources use xenon gas.	Operate at high voltage.	Power 2,000 W	Expensive	[15]
Excimer lamp	Monochromatic	Operate at low temperature.	Efficiency 10 to 35%	Tunable lamp	
Microwave UV lamp	Microwave energy	Similar working conditions of low-pressure mercury lamp.	Heat up faster	Three times electrode lamp.	[15]
Low-pressure amalgam lamp	UV 185 to 254 nm	Operates at 90 to 120°C.	Efficiency 95%	Long lifetime	
Carbon and plasma arc	Plasma forming gas (Argon)	300 to 500 nm	Emits a lot of radiation is a demerit.	–	[7]
Incandescent lamp	Tungsten filaments	Operating temperature ranges between 3,000 K and 3,400 K.	Mostly emits IR radiation.	–	
Halogen lamp	Halogen gas (I_2 or Br_2).	–	The greater output of UV light.	–	

3.2.2.1 PULSED UV LAMPS

Inert gases such as argon, xenon, and krypton, as well as mixes of these gases, are used in pulsed UV lamps. This pulsed UV is produced in two ways. The first is a flash lamp, which produces a pulse when gas travels over two confined electrodes, resulting in the range of wavelengths from 100 nm to 1,100 nm. Surface discharge lamp (SD) lamps, on the other hand, produce plasma as a result of a higher-intensity electric discharge within a xenon-confined fused silica tube. When compared to the flash lamp (600 W/ cm^2), the SD lamp has a higher UV intensity of 30,000 W/cm^2. Longer lamp life is possible with SD lamps [13, 15].

3.2.2.2 MERCURY UV LAMPS

The mercury vapor is the source of the irradiation which produces sufficient vapor pressure at low temperatures because of its inert behavior and low ionization energy. Based on the pressure of the mercury gas, the lamp was divided into three types: Low, medium, and high [7].

3.2.3 SHORT UV LAMP

The Low-pressure (LP) mercury UV lamp/short UV lamp works at a low pressure of about 0.1 to 10 Pa at a temperature ranging from 30°C to 50°C. The intensity ranges from 0.01 W/cm^2 and the lifespan is relatively higher, lasting for about 1 to 2 years. The germicidal effect is higher than the MP lamp, which is about 35 to 40%. The working is identical to a fluorescence lamp, but the ozone production happens at wavelengths below 260 nm, causing severe health issues. This radiation affects the genetic material of the microbes (DNA) through dimer formation. The air in the working environment should be less than 0.2 mg/l [28]. Low-pressure high-input lamps are available that provide greater germicidal effects compared with the other existing types.

3.2.4 MEDIUM UV LAMP

Medium pressure (MP) mercury lamp/medium UV lamp emits polychromatic light at the wavelength of 185 to 600 nm which operates at the pressure range of 50 to 300 kPa. The operating temperature is quite higher compared with the LP lamp with a range of 30°C to 50°C. When compared with the LP lamp, the germicidal effect on food is less (less than 20%). The lifetime is less than half a year. The increase in pressure yields more continuum light. These wavelengths are usually referred to as continuous light [20].

3.2.5 HIGH PRESSURE (HP) UV LAMP

High-pressure lamps are more efficient than low-pressure lamps because of their higher power (2,500 W) compared to 30 W for low-pressure lamps. Like fluorescent lamps, mercury lamps (medium, high, or low pressure) necessitate an electric generator for ignition and continuous operation [7]. The light from mercury vapor lamps can be filtered to eliminate the visible light and produce primarily UVA emissions [28].

3.2.6 SPECIAL TECHNOLOGIES

Traditional UV lamps have a non-uniform ultraviolet light intensity. Adoption of specific technologies is becoming increasingly important to enhance the efficiency of ultraviolet light treatment.

3.2.6.1 UV-LEDS

In the food industry, light-emitting diodes (LED) are becoming a popular alternative to ultraviolet lights. These diodes are gaining popularity owing to their economic cost and low energy demand, compactness, and sturdy construction. Furthermore, when compared to UV light lamps, the heat generated is less. Other lamp sources have a shorter average lifespan [1]. When a voltage is supplied between two terminals, a semiconductor called an LED bulb emits light. Under the phenomenon of electroluminescence, the wavelength difference is determined by the bandgap between semiconductors. The backbone of LEDs is made up of groups III and V elements in the periodic table. The emission of distinct UV wavelengths is caused by differences in the makeup of Group III and V. When the holes (positively charged) from the p-layer collide with electrons (negatively charged) in the n-type layer, energy is released in the form of photons. UV-emitting devices based on aluminum gallium nitride/aluminum nitride are commonly employed in the wavelength range from 210 nm to 400 nm [11].

3.2.6.2 MICROWAVE DRIVEN ULTRAVIOLET LAMP

A microwave-driven UV lamp is a novel technology that has emerged as a viable substitute for UV lamp processing of food products due to its numerous benefits in perspective of time and energy efficiency, consistent irradiance, and treatment effectiveness. The waveguide of four slots is mostly within the microwave electrodeless UV lamp configuration, and the agitated plasma due to the collision of negatively charged particles generates UV light in the waveguide. The ultraviolet light can be released via a tiny opening in the waveguide. Because water is not permitted inside the waveguide, a quartz column is installed on the exterior of the device to provide an enclosed condition. There is uniform energy coupled within the four slots of the waveguide [17].

3.3 EQUIPMENT FOR UV IRRADIATION OF FOODS

3.3.1 PHOTO REACTORS

Photochemical reactors or photo reactors are the chemical devices that consist of the photon source, photo catalyst, and reactants that are placed in contact with each other, and physiochemical transformations take place. Photo catalysts are involved in the excitation of absorbed light to a higher energy level, and they deliver energy to the reactants to perform various chemical reactions (Figure 3.1).

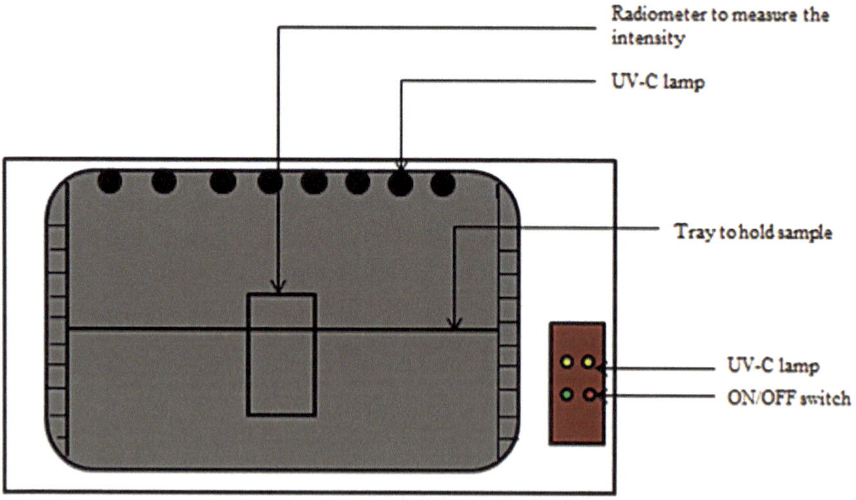

FIGURE 3.1 Schematic representation of UV light radiation setup.

Photochemical reactors can be used for irradiation purposes but depending on the operation mode, the reactors are classified into – (i) continuous reactors; (ii) discontinuous reactors; and (iii) semi-continuous reactors.

3.3.1.1 CONTINUOUS PHOTOREACTORS

When a large amount of mass needs to be treated and/or the reaction rate is particularly high, the continuous reactor is ideal. It necessitates numerous auxiliary devices; however, it is possible to accomplish excellent product quality control. There are different forms of continuous reactors based on the form and position of the lamp and the shape of the reactors. The light is placed in the central annulus space of a tubular annular photo reactor, which is made up of a cylinder with an annular section [21].

Cylindrical tank type and elliptical photo reactors are similar in construction with the discontinuous reactors, but the reaction continuously takes place. Another type is the parallel plate reactor, where the reactants are continuously flowing between the two parallel plates and where they are in close contact. The reactants with a high optical density can be treated with this type as the distance between the plates is very minimal and the radiation would reach every point of the reactants easily [6].

In a descendent film photo reactor, which is a tubular type, the lamp is situated on the axis of the center. The arrangement consists of the nozzle through which the reactants fall on the inner surface of the tube in the form of a film which would be treated without any contact with the UV source or the reactor wall [24]. The particle bed reactor consists of the layer of radioisotope source covered with fluorescent materials on the bed of glass particles. This produces high-energy visible light or UV radiation [24].

3.3.1.2 DISCONTINUOUS PHOTOREACTORS

The interaction of matter with the external surroundings will not occur in discontinuous reactors. Discontinuous reactors can be used to treat samples with higher viscosity. To drive out the reaction with low quantum efficiency, these reactors require a greater degree of a radiation source. The stirrer tank is the simplest and commonly used discontinuous reactor. It is composed primarily of a mixing tank and a set of UV light sources submerged inside the tank, resulting in homogeneous radiation. The elliptical photo reactor, for example, was constructed with an elliptical cross-section and a cylindrical contour, with the light focused and the reactor on the opposite end of the shell.

These reactors are quite simple and suitable for the study of experimental reaction kinetics [21]. This involves the treatment of fewer quantity samples.

In a semi-continuous reactor, the exchange of mass happens from the reactor to the outer environment. The semi-continuous reactor is more versatile but harder to quantify; it provides for better rate of reaction control because the reaction rate can be altered by the inclusion or exclusion of reagents and products [6].

3.3.1.3 DIFFERENT FORMS OF PHOTOREACTORS

Depending on the operating conditions, the reactors take on various shapes and configurations (temperature, pressure, products, the source used). There are two criteria to be considered in the design of the photo reactors:

- Source and form of illumination for the excitation of photocatalyst; and
- Utilization of the irradiation provided from the different sources by the photo reactors.

The main components of the photocatalytic system are – (i) radiation source; and (ii) photocatalyst. As the scaling of reactors is a complex and difficult process, some factors need to be taken into consideration for the effective irradiation process, such as:

- Mode of mass transfer;
- Reaction kinetics;
- Photo catalyst – type and particle size;
- Fluid flow (laminar or turbulent);
- Position of the photo catalyst; and
- Temperature.

Apart from the above-mentioned factors, other two factors that need to be paid attention to are,

- Type of irradiation source (Artificial or natural source); and
- Position of the irradiation source (External or immersed or distributed light source).

Based on the position and form of the photo catalyst, different forms can be classified, as Honeycomb Monolith, plate, annular and packed bed.

- In the honeycomb monolith, the photo catalyst was in the form of the honeycomb and the UV source was placed in between the photo catalyst.

- The photo catalyst was arranged at the bottom in the form of a plate and the sources are fixed on the top of the reactor in Parallel type.
- Annular: The photo catalyst and the UV source were in parallel with each other as a long tube form.
- Packed – The photo catalyst source was packed in between four to five UV lamps as a columnar.

Based on the mixing process, the photo reactors are of two types: (i) batch type; and (ii) plug type. The mixing of the reacting substances is promoted as much as possible in batch type to provide a thoroughly blended reactor. In Plug type, there is no mixing operation of the reacting substances for the formation of segregated reactors [6, 21].

3.4 DESIGN CONSIDERATIONS

The perfect UV reactor design is crucial, as it can eliminate the interference that occurs due to the inherent characteristics of the food and ensure effective deterioration of microbes and their toxins. One of the key elements that significantly affects the residence time and UV dosage to be applied is the flow pattern inside the UV reactor.

3.4.1 CHARACTERIZATION OF UV LIGHT SOURCE

The source must be cost-effective, easily available, and robust for operation. The radiation energy must be higher compared to natural sources. When selecting radiation sources, factors such as emission uniformity, ease of handling, longer lifespan, proper physical characteristics, and high efficiency are all important to consider.

3.4.2 LEVEL OF DOSAGE

The dose requirements are divided into two categories: (i) pre-filter; and (ii) post-filter applications. A UV dose of 3 logs is necessary for Giardia and Cryptosporidium inactivation, while a UV dose of 4 logs is required for viruses [3]. To enhance the procedure's efficacy, safety considerations should be developed before validating dose requirements. Previous research indicates that eliminating vegetative species requires a dose reduction of 1 to 4 logs, with dosage levels ranging from 10 to 80 Jm^{-2} and 25 to 200 Jm^{-2}. A radiometer is used to measure the UV irradiance on the target surface during calibration. The dose needed for inactivation can be determined using the microorganisms' D10 log time (10 log decimal decrease time). This method, generally referred to as bio-dosimetry [17], is used to determine the required UV dosage.

Chemical actinometry method can be used to calculate the number of photons produced per unit time by a light beam. The photochemical reaction of food products is directly correlated with the quantity of photons absorbed [17]. UV disinfection employs the UV radiation produced by the disinfecting unit to compute the inactivation ratio because it is an energy-intensive technology [18].

3.4.3 CONSIDERATION OF HYDRAULIC ANALYSIS AND FLOW PROFILE

The hydraulic analysis of the products was necessary to comprehend the flow dynamics in the reactor. Reynolds number is the basic equation for determining the type of flow through dynamic viscosity (μ), density (ρ), the velocity of the flow (v), and characteristic dimension (l) (length or diameter). The relationship between these parameters can be expressed by the equation:

$$\mathrm{Re} = \frac{\text{ñvl}}{\mu} \tag{2}$$

The Re number for the orange and pineapple liquid juices was found to be in the range of less than 2,000, implying that the flow in the reactor is laminar [15]. The viscosity varied for all the tested juices with the flow rate [9]. The Dean effect is one of the important factors to be considered as it occurs at the curvature of the tube at the inner and outer radius. This factor provides information about the secondary flow developed in the liquid foods, which involves additional mixing in the food produced. When the flow rate is higher due to mixing, the UV dosage level can be reduced. Insufficient irradiation occurs in the liquid at the radial position from the path of the incident light source [17].

Validation of the inlet and outlet pipes, which should be 10 pipe and 5 pipe diameters for upstream and downstream. The bend of 90°, 'S' or 'T' bend should be followed by a straight line in the inlet piping of the hydraulic system of the reactor. The velocity of the flow in the reactor should be within the range of less than 20% of the theoretical velocity, which can be calculated using the cross-sectional area of the flow rate [5].

3.4.4 GEOMETRIC CONFIGURATION OF THE REACTOR

UV reactors are available in a variety of configurations to suit a variety of purposes. Depending on the manufacturer and light orientation, reactors can be categorized as S-shaped, in-line, or U-shaped. Some of the restrictions that govern the UV reactor setup with the redundancy of the UV reactor should be assessed to determine the reactor's power modulation. For high energy efficiency in the reactor, the power modulation of lamps and the number of lamps that are required for the photochemical reaction must be evaluated. The orientation of the reactors influences the effectiveness of the reaction. The vertical orientation of the reactors, where the flow will be perpendicular, can minimize damage to the lamp due to the impact of dissolved particles in the sample. The reactor can be placed either in a perpendicular or parallel direction.

The control system is the central element of the UV reactor, where an allowable distance between the control system and the reactor is vital for the better treatment of food products. Some of the factors to be considered for the development of a UV reactor are:

- Piping configurations and the inlet and outlet connections need to be checked before the installation of the reactors.
- Availability of lifting facilities and provisions for handling chemicals (inert gas) and spare parts.
- Distance between the sources for lamp replacement, and space allowance for electrical systems such as generators, control panels, transformers, and power supply system [5].

3.4.5 LOCATION OF INSTALLATION

Before the installation of the UV reactors, it was recommended to filter the water in the treatment because of the presence of effluents, organic matter, and turbidity. The layout of the treatment should be easy to operate, easily accessible, and easy to maintain. The building must be under environmental stress as it may cause damage to the internal parts of the reactor due to bad weather conditions. The check for the availability of the provisions for conveying the water must be made before, during, and after the working of the reactor. UV reactors must be maintained according to the OCC procedure where the cleaning chemicals should be kept safe and isolated. As specified by the designer, appropriate drains, stockpiling, and safeguarding equipment (e.g., an emergency occlusive dressing station) should be installed. The quality sampling of water can be done by sample taps frequently for validation purposes. The installation of drain valves should be made at one or more spots for easy drainage during maintenance, which must be large enough to drain the reactor completely [5].

3.5 PARAMETERS INFLUENCING EFFICACY OF UV PROCESSING OF FOODS

The physiochemical properties and sensory properties of food products are diverse. Every property must be effectively evaluated to develop the keeping quality and maximize the efficiency of the UV reactor. The physical properties (Density, form of the products, flow pattern), and chemical properties (pH, TSS, water activity) are some of the properties that need to be analyzed before the selection of the UV source and UV reactor.

3.5.1 INTRINSIC PARAMETERS

3.5.1.1 EFFECT OF SUSPENDED SOLIDS

The presence of the suspended solids (SS) in the food products doesn't affect the UV scattering much but it interferes with the effective elimination of bacteria. The SS present in the food would undergo absorption, transmission, and scattering of the light. This would reduce the power of the light and even block the way for the UV light to flow. The larger particle size possibly leads to the backscattering of light. The suspended solids in the food products and its absorption coefficient should be considered for the calculation of the UV dosage for the elimination of bacteria [14].

3.5.1.2 EFFECT OF DISSOLVED SOLIDS

Dissolved solids refer to the inorganic and organic components in the liquid foods which are suspended within them. The presence of the DS dissolved solids greatly impacts the effectiveness of the Ultraviolet treatment. The presence of sugars like fructose, glucose, and organic acids absorbs the radiation at 200 nm whereas the organic acid is found to have a larger absorbance ranging from 220 to 300 nm. The presence of Vitamin C greatly reduces the incidence level of the UV dosage on targeted microbes [14]. The dirt and unwanted particles also hinder the path of the UV light by the formation of shadows that reduces the potency of the treatment in suppressing pathogens. Darker foods show higher absorption of UV light which causes a reduction in the energy of the incident light.

3.5.2 EXTRINSIC PARAMETERS

The growth conditions, temperature, UV dosage, and bandwidth are some of the external factors that influence the efficiency of the UV treatment. These parameters affect the keeping quality and safety of food production, and a better understanding of all these parameters is important. The phytochemical level has been decreased in the apples that are grown under the shade tree. The presence of bioactive molecules largely attenuates the UV radiation [10].

The temperature under which the UV treatment was carried out is one of the important criteria to be considered. UV-treated apples at 17°C were very effective compared with UV-treated samples at higher temperatures (27°C) [27], and the same behavior was contrasted in the sand pears [30]. This behavior may be due to the increased anthocyanin content [12].

The treatment of products with particular wavelengths Ultraviolet-A (315 nm–400 nm), Ultraviolet-B (280 nm–315 nm), and Ultraviolet-C (200 nm–280 nm) and combination causes changes in the physiochemical properties of the product. Different wavelengths cause different changes in the properties of the foods. The UV-B exposure is more effective in Vitamin D_2 formation in mushrooms [19]. UV-C treatment has effectively maintained the phenolic compound during the storage of blueberries [29] and green tomatoes [16].

The selection of the optimal dose is very critical to maintaining the sensory qualities of the food produced. Overdosing may lead to the browning reaction and formation of furfural compounds. The ripening of the produce gets accelerated along with the senescence effect, which are the consequences of the overdosing of UV radiation. The reduction in the germicidal effects and storage life causes serious losses [12].

3.5.3 UV LIGHT PROPAGATION

The UV light emitted from the source will propagate through the components of the reactor like the wall of the reactor, the sleeve of the lamp as well as the product to be treated. The light would undergo absorption, reflection, transmission, emission, and scattering. The absorption power of the food products and the microorganisms is related to Lambert's law. The transparency of the food products greatly influences the absorption behavior of the UV in the food products.

Absorption can be defined as the quantity of light observed by the food products from the incident light. The photons change from one form of energy to another when they travel through the product. The scattering phenomenon is very useful in the disinfection application that involves the deflection of the radiation from the path towards the absorber as the photon reacts with the food products [14, 26].

3.5.4 PENETRATION DEPTH

The penetration depth measures the depth through which UV radiation can pass through the product. It is the reciprocal of the absorption factor. The transmittance is mainly influenced by the thickness of the materials. If the thickness is more, the transmittance and penetration depth will be less [14, 26].

3.5.5 ABSORPTION COEFFICIENT

The absorption coefficient of water is higher compared with that of liquid fruit juices because of the higher optical density where the suspended particles are involved in both scattering and absorption of ultraviolet light. The viscosity of fruit juices greatly affects the absorption coefficient of UV light. A non-Newtonian fluid with a higher viscosity value has a high absorption coefficient [14].

3.5.6 EXPOSURE TIME

The exposure time depends on the intensity level of the UV dose applied to the surface of the food material. Depending on the thickness of the materials and intensity, the exposure time can range from a few seconds to several minutes. According to previous literature, the exposure time is short for higher intensity of UV light, lasting less than 2 to 3 seconds. The exposure time increases to 9 to 10 minutes with low-intensity UV dosage. Prolonged exposure of food products to higher intensity for extended periods can lead to undesirable changes in the food, notably browning and the formation of dark spots [25].

3.5.7 INTENSITY OF UV LIGHT

UV intensity/ultraviolet flux/ultraviolet irradiance, which is the quantity of radiation incident on the surface of the produce, is critical for the effective eradication of spoilage bacteria in fresh produce. It relies on the exposure duration and voltage. When the intensity increases as the gap between the source and sample decreases, the rate of bacterial cell suppression improves [25].

3.6 SUMMARY

UV light treatment is growing as a better alternative to non-thermal preservation compared to heat pasteurization. Although it is one of the most intriguing decontamination treatments, much more research is required in this domain. The effectiveness of the UV treatment relies on the treatment conditions like the design of the treatment chamber, extrinsic factors (temperature, UV dosage, and intensity), and intrinsic parameters (microbial resistance/sensitivity, pH, TSS, protein structure), light exposure duration which attributes to the changes in the physiochemical properties with great influences on the keeping quality of the products. Therefore, utilizing Ultraviolet processing on food surfaces necessitates technological solutions that can effectively process 100% of the food surface. Regrettably, current industrial approaches are insufficiently effective. Researchers can devise novel technologies by delving deeper into the information on the sensitivity mechanism and DNA repair pathways. Furthermore, the prospect of combining UV radiation with other physical or chemical inactivation methods to develop combination treatments for the effective inactivation of microbes can be explored. Shortly, issues such as the economic feasibility of ozone generation and workers' safety must be addressed, and the development of the 360° approach to food product treatment must be addressed.

KEYWORDS

- **absorption coefficient**
- **bio-dosimetry**
- **chemical actinometry**
- **Lambert's law**
- **light-emitting diodes**
- **Reynold's number**

REFERENCES

1. Akgün, M. P., & Ünlütürk, S. (2017). Effects of ultraviolet light emitting diodes (LEDs) on microbial and enzyme inactivation of apple juice. *International Journal of Food Microbiology, 260*, 65–74.
2. Bolton, J. R., Mayor-Smith, I., & Linden, K. G. (2015). Rethinking the concepts of fluence (UV Dose) and fluence rate: The importance of photon-based units–A systemic review. *Photochemistry and Photobiology, 91*(6), 1252–1262.
3. Cotton, C., & Passantino, L. (2005). Regulations in the United States: Requirements and guidance for ultraviolet disinfection of drinking water. *Journal of Environmental Engineering and Science, 4*(S1), S57–S63.
4. Delorme, M. M., Guimarães, J. T., Coutinho, N. M., Balthazar, C. F., Rocha, R. S., Silva, R., Margalho, L. P., Pimentel, T. C., Silva, M. C., & Freitas, M. Q. (2020). Ultraviolet radiation: An interesting technology to preserve the quality and safety of milk and dairy foods. *Trends in Food Science and Technology, 102*, 146–154.
5. Evrendilek, G. A., Tok, F. M., Soylu, E. M., & Soylu, S. (2008). Inactivation of Penicillium expansum in sour cherry juice, peach, and apricot nectars by pulsed electric fields. *Food Microbiology, 25*(5), 662–667.
6. Falguera, V., Pagán, J., Garza, S., Garvín, A., & Ibarz, A. (2011). Ultraviolet processing of liquid food: A review. Part 1: Fundamental engineering aspects. *Food Research International, 44*(6), 1571–1579.
7. Falguera, V., Pagán, J., Garza, S., Garvín, A., & Ibarz, A. (2011). Ultraviolet processing of liquid food: A review. Part 2: Effects on microorganisms and food components and properties. *Food Research International, 44*(6), 1580–1588.
8. Fan, X., Huang, R., & Chen, H. (2017). Application of ultraviolet C technology for surface decontamination of fresh produce. *Trends in Food Science and Technology, 70*, 9–19.
9. Guerrero-Beltrán, J. A., & Barbosa-Cánovas, G. V. (2005). Reduction of *Saccharomyces cerevisiae*, *Escherichia coli,* and *Listeria innocua* in apple juice by ultraviolet light. *Journal of Food Process Engineering, 28*(5), 437–452.
10. Hagen, S. F., Borge, G. I. A., Bengtsson, G. B., Bilger, W., Berge, A., Haffner, K., & Solhaug, K. A. (2007). Phenolic contents and other health and sensory-related properties of apple fruit (*Malus domestica* Borkh., cv. Aroma): Effect of postharvest UV-B irradiation. *Postharvest Biology and Technology, 45*(1), 1–10.
11. Hinds, L. M., O'Donnell, C. P., Akhter, M., & Tiwari, B. K. (2019). Principles and mechanisms of ultraviolet light emitting diode technology for food industry applications. *Innovative Food Science and Emerging Technologies, 56.*
12. Hui, Y., & Evranuz, E. (2012). Fruits and fruit products. In *Handbook of Plant-Based Fermented Food and Beverage Technology* (Second Edition, pp. 163–164), CRC Group of Taylor and Francis.
13. Keklik, N. M., Krishnamurthy, K., & Demirci, A. (2012). Microbial decontamination of food by ultraviolet (UV) and pulsed UV light. In *Microbial Decontamination in the Food Industry: Novel Methods and Applications* (pp. 344–369). Elsevier Ltd.
14. Koutchma, T. (2009). Advances in ultraviolet light technology for non-thermal processing of liquid foods. *Food and Bioprocess Technology, 2*(2), 138–155.
15. Koutchma, T. (2008). UV light for processing foods. *Ozone: Science and Engineering, 30*(1), 93–98.

16. Liu, C., Han, X., Cai, L., Lu, X., Ying, T., & Jiang, Z. (2011). Postharvest UV-B irradiation maintains sensory qualities and enhances antioxidant capacity in tomato fruit during storage. *Postharvest Biology and Technology, 59*(3), 232–237.
17. Lu, G., Li, C., & Liu, P. (2011). UV inactivation of milk-related microorganisms with a novel electrodeless lamp apparatus. *European Food Research and Technology, 233*(1), 79–87.
18. Matak, K. E., Sumner, S. S., Duncan, S. E., Hovingh, E., Worobo, R. W., Hackney, C. R., & Pierson, M. D. (2007). Effects of ultraviolet irradiation on chemical and sensory properties of goat milk. *Journal of Dairy Science, 90*(7), 3178–3186.
19. Mau, J. L., Chen, P. R., & Yang, J. H. (1998). Ultraviolet irradiation increased vitamin D2 content in edible mushrooms. *Journal of Agricultural and Food Chemistry, 46*(12), 5269–5272.
20. McDonald, K. F., Curry, R. D., Clevenger, T. E., Unklesbay, K., Eisenstark, A., Golden, J., & Morgan, R. D. (2000). Comparison of pulsed and continuous ultraviolet light sources for the decontamination of surfaces. *IEEE Transactions on Plasma Science, 28*(5), 1581–1587.
21. Molinari, R., Lavorato, C., & Argurio, P. (2015). Photocatalytic reduction of acetophenone in membrane reactors under UV and visible light using TiO_2 and Pd/TiO_2 catalysts. *Chemical Engineering Journal, 274*, 307–316.
22. Raeiszadeh, M., & Adeli, B. (2020). A critical review on ultraviolet disinfection systems against COVID-19 outbreak: Applicability, validation, and safety considerations. *ACS Photonics, 7*(11), 2941–2951.
23. Shama, G. (2014). Ultraviolet Light. In *Encyclopedia of Food Microbiology* (2nd ed., pp. 665–671). Elsevier.
24. Shin, J. Y., Kim, S. J., Kim, D. K., & Kang, D. H. (2016). Fundamental characteristics of deep-UV Light-Emitting Diodes and their application to control foodborne pathogens. *Applied and Environmental Microbiology, 82*, 2–10.
25. Turtoi, M. (2013). Ultraviolet light treatment of fresh fruits and vegetables surface: A review. *Journal of Agroalimentary Processes and Technologies, 19*(3), 325–337.
26. Ubi, B. E., Honda, C., Bessho, H., Kondo, S., Wada, M., Kobayashi, S., & Moriguchi, T. (2006). Expression analysis of anthocyanin biosynthetic genes in apple skin: Effect of UV-B and temperature. *Plant Science, 170*(3), 571–578.
27. Wambura, P., & Verghese, M. (2011). Effect of pulsed ultraviolet light on quality of sliced ham. *LWT – Food Science and Technology, 44*(10), 2173–2179.
28. Wang, C. Y., Chen, C. T., & Wang, S. Y. (2009). Changes of flavonoid content and antioxidant capacity in blueberries after illumination with UV-C. *Food Chemistry, 117*(3), 426–431.
29. Zhang, D., Yu, B., Bai, J., Qian, M., Shu, Q., Su, J., & Teng, Y. (2012). Effects of high temperatures on UV-B/visible irradiation induced postharvest anthocyanin accumulation in "Yunhongli No. 1" (*Pyrus pyrifolia* Nakai) pears. *Scientia Horticulturae, 134*, 53–59.

CHAPTER 4

EFFECT OF ULTRAVIOLET LIGHT ON MICROBIAL INACTIVATION: MECHANISM OF INACTIVATION AND IMPACT ON DIFFERENT MICROORGANISMS

MONISHA CHANDRAN, VIDYA D' CRUZ,
SANJAY KANDARATH GIRIJAN, and K. A. ATHMASELVI

ABSTRACT

Food safety has become a global concern in recent decades. One-third of processed food deteriorates due to microbial contamination and is considered to be lost and wasted. Contamination may occur at any stage of food processing, such as harvesting, processing, packaging, and storage. Foodborne pathogens are destroyed by both thermal and non-thermal techniques. However, thermal techniques have a few drawbacks due to the direct contact of heat, which leads to the formation of undesirable characteristics in food. Thus, non-thermal technology plays a vital role in food processing and safety. Ultraviolet (UV) light-based techniques are effective non-thermal disinfection methods for inactivating microorganisms to prevent food contamination. Despite UV disinfection on the surface, there are vast potential applications in the food sector. Photochemical alterations in the DNA strand are triggered by UV radiation, including damage to the same strand of DNA caused by the interbinding of adjacent pyrimidine nucleoside bases such as thymine and cytosine. This chapter mainly focuses on the mechanism of microbial inactivation caused by cellular deformation, which is influenced by structural changes in the DNA molecule that inactivate

Nonthermal Light-Based Technologies in Food Processing.
C. K. Sunil, Preeti Birwal, Megh R. Goyal, and R. Mahendran (Eds.)

microorganisms. The effect of UV light-based techniques on liquid and solid foods has been explored in this review. In addition, studies that depict the factors that govern UV light's potential to inhibit certain microorganisms such as bacteria, fungi, and spores have been discussed. In this study, the synergistic effectiveness of UV light-based microbe inactivation will be compared with other non-thermal techniques. This study will also look into UV light-based approaches, which are restricted in the food industry due to safety concerns regarding microbial reduction. To overcome the limitations of UV-based techniques, there are a few combination processing techniques that can be used to achieve complete microbial destruction. Thus, UV light-based techniques can be mass-commercialized in the food sector to inhibit microbial growth and enhance the shelf life of food products.

4.1 INTRODUCTION

Food safety is a serious concern confronting the food processors and service sectors, predominantly ready-to-eat foods. The potential of microbial contamination of the final product significantly increases [8]. Despite tremendous progress in the understanding of microbial transmission and virulence in food contamination, foodborne illnesses remain a global concern with serious public health consequences [4]. Recently, numerous foodborne outbreaks have emerged due to improper processing and handling of food products. As a result, regulations have been strictly made to strengthen the safety and quality of food products. According to the Food and Drug Administration, microbial reduction in the processed product should achieve a 5-log reduction compared to the unprocessed product [20, 49]. Thus, thermal techniques have prevailed for ages along with undesirable attributes to decrease the microbial load from food products. Some of the most important drawbacks of traditional thermal processing methods are mainly chemical reactions that result in flavor loss, nutrient destruction, and some product quality loss [42]. On the other hand, ultraviolet light-based techniques have emerged as a significant non-thermal tool to reduce or eliminate microorganisms, thereby sterilizing and disinfecting pathogenic organisms. Ultraviolet light is non-ionizing electromagnetic radiation, and its spectrum includes wavelengths from 100 nm to 400 nm [11]. Technology using ultraviolet light is a decontamination method used in the food industry as an alternative to thermal and chemical methods of food preservation [33, 35]. It has gained wide attention due to its cost-efficiency, eco-friendliness, and ease of use. UV technique has been effectively adapted for commercial applications like disinfection of food and drinking water, decontamination of the surface of packaging materials, wastewater treatment, and sterilizing of meat, fruit juices, and beverages [39, 41, 59]. There are many factors that influence the efficacy of the disinfection process [5]. The effectiveness of this technology against various microbes including bacteria, fungi, protozoans, viruses, etc., has been extensively studied by many researchers [2, 22, 61]. Recently, many advanced methods coupled with ultraviolet technique resulted in a significant approach with synergistic effects. By consistently reviewing all of these criteria, a comprehensive insight into the significance of such technology in the food sector can be acquired.

The chapter mainly focuses on different principles of UV light and various mechanisms by which microbial inactivation takes place and the impact of UV on different types of microbes. The impact of the nature of food and different process parameters on the efficiency of the treatment and comparison of UV with other similar light-based technologies are reviewed. Their limitations and future prospects are also covered in this chapter.

4.2 MECHANISM OF INACTIVATION OF MICROBES

Ultraviolet light is utilized in food applications for its preservative effect. The major objectives of food preservation are to obtain safe, stable, and quality products by reducing the microbial load without compromising their nutritional value. Most food preservation techniques inactivate microbes through the application of heat, which might cause detrimental effects on the sensory and nutritional attributes of food. Ultraviolet light (200–400 nm) is non-ionizing radiation having germicidal properties. The UV spectrum can be subdivided into three categories such as UV-A as a short wavelength of 320–400 nm, UV-B as a medium wavelength of 290–320 nm, UV-C as a long wavelength of 200–290 nm, and vacuum UV wavelength of 100–200 nm. UV-A is further split into UVA2 (320–340 nm) and UVA1 (340–400 nm), while UV-B is further classified into broadband UVB (290–320 nm), 380 nm man-made excimer laser, and narrowband UVB (NB-UVB), which ranges from 311 to 313 nm. The UV-C region having wavelengths of 200 to 280 nm is considered the germicidal spectrum. It induces a lethal effect on bacteria, fungi, and other microorganisms [46]. The type of lesion produced in the microorganism depends on the radiation band used. A sigmoidal curve of destruction results from the inactivation of microbes. The food undergoing treatment absorbs the energy of ultraviolet radiation leading to alterations in the food constituents. Distinct UV spectral areas have different biological effects in terms of the cellular targets they prefer and the mode of action through which they cause damage (Figure 4.1).

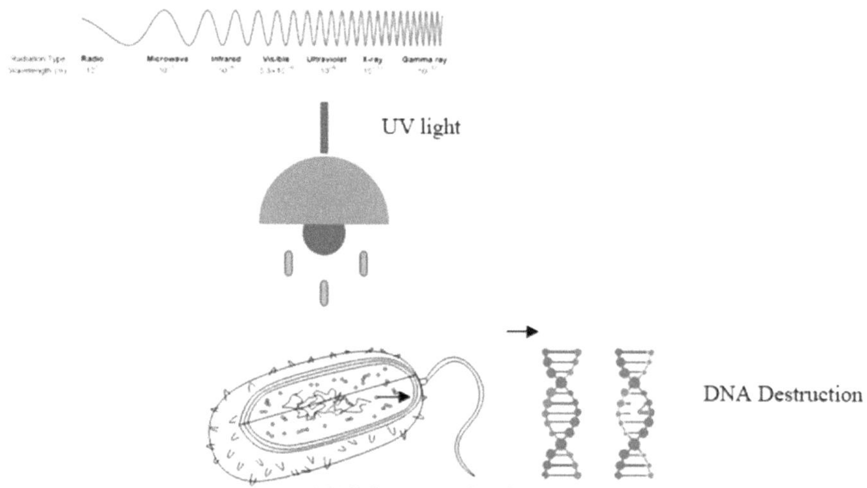

FIGURE 4.1 Mechanism of UV inactivation of microorganisms.

4.2.1 DNA DAMAGE

The most potent wavelength for inactivation of microorganisms ranges between 250 and 260 nm (peak at 254 nm) since nitrogenous bases in DNA intensely absorb these photons due to their aromatic ring structure. These UVC photons absorbed in the food can rupture the genetic material of microbial cells leading to alteration of cell metabolisms and enzyme activities, damage to the integrity of the cytoplasmic membrane, and inhibition of reproduction followed by cell death [17]. Potentially toxic DNA lesions are scattered randomly over the cells of microbes, affecting the population as a whole. Lethal mutation to DNA or RNA occurs due to hydration of base or base dimerization by cross-linking between neighboring pyrimidine nucleoside bases thereby disrupting its normal assembly.

UV light can produce six different photoproducts. This includes dimers between two uracils, two cytosines, two thymines, cytosine and thymine, uracil and thymine, and uracil and cytosine [39]. The photoproduct with the maximum quantum efficiency in DNA is the thymine dimer. Dimer formation occurs by the break of the hydrogen bonds linking the thymine bases, followed by cross-linking of the 5th and 6th carbon atoms [39]. The photoproducts produced also include CPDs and 6–4PPs. Among these, the former is usually formed between thymines (DNA), cytosines (DNA and RNA), and uracil (RNA), depending on nucleotide composition. They interfere with DNA replication and RNA transcription, leading to mutagenesis [56]. Apart from the development of these photoproducts, bacterial spore photoproducts (SPs) are formed in UV-irradiated bacterial spores [55]. Furthermore, indirect photochemical actions such as the production of free radicals can cause ultra-structural mutations [15]. The amount of mutations that can be recovered is determined by the target microorganism's UV repair mechanism. In bacteria, dimer formation can be reversed by absorption of wavelengths between 300 and 500 nm; it is termed photo-reactivation by photolyase enzymes dividing the dimers. The quantity of cross-links is beyond repair once the cross-linking threshold is surpassed, and cell death ensues.

4.2.2 *LIPID OXIDATION*

UV-A has largely indirect effects, which are driven by ROS (reactive oxygen species) produced by photodynamic processes that involve photosensitizers. These highly reactive species can react with cellular components, particularly lipids and proteins, causing changes in the permeability of the cell membrane and/or modification of transmembrane ion gradients, leading to cell death [10]. Irradiation with long-wave UV causes hydroperoxide radicals to develop in the membrane's unsaturated fatty acids, causing alterations in membrane permeability [40]. According to Santos et al. [53], UV exposure increased the number of methyl groups attached to triglycerides, oxidation of fatty acids, and compositional changes in lipids. These changes were confirmed by using gas chromatography. The lipid peroxidation in the liposomal membrane increases with an increase in UVA fluence (Figure 4.2).

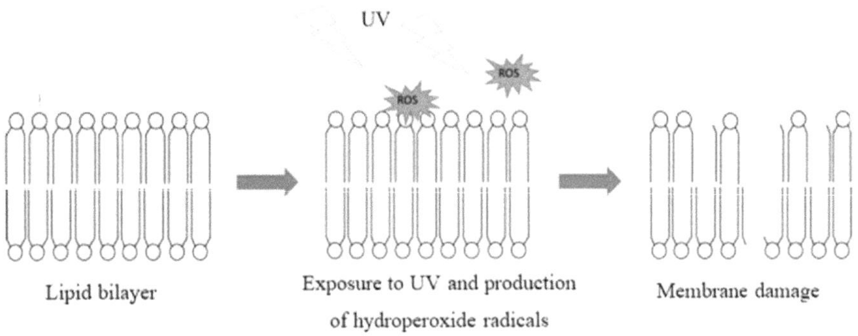

FIGURE 4.2 Lipid oxidation.

4.2.3 PROTEIN ALTERATION

Apart from DNA damage, photooxidation (especially UVA) may cause damage to proteins, lipids, and sterols. Because of the enormous availability of endogenous chromophores inside their cellular structures, proteins are important biological targets for photooxidation. The majority of UV-penetrated protein destruction is caused during the formation of reactive oxygen species (ROS) [56]. Furthermore, the primary, secondary, and tertiary structures of microbial proteins replicate on their nitrogen base (amino groups) components. One out of every 10 amino acids is known to be sensitive to photochemical reactions, and any of these amino acids might hypothetically impact the structure and function of the protein [30]. Aromatic amino acids present in microbial proteins like phenylalanine and tryptophan absorb UV light. UV radiation causes them to cross-link at their carbon-carbon double bonds. This induces protein denaturation leading to membrane depolarization and irregular ionic flow [40]. When bacterial protein absorbs UV, it induces the development of protein oxidation products, and changes in protein post-translational modifications including phosphorylation, glycosylation, and propionylation levels [53].

4.3 EFFECTS OF ULTRAVIOLET LIGHT TREATMENT

The usage of chemicals to preserve food and prevent spoilage has led to a negative public reaction, with consumers being much more aware and informed about the situation prevailing in the food industry. In the food processing industry, currently, ultraviolet light treatment holds considerable promise. UV light has shown potential for pasteurization of juices and beverages, as well as controlling microbial contamination on egg and meat surfaces. Apart from being a non-thermal alternative to conventional heat processing, it has the potential to increase the shelf life of raw fresh products. Even selected solid and liquid meals can be made safer by UV light processing without sacrificing quality or nutrient content [35]. Contamination of fresh produce can occur at any time during the product's processing line, which can lead to spoilage. Microbial contamination on fruits and vegetables occurs due to animal and human fecal matter, contaminated manure, polluted dust, industrial additives, irrigation water, water used for agrochemicals, and other agricultural reasons [9].

Decontamination of water was earlier done using UV radiation, now it is also used for disinfection of liquids other than water, air sanitation, and surface sterilization [11]. Contaminations and their consequences, which further lead to spoilage and pathogenic after-effects, are the results of the action of pathogenic and spoilage-causing microorganisms that can thrive in the conditions provided. Conditions that can make their habitat unfavorable can help prevent their growth and the detrimental outcome that follows. UV light treatment is one such method we need to emphasize to bring down the worsening effects caused by spoilage and pathogenic microorganisms.

4.3.1 EFFECT ON DIFFERENT MICROORGANISMS

UV light has been shown to be effective against viruses and protozoa, as well as a number of food pathogens and spoilage bacteria. UV light has a wide range of wavelengths between 100 and 400 nm, although the fatal impact on microbes was primarily due to the short-wavelength of UV-C light, which ranges between 200 and 280 nm and causes organisms to inactivate as a result of DNA damage. Inter-linking in the same DNA strand between thymine and cytosine, which are pyrimidine nucleoside bases, was caused by the photochemical process, disrupting DNA transcription and replication, and ultimately resulting in cell death [11]. Psychotropic aerobic Gram-negative bacteria, yeasts, molds, heterofermentative lactobacilli, and spore-forming bacteria are some of the spoilage microorganisms that might cause issues. The amount of microbial recontamination of pasteurized fluid milk products is a major determinant of their shelf life. The presence of a wide range of metabolic by-products, which create off-odors and scents, as well as visible color and texture changes, characterizes fungal spoiling of dairy items. Extracellular hydrolytic enzymes are produced in great quantities by psychotropic bacteria, and the amount of microbial recontamination of pasteurized fluid milk products is a major determinant of their shelf life. The presence of a wide range of metabolic by-products, which create off-odors and scents, as well as visible color and texture changes, characterizes fungal spoiling of dairy items [51].

4.3.1.1 BACTERIA

Spores can survive high processing temperatures, thus spore-forming bacteria are generally connected with the deterioration of heat-treated goods. Gram-positive bacteria can be either strict anaerobes or facultative anaerobes (able to grow in the presence or absence of oxygen). Some spore-formers are thermophilic (up to 55C). Some anaerobic thermophiles create hydrogen sulfide (Desulfotomaculum), and others produce hydrogen and carbon dioxide during the development of Thermo-anaerobacterium in canned/hermetically sealed foods maintained at high temperatures, such as soups served in vending machines [51]. The susceptibility of the microorganisms in product to UV radiation varies greatly. Gram-negative bacteria, yeast, bacterial spores, molds, and viruses are all more susceptible than gram-positive bacteria. Variations in cell structure may limit the effect of ultraviolet rays' penetration into cells, which leads to affects media turbidity. The genus' DNA repair capabilities, as well as the helix structure of DNA constituents and cell size, assess the sensitivity of microorganisms to UV irradiation. Different species and strains have diverse sensitivities. The physiological status of the cells can also influence resistance to microbial activity upon ultraviolet light techniques [23]. A measurement known as the D value was widely implemented to quantify UV effect on microbial growth. UV doses required to lower microbial populations by one order of magnitude. The vulnerability of a certain organism to varying levels of UV radiation is demonstrated using survival curves (UV dose-response). UV disinfection experiments display the average D values for distinct bacteria groupings [37]. The bacteria spectrum does not include *Deinococcus radiodurans,* the bacteria that has shown to be the most UV-resistant to date. *Deinococcus* D values range from 19.7 to 145 mJ/cm^2. UV inactivation doses, D value (mJ/cm^2) for various microbial species measured at 253.7 nm [37]. Cocci and micrococci showed a D value ranging from 1.5 to 20, Spore formers 4 to 30, Yeast 2.3 to 8, and enteric bacteria and enteric viruses having values from 2 to 8 and 5 to 30, respectively. Fungi, protozoa, and algae showed D values ranging from 30 to 300, 60 to 120, and 300 to 600, respectively.

4.3.1.2 FUNGI

Fungi generally include both yeast and mold, which play a vital role in food decomposition. Molds are multicellular fungi that reproduce by spore production (single cells that can grow into a mature fungus). Spores grow in enormous quantities and disperse quickly through the air. If conditions are favorable, these spores can grow and multiply after they fall on a feeding substrate. Yeasts are unicellular fungi, which are much larger cells than bacteria. Cell division (binary fission) or budding are the two ways they replicate. Yeast and mold being the major cause of food spoilage, the effectiveness of UV radiation on these fungi has been the subject of numerous investigations and research projects. In a study conducted on the action of various species of yeast in winemaking, different species of yeast were introduced, and it was found that different quantities of yeast occur during stages of winemaking. The samples were then treated with UV light to determine their tolerances at various dosages. Among all the species studied, *M. pulcherrima* had the maximum resistance to ultraviolet radiation. There were still 10^2 countable colonies at the maximum exposure of 1.0 kJ/L. UV-C sensitivity was much higher in *S. cerevisiae* and *H. uvarum*. In both yeasts, UV-C treatments surpassing 0.6 kJ/L resulted in the inactivation of 6-log ranges. The UV-C dose of 1.0 kJ/L, *P. fermentans* and *Candida* sp. depicted 6-log microbial inactivation, putting them between *S. cerevisiae* and *M. pulcherrima* [18]. Similarly, Marquenie et al. [44] concluded with the results that *Botrytis cinerea* and *Monilinia fructigena*, ultraviolet irradiation ranges in 1.00 J/cm^2 and 0.50 J/cm^2 rendered them totally inactive. The two most common post-harvest rotting fungi found in strawberries and cherries are these two [44]. Also, it was found UV dosage of minimum 4,800 and maximum of 7,500 J/m^2 which has the highest impact on resistance to the formation of surface culture of the fungal growth. The ultraviolet (UV-C) light has been shown to lower the frequency of disease in storage rot peach caused by *M. fructicola*, tangerine green mold (*Penicillium digitatum*), tomato soft and firm rot, and the occurrence of storage rot in sweet potato [58].

4.3.1.3 SPORES

Spores are the reproductive means of fungi, bacteria, or any plant material. When the spores come in contact with a favorable growth surface, they are able to produce their parent organism. Later, they can turn into colonies, which upon maturation produce spores and hence the cycle of reproduction continues [54, 55]. Spore contamination is particularly common in the baking business, where spores can infect newly baked items, leading to the growth of molds. Possible molds that lead to loss and spoilage in bakery products include *Eurotium, Aspergillus*, and *Penicillium* species [7]. In baked foods, ultraviolet light lamps are commercially applied on bread slicing machines to decrease the contaminated air and food packages. UV irradiation will not liberate thermal energy, though it might scorch and induce condensation on films. Bakery products that are packed in packaging film might show significant increases in mold-free shelf life after administering UV at intensity ranges of 12 mJ/cm^2 for 30–60 seconds. After baking, the loafed mare was transported into ultraviolet fixed tunnels to inhibit any microbial activity that may have surface contamination. The use of an ultraviolet chamber to treat baked loaves resulted in a significantly longer shelf life [7]. Rice products, starch-rich foods, cooked meat, cereals, and cereal by-products, milk products, spices, and condiments are all common habitats of *B. cereus* spores. Through the synthesis of different toxins, *B. cereus* may lead to either diarrheal or emetic food poisoning. In immunologically impaired and fragile people, *B. cereus* may lead to local disease [34]. The use of UV-TiO_2 photocatalysis as a physical disinfection strategy in the reduction of bacteria by non-thermal techniques has been reported to be successful [62].

4.3.2 EFFECT ON DIFFERENT FOODS

4.3.2.1 LIQUID FOODS

UV light has a lot of promise when it comes to minimizing microbial growth in a wide range of food products, specifically in fruit juices and beverages. Thus, the final product contains fewer color components, organic compounds, and suspended debris. Liquid foods may absorb less ultraviolet light, reducing the efficacy of UV pasteurization procedures. UV radiation has the potential to be used for pasteurization of juices and drinks as a non-thermal alternative to typical thermal processing [35].

Liquid foods such as fruit juices and beverages have a variety of optical and physical properties, as well as chemical compositions, all of which affect UV light transmission (UVT), dosage administration, momentum transfer, and, as a result, microbial inactivation when compared to potable water. In contrast, ultraviolet irradiation has an optimistic consumer perception in physical preservation methods and is appealing to the food sector as an economical non-thermal method. UV light-based technology has the potential to be a viable substitute for pasteurization (thermal technique) in a variety of liquid foods [49]. Recent trends in the application of ultraviolet light-based science and engineering reports that there is sufficient data supporting the literature. Ultraviolet light transmission through liquid foods is limited when compared to UV-treated drinking water due to their high optical density and substantial absorption and dispersal properties of UV light [35].

4.3.3.2 SOLID FOODS

The relevance of ultraviolet light treatment to raw fruits, vegetables, and roots before storing consists of two applications. UV radiation has the ability to reduce the number of germs on the product's surface as well as induce host resistance to the bacteria. Hormesis refers to the beneficial effect of UV-C light on fresh foods, and the agent (UV light) is referred to as 'hormetin' or 'hormetic effect' [14, 25]. UVC light has been demonstrated to inhibit the occurrence of peach storage rot disease caused by *M. fructicola*, tangerine green mold (*Penicillium digitatum*), tomato soft rot, and sweet potato storage rot [58]. Before refrigeration, certain varieties of beef can be UV-C treated on the surface to reduce microbial load. Depending on the dose, UV-C radiation decreases the microbial burden in fresh beef by two or three log cycles. The microbial decrease improves as the dose is increased. Opaque materials, on the other hand, do not allow the radiation to pass through [58]. Fresh meat does not change color or appearance due to the UV-C radiation used for disinfection.

4.4 EFFECT OF PROCESS PARAMETERS OF ULTRAVIOLET LIGHT ON MICROORGANISMS

The rate of reduction in microbial load in the food being treated by UV depends on its different process parameters. This includes the wavelength to which food is exposed, the light source of UV, the distance of the subject from the source, and the time of exposure. There is a need to understand the effect of each parameter so that an optimum microbial reduction in less time with more efficiency can be achieved using specific process parameters. Hence, the effect of each parameter is being discussed here.

4.4.1 UV LIGHT SOURCE AND WAVELENGTH

The light source is the most critical component of a UV disinfectant system. It is important to choose a proper UV source with an optimum wavelength to achieve increased penetration of radiation in food and maximum efficiency of the process. The mechanism of inactivating microbes depends on the wavelength range used. As previously discussed, UVC is the germicidal range that directly affects microbial DNA, while the UVA and UVB ranges act on microbes by altering the lipids and proteins in their structure. The effectiveness of different UV lamps in destroying microbes depends on their ability to produce wavelengths close to the absorption peak of the target DNA.

The emission of wavelength entirely depends on the lamp used for treatment. The major two sources of UV are the sun in the solar system (180–400 nm) and man-made UV generating lamps. The latter include arc lamps, excimer lamps, fluorescent lamps, UV gas-discharge lamps,

UV light-emitting diodes (UV-LED), microwave discharge electrodeless lamps, TL-01 Ultraviolet lamps, Ultraviolet lasers, TL-12 Philips Lamp, black-light ILTs, short wave UV lamps, etc. [1]. These can be categorized as low-pressure (LP) lamps (UVC), medium-pressure (MP) lamps (UVC/UVB), and pulsed xenon lamps used in Pulsed UV technology (PUV). MP lamps have a broad emission spectrum (185–400 nm) whereas LP lamps emit a single wavelength peak (254 nm) in the UVC band [31]. Hence, UV generated from LP lamps affects only DNA and there is a chance of photoreactivation and repairing of affected DNA. In the case of MP lamps, their broad polychromatic spectrum of UV irreversibly affects not only DNA but other microbial cell constituents such as lipids and enzymes as well. Kalisvaart et al. [31] observed that low-pressure lamp UV damage was repaired by pathogenic *E. coli* O175:H7, whereas repair was not detected for the same after being exposed to UV from MP lamps. Therefore, MP lamps can be considered more effective for use in disinfectant systems. However, Gayan et al. [22] reported that both lamps show the same microbial survival rate, i.e., the UV dose required to inactivate a specified number of microorganisms using LP and MP lamps is not significantly different. Even though it reduces the microbial load in the same manner, chances of repair are still possible. Mbonimpa et al. [46] observed that bacterial inactivation does not occur beyond 320 nm in the UV range. And it is possible for photoreactivation of bacteria, which restored their viability after inactivation at 297 nm (UVA). Hence, to create permanent damage, multiple dosages or destruction mechanisms over inactivation thresholds may be necessary. Hence, MP lamps may be preferred over LP lamps.

4.4.2 DISTANCE AND TIME OF EXPOSURE

Other important parameters are how long you expose the food to UV and at what distance the food is being kept from the UV source. Less distance with more time of exposure yields more disinfection efficiency. Fahim et al. [19] observed that the time of exposure and distance from the source are directly proportional to each other. This means lesser exposure time is required if food is kept close to the source and vice versa, provided that the whole area should be thoroughly exposed to the source, without any obstructions in between the samples. Katara et al. [32] found that good disinfection efficiency can be obtained when UV is applied from a distance of eight feet (2.4384 m) on either side of the tube. Also, an adequate exposure time was found to be 30 minutes under this condition. Application of these optimized conditions inactivated pathogenic aerobic and anaerobic bacteria in the order of four logs. Similar conditions like a distance of 2 m from the source for 30 minutes were found effective [16]. *Candida auris* was the micro-organism under study. If the time is halved or the distance is doubled, the efficacy of the treatment diminishes by 10-fold and 50-fold, respectively. Hence, the exposure time and distance are two critical parameters to resist microbes using UV.

4.4.3 UV DOSE

The intensity of the consequences of UV radiation on microbes depends on the dose of UV applied. UV dose, also referred to as UV irradiance or intensity flux, is a function of the strength or intensity of UV and exposure time. It is the amount of energy of UV put into a given surface over a specific period. It can be determined by the following equation [45].

UV dose, D (mJ/cm^2) = irradiance (mW/cm^2) × exposure time (s)

$$1\ mJ/cm^2 = 1{,}000\ mWs/cm^2$$

Irradiance is the intensity of UV radiation on the specified area of surface, expressed in W/cm^2. Exposure time is a factor of the flow rate of the product in the case of liquids, whereas for other foods, it depends on the speed of the conveyor on which the product is being carried. The relationship between dosage and destruction of a target microorganism can be given by,

$$N/N_0 = e^{-KD}$$

where; 'N' is the initial population of target organisms; 'N_0' is the population of target organisms after UV exposure; 'K' is a constant related to the sensitivity of the organism; and 'D' is the UV dose [52]. According to this relationship, doubling the dosage will result in a tenfold increase in the destruction of the target organism. Doubling the dosage required for 90% destruction will result in 99% destruction in the target organism. If you triple the dose, the target organism will be destroyed by 99.9%, and so on. The dose required to kill the microorganisms depends on their characteristics. It varies for different bacteria, fungi, viruses, vegetative cells, and spores. The sensitivity of different microbes concerning UV dose can be given as bacteria ~ protozoa > yeast > molds > bacterial spores. Molds have a more complex structure and hence are more resistant to UV than bacteria. The

UVC energy dosage of 12,000 mWs/cm^2 was required to inhibit vegetative bacteria. However, 22,000 $\mu Ws/cm^2$ was needed against all pathogens [43].

4.5 SAFETY ASPECTS OF UV LIGHT IN MICROBIAL INACTIVATION

The ultraviolet light technique is a novel non-thermal method used as an alternative technique to thermal processes to retard microbial growth and enhance the shelf life of processed food products. The major concern in UV-treated food products is safety regulation limits. Currently, many countries have approved UV treatments on food products at recommended doses. The required UV irradiation is assessed based on the type of food, initial concentration of microorganisms, and UV equipment design setup. According to FDA regulations, the radiation source should be a low-pressure mercury (LPM) lamp which emits a wavelength of 253.7 nm. Table 4.1 shows national and international standards for radiation in food and potable water [20].

4.6 LIMITATIONS OF ULTRAVIOLET LIGHT IN MICROBIAL INACTIVATION

Few studies have shown that microbial inactivation using UV light is more effective in gram-negative bacteria than in gram-positive bacteria, yeast, fungi, spores, and viruses [5]. The mechanism behind the antimicrobial effect on bacteria is the impact on the pyrimidine dimers. Furthermore, intrinsic factors and the thick peptidoglycan layer in the cell wall influence the penetration of UV light into the bacterial cell structure. Additionally, spore cells in a dormant state reduce pyrimidine dimerization. Spores are more resistant to microbial inactivation compared to vegetative cells due to the presence of a thick protein layer. The inability of UV radiation to penetrate opaque and uneven surfaces results in poor microbial inactivation efficacy. Light properties, food characteristics, transmittance, color, dose, and thickness of the food substance are all important aspects to consider [17]. Low- and medium-pressure mercury lamps (LPM and MPM), pulsed light (PL), and light-emitting diodes (LEDs) are all commercially available as UV sources which significantly play a role in microbial inactivation [28, 37]. Because the action has a better impact on their direct contact with the sample positioned under

TABLE 4.1 Safety Regulations of UV Light in Treated Foods

Regulation	Commodity	Parameters		References
FSSAI	Food additives – Microbial decontamination.	Dose – minimum 5–10 maximum (kiloGray).		[36]
	Packaging substances and other related materials.	Dose – minimum 5–10 maximum (kiloGray).		
	RTE foods, traditional foods, space foods, minimal manufactured foods.	Dose – minimum 2–10 maximum (kiloGray).		
US FDA	UV pasteurized water for dairy sector.	LP UV at 254 nm at 186 $mJcm^{-2}$.	Reduction of pathogenic microorganism in water.	[20]
	Drinking water	Irradiation wavelength – 2537A, operating temperature of lamp was 36 to 46°C, penetration of radiation in water bath should be 1 cm or less than 1 cm, absorption coefficient was 0.19/cm.	Complete sterilization of potable water.	
	Food and processed food products.	Irradiation wavelength should be 2,537 A with 1 W intensity. It has drawback on high fat foods.	Complete destruction of microorganism present in surface area.	
	Beverages (juices)	The flow should be maintained turbulent within tubes holding minimum 2,200 of Reynolds number.	Destruction of pathogenic microorganism.	
EU regulation	UV treatment on baker's yeast.	Irradiation wavelength – 254 nm, concentration of vitamin D in yeast ranges between 200 and 875 μg/g.	Inactivation of yeast and reduction of microbial growth.	[21]

the germicidal lamps, the positioning of the samples may alter the rate of microorganism suppression. Photoreactivation and dark repair are two mechanisms for reversing radiation-induced dimer generation. Using photolyase enzymes, photoreactivation is the major approach for reducing damage to irradiated microorganisms. The ability to recover increases the likelihood of microbe survival and extends the shelf life of ultraviolet light-treated products [25]. As a result, proper storage conditions are essential to prevent microbial photoreactivation in foods that are subjected to UV irradiation.

The application of the appropriate wavelength and dose of UV light should be a top priority in order to ensure the delivery of safe food products and reduce the risk of deterioration due to photoreactivation [25]. Few studies have reported that human eyes, burns, and skin care have all been reported to be damaged by continuous exposure [17]. Another major drawback of UV light-based techniques is that they are photosensitive to a few water-soluble and fat-soluble vitamins. In general, the photosensitive components of water-soluble vitamins include Vitamin C, B12, B6, B2, and folic acid, and similarly for fat-soluble vitamins include A, E, K, and carotene [38]. Furthermore, the design of UV equipment and the flow pattern of UV light will have an impact on the inactivation of microbes [17].

4.7 COMPARISON OF MICROBIAL INACTIVATION WITH OTHER TECHNOLOGIES

The microbial inactivation with light-based technologies has been given in Table 4.2. The combination treatment will give a synergistic effect and explore novel processing alternative techniques on the sample. The comparative study on the effect of microbial inactivation of UV with other combinations has been discussed in Table 4.2. The UV light-based technique will inhibit the growth of microorganisms, but is limited due to a few drawbacks like dark action and photoreaction. The sample has been treated with a combination of other light-based/other techniques such as pulsed, LED, ultrasound, PEF, and plasma. According to FDA regulations, the microbial reduction should be 5 log CFU/ml [20].

4.8 FUTURE SCOPE

Despite technological advancements, UV light-based approaches have emerged as optimistic non-thermal techniques with remarkable features such

TABLE 4.2 Comparisons of Microbial Inactivation of UV Light with Other Novel Techniques

Treatments	Parameters	Microbial Reduction (CFU/ml)	References
Pulsed and continuous UV light.	Surface sterilization Lamp – low pressure mercury lamp (253.7 nm) and pulse frequency was adjustable.	*E. coli* – Initial population 9.47 log reduced to 6.72 log (2 Hz) and 6.91 log (8 Hz). *Salmonella* spp. – Initial population 9.01 log reduced to 6.15 log (6 Hz) and 6.45 log (8 Hz).	[42]
Ultraviolet light	UV irradiation of apple juice – low pressure mercury lamp (254 nm).	Microbial population reduced to under detection limit of 1 log CFU/ml.	[11]
UVC radiation	UV-C radiation effect on fish meal (radiator power 30 W and distance 1 m).	Microbial population reduced in *Salmonella* to 3 log and *E. coli* has more than 4 log reduction.	[57]
UV irradiation	UV irradiation of apple cider, low mercury lamp with exposure of 14.32 mJ UV for 1.2 to 1.9s.	Microbial reduction to 5 logs 10 in *C. parvumoocysts*.	[26]
UV irradiation	Apple cider – Low mercury lamp, UV wave-length of 254 nm. Dose 6,500 $\mu J/cm^2$.	*E. coli* has initial population of 6–7 log CFU/ml which reduced by 5 log CFU/ml after treated.	[50]
UV-light emitting diodes	Mixed beverage – four UV-LED lamps with 8.33 mm diameter; 2 lamps at 280 nm and other at 365 nm.	Microbial reduction of *E. coli* has $<$ 4 log after UV treated from 4.09 log CFU/ml. Yeast and mold were reduced by 2.59 log (5.69 log) and 0.17 log (3.28 log).	[6]
Ultraviolet and pulsed electric field	UV and PEF exposed 50 cm chamber length in apple juice with a treatment time of 2.94s.	*E. coli* was reduced by 5.35 and 5.30 CFU/ml PEF following by UV treatment (no synergistic effect).	[24]
Ultraviolet and pulsed electric field	UV and PEF treatment in apple juice.	The microbial reduction in UV and PEF resulted in 2.2 and 5.4 log; In combination resulted in 7.1 log cycles.	[48]

TABLE 4.2 *(Continued)*

Treatments	Parameters	Microbial Reduction (CFU/ml)	References
UV-C irradiation	UV-C irradiation in RTE salads.	UV-C irradiation at 8,000 J/m^2 declined to microbial population of *E. coli* with 2.16 log CFU/g and *L. monocytohenes* with 2.57 log CFU/ml.	[13]
Ultrasound and UV-C light treatment.	Fruit juices – Ultrasound treatments with 20 kHz, 95 µm-wave amplitude and UV-C has a lamp (100 W).	USc treatment in *E. coli* provoked reduced up to 2.2 log CFU/ml.	[12]
Ultrasonication, ultraviolet, and atmospheric plasma processing.	Tomato based beverage – UV (254 nm) ultrasonication (Elmasonic S 100 H, 240 V, 37 kHz), ACP (260 V, 60 Kv, 50 Hz).	TBC of initial population is 2.3×10^3, yeast is 4.5×10^3, respectively, the microbial reduction after treatment was reduced as 1.5×10^2 CFU/ml at pasteurization, following in ultrasonication – 2.0×10^2 and ACP – 2.5×10^2 CFU/ml.	[47]
Thermal, HPP (high hydrostatic pressure), ultraviolet-C processing techniques.	Wheatgrass juice – thermal treatment (70C), HPP at 500 MPa for 60s, and UV-C fluence of 69.2 $mJcm^{-2}$.	The treatments support reduction by 5 log CFU in wheatgrass juice.	[3]
UV multiwave-length with light emitting diode (LED) based technology.	Synergistic effects observed at treatment at 285/405 nm combinations.	*B. subtilis* microbial reduction of 6.82 $\log_{10}$ at 285 nm, but the combined treatment resulted in lower reduction of 3.9 $\log_{10}$ CFU/ml.	[27]
Ultrasound combined with ultraviolet treatment.	Mango juice – US (600 W/10 min; 20 kHz); UV (8 W, 254 nm) were the optimal parameters for treatment.	The study depicts complete reduction of pathogenic by collapse induced due to microscopic bubbles and great shear rates due to HPP followed by DNA damage caused by UV light.	[60]

as rapidity, ease, less energy, and adaptability. For several years, ultraviolet radiation has been exploited as a physical disinfectant for air particles, surfaces, contact materials, and drinking water. It has been used in buildings, hospitals, institutions, and enterprises for a variety of reasons. Freshwater, material rinse water, industrial effluent, and water used in breweries are among the liquids that have been UV-C treated to reduce the concentration of bacteria. Furthermore, UV light is used to disinfect surface contamination. UV radiation is a cold disinfection technique; thus, it could be utilized for applications other than water disinfection in the food sectors. In the dairy and fruit juice industry, it can be utilized as an alternative technique for pasteurization. Additionally, all food packaging materials have been surfaced sterilized, including RTE foods (ready to eat), in order to safeguard against microbial contamination.

4.9 SUMMARY

The food safety is an emerging concern in the global population in this pandemic era. The production of microorganisms due to various intrinsic and extrinsic factors plays a vital role in the critical aspect of food safety. Thus, to preserve the food and increase the shelf life, processing techniques adopted like conventional thermal treatments. However, these thermal techniques impart changes in undesirable attributes due to heat induced in the material. Thus, emerging non-thermal light-based techniques eliminate and reduce microbial growth rapidly. The UV light technique has a prominent feature in cell destruction without altering the nature of the food component. However, there are a few constraints which can be overcome by a combination of other techniques. Thus, utilizing low-cost, easy handling equipment which inhibits the growth of pathogenic organisms.

ACKNOWLEDGMENTS

The authors would like to express sincere thanks to the National Institute of Food Technology, Entrepreneurship, and Management, Thanjavur (NIFTEM-T), for their supervision and guidance to accomplish this review work.

KEYWORDS

- **food borne pathogens**
- **microbial inactivation**
- **non-thermal light-based techniques**
- **ultraviolet**
- **UV light-emitting diodes**
- **UV-PEF**

REFERENCES

1. Ahmad, S.I., Christensen, L., & Baron, E. (2017). History of UV Lamps, Types, and Their Applications. In: Ahmad, S. (eds.). Ultraviolet Light in Human Health, Diseases and Environment. *Advances in Experimental Medicine and Biology, 996*. Springer, Cham.
2. Aisha, M. A., & Maznah, W. O. (2018). Effect of ultraviolet radiation on the growth of microorganisms developing on cave wall paintings. *AIP Conference Proceedings, 1994*, 070006. AIP Publishing LLC.
3. Ali, N., Popovi, V., Keith, K., & Yan, W. (2019). Effect of thermal, high hydrostatic pressure, and ultraviolet-C processing on the microbial inactivation, vitamins, chlorophyll, antioxidants, enzyme activity, and color of wheatgrass juice. *Journal of Food Process Engineering, 43*(1), 1–11.
4. Anderson, J. G., Rowan, N. J., MacGregor, S. J., Fouracre, R. A., & Parish, O. (2000). Inactivation of food-borne enteropathogenic bacteria and spoilage fungi using pulsed-light. *IEEE Transactions on Plasma Science, 28*(1), 83–88.
5. Barbosa-Cánovas, G. V., Tapia, M. S., Cano, M. P., Martín-Belloso, O., & Martínez, A. (2004). *Novel Food Processing Technologies* (1st ed.). CRC press.
6. Baykuş, G., Akgün, M. P., & Unluturk, S. (2021). Effects of ultraviolet-light emitting diodes (UV-LEDs) on microbial inactivation and quality attributes of mixed beverage made from a blend of carrot, carob, ginger, grape, and lemon juice. *Innovative Food Science and Emerging Technologies, 67*, 102572.
7. Begum, M., Hocking, A. D., & Miskelly, D. (2009). Inactivation of food spoilage fungi by ultraviolet (UVC) irradiation. *International Journal of Food Microbiology, 129*(1), 74–77.
8. Bintsis, T., Litopoulou-Tzanetaki, E., & Robinson, R. K. (2000). Existing and potential applications of ultraviolet light in the food industry–A critical review. *Journal of the Science of Food and Agriculture, 80*(6), 637–645.
9. Birmpa, A., Sfika, V., & Vantarakis, A. (2013). Ultraviolet light and ultrasound as non-thermal treatments for the inactivation of microorganisms in fresh ready-to-eat foods. *International Journal of Food Microbiology, 167*(1), 96–102.

10. Bose, B., & Chatterjee, S. N. (1995). Correlation between UVA-induced changes in microviscosity, permeability, and malondialdehyde formation in liposomal membrane. *Journal of Photochemistry and Photobiology B: Biology, 28*(2), 149–153.
11. Caminiti, I. M., Palgan, I., Muñoz, A., Noci, F., Whyte, P., Morgan, D. J., & Lyng, J. G. (2012). The effect of ultraviolet light on microbial inactivation and quality attributes of apple juice. *Food and Bioprocess Technology, 5*(2), 680–686.
12. Char, C. D., Mitilinaki, E., Guerrero, S. N., & Alzamora, S. M. (2010). Use of high-intensity ultrasound and UV-C light to inactivate some microorganisms in fruit juices. *Food and Bioprocess Technology, 3*(6), 797–803.
13. Chun, H. H., Kim, J. Y., & Song, K. B. (2010). Inactivation of foodborne pathogens in ready-to-eat salad using UV-C irradiation. *Food Science and Biotechnology, 19*, 547–551.
14. Civello, P. M., Vicente, A. R., & Martínez, G. A. (2006). UV-C technology to control postharvest diseases of fruits and vegetables. In *Recent Advances in Alternative Postharvest Technologies to Control Fungal Diseases in Fruits and Vegetables* (pp. 71–102), Transworld Research Network, 37/661(2), Fort P.O., Trivandrum–695023, Kerala, India.
15. Cutler, T. D., & Zimmerman, J. J. (2011). Ultraviolet irradiation and the mechanisms underlying its inactivation of infectious agents. *Animal Health Research Reviews, 12*(1), 15–23.
16. De Groot, T., Chowdhary, A., Meis, J. F., & Voss, A. (2019). Killing of Candida auris by UV-C: Importance of exposure time and distance. *Mycoses, 62*(5), 408–412.
17. Delorme, M. M., Guimarães, J. T., Coutinho, N. M., Balthazar, C. F., Rocha, R. S., Silva, R., Margalho, L. P., & Cruz, A. G. (2020). Ultraviolet radiation: An interesting technology to preserve the quality and safety of milk and dairy foods. *Trends in Food Science and Technology, 102*, 146–154.
18. Diesler, K., Golombek, P., Kromm, L., Scharfenberger-Schmeer, M., Durner, D., Schmarr, H. G., & Fischer, U. (2019). UV-C treatment of grape must: Microbial inactivation, toxicological considerations, and influence on chemical and sensory properties of white wine. *Innovative Food Science and Emerging Technologies, 52*, 291–304.
19. Fahim, Q., Uddin, N., Hassan, N. U., Sabir, N., Tariq, A., Bashir, A., & Ahmad, S. (2021). Evaluation of germicidal UV-C light for surface disinfection in a tertiary care hospital. *Pakistan Armed Forces Medical Journal, 71*(3), 889–894.
20. FDA. (2000). Irradiation in the production, processing, and handling of food: Ultraviolet radiation for the processing and treatment of food. *Federal Register, 65*, 71056–71058. https://www.federalregister.gov/documents/2000/11/29/00-30453/irradiation-in-the-production-processing-and-handling-of-food (accessed on 03 June 2024).
21. EFSA Panel on Nutrition, Novel Foods, Food Allergens (NDA), Turck, D., Castenmiller, J., De Henauw, S., Hirsch-Ernst, K. I., Kearney, J., & Knutsen, H. K. (2021). Safety of extended uses of UV-treated baker's yeast as a novel food pursuant to Regulation (EU) 2015/2283. *EFSA Journal, 19*(6), e06602. https://doi.org/10.2903/j.efsa.2021.6602.
22. Gayan, E., Condón, S., & Álvarez, I. (2014). Biological aspects in food preservation by ultraviolet light: A review. *Food and Bioprocess Technology, 7*(1), 1–20. https://doi.org/10.1007/s11947-013-1168-7.
23. Gayan, E., Monfort, S., Álvarez, I., & Condón, S. (2011). UV-C inactivation of Escherichia coli at different temperatures. *Innovative Food Science and Emerging Technologies, 12*(4), 531–541. https://doi.org/10.1016/j.ifset.2011.07.006.

24. Gachovska, T. K., Kumar, S., Thippareddi, H., Subbiah, J., & Williams, F. (2008). Ultraviolet and pulsed electric field treatments have an additive effect on the inactivation of *E. coli* in apple juice. *Journal of Food Science, 73*(9), 412-417. https://doi.org/10.1111/j.1750-3841.2008.00909.x.
25. Guerrero-Beltrán, J. A., & Barbosa-Cánovas, G. V. (2004). Advantages and limitations on processing foods by UV light. *Food Science and Technology International, 10*(3), 137–147. https://doi.org/10.1177/1082013204044359.
26. Hanes, D. E., Worobo, R. W., Orlandi, P. A., Burr, D. H., Miliotis, M. D., Robl, M. G., & Jackson, G. J. (2002). Inactivation of Cryptosporidium parvum oocysts in fresh apple cider by UV irradiation. *Applied and Environmental Microbiology, 68*(8), 4168–4172. https://doi.org/10.1128/AEM.68.8.4168-4172.2002.
27. Hinds, L. M., Charoux, C. M. G., Akhter, M., O'Donnell, C. P., & Tiwari, B. K. (2020). Effectiveness of a novel UV light emitting diode-based technology for the microbial inactivation of Bacillus subtilis in model food systems. *Food Control, 114*, 106910. https://doi.org/10.1016/j.foodcont.2020.106910.
28. IJpelaar, G. F., Harmsen, D. J., Beerendonk, E. F., van Leerdam, R. C., Metz, D. H., Knol, A. H., & Krijnen, S. (2010). Comparison of low pressure and medium pressure UV lamps for UV/H2O2 treatment of natural waters containing micropollutants. *Ozone: Science and Engineering, 32*(5), 329–337. https://doi.org/10.1080/01919512.2010.5148 83.
29. FSSAI, 2011: Food Safety and Standards (Food Products Standards and Food Additives) Regulations, 2011. (Available online at https://faolex.fao.org/docs/pdf/IND132615.pdf).
30. Kalisvaart, B. F. (2001). Photobiological effects of polychromatic medium pressure UV lamps. *Water Science and Technology, 43*(4), 191–197. https://doi.org/10.2166/wst.2001.0294.
31. Jagger, J., Stafford, R. S., & Mackin Jr., R. J. (1967). Killing and photoreactivation of Streptomyces griseus conidia by vacuum-ultraviolet and far-ultraviolet radiation (1500 to 2700 Å). *Radiation Research, 32*(1), 64–92. https://doi.org/10.2307/3572352.
32. Kalisvaart, B. F. (2004). Re-use of wastewater: Preventing the recovery of pathogens by using medium-pressure UV lamp technology. *Water Science and Technology, 50*(6), 337–344. https://doi.org/10.2166/wst.2004.0350.
33. Katara, G., Hemvani, N., Chitnis, S., Chitnis, V., & Chitnis, D. S. (2008). Surface disinfection by exposure to germicidal UV light. *Indian Journal of Medical Microbiology, 26*(3), 241–242. https://doi.org/10.4103/0255-0857.42060.
34. Keklik, N. M., Krishnamurthy, K., & Demirci, A. (2012). Microbial decontamination of food by ultraviolet (UV) and pulsed UV light. In Demirci, A. & Ngadi, M. O. (Eds.), *Microbial Decontamination in the Food Industry: Novel Methods and Applications* (pp. 344–369). Woodhead Publishing. https://doi.org/10.1533/9780857095756.3.344.
35. Kim, J. U., Shahbaz, H. M., Cho, J., Lee, H., & Park, J. (2021). Inactivation of *Bacillus cereus* spores using a combined treatment of UV-TiO_2 photocatalysis and high hydrostatic pressure. *Innovative Food Science and Emerging Technologies, 70*, 102676. https://doi.org/10.1016/j.ifset.2021.102676.
36. Koutchma, T. (2008a). UV light for processing foods. *Ozone: Science and Engineering, 30*(1), 93–98. https://doi.org/10.1080/01919510701799222.
37. Koutchma, T. (2018b). Status of international regulations for ultraviolet treatment of foods. *IUVA News, 20*, 13–16.

38. Koutchma, T. (2009). Advances in ultraviolet light technology for non-thermal processing of liquid foods. *Food and Bioprocess Technology, 2*(2), 138–155. https://doi.org/10.1007/s11947-008-0187-6.
39. Koutchma, T., Popović, V., Ros-Polski, V., & Popielarz, A. (2016). Effects of ultraviolet light and high-pressure processing on quality and health-related constituents of fresh juice products. *Comprehensive Reviews in Food Science and Food Safety, 15*(5), 844–867. https://doi.org/10.1111/1541-4337.12210.
40. Kowalski, W. (2009). UVGI disinfection theory. In Kowalski, W. (Ed.), *Ultraviolet Germicidal Irradiation Handbook* (Vol. 2, pp. 17–50). Springer Berlin Heidelberg. https://doi.org/10.1007/978-3-642-01999-9_2.
41. Lado, B. H., & Yousef, A. E. (2002). Alternative food-preservation technologies: Efficacy and mechanisms. *Microbes and Infection, 4*(4), 433–440. https://doi.org/10.1016/S1286-4579(02)01557-5.
42. Lin, L., Cai, C., Gilbert, R. G., Li, E., Wang, J., & Wei, C. (2016). Relationships between amylopectin molecular structures and functional properties of different-sized fractions of normal and high-amylose maize starches. *Food Hydrocolloids, 52*, 359–368. https://doi.org/10.1016/j.foodhyd.2015.07.032.
43. Luo, W., Chen, A., Chen, M., Dong, W., & Hou, X. (2014). Comparison of sterilization efficiency of pulsed and continuous UV light using tunable frequency UV system. *Innovative Food Science and Emerging Technologies, 26*, 220–225. https://doi.org/10.1016/j.ifset.2014.10.005.
44. Mahida, N., Vaughan, N., & Boswell, T. (2013). First UK evaluation of an automated ultraviolet-C room decontamination device (Tru-D™). *Journal of Hospital Infection, 84*(4), 332–335. https://doi.org/10.1016/j.jhin.2013.05.005.
45. Marquenie, D., Lammertyn, J., Geeraerd, A. H., Soontjens, C., Van Impe, J. F., Nicola, B. M., & Michiels, C. W. (2002). Inactivation of conidia of *Botrytis cinerea* and *Monilinia fructigena* using UV-C and heat treatment. *International Journal of Food Microbiology, 74*(1–2), 27–35. https://doi.org/10.1016/S0168-1605(01)00736-1.
46. Matak, K. E., Sumner, S. S., Duncan, S. E., Hovingh, E., Worobo, R. W., Hackney, C. R., & Pierson, M. D. (2007). Effects of ultraviolet irradiation on chemical and sensory properties of goat milk. *Journal of Dairy Science, 90*(7), 3178–3186. https://doi.org/10.3168/jds.2006-780.
47. Mbonimpa, E. G., Blatchley, E. R., Applegate, B., & Harper, W. F. (2018). Ultraviolet A & B wavelength-dependent inactivation of viruses and bacteria in water. *Journal of Water and Health, 16*(5), 796–806. https://doi.org/10.2166/wh.2018.021.
48. Mehta, D., Sharma, N., Bansal, V., Sangwan, R. S., & Yadav, S. K. (2019). Impact of ultrasonication, ultraviolet and atmospheric cold plasma processing on quality parameters of tomato-based beverage in comparison with thermal processing. *Innovative Food Science and Emerging Technologies, 52*, 343–349. https://doi.org/10.1016/j.ifset.2018.12.013.
49. Noci, F., Riener, J., Cronin, D. A., Morgan, D. J., & Lyng, J. G. (2008). Ultraviolet irradiation and pulsed electric fields (PEF) in a hurdle strategy for the preservation of fresh apple juice. *Journal of Food Engineering, 85*(1), 141–146. https://doi.org/10.1016/j.jfoodeng.2007.07.021.
50. Oteiza, J. M., Peltzer, M., & Gannuzzi, L. (2005). Antimicrobial efficacy of UV radiation on *Escherichia coli* O157 (EDL 933) in fruit juices of different absorptivities. *Journal of Food Protection, 68*(1), 49–58. https://doi.org/10.4315/0362-028X-68.1.49.

51. Quintero-Ramos, A., Churey, J. J., Hartman, P., Barnard, J., & Worobo, R. W. (2004). Modeling of *Escherichia coli* inactivation by UV irradiation at different pH values in apple cider. *Journal of Food Protection, 67*(6), 1153–1156. https://doi.org/10.4315/0362-028X-67.6.1153.
52. Rawat, S. (2015). Food spoilage: Microorganisms and their prevention. *Asian Journal of Plant Science and Research, 5*(4), 47–56.
53. Riley, R. L., & Kaufman, J. E. (1972). Effect of relative humidity on the inactivation of airborne *Serratia marcescens* by ultraviolet radiation. *Applied Microbiology, 23*(6), 1113–1120. https://doi.org/10.1128/am.23.6.1113-1120.1972.
54. Santos, A. L., Moreirinha, C., Lopes, D., Esteves, A. C., Henriques, I., Almeida, A., & Cunha, A. (2013). Effects of UV radiation on the lipids and proteins of bacteria studied by mid-infrared spectroscopy. *Environmental Science & Technology, 47*(12), 6306–6315. https://doi.org/10.1021/es4005082.
55. Sastry, S. K., Datta, A. K., & Worobo, R. W. (2000). Ultraviolet light. *Journal of Food Science, 65*(3), 90–92. https://doi.org/10.1111/j.1750-3841.2000.tb00781.x.
56. Setlow, P., & Li, L. (2015). Photochemistry and photobiology of the spore photoproduct: A 50-year journey. *Photochemistry and Photobiology, 91*(6), 1263–1290. https://doi.org/10.1111/php.12505.
57. Singh, H., Bhardwaj, S. K., Khatri, M., Kim, K. H., & Bhardwaj, N. (2021). UVC radiation for food safety: An emerging technology for the microbial disinfection of food products. *Chemical Engineering Journal, 417*, 128084. https://doi.org/10.1016/j.cej.2021.128084.
58. Skowron, K., Bauza-Kaszewska, J., Paluszak, Z., & Skowron, K. J. (2014). UV-C radiation as a factor reducing microbiological contamination of fish meal. *The Scientific World Journal, 2014*, 928094. https://doi.org/10.1155/2014/928094.
59. Stevens, C., Khan, V. A., Lu, J. Y., Wilson, C. L., Pusey, P. L., Igwegbe, E. C. K., & Droby, S. (1997). Integration of ultraviolet (UV-C) light with yeast treatment for control of postharvest storage rots of fruits and vegetables. *Biological Control, 10*(2), 98–103. https://doi.org/10.1006/bcon.1997.0552.
60. Wambura, P., & Verghese, M. (2011). Effect of pulsed ultraviolet light on quality of sliced ham. *LWT – Food Science and Technology, 44*(10), 2173–2179. https://doi.org/10.1016/j.lwt.2011.07.018.
61. Wang, J., Liu, Q., Xie, B., & Sun, Z. (2020). Effect of ultrasound combined with ultraviolet treatment on microbial inactivation and quality properties of mango juice. *Ultrasonics–Sonochemistry, 64*, 105000. https://doi.org/10.1016/j.ultsonch.2020.105000.
62. Zimmer, J. L., & Slawson, R. M. (2002). Potential repair of *Escherichia coli* DNA following exposure to UV radiation from both medium- and low-pressure UV sources used in drinking water treatment. *Applied and Environmental Microbiology, 68*(7), 3293–3299. https://doi.org/10.1128/AEM.68.7.3293-3299.2002.
63. Zhu, Z., Cai, H., & Sun, D. W. (2018). Titanium dioxide (TiO_2) photocatalysis technology for nonthermal inactivation of microorganisms in foods. *Trends in Food Science and Technology, 75*, 23–35. https://doi.org/10.1016/j.tifs.2018.02.010.

CHAPTER 5

KINETIC MODELING AND EVALUATION OF PHYSICOCHEMICAL AND MICROBIAL CHANGES BY ULTRAVIOLET LIGHT EXPOSURE

NIKITHA MODUPALLI, MALINI BUVANESWARAN, C. K. SUNIL, and VENKATACHALAPATHY NATARAJAN

ABSTRACT

The increase in the concept of food safety and quality has increased the processing of food products using different techniques. The deteriorative effects of thermal processing techniques have caused a paradigm shift in attention towards non-thermal technologies. Ultraviolet (UV) light, being a non-thermal technique, has been studied highly. The photochemical exposure to UV light has shown several chemical and microbial changes in different food substrates. Also, the inactivation of microorganisms using UV exposure has shown the different resistances of different strains against UV light, leading to needing exposure at certain wavelengths for a certain period to achieve the desired log reduction of the microbial population. However, the practical applications of UV light on the industrial front have been hindered, probably due to the lack of empirical data on factors affecting the UV treatment efficiency in food processing. Kinetic modeling of thermo-chemical and microbial activities caused by UV light exposure in food products helps understand the factors affecting the efficacy of UV treatment for food processing applications. This chapter elaborates on kinetic modeling and its importance in food processing and modeling thermo-chemical and microbial interactions by UV light treatments in selected food products.

Nonthermal Light-Based Technologies in Food Processing.
C. K. Sunil, Preeti Birwal, Megh R. Goyal, and R. Mahendran (Eds.)

5.1 INTRODUCTION

UV irradiation technique has a promising role in the food industry because of its application in low-cost non-thermal preservation techniques and consumer trends on 'green' food products. It tends to replace thermal pasteurization in liquid food like fresh juice, soft drinks, and beverages. In food processing, UV light from 100 to 400 nm is used for different purposes, including disinfection, microbial inactivation, and preservation. UV rays penetrate the entire volume of food substances and are absorbed by the food matrix. The food sample is a mixture of several organic compounds with different optical and physical characteristics, chemical reactions, momentum transfer, dose delivery, degradation kinetics, and microbial inactivation kinetics to UV light exposure. In most food applications, UV light is used to increase the shelf life of food products by reducing pathogenic microorganisms. On the other hand, UV exposure leads to some physiochemical modifications in organic compounds like degradation of color and lipid oxidation. Especially in milk processing, UV exposure leads to the oxidation of fats, which highly deteriorates the quality parameters by producing foul-smelling compounds and a cabbage-like taste. Most vitamins are sensitive to UV exposure. During processing, considerations should be taken to ensure the security of other UV degradable compounds in food products like carotenoids.

The kinetic study can provide information about the activation energy required for the degradation and inactivation of specific compounds, which helps to determine the process variables such as UV dosage and exposure time. Also, the kinetic model helps predict the change in the food matrix in a particular period using experimental data values. It is necessary for kinetic study while UV light exposure in food products. Food quality is a multidimensional aspect, including both subjective and objective constraints and can be dynamic over time. Several elements play a role in expressing the quality of the product, so make it clear that the quality parameters of food items were divided into intrinsic and extrinsic attributes. In modeling, we can either focus on the consumer or process perspective. In some cases, the integrated models are also studied (e.g., Quality function deployment). This chapter focuses on the effect of intrinsic changes occurring in food products by UV light exposure on quality attributes and the availability and application of kinetic modeling in the same area.

5.2 IMPORTANCE OF KINETIC MODELING IN FOOD PROCESSING

Modeling can be used in two different ways. One is mathematical modeling, and the other is kinetic modeling. Mathematical modeling can be used based on a set of reactions and is mainly used for the optimization and design of a process. However, kinetic modeling is used to define the chemical reactions, for example, the effect of process variables on the chemical properties of the food material. In food processing, mathematical models are used to minimize product deterioration and maximize financial profits.

The modeling application will vary based on the goal of the research work. Modeling is used to analyze the property/output of the system based on the process variables. It helps to predict the actual changes of the system based on the process variables using simulations. It helps arrive at the real-world outcome in complex situations, optimize, and study the changes in a system based on variables. In studies, changes in process and variables can be called decision variables. The decision variables are fixed based on the final quality requirement of the product. Here, to predict the effect of the process variables (light intensity, exposure time) on the food product quality (physicochemical and microbial changes) are predicted by kinetic modeling. Two main terms were used in kinetic modeling: model fit and prediction. Model fit is used to analyze the correlation between the experimental data with a model performance and prediction used to predict the changes in quality of the food products based on the process variables. For optimization and design of a process, mathematical modeling will be used.

This chapter aims to provide a detailed view of the application and implementation of kinetic models to predict the physicochemical and microbial changes by UV light exposure. Here, kinetics refers to the change in product quality in a time-dependent physicochemical and microbial reaction. The mathematical equations describe these kinetics, and the equations are used to form models to predict the quality changes. To create these models, clear knowledge about chemical and microbial changes by UV light exposure, composition, and structural characteristics of food are required to build a kinetic model. Prediction of the activation energy required for either chemical or mechanical degradation by kinetic modeling will create a new revolution in food processing when considering quality as a primary factor. Most of the studies have been carried out to find the microbial degradation kinetics, and modeling has been successfully conducted, but only a few studies have concentrated on studying physicochemical degradation by UV light exposure.

5.3 KINETIC MODELING: TYPES AND APPLICATIONS

In kinetic modeling, certain chemical reactions, not quality attributes, are used to determine the quality, and these chemical reactions are indirectly related to some quality attributes [12]. So, to express quality, we need to find the kinetics of chemical reactions closely related to the kinetics of physical or chemical changes. Some intrinsic factors like the presence of toxic and pathogenic compounds cannot be perceived directly by the consumer, but kinetic modeling helps to include such hidden attributes in modeling. When considering intrinsic attributes for kinetic modeling, it includes safety attributes (microbial, toxic, and mutagenic) and physicochemical attributes (nutritional value, texture, color, appearance). The sensory attribute explains the quality by processing stimuli by the consumer, so exact sensory analysis cannot be replicated using kinetic modeling. Instead of finding the flavoring and tasting compounds in food, based on whether the product is quality or not is analyzed. To analyze the physicochemical changes in a food product by UV light physical, chemical, biochemical, and microbial quality indicators to provide information about the product, these indicators results will change in time, so this approach is characterized by the kinetic approach [38].

Normally, the reaction rate of a particular process is calculated by the zero, first, and second-order models or fractional order can be used based on the reaction behavior. Chemical and biological reactions normally follow any one of the:

$$V = -\frac{dc}{dt} = kc^n \tag{1}$$

where; v, c, k, and n represent the rate of the reaction, the concentration of the reactant, the rate constant, and the order of the reaction, respectively. Nevertheless, the order of the reaction is the mathematically derived empirical value, and it does not contain any relation to the reaction mechanism. In most studies, the ultimate aim is to find the concentration (C) of the particular compounds after some time (t) period. The required model to calculate the concentration based on the time can be obtained by integrating basic models. For the zero-order reaction based on the process time t in linear form:

$$c = c_0 - kt \tag{2}$$

For first order, $c = c_0 \exp(-kt)$ (3)

$$\text{For a second order, } \frac{1}{c} = \frac{1}{c_0} + kt \tag{4}$$

where; c_0 represents the initial concentration.

In some cases, the reaction rate will depend on the concentration of one or two reactants, and then molecularity helps indicate the number of reactants involved. However, in the case of a composite reaction, where the reaction rate involves more than one reaction mechanism, the usage of molecularity has no meaning. Most of the chemical degradation mechanisms in food products follow composite reactions. For modeling shelf life or chemical changes in the food system, simple reaction orders are more than enough, but for explaining reaction kinetics by a particular treatment method in detail, reaction mechanisms and their differential form need to be analyzed. For example, during milk heating, the lactose present in the milk isomerizes to lactulose, but this formation is not a monomolecular reaction because it can further degrade to galactose.

On the other hand, lactulose can reverse isomerize to lactose. Based on this, two or more differential equations are formed to study the lactose degradation during heating using numerical or another integration method. When numerical integration is widely accepted and used, the data fitting is flexible. Different models can be tested by adjusting the process-related differential equations, and the experimental data can directly fit into them. A suitable model replicating the real reaction kinetics can be selected from the different models based on the goodness of fit. When the reaction is carried out in an unsteady state, the kinetic models are required to predict the changes in the reactants over a period.

Apart from the kinetic reactions, studying activation energy for particular chemical reactions is important during the kinetic study, which denotes the energy required to start a particular reaction. In most cases, the activation energy is derived from the Arrhenius equation with simple process-related modifications. Badin et al. [7] described the Arrhenius equation to study the activation energy for the degradation of lycopene and ascorbic acid based on temperature [7]:

$$k = K_0 \exp\left(-\frac{E_a}{RT}\right) \tag{5}$$

In which k is the temperature dependence reaction rate constant; k_0 is the frequency factor; E_a is the activation energy; R is the universal gas constant (0.00831 kJ mol^{-1} K^{-1}); and T is the absolute temperature. Here, the k is obtained from any kinetic models used to calculate the activation energy. Based on the specific process factor (e.g., temperature), the minimum energy required to initiate a chemical reaction can be calculated here if one is more parameters involved in the chemical reaction. In that situation, the number of kinetic rates constant obtained is equal to the parameters involved in the chemical reaction, so several activation energies based on process parameters are calculated. This helps set the critical process point where the photodegradation of nutritive compounds or microbial inactivation.

5.4 UV LIGHT: MODELING OF MICROBIAL INACTIVATION MECHANISMS

For the past two decades, UV light-assisted inactivation of microorganisms has been a go-to solution alternative for thermal processing, especially for liquid foods. UV light has also been actively applied for achieving surface decontamination, water sterilization, milk processing, etc. The food processing sector can prominently benefit from utilizing UV light in its different forms (UV-A, B, and C), which can be an efficient way for microbial inactivation [26]. It has been widely observed that UV-C has prominent germicidal properties due to its ability to induce photocatalytic chemical changes in the DNA of microorganisms [35]. Though UV light has been highly effective in microbial destruction, the level of UV light penetration into the target product depends on the color and transparency of the substrate. This is one of the major drawbacks and factors to be considered when designing and adapting UV treatment for food decontamination.

UV light is a non-ionizing radiation and does not cause ionization of particles in substrate. When UV radiation (especially UV-C and vacuum UV) hits the surface of the target product, depending on the transparency of the product, it will be partly absorbed and partly transmitted. The absorbed light radiation can directly interact with the biological compounds present and can initiate the chemical reaction. The nucleic acids present in the DNA of the microorganisms in the target product can absorb the radiation between 200 and 310 nm and can cause breakage and realignment of pyrimidines with the adjacent cytosine and adenosine. Thymine, the most sensitive nitrogen base, reacts immediately after exposure to radiation [1]. These pyrimidine dimers can prevent cell replication, making it impossible for the microbes to proliferate. The photon energy absorbed by the microbial DNA can also cause the formation of cross-linking chemicals that can arrest the transcription process, eventually leading to the death of the cells [30, 33].

The antimicrobial effect exerted by the UV light exposure can be attributed to chemical changes due to DNA splicing and the formation of cis-syn pyridine dimers of cyclobutene [18]. Microbial strains like *E. coli, Listeria innocua, Yersinia pseudotuberculosis, Bacillus subtilis, Staphylococcus aureus*, yeast *Saccharomyces cerevisiae* and protozoans like *C. parvum* have been extensively studied and reported for the efficacy of UV light in the destruction of these microorganisms in several food systems [25]. The application of UV light has been multidimensional in the food industry, for disinfection of air and surfaces, decontaminating liquid foods like fruit juices, liquid egg and milk products, and solid food surfaces like fresh foods, cut flesh foods, and high moisture foods.

Preservation of the quality of food products directly implies arresting microbial growth for elongation of shelf stability. Foods, the carriers of sufficient nutritious substances required for growth, often become highly vulnerable to microbial infestation. The knowledge about the growth of various microorganisms in food systems can help in the prediction of the shelf-life of the product in terms of microbial growth. It can pave a base for risk assessment and hazard analysis. The growth of microorganisms can be affected by various extrinsic factors (temperature, packaging, gas composition, etc.), intrinsic factors (acidity, ionic character, nutritional concentrations, enzymes, etc.), and their interactions. Unlike chemical reaction kinetics, microbial kinetics deal with the increase in the number of cells, not the molecular changes or the population dynamics in short [40, 41]. Prediction of microbial growth studied as Predictive microbiology has been a topic of interest in the past two decades, as it explains the growth kinetics of different microorganisms in food systems concerning the changes in one or more extrinsic and intrinsic parameters [42].

In the case of UV or pulsed UV treatment, the knowledge about the growth kinetics of microbes can give a distinctive idea about the parameters to be adapted. However, this chapter elaborates on the kinetics of cell death that occurs after the UV light treatment to various foods, which is the most obvious effect of light exposure. The concept of 12D refers to the process of sterilizing the foods to a 12-log reduction of *Clostridium botulinum*, considered the most lethal food pathogen. Inactivation kinetics of microbial cells has been studied since the early 1920s after the coining of the terms *D, Z*, and Q_{10} values. The *D* value is a decimal value for microbial reduction, which defines the amount of time required for reducing the microbial concentration by a factor of 10 or one decimal log cycle. The *Z* value is representative of temperature dependency, which explains the temperature increase required to inactivate the microorganisms by a factor of 10. Q_{10} refers to the change in reaction rate when the temperature is raised by 10°C. The *D*, *Z*, and Q_{10} values follow the first-order kinetics in most cases and are very important in explaining the Thermal Death Time (TDT) curve [2, 20, 28].

5.4.1 KINETIC MODELING OF UV INACTIVATION OF MICROORGANISMS

Inactivating foodborne pathogens is crucial for the safety and quality of food products and ingredients. Several processes can achieve the inactivation of pathogens, both thermal (heating using various sources of sterilization) and non-thermal (by UV light, electromagnetic pulses and radiations, the addition of preservatives, etc.) [3]. The level of inactivation achieved by each technique can vary with several factors like the type of technique, type of food substrate, intensity, time of treatment, etc. The dependency of inactivation on these factors makes it very important to perform modeling of the kinetics of inactivation to study the effects of different factors on the inactivation of different pathogens. In a general sense, the inactivation of vegetative cells follows the first-order kinetics [9], which means the reduction in the number of microbes *N*, for time *t* follows the given equation:

$$-\frac{dN}{dT} = kN \tag{6}$$

It can be simplified to, $In\,N = In\,N_0 - kt$ (7)

where; *N* and N_0 are the populations of microorganisms in the substrate at time *t* and 0 and *k* is the first-order rate constant, respectively. This equation can be explained in terms of the survival ratio at a given time *t* is:

$$S(t) = \frac{N}{N_0} = \exp(-\frac{t}{D}) \text{ or } S(t) = -\frac{t}{D} \tag{8}$$

where *D* denotes decimal reduction time, which can be calculated using:

$$D = \frac{In10}{k} = \frac{2303}{k} \tag{9}$$

The survival rate versus time plot is called a survival curve and needs to be linear if the system follows first-order kinetics. The slope of this plot should be D^{-1}. This implies that first-order inactivation kinetics is mainly characterized by the *D* value or first-order rate constant. However, in practical applications such as enzyme inactivation or heat degradation, microbial inactivation does not usually follow first-order kinetics. Several studies on the inactivation of different microbes in various food systems have shown frequent non-linear survival curves.

Factors such as the presence of subpopulations, interference with the composition of matrices, uneven treatment exposure, etc., can cause non-linearity in the curves. Additionally, the multitude of chemical reactions required for microbial inactivation, which would be considered a single event in explaining first-order kinetics, do not align with survival kinetics. These factors may be major reasons for the non-linearity of microbial survival kinetics and for reconsidering views on microbial kinetics. These reasons justify the need to adopt alternative models to explain survival curves that exhibit non-linear patterns [5, 13, 41].

5.4.1.1 ALTERNATIVE MODELS

The inactivation of food pathogens needs to be considered a stochastic series of events rather than a single holistic occurrence. Overlooking the mechanism of death, the individual microbe among the whole organism can survive or die under applied stress when cell death is considered. This can be called a probabilistic lethal event. To put it simply, it is not possible for all cells to die at the same given time t, which can be given as t_c is the time taken for individual cell death ($t > t_c$). The survival of any single pathogen (i) can be given as a step function $Si(t) = 0$ if the cell dies and $Si(t) = 1$ if the cell lives. The consequent result of the survival rate of the whole microbial population is formed by the cumulation of the step functions distribution. However, the underlying molecular reactions leading to each cell's death might vary, making it easier to explain in probabilistic reaction kinetics than deterministic kinetics. Hence, the survival curves can be looked at using the statistical distribution of individual organisms among the population at time t_c (a critical time for the death of each microorganism). Most curves following this distribution were concave (upward or downward) curves [23].

Weibull model is one of the most effective models representing nonsymmetrical distribution kinetics. Though it is used widely in describing the time-dependent failure of electric and mechanical systems, it also appropriately fits to analyze the survival kinetics data. The survival kinetics using the Weibull model explains the time-dependent cell death in the food matrix after applying external stress. The Weibull function for the asymmetrical probability distribution of survival rate $S(t)$ is given as:

$$S(t) = \exp(-(\frac{t}{a_w})^{\beta_w}) \tag{10}$$

which can be simplified further deducted using the logarithm of the survival ratio as:

$$S(t) = \exp(-(\frac{t}{a_w})^{\beta_w}) \tag{11}$$

where; α_w and β_w are the distribution parameters, time scale and shape parameter, respectively, the distribution curve follows an upward bend when $\beta_w < 1$ and a downward bend when $\beta_w > 1$.

When the shape parameter of the distribution is equal to 1, the distribution follows the first-order kinetics and gives a linear curve [9, 21]. The characteristics of survival curves largely depend on the technique used, type of food matrix, and target microorganism, which dictate the *D* and t_c values. Several studies have examined the effect of different parameters on microbial inactivation from different food products for specific microorganisms [17].

5.4.2 KINETICS OF UV-ASSISTED MICROBIAL INACTIVATION

Using kinetic mathematical modeling can be an excellent tool for estimating the UV inactivation of food pathogens. It can also effectively explain the phenomena of the shoulder effect (resistance by the microorganisms during the start of the treatment) and tailing effect (explains the clumping of subpopulations, protection effects offered by the continuous phase, and occurrences of mixed populations). In most of the microorganism inactivations using UV, the following models have been used to fit in and analyze the experimental data: Log-linear model; log-linear with shoulder and tailing effects, each separately and cumulatively; Weibull model; Weibull model with combinations of shoulder and tailing effects with biphasic models and fixed-parameter p [8]. In Weibull distribution of inactivation kinetics, the upward concave curve explains that the surviving cells endured severe damage with time t, and the microbial inactivation rate is elevated. The downward concave plot elaborates on removing the most sensitive population first, then elimination according to the weakness. Similarly, the biphasic plot explains the death of a more sensitive population among the subpopulations followed by the other [40, 41].

A study on the effect of UV exposure on seven pathogens, namely *Listeria monocytogenes, Serratia marcescens, Salmonella Senftenberg, Yersinia enterocolitica, Aeromonas hydrophila, Escherichia coli*, and *Staphylococcus aureus*, in whole milk [11]. The UHT whole milk was inoculated with the selected food pathogens separately, and each sample was treated with UV-C radiation dosages between 0 and 5,000 J/L at a flow of 4,300 L/h. *L. monocytogenes* has been most effectively killed to a 5-log reduction at 2,000 J/L dosage, whereas *S. aureus* has shown a 5-log reduction at 1,450 J/L dosage. The microbial inactivation kinetics of all the selected organisms followed a non-linear pattern, showing sigmoid inactivation curves. Another study on modeling the efficacy of UV-emitting LED light on the inactivation of *E. coli* in potable water revealed the use of a model based on the Monte Carlo method, adapting an incremental ray tracing algorithm [46]. The UV light emissions at 254 nm have been selected to effectively study and predict the optical fluency rates for on-point use of potable water, considering the effect of chamber geometry.

Similarly, Lebik et al. [27] studied the effect of UV radiation on two strains of microorganisms, *E. coli* and *Lactobacillus helveticus*, in water. Two empirical kinetic models, Chick-Watson and Hom, were used to mathematically analyze the real-time experimental data. It was revealed that the models fit the experimental data of UV inactivation of *E. coli* perfectly but failed to explain data from inactivation of *L. helveticus*, thus concluding that the application and success of the kinetic models cannot be the same for different microorganisms and treatment conditions.

Similarly, a previous study evaluating the effect of UV light of varying wavelengths for inactivation of isolated *Campylobacter jejuni* in sterile broth has shown that UV light with a 280 nm and 300 nm combination effectively achieved a 4-log reduction for an exposure time of 5 seconds [47]. The inactivation kinetics were examined to fit in four models, Log-linear, log-linear-shoulder, Weibull, and double Weibull, using the parameters for the goodness of fit. The experimental inactivation data fit the log-linear model, following a linear symmetric pattern. Gouma et al. studied the effect of combination treatment of UV light with mild temperatures in apple juice for *E. coli* and *L. monocytogenes* inactivation [15]. It was observed that *E. coli* inactivation occurred with a UV-C dose of 13.81 J/mL for 12.71 min and 5.20 J/mL for 4.78 min to reach a 5-log reduction in the pathogen when the heat applied was 44°C to 54°C. Similarly, for *L. monocytogenes* inactivation, the temperatures between 54°C and 60°C required a UV-C dosage of 5.20 J/mL for 4.78 min, whereas 2.11 J/mL for 1.93 min. The inactivation kinetics of both pathogens in given conditions were explained using tertiary models given by Gerard et al. which elaborates the temperature dependency and logarithm of exposure time and dosage in achieving a 5-log reduction of *E. coli* and *L. monocytogenes* in apple juice.

A similar study on establishing a kinetic model on UV inactivation of microbes on solid surfaces, in which the surface has been divided into several zones, works on the ratio of UV light intensity exposed to the microorganisms to that on the whole surface area [14]. The model utilizes the inactivation data of the individual microorganisms under full UV intensity ($\alpha=1$) with that of microorganism inactivation on the targeted surfaces. The model from standard non-linear programming was used to explain the data of microbial inactivation, and the equation obtained is:

$$y' = \frac{N}{N_0} = \sum_{j=1}^{n_z} f_j . g(F_s . a_j) \tag{12}$$

where; j is the distinguished zone on the solid surface; f_j is the fraction of organisms in zone j of the surface; α_j is the light exposure in zone j; and n is the number of target zones.

It has been stated that this model would be very useful in occurrences of heavy reductions in log numbers. However, the applicability of this model in real-time can be quite challenging without knowing the inactivation dynamics of the target microbe and the microbial load on the target surfaces. Severin et al. [45] studied UV inactivation of *E. coli* and *Candida parapsilosis* in water disinfection using batch type and mixed flow through an annular reactor. Two simple kinetic models were initially proposed and tested for their ability to fit the scale-up process's data, which was not completely successful. This has further been elaborated by developing comparative data between two kinetic models, a multi-target kinetic model and series event kinetics, especially for the inactivation in a mixed-flow annular reactor.

Brahmi et al. studied the effect of suspended solid particles on UV-assisted microbial inactivation in secondary treated wastewater [10]. The work seeks to determine the disinfection kinetics of *Pseudomonas aeruginosa* in two phases: the initial rapid phase of inactivation followed by the lag phase, with an array of models from the simplest Chick-Watson first-order kinetics to the multidimensional Collins-Selleck model. It was found that the initial delay in the start of inactivation on most of the microbial strains rendered the simple first-order kinetics unsuitable for the obtained data. The rapidly changing dynamics of the UV light disinfection process have been found to fit well in the Collins-Selleck model. Another study on the inactivation of ascospores of microbial strains *Aspergillus fischeri* and *Paecilomyces niveus* in apple juice used different non-symmetric kinetic models [34]. Weibull and first-order biphasic models were used to fit the data of ascospores survival to analyze the inactivation kinetics. The survival ratio is given by:

$$S(d) = \frac{\mathrm{Nd}}{N_o} \tag{13}$$

where; N_d and N_0 denoted the ascospore population in the substrate after and before UV exposure. This model also includes the fluence (*F*) of UV light given as:

$$F = \frac{60tI}{1000} \tag{14}$$

where; *t* is the exposure time; and *I* is the intensity of UV light. This model considers the inactivation rates of both sensitive and resistant populations as k_1 and k_2 (in terms of m^2/kJ or min^{-1}), respectively.

The biphasic decimal logarithmic model using k_1 and k_2, along with *f*, which is a fraction of the sensitive population, is given as:

$$log\frac{(N(d)}{N_0} = log(f10^{-k_1d} + (1-f)10^{-k_2d})) \tag{15}$$

The use of fluence has simplified the model to be estimated in primary model parameters to explain the entire experimental data.

A study by Ngadi et al. on the inactivation of *E. coli* using UV light in cultured liquid broth has shown that though the kinetics of inactivation followed first-order equations, a tailing effect has been observed in the inactivation due to a slowdown in death rate [36]. The UV light intensity I play a crucial role in inactivation kinetics due to direct proportionality and can be derived using:

$$I = I_0 e^{-az} \tag{16}$$

where; *I* and I_0 are the light intensities (mWcm^{-1}) on the top and within the substrate surface, z and α are the equation constant.

The microbial inactivation by first-order kinetics can be put in the following equation.

$$\frac{dX}{dt} = -kIX \tag{17}$$

where; X is the number of surviving cells; and t is the time; and k is the inactivation rate constant.

It has been observed in several studies that microbial inactivation starts rapidly initially and slows down subsequently due to the tailing effect. X_r is the ratio of the surviving population out of the total population X is explained using:

$$X_r = X_c e^{-k_1 d} + (1 - X_c) e^{-k_2 d} \tag{18}$$

where; d is the UV dosage; and X_c is the ratio of critical survival.

5.5 MODELING OF THERMO-CHEMICAL REACTIONS IN THE SUBSTRATE

In most food applications, UV light is used to inactivate pathogenic microorganisms, but coincidentally the organic compounds present in the food get affected by the UV light. The major drawback is the formation of free radicals during light exposure, which leads to the digestion of protein, lipid oxidation, vitamin degradation, antioxidant destruction, color degradation, and foul smell formation.

$$A + hv \rightarrow A_+^+ \rightarrow Products \tag{19}$$

When we speak about thermochemical reactions, the absorption of UV photons by organic compounds and the photon energy to start the reaction are crucial factors determining the level of photodegradation [52]. The photodegradation reaction kinetics of organic compounds, including vitamins, fats, and pigments during UV treatment and the storage period, help to measure the effect of UV light on foodstuffs, which is highly necessary for organoleptic studies. The degradation kinetics largely depend on the process conditions such as UV wavelength, photon energy, treatment time, the composition of food, and concentration of photodegradation compounds, so kinetic modeling provides the necessary safe limits for different process variables.

The activation energy required for the food-related chemical reactions usually falls under 30–150 kJ/mol [48]. A UV photon (254 nm) with the energy of 472 joules is enough to loosen the bond between O-H, C-C, C-H, C-N, H-N, and S-S; to activate several chemical reactions [12]. When we discuss the energy for degradation reaction, the wavelength of UV light comes first, which will decide the photon's energy. The second thing is the organic matrix to be treated and its light absorption characteristics. The Lambert-Beer law calculates the amount of light absorbed by the food system by using the relationship between absorbance (A), the concentration of absorbent (c, mol/L) in the food matrix, molar absorptivity (ε, L/mol/cm), and treatment path length (d, cm),

$$A = \varepsilon c d \tag{20}$$

If more than one UV light-absorbing compound is present, the cumulative absorbance will be considered. The absorbance rate of UV varies based on the number of suspended solids in the food matrix [24]. The photosensitive compounds create intermediate freely reactive compounds during UV exposure, which cause a series of chemical reactions. Proteins with aromatic rings and disulfide bonds, Vitamins A, B12, E, unsaturated fatty acids in oil, and phospholipids are very sensitive to UV light exposure. The photodegradation level will vary based on the concentration of sensitive products. The level of unsaturated bonds in the food matrix determines the degradation of organic compounds. Hence, unsaturated bonds can absorb UV light more strongly than saturated bonds.

Compounds like vitamin A, C, riboflavin, and coloring agents can absorb UV light. Vieira et al. [49] formulated the reversible kinetic model to analyze the degradation of ascorbic acid by thermal treatment using the concentration of ascorbic acid at the initial (C_{AA0}), at time t (C_{AAt}), and at infinity ($C_{AA\infty}$),

$$\frac{C_{AAt} - C_{AA\infty}}{C_{AA0} - C_{AA\infty}} = e^{-k_1 t} \tag{21}$$

For specific treatment time (t), the kinetic model is modified to:

$$\frac{C_{AA\infty}}{C_{AA\infty}} = e^{-k_1 t} \tag{22}$$

The reaction rate constant (k_1) is calculated from this equation, and the activation energy required to activate this ascorbic degradation reaction (73 kJ/mol) was calculated using the Arrhenius law. The rate of reaction is influenced by the product conditions like pH, oxygen, cupric ion, sucrose, enzymes, amino acids, and the concentration of ascorbic acid in a product) [49]. So, this product condition can be taken as an independent variable to study the response of degradation behavior. Some studies pointed out that vitamin C degradation follows zero-order kinetics in apple and apple juice. It shows that the reaction kinetics vary based on the type of processing applied and product composition [39, 43]. Some kinetic degradation models of photodegradation compounds in the food system concerning their treatment method are shown in Table 5.1.

Discoloration of pigments and chemical substances (dye) follows the first-order kinetic model, but this reaction was highly sensitive to the pH of the substance. At neutral pH, the degradation rate is higher than in acid and alkali conditions [31]. Xu et al. [52] described the kinetic model to express the degradation kinetics of *p*-arsanilic acid (As) using UV light exposure. The reaction was highly dependent on the intermediate inorganic compounds like As(III) and As(IV) generation by chemical bond cleavage [52].

Liu et al. [29] noticed the aflatoxin B_1 degradation under UV irradiation in an aqueous medium follows the first-order kinetics $(\text{In}\left(\frac{c_t}{c_0}\right) = -0.07325t)$, the half-life of aflatoxin and reaction constant does not affect by the initial concentration of aflatoxin B_1, and the photo degradation rate follows 800 > 400 > 200 μw/cm² based on the intensity of UV irradiation. Three reactive substances such as $C_{17}H_{14}O_7$, $C_{16}H_{14}O_6$, and $C_{16}H_{12}O_7$ were identified. Different studies have calculated the activation energy required for the photodegradation of different compounds like lycopene, ascorbic acid, and polyphenols, which are listed in Table 5.2. The study of activation energy helps to control the degradation kinetics using UV light exposure in a positive way.

Haddouche et al. worked on UV light inactivation of polyphenol oxidase (PPO), an enzyme mostly occurring in fresh produce like fruits and vegetables [16]. PPO is known to cause enzymatic browning, especially in cut fruits and juices. Researchers have adopted several methods to inactivate PPO and increase the shelf stability of the products. The study explains the effect of UV light at 254 nm with an incidence intensity of 20 mW/cm² on reducing PPO activity by 90% when exposed for 5 minutes. The study also analyzed the effect of glucose, fructose, and sucrose interference on PPO inactivation. It has been observed that though sucrose and glucose had no significant effect on the process, the presence of fructose quickened the inactivation, which might be due to the development of reactive oxidation species. The study also concluded that the inactivation kinetics were profound when the pH was more acidic, and that the PPO was not reactivated during storage.

5.6 FUTURE PROSPECTS AND CHALLENGES

UV light is a non-thermal technology that depends on several extrinsic and intrinsic factors. Extensive knowledge of chemical reactions and microbial kinetics in food systems can help in practical applications of UV light for food processing. The dissimilarities of product characteristics, subpopulations of microorganisms, the difference in the level of resistance against external stress for different cells, penetration of UV light in the substrate, etc., pose serious trouble in attaining a perfect fit using mathematical modeling. Though lethal germicidal effects of UV and pulsed UV light have been extensively studied, there are certain gaps in the literature regarding the resistance of certain microbial strains and spores against UV, lethal doses of different food pathogens, and biological variability in both pathogens as well as food substrates, which need to be studied to envision a holistic picture of UV light processing. Also, there have not been more studies on the effect of UV light on the nutritional properties of solid and liquid foods, which need to be widely researched. The development of resistance against UV light by food pathogen strains can also be a very interesting branch to explore. Predictive microbiology, which has been quite interesting in recent years, can be applied to UV light processing for the inactivation of different microorganisms using various treatment parameters. This work can create a database of different pathogens at various wavelength-dosage-time combinations in many food products to improve the practical applicability of UV and pulsed UV light processing in food preservation.

Kinetic modeling can be used as a powerful tool in food product development. By considering food safety issues and risk analysis as a reference, we can ensure that the product meets quality standards. Applying UV treatment to food products for microbial inactivation causes minimal degradation of nutritional organic compounds and may alter the palatability of the food. Most of the photodegradation of organic compounds follows first-order kinetics. The reaction rate depends on process variables and food composition. There have been very few studies analyzing the physicochemical degradation kinetics of food systems. In contrast, the effects of UV light exposure and the lack of studies on the generation of free radicals and intermediate products on food quality need to be further researched.

TABLE 5.1 Application Kinetic Models to Describe the Degradation of Food Quality by the Thermo-Chemical Reaction

Food Product	Process Type	Respective Thermo-Chemical Reaction	Model	Model Equation	Rate Constant, k	Correlation Coefficient (R^2)	References
Fermented turnip juice	UV treatment	Anthocyanin degradation	Weibull model	$C_t = C_0 \exp[-(\frac{t}{a})^{\beta}]$	0.016	0.982–0.998	[22]
Lychee juice	UV treatment	Antioxidant activity	First-order kinetic model.	$\ln[C]=\ln[C_0]-k_1t$	0.0237	0.9876	[50]
		Total phenolic compound.			0.0012	0.9814	
		Total flavonoid content			0.0008	0.9827	
Wastewater from a poultry farm.	UV treatment	P-arsanilic acid photolysis.	pH based kinetic model	–	0.20	–	[52]
Crushed tomato	Thermal treatment	Degradation of color Lycopene L-ascorbic acid	First-order kinetic model.	$\ln[C]=\ln[C_0]-k_1t$	0.249 0.1667 0.0423	>0.96	[7]
Apple peel by-product	Drum drying	Total phenolic content degradation.	First-order kinetic model.	$\ln[C]=\ln[C_0]-k_1t$	–	0.9784	[19]
Orange juice	Pasteurization	Color and L-ascorbic acid degradation.	Arrhenius model	–	–	0.945	[43]
Carotenoids	Thermal degradation	lutein, zeaxanthin, β-cryptoxanthin, β-carotene.	First-order kinetic model by light-induced model.	$\ln[C]=\ln[C_0]-k_1t$	–	–	[51]
Blueberry	Thermal degradation	Anthocyanin	Zero kinetic model.	–	–	0.9635	[32]
Aqueous solution	UV irradiation	Gossypol	Monophasic exponential kinetics.	–	–	–	[37]
Apple juice and apple cider.	UV irradiation	Patulin	First-order kinetics.	$\ln[C]=\ln[C_0]-k_1t$	0.0294	0.9950	[6]

*Scale parameter (s); β shape parameter; 't' time of exposure; and 'k_1' first-order reaction constant.

TABLE 5.2 Energy Required for Degradation of Some Nutritional Compounds

Food System	Treatment Parameter	Compounds	Activation Energy Ea (kJ mol^{-1})	References
Crushed tomato	Heat treatment from 70 to 100°C.	Ascorbic acid	25.69	[7]
		Lycopene	45.96	
Apple peel by product	Drum drying at 140°C	Phenolic content	1.026	[19]
Carotenoids	Thermal treatment	Zeaxanthin	65.6	[51]
		Lutein	38.9	
		β-cryptoxanthin	33.9	
		β-carotene	8.6	
Blueberry	Hydro thermodynamic process 70 to 105°C.	Anthocyanin	66.37	[32]
Orange juice	Pasteurization	Vitamin C	12.192	[39]
		β-carotene	19.097	

KEYWORDS

- **anthocyanin**
- **kinetics**
- **microbial inactivation**
- **pasteurization**
- **thermo-chemical reaction**
- **thermodynamic process**
- **UV light**

REFERENCES

1. Adams, M. R., Moss, M. O., & Moss, M. O. (2000). *Food Microbiology*. Publisher: The Royal Society of Chemistry, pp. 65–117.
2. Aisami, A., Yasid, N. A., Johari, W. L. W., & Shukor, M. Y. (2017). Estimation of the Q10 value: The temperature coefficient for the growth of Pseudomonas sp. AQ5-04 on phenol. *Bioremediation Science and Technology Research, 5*(1), 24–26.
3. Alvarez, I., Virto, R., Raso, J., & Condón, S. (2003). Comparing predicting models for the *Escherichia coli* inactivation by pulsed electric fields. *Innovative Food Science & Emerging Technologies, 4*(2), 195–202.
4. Andrews, G. R. (1986). Formation and occurrence of lactulose in heated milk. *Journal of Dairy Research, 53*(4), 665–680. https://doi.org/10.1017/S0022029900033197.
5. Aragao, G. M. F., Corradini, M. G., Normand, M. D., & Peleg, M. (2007). Evaluation of the Weibull and log normal distribution functions as survival models of *Escherichia coli* under isothermal and non-isothermal conditions. *International Journal of Food Microbiology, 119*(3), 243–257.
6. Assatarakul, K., Churey, J. J., Manns, D. C., & Worobo, R. W. (2012). Patulin reduction in apple juice from concentrate by UV radiation and comparison of kinetic degradation models between apple juice and apple cider. *Journal of Food Protection, 75*(4), 717–724. https://doi.org/10.4315/0362-028X.JFP-11-429.
7. Badin, E. E., Quevedo-Leon, R., Ibarz, A., Ribotta, P. D., & Lespinard, A. R. (2021). Kinetic modeling of thermal degradation of color, lycopene, and ascorbic acid in crushed tomato. *Food and Bioprocess Technology, 14*(2), 324–333. https://doi.org/10.1007/s11947-021-02579-1.
8. Baysal, A. H. (2018). Short-wave ultraviolet light inactivation of pathogens in fruit juices. In *Fruit Juices* (pp. 463–510). Elsevier.
9. van Boekel, M. A. J. S. (2002). On the use of the Weibull model to describe thermal inactivation of microbial vegetative cells. *International Journal of Food Microbiology, 74*(1–2), 139–159.

10. Brahmi, M., Belhadi, N. H., Hamdi, H., & Hassen, A. (2010). Modeling of secondary treated wastewater disinfection by UV irradiation: Effects of suspended solids content. *Journal of Environmental Sciences, 22*(8), 1218–1224.
11. Crook, J. A., Rossitto, P. V., Parko, J., Koutchma, T., & Cullor, J. S. (2015). Efficacy of ultraviolet (UV-C) light in a thin-film turbulent flow for the reduction of milk borne pathogens. *Foodborne Pathogens and Disease, 12*(6), 506–513.
12. Csapo, J., Prokisch, J., Albert, C., & Sipos, P. (2019). Effect of UV light on food quality and safety. *Acta Universitatis Sapientiae, Alimentaria, 12*(1), 21–41. https://doi.org/10.2478/ausal-2019-0002.
13. Fernandez, A., Salmeron, C., Fernández, P. S., & Martinez, A. (1999). Application of a frequency distribution model to describe the thermal inactivation of two strains of *Bacillus cereus*. *Trends in Food Science & Technology, 10*(4–5), 158–162.
14. Gardner, D. W. M., & Shama, G. (2000). Modeling UV-induced inactivation of microorganisms on surfaces. *Journal of Food Protection, 63*(1), 63–70.
15. Gouma, M., Álvarez, I., Condón, S., & Gayán, E. (2015). Modelling microbial inactivation kinetics of combined UV-H treatments in apple juice. *Innovative Food Science & Emerging Technologies, 27*, 111–120.
16. Haddouche, L., Phalak, A., & Tikekar, R. V. (2015). Inactivation of polyphenol oxidase using 254nm ultraviolet light in a model system. *LWT, 62*(1), 97–103. https://doi.org/10.1016/j.lwt.2015.01.039.
17. Halder, A., Datta, A. K., & Geedipalli, S. S. R. (2007). Uncertainty in thermal process calculations due to variability in first-order and Weibull kinetic parameters. *Journal of Food Science, 72*(4), E155–E167.
18. Hallmich, C., & Gehr, R. (2010). Effect of pre-and post-UV disinfection conditions on photoreactivation of fecal coliforms in wastewater effluents. *Water Research, 44*(9), 2885–2893.
19. Henriquez, C., Cordova, A., Almonacid, S., & Saavedra, J. (2014). Kinetic modeling of phenolic compound degradation during drum-drying of apple peel by-products. *Journal of Food Engineering, 143*, 146–153. https://doi.org/10.1016/j.jfoodeng.2014.06.037.
20. Huang, K., Tian, H., Gai, L., & Wang, J. (2012). A review of kinetic models for inactivating microorganisms and enzymes by pulsed electric field processing. *Journal of Food Engineering, 111*(2), 191–207.
21. Jagannath, A., Tsuchido, T., & Membre, J. M. (2005). Comparison of the thermal inactivation of *Bacillus subtilis* spores in foods using the modified Weibull and Bigelow equations. *Food Microbiology, 22*(2–3), 233–239.
22. Karaoglan, H. A., Keklik, N. M., & Develi Isıklı, N. (2019). Degradation kinetics of anthocyanin and physicochemical changes in fermented turnip juice exposed to pulsed UV light. *Journal of Food Science and Technology, 56*(1), 30–39. https://doi.org/10.1007/s13197-018-3434-1.
23. Keklik, N. M., Demirci, A., Puri, V. M., & Heinemann, P. H. (2012). Modeling the inactivation of *Salmonella typhimurium*, *Listeria monocytogenes*, and *Salmonella enteritidis* on poultry products exposed to pulsed UV light. *Journal of Food Protection, 75*(2), 281–288.
24. Koutchma, T. (2009). Advances in ultraviolet light technology for non-thermal processing of liquid foods. *Food and Bioprocess Technology, 2*(2), 138–155. https://doi.org/10.1007/s11947-008-0178-3.

25. Koutchma, T. (2019). *Chapter 4: Ultraviolet Light in Food Technology: Principles and Applications*. CRC Press.
26. Koutchma, T., & Popović, V. (2019). UV Light-Emitting Diodes (LEDs) and food safety. In *Ultraviolet LED Technology for Food Applications* (pp. 91–117), Elsevier.
27. Lebik, H., Madjene, F., Aoudjit, L., & Igoud, S. (2014). Modelling the kinetics of UV water disinfection. *International Journal of Scientific Research and Management Studies, 1*, 60–64.
28. Ling, B., Tang, J., Kong, F., Mitcham, E. J., & Wang, S. (2015). Kinetics of food quality changes during thermal processing: A review. *Food and Bioprocess Technology, 8*(2), 343–358.
29. Liu, R., Jin, Q., Tao, G., Shan, L., Huang, J., Liu, Y., Wang, X., Mao, W., & Wang, S. (2010). Photodegradation kinetics and byproducts identification of aflatoxin B1 in aqueous medium by ultra-performance liquid chromatography-quadrupole time-of-flight mass spectrometry. *Journal of Mass Spectrometry, 45*(5), 553–559. https://doi.org/10.1002/jms.1741.
30. López, M. A., Palou, E., Barbosa, C. G., Tapia, M. S., & Cano, M. P. (2005). Ultraviolet light and food preservation. In *Novel Food Processing Technologies* (pp. 405–421), CRC Press.
31. Mahmoodi, N. M., Arami, M., Limaee, N. Y., & Tabrizi, N. S. (2005). Decolorization and aromatic ring degradation kinetics of Direct Red 80 by UV oxidation in the presence of hydrogen peroxide utilizing TiO_2 as a photocatalyst. *Chemical Engineering Journal, 112*(1–3), 191–196. https://doi.org/10.1016/j.cej.2005.07.008.
32. Martynenko, A., & Chen, Y. (2016). Degradation kinetics of total anthocyanins and formation of polymeric color in blueberry hydrothermodynamic (HTD) processing. *Journal of Food Engineering, 171*, 44–51. https://doi.org/10.1016/j.jfoodeng.2015.10.008.
33. Matak, K. E., Churey, J. J., Worobo, R. W., Sumner, S. S., Hovingh, E., Hackney, C. R., & Pierson, M. D. (2005). Efficacy of UV light for the reduction of *Listeria monocytogenes* in goat's milk. *Journal of Food Protection, 68*(10), 2212–2216.
34. Menezes, N. M. C., Longhi, D. A., Ortiz, B. O., Junior, A. F., & de Aragão, G. M. F. (2020). Modeling the inactivation of *Aspergillus fischeri* and *Paecilomyces niveus* ascospores in apple juice by different ultraviolet light irradiances. *International Journal of Food Microbiology, 333*, 108773.
35. Miller, R. V., Jeffrey, W., Mitchell, D., & Elasri, M. (1999). Bacterial responses to ultraviolet light. *ASM News–American Society for Microbiology, 65*, 535–541.
36. Ngadi, M., Smith, J. P., & Cayouette, B. (2003). Kinetics of ultraviolet light inactivation of *Escherichia coli* O157 in liquid foods. *Journal of the Science of Food and Agriculture, 83*(15), 1551–1555.
37. Nomeir, A. A., & Abou-Donia, M. B. (1985). Photodecomposition of gossypol by ultraviolet irradiation. *Journal of the American Oil Chemists' Society, 62*(1), 87–89. https://doi.org/10.1007/BF02541497.
38. Ochoa-Velasco, C. E., Avila-Sosa, R., Hernandez-Carranza, P., Ruiz-Espinosa, H., Ruiz-Lopez, I. I., & Guerrero-Beltran, J. A. (2020). Mathematical modeling used to evaluate the effect of UV-C light treatment on microorganisms in liquid foods. *Food Engineering Reviews, 12*(3), 290–308. https://doi.org/10.1007/s12393-020-09219-y.
39. Ogori, A. F., Kurazheva, M., Demidov, A., Bobruyko, E., Tsvelikhovskaya, A., Sirotkina, L., Ekaterina, P., Căsărovă, M., Hleba, L., Ishaku, S. O., & Amove, J.

(2020). Degradation kinetics of vitamin C and β-Carotene in mango juice. *Journal of Microbiology, Biotechnology and Food Sciences, 9*(4), 861–864. https://doi.org/10.15414/jmbfs.2020.9.4.861-864.
40. Peleg, M. (1995). A model of temperature effects on microbial populations from growth to lethality. *Journal of the Science of Food and Agriculture, 68*(1), 83–89.
41. Peleg, M. (2006). *Advanced Quantitative Microbiology for Foods and Biosystems: Models for Predicting Growth and Inactivation.* CRC Press.
42. Perez-Rodriguez, F., & Valero, A. (2013). Predictive microbiology in foods. In *Predictive Microbiology in Foods* (pp. 1–10). Springer.
43. Remini, H., Mertz, C., Belbahi, A., Achir, N., Dornier, M., & Madani, K. (2015). Degradation kinetic modelling of ascorbic acid and color intensity in pasteurized blood orange juice during storage. *Food Chemistry, 173*, 665–673. https://doi.org/10.1016/j.foodchem.2014.10.069.
44. Ruiz-Hernández, K., Ramírez-Rojas, N. Z., Meza-Plaza, E. F., García-Mosqueda, C., Jauregui-Vázquez, D., Rojas-Laguna, R., & Sosa-Morales, M. E. (2021). UV-C treatments against *Salmonella typhimurium* ATCC 14028 in inoculated peanuts and almonds. *Food Engineering Reviews, 13*(3), 706–712. https://doi.org/10.1007/s12393-020-09272-7.
45. Severin, B. F., Suidan, M. T., & Engelbrecht, R. S. (1983). Kinetic modeling of UV disinfection of water. *Water Research, 17*(11), 1669–1678.
46. Simons, R., Gabbai, U. E., & Moram, M. A. (2014). Optical fluence modelling for ultraviolet light emitting diode-based water treatment systems. *Water Research, 66*, 338–349.
47. Soro, A. B., Whyte, P., Bolton, D. J., & Tiwari, B. K. (2021). Modelling the effect of UV light at different wavelengths and treatment combinations on the inactivation of *Campylobacter jejuni. Innovative Food Science & Emerging Technologies, 69*, 102626.
48. Valdramidis, V. P., Taoukis, P. S., Stoforos, N. G., & Van Impe, J. F. M. (2012). Modeling the kinetics of microbial and quality attributes of fluid food during novel thermal and non-thermal processes. In *Novel Thermal and Non-Thermal Processing of Fluid Foods* (pp. 365–382). Elsevier Inc. https://doi.org/10.1016/B978-0-12-381470-8.00014-1.
49. Vieira, M. C., Teixeira, A. A., & Silva, C. L. M. (2000). Mathematical modeling of the thermal degradation kinetics of vitamin C in cupuacu (*Theobroma grandiflorum*) nectar. *Journal of Food Engineering, 43*(1), 1–7. https://doi.org/10.1016/S0260-8774(99)00121-1.
50. Visuthiwan, S., & Assatarakul, K. (2021). Kinetic modeling of microbial degradation and antioxidant reduction in lychee juice subjected to UV radiation and shelf life during cold storage. *Food Control, 123*, 107770. https://doi.org/10.1016/j.foodcont.2020.107770.
51. Xiao, Y. D., Huang, W. Y., Li, D. J., Song, J. F., Liu, C. Q., Wei, Q. Y., Zhang, M., & Yang, Q. M. (2018). Thermal degradation kinetics of all-trans and cis-carotenoids in a light-induced model system. *Food Chemistry, 239*, 360–368. https://doi.org/10.1016/j.foodchem.2017.06.107.
52. Xu, J., Shen, X., Wang, D., Zhao, C., Liu, Z., Pozdnyakov, I. P., Wu, F., & Xia, J. (2018). Kinetics and mechanisms of pH-dependent direct photolysis of p-arsanilic acid under UV-C light. *Chemical Engineering Journal, 336*, 334–341. https://doi.org/10.1016/j.cej.2017.12.037.

CHAPTER 6

APPLICATION OF ULTRAVIOLET LIGHT IN SOLID AND SEMI-SOLID FOODS

SIVASHANKARI MANICKAM,
SADVATHA RAMANNA HAROMUCHADI, and RAVINDRA NAIK

ABSTRACT

Modern food production would clearly benefit from ultraviolet (UV) light processing methods that do not appreciably change the solid and semi-solid foods' organoleptic or nutritional properties or produce any unwanted chemical components. The extensive application of high-tech non-thermal or low-heat techniques that could eventually replace traditional thermal preservation techniques and produce products with better quality, longer shelf lives, and perhaps even higher nutritional values was sparked by interest in UV light processing technologies. Recent years have seen a lot of research on non-thermal food processing techniques in an effort to create food items with increased preservation of nutritional and organoleptic qualities as well as shelf life to meet shifting consumer preferences [90, 127, 263]. Ultraviolet (UV) light technology has the possibility to inactivate harmful microorganisms. The use of ultraviolet light in solids and semi-solids could enhance shelf life by preserving nutritional value and preventing the growth of microorganisms. Critical UV processing factors and their effects on physicochemical and sensory properties are being monitored. UV technology combined with other techniques could increase the efficacy of decontamination, scope, future trends, and industrial applications of UV light in solid and semi-solid foods while also addressing the limitations of UV light.

Nonthermal Light-Based Technologies in Food Processing.
C. K. Sunil, Preeti Birwal, Megh R. Goyal, and R. Mahendran (Eds.)

6.1 INTRODUCTION

A cost-effective, chemical-free, and effluent-free physical, non-thermal way of food processing, UV light is an eco-friendly and long-lasting technological advancement. Therefore, caution must be exercised to limit UV exposure to people and get rid of the ozone produced. This method uses vacuum and far UV wavelengths, is risk-free, and produces no byproducts [256].

UV treatment is applied to solid and semi-solid products having the main objective of increasing the product's shelf life and making it safer for consumers by reducing the presence of pathogenic bacteria [135]. In order to cause a photochemical change, a wavelength change, and DNA damage in an inert, dangerous bacterium, UV light exposure must be seen. To lessen the microbial growth, surfaces can be disinfected using UV light [33]. When UV-C exposure caused the formation of viniferins and resveratrol in grape vines in 1977, the earliest research on fruits and veggies was released. UV-C radiation was first used to preserve food in the 1930s [87]. Later, it was used in semi-solids like cheese and yogurt as well as solid fruits, vegetables, meat, fish, and RTE or RTC items. The efficiency of UV-C radiation for decontaminating produce is greatly influenced by the kind of purifying systems, method of application, pulse intensity, form and site of microorganisms on fruits and vegetables, distance from the source of irradiation, and type of microbe [87, 163]. Since a small amount of light penetrates the interior. Depending on the composition of the food, UV light can only penetrate solid and semi-solid materials up to a few hundredths of a millimeter. The surface of solid food products is affected by UV exposure. UV radiation can effectively kill surface bacteria, ensuring product safety, extending product shelf life, and having minimal effects on chemical and nutritional properties due to its limited penetration depth. The limited use of UV light may be due to the limited availability of equipment for disinfecting solids commercially [156].

On the other hand, it raises the question of how UV exposure may affect the composition of food. What changes in dietary composition could result from photochemical processes that may ultimately have a negative impact on nutritional value and quality? The structural diversity of vitamins is particularly important because many of them are sensitive to UV light due to their structure [163].

Solid and semi-solid food surfaces that have been treated with UV light may have commercial applications. UV light irradiation is a germicidal treatment that may be more effective than ozone, hydrogen peroxide, or chlorine. Recent advancements in UV light irradiation science and engineering have demonstrated that both freshly cut fruits and vegetables, as well as whole fresh fruits and vegetables, could benefit significantly from UV treatment to extend their shelf life [272]. Surprisingly, there is still limited understanding of how substances, especially complex food matrices, respond to UV light, despite the importance of this interaction [250].

In comparison to thermal processing, ultraviolet (UV) has the potential to offer consumers products that are minimally processed, microbiologically secure, and fresh-like items with only negligible effects on the product's sensory and nutritional qualities [156]. However, this technology must not take the place of sound agricultural, manufacturing, or hygiene practices. UV light application is another potential substitute for the use of chemicals in the food business. Additionally, UV light usage leaves no chemical traces behind. It also has some technological benefits for small-scale production, particularly in developing nations, because of its low operational, installation, and maintenance costs and low energy consumption. The treatment is fairly simple to use and clean. Despite having several benefits, its low penetration power limits its application in the food business. High doses may have a harmful impact on some vitamins and quality. Applications should be made with these circumstances in mind in order to get effective outcomes. Examining the process variables that affect UV light treatment and processing for food safety is the goal of this chapter [272]. It provides a basic assessment of the applications and efficiency of UV light treatment for solid and semi-solid meals in addition to future suggestions on this approach and its scalability [161].

This chapter explores some of the most current findings that may lead to the application of UV light for the post-harvest treatment of fruits and vegetables, not to illustrate completely and in detail all the research on the effects of UV radiation on vegetables and fruits [169, 250].

6.2 CURRENT TRENDS IN SURFACE DECONTAMINATION OF SOLID AND SEMI-SOLID FOODS

Governments, international organizations, and researchers around the world are increasingly focused on food safety and the prevalence of microorganisms that cause serious illness. Therefore, engineers now confront difficulties in developing novel technologies to handle microbial pollution, preserve public health, and avoid the proliferation of infections and pandemics, as well as industrial solutions.

Due to these difficulties, engineers must work to develop and improve appropriate decontamination techniques [82, 100, 159, 229, 300]. Cleaning, sanitizing, disinfection, and sterilization are some of the types of decontamination techniques based on the level of microbiological decontamination. According to the decontamination process, they can also be separated into physical and chemical cleaning techniques [82]. Numerous researchers have noted the development of new decontamination techniques for solid and semi-solid materials, including the use of vaporized hydrogen peroxide, cold plasma, ozonation, high hydrostatic pressure (HHP), and sterilization. The technologies that were created, however, have advantages and disadvantages. A cost-effective, chemical-, and waste-free, long-lasting, non-thermal physical method of processing food, UV light is devoid of waste effluents and chemicals. It doesn't generate waste products. Although measures must be made to keep people from being exposed to UV radiation and to eliminate the ozone created by distant UV rays and vacuum, allowing for safe use [165].

Compared to thermal and chemical sterilizing techniques, which are typically carried out at ambient temperature and pressure, UV radiation is thought to be cheaper and more effective [20]. With a germicidal effect on air or surfaces between 200 and 365 nm, it is also known that type C ultraviolet radiation (UVC) kills germs and viruses [133, 326]. UVC kills bacteria by inactivating DNA/RNA through the generation of pyrimidine dimers and cytosine, which are mutagenic DNA lesions caused by UVC absorption [149, 169].

6.3 MONITORING UV CRITICAL PROCESS FACTORS

6.3.1 *THICKNESS, SHAPE, DURATION OF TREATMENT*

When the right amount of UV radiation is applied to the solid and semi-solid food surfaces, a sufficient decrease in microbial load can be obtained. The topography of the skin and the method of application, however, can have an impact on how much UV radiation reaches the fruit, thus this must be carefully managed.

Even with low levels of UV-C radiation doses, adjustments to the typical. It has been demonstrated that self-contained chambers can increase microbial reductions and maintain quality attributes [220]. The study found that a higher efficiency was attributed to a dual action of immersion irradiation and simultaneous operations with process water resulting from the replacement of the conventional treatment chambers with pressurized water sprinklers [87]. In disinfection systems, [311] showed that it is possible to enhance UV-C radiation exposure on fruit surfaces while minimizing major fluctuations. In this investigation, conveyor belts were positioned 14 cm from UV-C lights, and fruits were moved and rotated between two treatment chambers that were joined by an inclined belt. As a result of this modification to the typical UV-C lamp self-contained chambers, fruits received an average dose of 1 kJ/m^2, as opposed to 0.2 kJ/m^2 for fruits that were not rotated.

Escherichia coli O157:H7 (*E. coli* O157:H7) and *E. coli* ATCC 25922 both saw reductions in log CFU/fruit of 1.3–1.8 and 1.0–1.2, respectively, thanks to the modified rotating device, as opposed to 0.5–0.7 log CFU/fruit reductions obtained without the rotation device. This demonstrates the significance of dose consistency for maximizing UV-C radiation's effectiveness in disinfection systems [220]. A tabletop UV-C radiation system was also employed by Hakguder et al. [127] to offer uniform intensity, allowing fruits to move up and down from a rotating roller. They found decreases in the overall amount of aerobic mesophilic bacteria, yeast, and mold of 3.0 and 2.38 log CFU/g at dosages of 31.01 and 7.75 kJ/m^2, respectively. It was also observed that the lamp's midsection had approximately constant light intensity, but the light intensity at either end was decreased. Due to UV-C radiation's low penetration capacity, the main challenge to commercialization is ensuring uniform exposure on food surfaces.

Experiments were conducted where UV rays were projected onto the entire fruit surface to lower the microbial population and strengthen post-harvest resistance to deterioration [256]. This was accomplished by manually turning the products during UV treatment two or four times [282, 312]. However, Stevens et al. [282] noted that such procedures are quite problematic and could negatively impact the marketing of fresh produce's post-harvest UV treatments.

The potential impact of fruit rotation on the deterioration of the rots that cause bitterness (*Colletotrichum gleosporioides*), brownness (*M. fructicola*), and greenness (*P. digitatum*) in apples, peaches, and tangerines was examined [256]. In comparison to when the fruits were rotated four times, apples, peaches, and tangerines that were exposed to UV-C light at their stem ends for 7.5 kJ/m^2, 7.3 kJ/m^2, and 1.3 kJ/m^2, respectively, generated a similar or marginally superior resistance to mold deterioration [282]. Fruits were most susceptible to rotting and degradation when only one or two surfaces were exposed to UV radiation.

The differences in how different fruits react to various treatments. There was a connection between the regions of UV-C photoreception and potential transmission mechanisms to the fruits' phloem vascular tissue has a transduction [282]. Obande & Shama [217] used biodosimetry to assess the

UV-C dose given to a polystyrene sphere that might mimic fruits like apples, peaches, tomatoes, etc. *Bacillus subtilis* spores were used to inoculate the spheres, which were then exposed to UV-C light using static and rotational techniques. According to the authors, spore biodosimetry under UV irradiation settings produced 9.1 ± 0.9 J for a single exposure to UV-C for 80 s, 10.7 ± 1.0 J for two rotations by 180° (2 × 40 s), and 6.1 ± 0.6 J for a sphere rotated four times by 90° (4 × 20 s) in their studies. The scenario of a rotating sphere constantly for 80 s produced the lowest UV dose, or 3.5 J. There isn't much conflict when comparing the outcomes from Stevens et al. with Obande & Shama [217, 282]. The rotation for two times produced the most UV dosage for the polystyrene sphere. When the same approach was used with fruits, the decay inhibition was the lowest. Undoubtedly, accurate measurement of the UV dosage given to the fruits is crucial for future commercialization. To determine the relationship between the administered UV radiation, its distribution throughout the solid fruit surface, and the physiological processes brought on by UV hermetic processing, more research must be done.

In less than 5 minutes, a mercury UV light source with a 254 nm spectral line demonstrated a 99.99% reduction. In experiments using *Escherichia coli*, the thallium-antimony light source displayed a decreased but comparable efficiency. Consequently, irradiating items and spaces with UV-C utilizing mercury-free light sources could be a useful cleaning technique [276].

When carried out correctly, UV treatment can be a competitive substitute for heat treatments for reducing the number of harmful microorganisms in food. It is evident that UVC has the biggest effect on disease control [60]. The second critical process variable in the method is the radiation dose (energy per unit area). A wide range of doses (between 0.01 and 50 kJ/m^2) have been researched for the UVC region. However, one study made an effort to deliver more than two or three dosages, and full optimization research is now being conducted. Using strawberries as a test subject, Cote et al. [58] discovered that the most crucial process component impacting the efficacy of treatment is not radiation intensity (energy per unit time and unit area). Surprisingly, this variable has gone unnoticed in certain studies that omit mentioning the illuminant's radiation intensity [65].

When maximizing UV therapies, radiation dose and intensity should be taken into account simultaneously [58]. Low UVB levels (less than 5 kJ/m^2) and intensities can help refrigeration work more effectively by delaying the destruction of chlorophyll [194]. The freezing sector may be better suited for high-intensity UVB exposure before further processing as a preventative measure to boost antioxidant capability [64]. Another factor to take into account while choosing an appropriate treatment schedule is the radiation pattern. For pulsed therapies using xenon lamps, this has been explored to a fair extent [299]. The importance of the radiation exposure pattern has not been considered when it comes to therapies using traditional single-UV radiation sources. Araque et al. [10] demonstrated that when compared to a single pre-storage irradiation at the same total dose, treatments that were fractionated (two-step and five-step) were substantially better at regulating intensity (36 W/m^2) and softness and degradation (4 kJ/m^2). During storage, fractionating the treatments prevented pectin from debranching and prevented polyuronides from becoming soluble [328].

6.3.2 FREQUENCY, DISTANCE FROM LIGHT SOURCE

Due to its germicidal effects on microorganisms, UV-C light, or more particularly, light with a wavelength of 254 nm, is the only light source allowed [99]. In order to achieve the necessary effect on the microbial load in solid and semi-solid foods to offer microbiologically safe food items, exposure time, dosage, and product surface are crucial [69].

The effects of various UV bands, including UV-C (200–280 nm), UV-B (280–315 nm), and UV-A (315–400 nm), both individually and in conjunction with visible light, were investigated. For pears and apples, combined UV-B and visible light treatments have positive effects on the plants' functional qualities [72, 129, 180, 186]. In comparison to UV-B or visible light treatment alone, UV-B/visible light treatment was more successful at accumulating apple phytochemicals [63, 158]. Researchers looked into how common (*Agaricus bisporus*) mushrooms convert ergosterol to vitamin D2 [118, 134]. Although vitamin D2 was produced in response to both tested treatments, it was discovered that UV-B radiation was more effective [148, 261]. Vitamin D2 increased by 387% in response to UV-B exposure (4.93 kJ/m^2) and by 173% in response to UV-C (6.06 kJ/m^2) [165]. Jasinghe & Perera [170] examined how vitamin D2 production in edible mushrooms was affected by UV-C (23.0 kJ/m^2) and UV-A (25.2 kJ/m^2) exposure. In another study, Shiitake, oyster, button, and abalone mushrooms were all evaluated, and all of them had higher amounts of vitamin D2 as a result of the UV-C exposure [141]. Various other fruits have also been effectively exposed to UV-C light. In the case of mangoes [15, 104, 237, 288], blueberries [112, 262, 330], pepper fruits [87, 192], and green tomato fruits [184, 195, 226], it was discovered that UV-C irradiation increased and/

or enhanced the preservation of the phenolic compounds during storage [77, 308]. Mangoes exposed to 4.93 kJ/m^2 had the maximum accumulation of phytochemicals, but treatments with energy densities of 2.46 kJ/m^2 or 9.86 kJ/m^2 produced fewer phenols and polyamine compounds [167]. Since browning and drying of the sepals occurred with higher treatments, Lammertyn et al. and Koutchma & Orlowska [165, 175] advised 1.0 kJ/m^2 as the appropriate fluence for treating strawberries with UV-C [165, 175]. Additionally, an overdose can speed up the ripening and senescence processes and reduce the ability to resist microbial and/or fungal deterioration, resulting in decreased fruit storage and financial loss [167]. Therefore, the ideal UV treatment settings must be used in order to acquire the highest quantities of nutraceuticals without adversely compromising a certain fruit commodity's look and shelf life.

The UV-C dose of a light is commonly measured using UV sensors in W/m^2 units. UV irradiance is measured with radiometers, which can be thermal or photonic [120]. The amount of the dosage is also measured via chemical actinometry. Actinometers are used to measure the concentrations of products from photochemical reactions that have well-defined energy; these concentrations are directly correlated with how much UV light the treated product absorbed [304]. To get repeatable measurements, all sensors must be calibrated for irradiance while accounting for spectral selectivity radiation (240–290 nm) [305]. However, biodosimetry, which involves calculating the log reduction after the fluid has undergone specific conditions been treated with UV-C after the inoculation of a surrogate bacterium is the most reliable method for monitoring irradiance [119]. Factors like light source, product composition, and have an impact on how well UV rays penetrate solid and semi-solid foods [69]. The product ripening stage is an important variable in determining the effectiveness of UV irradiation [224]. In general, treating immature fruit may result in stronger ripening-delay phenotypes than treating fruit at advanced maturity stages [65]. The ripening stage affects post-harvest decay control as well as how UV irradiation impacts ripening changes [2]. In fresh-cut bell peppers, UV light proved more effective at reducing soft rot in red mature fruit [251, 252]. The level of commodities processing also significantly affects antioxidant elicitation and quality maintenance [65]. This might be because processing exposes more of the product's surface area, increasing the surface area. However, it has also been proposed that synergistic effects brought on by concurrent UV and wounding stressors are also at play [285]. The nature of the phase between the product and the UV light can also have a significant impact.

The evaluation of UV light as an alternative food processing technique takes into account the physical and chemical composition of the food product to be treated, the UV equipment's qualities (such as UVC intensity and exposure length), microbiological characteristics, and other commercial issues [272].

To maximize process effectiveness and UVC light penetration into food products, especially solid and semi-solid food products, for microbial inactivation, it's crucial to select a UVC source with the right wavelength [69, 272]. Low-pressure (254 nm line), high-pressure (broadband fine lamps), and excimer lamps (207, 222 nm) UVC germicidal lamps have tremendous potential for use in food applications [272]. In comparison to traditional 254 nm mercury lamps, a krypton-chlorine excimer lamp had better bactericidal efficacy against gastrointestinal microorganisms. Additionally, UVC radiation produced by pulsed xenon lamps, which release light at regular intervals, can offer a greater immediate energy intensity than traditional mercury lamps. The wavelength has a significant impact on microbial inactivation during UVC food processing. In close proximity to the maxima of DNA absorption, UVC is more deadly to cells. More than 6 log reductions in contaminating bacteria were seen in a study on the effects of different UV wavelength combinations on model food systems (produced with UVC LEDs). To achieve the necessary amount of microbial inactivation, UVC light applications for solid and semi-solid food products use various tools and apparatus. Simple UV chambers can be used to disinfect solid meals like fruits and fresh produce, but specific reactors are needed to apply UVC light to liquid consumables like milk or alcoholic beverages [272].

6.3.3 TEMPERATURE

Apples from the varieties Jonathan, Gala, and Royal Gala produced more anthocyanins when exposed to radiation at higher temperatures (20–25°C) than when exposed to lower temperatures (10–15°C), according to studies [283]. Similarly, Red Chinese sand pears responded better to UV-B/visible irradiation at 27°C than at 17°C [322]. Higher amounts of anthocyanins were seen in apples after postharvest UV-B/visible light exposure at –0.5/–0.5°C (day/night), 20/20°C, and 20/6°C [223]. Therefore, it is necessary to optimize the best temperature ranges for postharvest UV treatment for a particular commodity and cultivar.

6.4 DECONTAMINATION APPROACH FOR SOLID AND SEMI-SOLID FOODS

Numerous reviews have been published to explain how UV technology is used to process liquid meals including dairy products and fruit juices. There is not a lot of information available about the cleanliness of the surfaces of solid foods, meat, and fresh items because it was limited to such liquid food products [161, 272].

UV and its uses in the preparation of solid foods. UV radiation is primarily used to achieve two main goals for fresh fruits and vegetables [29, 119, 163]. The primary goal is to reduce the number of microorganisms on the product surface, and the secondary goal is to increase host tolerance to microbes. In a process known as hormesis, UV light—also known as hormetin—proves to be helpful in processing a variety of different food products. The entire procedure is generally referred to as the Hormetic effect [27, 173, 178]. A phenolic chemical known as phytoalexins is created as a result of this effect, which also causes the creation of phenylalanine ammonia-lyase [257]. Fresh, pulpy fruits and vegetables become resistant to a variety of microorganisms mostly due to phytoalexin production [83]. The fundamental underlying idea behind this is that phenolic chemicals are known to suppress ethylene synthesis, which prolongs the shelf life of food by slowing the ripening process [182].

According to research, fruits and vegetables can help prevent some cancers, diabetes, obesity, and cardiovascular diseases [303]. Fruits and vegetables must be safe and free of microorganisms whether they are eaten raw or cooked [138, 146, 307]. Numerous fruits, vegetables, and other solid food items belonging to different genera can carry a variety of diseases [21]. *Salmonella, Listeria, Staphylococcus, Clostridium*, and *Bacillus* are a few examples of such pathogens [113]. Since all of these are regarded as hazardous bacteria, food that has been contaminated with them is not at all fit for ingestion, and their removal from food is necessary [101, 307]. Food that has been contaminated with these dangerous pathogens is regarded as ruined, and it can be challenging to remove the microorganisms from food just by washing it with water [277]. However, on occasion, hot water with a higher temperature could be helpful to get rid of some of these bacteria [57, 247]. Therefore, a solution is needed that can process these solid food items effectively and efficiently, and UV plays a crucial part in that. Varying bacteria require different amounts of UV radiation exposure, which can range from a fraction of a second to several minutes [7, 32]. Since 1997, solid food products have routinely been exposed to UV radiation (mostly 180–280 nm with a maximum at 254 nm) in place of chemical fungicides and bactericides to prevent various postharvest illnesses [241, 292]. After that, numerous studies were conducted to support it, and it was established that UV helps to induce fruit and vegetable resistance to postharvest storage rots [79, 279]. Additionally, it was demonstrated experimentally that it prolongs food life by delaying ripening [136, 147, 238, 266]. Since most pathogenic bacteria are killed by UV, it is employed widely because an ideal UV intensity causes phenolic compounds to develop and accumulate, which in turn activates genes involved in pathogenesis [213, 298]. When UV radiation is utilized at a certain wavelength of 254 nm, pyrimidine dimers are formed. This causes a mutation in the helical helix of the bacteria's DNA, which prevents future microbial reproduction and multiplication [274]. There are two ways that UV-C in general can function and demonstrate its germicidal efficacy. The first is by a process that causes DNA damage, and the second is through the development of resistant defenses against bacteria in different fruits and vegetables [219]. In this approach, UV demonstrates that it may be used to extend the shelf life of solid goods like fruits and vegetables and to eradicate germs [65, 294].

6.4.1 FRESH PRODUCE-FRUITS – CHERRY TOMATO, STRAWBERRY, BLUE BERRY

Owing to fruits' positive effects on health, production of fresh fruit and fruit-related goods has increased in recent years [59, 280, 284]. There are various non-thermal pathogen control methods for fresh produce. These non-traditional therapeutic options can be chemical in nature or vice versa. Because UV is safe and has no aftereffects on the food it treats, it is a potential surface decontamination solution [97, 128]. According to research, UV treatments can boost the nutritional value and shelf life of fruits and vegetables, as well as their antioxidant activity, in addition to being germicidal [41, 236]. UV systems are inexpensive because they have a cheap initial investment and low ongoing treatment costs [55]. For the purpose of extending shelf life and enhancing product quality, numerous studies on the effects of UV irradiation on fruits and vegetables have been conducted [7, 67, 154]. The shelf life of peaches and tomatoes was researched [2, 201, 328], which discovered that low UVC light (130–4,000 mJ/cm^2) delayed ripening and prevented post-harvest rots [55].

Fresh fruits, vegetables, and grains make up fresh produce, which is a significant source of nutrition for people. Fresh produce can become contaminated by microbes anywhere, including during harvest, storage, transit, retailing, and handling [55, 272]. Recent years have seen an upsurge in *E. coli*, *L. monocytogenes*, norovirus, or hepatitis virus, usually as a result of people eating contaminated fresh produce [55, 165].

Chlorine dioxide, sodium chlorite, organic acids, cold plasma, ozone, and electrolyzed water are a few examples of conventional surface sanitation techniques for fresh produce [34, 54, 76, 272]. To minimize the amount of bacteria on the product surface, fresh fruits and vegetables are currently frequently exposed to UV light [28, 206, 209]. Numerous studies have found that UV treatment increases the commercial potential of the surface treatment of newly cut fruits. Additionally, it has been claimed that using LVC technology increases the antibacterial capabilities of natural substances [76, 176].

Apart from inactivating microorganisms, UV light assists in identifying skin imperfections or wounds and aids in sorting/handling. Separating fruit with these flaws is one of the most time-consuming tasks carried out by packing houses. The fruit is typically examined visually or sorted optically by a machine when it is appropriately lighted with white light [61, 189, 196, 227]. The initial classification of citrus using UV lasers may make it easier to detect physical deterioration [181, 191, 290]. Fruit is illuminated by UVA "black light," which reveals that microscopic peel cracks glow intensely and enable segregation at the initial stages of categorization. These rooms contain the fruit-transporting conveyor belts, and the workers here are required to put on safety goggles and gloves [196, 273].

Cherry tomatoes were given a stand-alone and barrier treatment of UV-C irradiation and edible coating. The white dragon fruit's mucilage was used to make the coating, which is edible. Quality indicators during 21 days of storage at 4°C such as antioxidant analysis, ascorbic acid, color, total soluble solids, titratable acidity, and weight loss (total phenolic content and flavonoid content), and microbiological analysis were measured [327]. The hurdle therapy outperformed stand-alone therapies by 21 days, extending the shelf life, reducing weight loss (0.87 ± 0.05%) and color changes (11.61 ± 0.95 DE), and inhibiting microorganisms [249]. Additionally, fruits treated with UV-C and edible coating had increased levels of ascorbic acid (1.07 ± 0.06 mg/100 mL), total flavonoids (13.179 ± 0.002 mg CE/100 mL), and total polyphenols (0.132 ± 0.003 mg GAE/100 mL). These findings suggest that an inventive way to extend the shelf life and quality of fruits is to use an edible covering combined with UV-C as a hurdle treatment [72].

Fresh strawberries from the "Kurdistan" variety (*Fragaria ananassa*, Duch. cv. Kurdistan) were subjected to UV-C at different intensities (254 nm) radiation and kept at –1 to –5°C for up to 7 days [66, 281]. All UV-C radiation dosages slowed yeast growth, but only the highest dose produced effects that were statistically significant. A suitable amount of UV-C radiation could lower microbial burdens without impairing the Kurdistan strawberry's sensory quality. On day 7, fruit treated with greater doses (0.5 kJ/m^2) is much firmer ($p<0.05$), and this dose enhanced the product's sensory quality. Instead of using antibacterial agents, this UV technique could be a substitute [292].

When it comes to fruit ripening that is impacted by UV-C and softening. Among other fruits, the food industry needs strawberries [26, 244]. One of the key determinants of approaches to meet customer expectations for fruit postharvest life is how to postpone fruit softening through exposure to UV-C, a novel alternative in processing and preservation [37, 77]. With safer and more recent horticultural products, this softening delay might not occur. The majority of the alterations in enzyme activity and these new technologies (modified atmospheric proteins implicated in cell wall disintegration) are to blame [19].

According to Guo et al. [121], blueberries can be cleaned of *Salmonella* enterica and *E. coli* utilizing UV treatments. After 60 seconds of pulsed UV irradiation, they reported a maximum reduction of 4.3 and 2.9 log CFU/g for *Salmonella* and *E. coli*, respectively. Continuous-wave UV is less expensive than pulsed UV. Those with little cash to spend, which includes the majority of commercial blueberry packinghouses, may profit from continuous-wave UV's low starting cost and the lack of costly safety equipment [56].

Several outbreaks involving fresh and frozen berries have been linked to HAV and NoV [49, 211]. Cleaning fresh, frozen, and HAV-contaminated strawberries, blueberries, and raspberries, as well as MNV-1, has been attempted [36, 221]. Although viral titers on berry surfaces could occasionally be decreased by up to 3 log cycles by UV-C, its impact was limited, and treatment durations greater than 1.3 J/cm^2 had no positive impact on inactivation [103]. Despite the fact that the UV reactor was constructed with mirror-finish reflectors on both sides to increase exposure, it's probable that shielding from the uneven surface of berries is to blame. Due to its minimal impact on fruit sensory quality and lack of harmful by-product furan, however, as part of a barrier-building strategy, UV-C may be useful for berries [109]. With a range of 1.9 to 2.6 log TCID50/mL for FCV, HAV, and Aichi virus, strawberries posed a significant barrier to viral inactivation

[6, 89], which was achieved at a rate comparable to foods with smoother surfaces, including lettuce and green onions, but several log cycles lower. Safety of lettuce, another food that is ingested fresh. For HAV, Aichi virus, and FCV on Romaine lettuce [143]. The effectiveness of UV-C is high, at least 4 log TCID50/mL, but not for phage MS2 on Iceberg lettuce [106]. In each case, inactivation started out quickly before slowing down. For MNV-1, a low level of inactivation was likewise seen, but this time with a log-linear drop. Depending on the virus kind, disinfecting green onions may or may not be necessary. While HAV, FCV, Aichi virus, human adenovirus type 41, and MNV-1, in decreasing sequence of inactivation, have not been linked to large levels of inactivation (>5 log TCID50/mL) [35, 62]. Inactivation was limited when viruses were absorbed, which can occur during hydroponic cultivation by uptake through roots. These results are to be anticipated given that UV-C is a topical treatment.

6.4.2 FRESH-CUT VEGETABLES: MUSHROOMS AND CARROTS

Due to the numerous health advantages linked to eating these foods, the market for fresh fruit and ready-to-eat veggies has developed considerably [243, 271]. Conversely, some fresh fruits and vegetables have had their shelf life extended by ultraviolet (UV) light [7, 179]. The UV wavelength band is made up of three sub-bands with wavelengths of UV-A (320–400 nm), UV-B (280–320 nm), and UV-C. The UV wavelength band encompasses the range of 100 to 400 nm (200–280 nm) [56]. It has been discovered that the shortwave ultraviolet light range of (UV-C) (200–280 nm), which is referred to as a germicidal range, is effective for controlling postharvest ripening and diseases in a variety of plants, including tomatoes, strawberries, baby spinach, broccoli, peppers, and blueberries [79, 209, 212]. In addition, it has been established that exposure to UV-C light stresses plant tissues, which triggers defense mechanisms in plants such as the accumulation of antimicrobial compounds, a reduction in the activity of enzymes that break down cell walls, an increase in the activity of defense enzymes, and an increase in antioxidant activity [188, 214, 295]. These substances are highly sought after because they can help maintain the quality and extend the life of vegetables and fruits by postponing senescence and fruit ripening, as well as by boosting the production of natural defenses against fungus and bacteria [153, 212, 323]. According to studies, exposure to UV rays increased the levels of total phenols and polyamine compounds in mangoes [124, 193], anthocyanin, phenolic content, and antioxidant capacity in strawberries [204, 253], flavonoids in blueberries [313, 314], and ascorbic acid, phenolic compounds, and antioxidant capacity in tomatoes [13, 80]. High UV-C exposure levels, nevertheless, can occasionally cause harm. Strawberry calyx browning, soft patches, and loss of anthocyanin and phenolic content were all caused by dosages of 1 to 4.1 kJ UV-C/m^2 [45, 122, 267, 310]. The purpose of this study was to ascertain how three pre-packaging UV-C treatments under passive MAP affected the fresh-cut rocket leaves' post-harvest quality, total phenolic content, and antioxidant activity. Fruits can be contaminated by microbes from the air, water, and surfaces during the manufacturing process [68, 218]. Targeting certain locations within the manufacturing environment with cleaning and disinfection regimes has been the usual strategy to reducing such pollution.

An affordable step toward better hygiene control methods in the food business is UV light. In order to prevent microbiological contamination and thereby increase the safety of fruits, sanitation, disinfection, and oxidation using UV light is a flexible, environmentally friendly technology that may be employed in the processing and storage facilities for fruits [67]. The ability of 254 nm-wavelength UV light to lessen *Salmonella* contamination on tomatoes and food contact surfaces was assessed. Tomatoes that had been inoculated underwent varying UV-C exposure. *Salmonella* populations were dramatically reduced by all UV treatments ($p<0.05$). It was also investigated how well UV-C light reduced *Salmonella* contamination on various tomato surface areas and at various UV dosages. UV therapy proved successful in reducing *Salmonella* numbers wherever it was used [39]. Studies after that examined any potential photoreactivation or dark healing of UV-damaged *Salmonella*. Both photoreactivation and dark repair were not found. We also looked at how *Salmonella* contamination on surfaces in touch with food could be reduced by UV light (stainless steel, HDPE, waxed cardboard, and PVC). On waxed cardboard, *Salmonella* populations reduced the least. It is possible to use UV-C light in tomato handling facilities [183]. Fresh mushrooms could potentially be sterilized with UV-C radiation, which could also be a practical non-chemical technique to maintain mushroom quality and lengthen their postharvest life [302].

It was investigated the impact of UV-C light, applied to both sides of mushrooms, on microbial loads and product quality during 21 days of storage at 4°C [117]. At 1, 7, 14, and 21 days of storage, ascorbic acid, color, antioxidant activity, total phenolics, and microflora populations were all assessed. *Escherichia coli* O157:H7 inactivation by UV-C was also identified. Results indicated that *E. coli* O157:H7 inoculated on mushroom cap surfaces decreased by 0.67–1.13 log CFU g^1 at UV-C dosages of 0.45–3.15 kJ/m^2. On the surface of mushrooms, UV-C radiation also decreased total aerobic plate counts by 0.63–0.89 log CFU g^1. Additionally, it appears that the UV-C treatments prevented lesion growth on the mushroom surface. Irradiated mushrooms displayed decreased antioxidant activity, total phenolics, and ascorbic acid concentration over the first seven days than non-irradiated samples [308].

Shiitake mushrooms (Lentinus edodes) were treated to UV-C light (4 kJ/m^2) and stored in modified atmosphere packaging (MAP) for 15 days at 11C and 95% relative humidity, followed by 3 days at 20°C. A high degree of firmness was maintained for 15 days at a low temperature thanks to UV-C treatment, which also lessened the firmness loss during the course of the product's shelf life. Additionally, samples that had been treated exhibited greater levels of total flavonoids and ascorbic acid and delayed increases in the generation rate of superoxide anion and H_2O_2 [172].

6.4.3 ANIMAL SOURCE PRODUCTS: MEAT, FISH, EGGS, ETC.

There is evidence in the literature that UV radiation can clean meat products' surfaces. Several research studies have examined the use of UV irradiation to lower levels of *Salmonella* and *Escherichia coli* on hog skin and muscle [157]; *Escherichia coli* and *Salmonella* on fresh beef [314]; *Campylobacter jejuni, Enterobacteriaceae*, and total aerobic bacteria [208]; *Listeria monocytogenes* on chicken meat [310]; *Salmonella, Listeria monocytogenes*, and *Enterohemorrhagic Escherichia coli* on eggs [132]; *Escherichia coli* O157:H7, *Listeria monocytogenes, Salmonella Typhimurium, Staphylococcus aureus*, and *Staphylococcal enterotoxin* A [245]; and *Salmonella* populations in ground beef [317]. Despite the known restricted ability of UV light to penetrate hard food surfaces like meats, these investigations demonstrated that UV light has the potential to minimize bacterial contamination on food surfaces. As a result, *L. monocytogenes* and other infections of concern could potentially be controlled in meat and poultry processing facilities using UV radiation as a post-lethality treatment [205].

Pathogenic bacteria may be present in fish meals, a source of protein added to feeds. To reduce the epidemiological risk, efficient cleaning techniques like UV-C radiation are required [88, 205]. This study set out to determine how UV-C radiation affected how sanitary fish meals were. Meals of cod and salmon were included in the research supplies. *Salmonella, E.*

coli, *enterococci*, and *C. sporogenes* spore suspensions were used to inoculate samples of the fish meals, which were then subjected to surface UV-C fluencies of 0–400 Jm^2 for bacteria and 0–5,000 Jm^2 for spores. According to the kind of meal, the maximal theoretical lethal UV-C dose for *Salmonella*'s vegetative forms ranged from 670.99 to 688.36 Jm^2. The lowest UV-C fluency of 363.34–363.95 Jm^2 was required to inactivate Enterococcus species. Spores were far more resistant, and the UV-C dosages required to inactivate them in salmon meal and cod meal, respectively, were 159571.1 Jm^2 and 66836.9 Jm^2, respectively. Using UV-C light to sterilize fish meals turned out to be a technology that was largely efficient against vegetative forms of bacteria but almost completely worthless against spores [275].

Although the results are modest, it is possible to inactivate viruses in meat products using UV light alone. This is likely because some light can be absorbed by animal proteins and the uneven surface protects virus particles from exposure [5]. In all instances, the inactivation curve advanced without showing any signs of tailing, indicating that greater fluencies (up to 3.6 J/cm^2) might boost inactivation. However, the chicken meat's sensory quality declined at fluencies over 1.2 J/cm^2, where only 0.5 log declines were observed for both viruses. Results would probably be better if UV-C therapy were paired with another decontamination method [82]. On pork liver, ham, and sausage treated with pulsed light at fluencies of 45–60 J/cm^2, a still negligible but larger inactivation of FCV and *Escherichia coli* phages MS2 and 'X174 (0.97–2.8 log decrease) was noted [207].

6.4.4 BAKERY-CHEESE

To stop mold from growing, the cheese surface can be applied right before packaging. Mold on the surface is one of the most typical issues with cheese technology. A good way to stop mold growth on cheese surfaces is to apply UV light right before packaging. UV light was observed to reduce *Pseudomonas* spp. and *Enterobacteriaceae* by 1–2 logs on the surface of cheese without changing the cheese's color, texture, or appearance [155, 156, 174]. The authors came to the conclusion that this treatment demonstrated an intriguing surface microbial cleansing and prolonged shelf-life of cheese with the least amount of transmission inside the product. The effectiveness of pulsed UV light for inactivating infected *Listeria monocytogenes* and *Penicillium roqueforti* of hard cheeses that were packaged and unpackaged was looked into [40, 42, 151]. In packaged and unpackaged cheeses, *P.*

roqueforti reductions were 1.32 log and 1.24 log, respectively. About 2.8 log less *L. monocytogenes* was detected in both packaged and unpackaged cheeses [142, 325]. It was claimed that *P. roqueforti* and *L. monocytogenes* on the surface of hard cheeses may be rendered inactive by pulsed UV light. The efficiency of pulsed-light (PL) therapy on the inactivation of the spoilage bacteria on cheese surface was investigated in order to identify the effects of inoculum level, cheese surface topography, and the presence of clear polyethylene packaging [240]. Samples of inoculated cheese were exposed to PL dosages ranging from 1.02 to 12.29 J/cm^2 [156]. With a maximum inactivation level of 3.37 log, *Listeria innocua* was the least sensitive, followed by *P. fluorescens* and *Escherichia coli* ATCC 25922 with maximum inactivation levels of 3.74 log and 5.41 log, respectively. After three pulses, the inactivation reached a plateau (3.07 J/cm^3). The authors came to the conclusion that both PL treatments with and without packaging under UV light produced levels of inactivation that were consistently identical.

Applying UV-C after cheese packaging might be a useful safety procedure to render dangerous germs on cheese surfaces inactive [93, 125, 199]. In this application, it is essential that the packing film be thick, and that UV light can pass through it to inhibit the formation of foodborne bacteria on the surfaces. Sliced cheese was wrapped in various types and thicknesses of plastic sheets, and foodborne bacteria were killed by UV-C radiation. The authors' findings show that foodborne pathogens like *E. coli* O157:H7, *S. Typhimurium*, and *L. monocytogenes* can be successfully controlled using UV-C radiation and adjusted 0.07 mm thick PP or PE film packaging [125].

Very little study has been conducted on the use of UV light to clean the surfaces of other dairy products. Similar to cheese, post-processing mold infestation reduces the shelf life of set-type yogurt. In a batch UV light cabinet at Ege University, the surface of set-type yogurt samples that were naturally mold-contaminated was exposed to UV light at varying doses [156], in order to eradicate the mold. They came to the conclusion that greater UV light doses slightly increased the oxidation levels in yogurt and would be effective in killing mold on surface-set varieties of yogurt [156].

6.5 UV LIGHT EFFECTS ON SOLID AND SEMI-SOLID FOODS

6.5.1 *MICROBIAL INACTIVATION*

In response to prior instances of the bacteria in such meats, the USDA adopted a restriction to restrict Listeria in facilities producing Ready-to-eat 65 (RTE) items [94]. The use of an antimicrobial agent and/or a post-lethality treatment can be combined or done independently to diminish or remove the bacterium on the product in compliance with this regulation's alternatives [278]. Surfaces of meat products can be cleaned by UV light [161]. In several investigations, UV irradiation was used to reduce the levels of *Salmonella Typhimurium, Listeria monocytogenes*, and *Escherichia coli* in chicken flesh, pig skin, and muscle [157, 254]. Despite the known limited ability of UV light to penetrate hard food surfaces like meats, these investigations demonstrated that UV light has the potential to minimize bacterial contamination on food surfaces [163]. In order to manage *L. monocytogenes* and other dangerous infections in meat- and poultry-processing facilities, UV light may be employed as a post-lethality treatment [16].

Additionally, UV treatment has been used to decrease post-baking contamination and lengthen the shelf life of baguettes that haven't finished baking [12, 75, 161]. Numerous bacterial populations on eggshell surfaces, including the total aerobic plate count [51], *S. Typhimurium*, *E. coli*, and *Yersinia enterocolitica* [96, 293], have been shown to be significantly reduced by UV radiation. Although UV treatment of 120 eggs is urgently needed to increase egg safety, it has not yet been commercially adopted.

The photochemical breakdown of nucleic acids, which prevents cell division, is the primary cause of UV radiation's bactericidal effects [139]. This impact is frequently used in a variety of business and economic sectors. Based on the production of UV light with the proper wavelength, specialized equipment has been created for the efficient disinfection of air, water, and sewage [105, 109].

Numerous reports also support its potential applications, such as in the food business to lessen the quantity of germs that grow on the surfaces of solid foods viz fruits, vegetables, eggs, and meat [50]. This applies to both pathogens and saprophytes, the latter of which are unimportant from an epidemiological perspective but are in charge of the microbiological breakdown of goods [18, 169]. In cases where *Salmonella Senftenberg* and *Escherichia coli* were inoculated on the surface of fresh pork meat and pig leather [275]. The effectiveness of UV light in inactivating the bacteria was examined, and the results showed that *Salmonella* had a higher vulnerability to UV radiation than *E. coli*, regardless of the radiation dose that was used [321]. Similar relationships were observed when radiation-exposed indicator bacteria were present in sewage. The administered dose of 100 J/m^2 resulted in a reduction of more than 2 logs in the number of *S. enteritidis* and more than 1 log in the number of *Enterococcus fecalis*, but only about 0.5 logs in the number of *E. coli* [301].

According to the findings of the most recent research, *Salmonella* had the highest level of vegetative forms' resistance to the effects of UV-C radiation [137]. The Enterococcus genus was most vulnerable, and a dose of 400 J/m^2 completely rendered it inactive [145]. The UV radiation dose required for completely inactivating indicator bacteria is higher for *Salmonella* than for *E. coli* [115]. Radiation levels of 320 J/m^2 and 110 J/m^2 were needed to eradicate these germs from sewage, respectively. Enterococci did not fare as well in sewage exposed to UV radiation as *E. coli* did [53]. Numerous studies examining the impact of UV-C radiation on bacteria of fecal origin found similar vulnerability [102].

In the literature, there is relatively little information about how UV-C impacts bacterial spores in foods and animal feeds. UV-irradiated sewage spores displayed a resistance 9 times greater than that of bacterial cells in the vegetative stage [46, 130]. The effectiveness of UV light in lowering the quantity of *Clostridium difficile* spores on various surfaces in hospital rooms was documented [215]. The authors approached the phenomena from the perspective of an application, specifically how to improve the sporicidal effectiveness of UV radiation by first stimulating the germination of mature spores. The origin of the spores may also affect susceptibility to UV light. Laboratory-grown *Bacillus subtilis* spores were noticeably less UV-resistant than those that were taken straight from the environment [22]. Their numbers were reduced by 1.5 and 2.5 logs, respectively, by a dosage of 400 J/m^2. The initial concentration and level of bacterial species were cited by the authors as additional factors that may affect the effectiveness of spore inactivation.

The latter component defines the specific spore structure and regulates factors such as how deeply UV radiation may penetrate spores and an organism's capacity to heal UV damage [109]. Studies done at the molecular level have shown that the cortex's peptidoglycan structure and spore dehydration levels, which significantly affect thermoresistance, have an insignificant influence on the development of spore resistance to UV radiation [264]. The SASP—small acid-soluble proteins/-type—saturation of spore DNA is said to be of utmost significance in this process [265]. Expanding the scope of this agent's applications is encouraged by further understanding of the mechanisms behind the bactericidal impact of UV radiation and its connection to its inactivation efficiency, which has been empirically demonstrated in a variety of situations [92].

UVC is the most effective at eradicating germs because of its higher energy [84]. Bacteria, viruses, protozoa, and fungi can all be killed by UV radiation. UV light, ozonation, and chlorination all remove viruses [160].

Multiple studies have demonstrated that UV light effectively inactivates Gram-negative bacteria more quickly than Gram-positive bacteria. The variation in the peptidoglycan structure of the cell wall, which can impact radiation penetration, has been linked to this effect [69]. Additionally, due to their larger, more sophisticated, and genetically redundant cells than bacteria, eukaryotic organisms are typically more UV-resistant [234]. Because yeast has a lower DNA pyrimidine concentration than bacteria, it may be more likely for photons to be absorbed by other substances, contributing to its comparatively strong UV radiation resistance [9]. Studies on the majority of prevalent post-harvest fungal infections and the impact of postharvest pre-storage single exposure, including *Rhizopus, Penicillium digitatum, P. expansum,* and *Penicillium italicum, Monilinia* sp., *Botrytis cinerea, Colletotrichum* sp., and *Fusarium* sp., among others, have been conducted [65]. These studies have found that these exposures have positive effects. Numerous reports of direct germicide activity impairing microbial viability have been made, but less severe effects, including slower germination of viable conidia, have also been noted. Direct UV irradiation of fresh food decreased the survivability of *E. coli*, *Salmonella*, and *Listeria* in terms of human pathogens [268].

The majority of this research has thus far been carried out in a lab setting or, in the best case, on a pilot scale. An analysis of the literature suggests that some crucial issues need to be resolved before they can be used commercially on a large scale, especially in relation to their usefulness for limiting radiation exposure, worker safety procedures, mechanical damage—particularly to fruits and vegetables—for continuous processing lines (where the treatment duration could last minutes at low irradiances). Furthermore, the long-standing use of wet cleaning techniques for fresh produce may hinder the rapid adoption of a new technology [24].

6.5.2 NUTRITIONAL/PHYSIOCHEMICAL

Consumer interest in functional food products that could support sustaining peak performance, health, and well-being has increased in recent years [81, 108, 297]. Functional foods are those that have been scientifically shown to have health advantages above and beyond their basic nutritional content and/or to lower the risk of developing chronic diseases because physiologically bioactive substances are present [30]. Natural foods (such as fruits and vegetables) and processed foods that have acquired vitamin fortification or supplements, phytochemicals, or botanicals are both considered functional foods [44]. Abiotic stressors like UV radiation or specific growing conditions can be exposed to after harvest to increase the nutraceutical potential of plant meals [188].

Fruits' hormetic response is an intricate, still unknown process. UV radiation has been shown to trigger cellular defense mechanisms, changing metabolic activity by turning on particular genes and enzymes [230]. This includes: (i) glucanases and chitinases with lytic activity against the main components of fungal cell walls; (ii) enzymes of the peroxidases and reductases family that cause an oxidative burst and the production of lignin polymers in order to provide structural barriers against invasive diseases; and (iii) phenylalanine ammonia lyase (PAL) – engaged in the production of phenolics, which exhibit high antioxidant properties [110]. Chlorogenic acid, gallic acid, epicatechin, and quercetin are examples of plant phytochemicals that absorb UV light [171].

It was reported 21 patulin-positive samples of rotten sections of apples [25]. Patulin has been found in concentrations as high as 130 mg kg^{-1}. Considering how stable it is and how long it may stay in juice, like the majority of mycotoxins. Even though washing and removing rotting apples can lower patulin concentration by 90% [177], apple juice samples showed up to 733 µg/L of patulin contamination (2001) [144, 260].

For baby food and formula, the European Union has gone a step further and set a maximum limit of 10 g/L^{-1} (10 ppb). There isn't a single, commercially viable approach for lowering patulin while maintaining produce quality, despite numerous methods for controlling and eliminating patulin having been presented [328]. A few recent studies assessed the viability of UV light as a potential commercial substitute for the elimination of patulin and Penicillium spores that produce patulin in apple juice [256]. Patulin levels decreased significantly and virtually linearly in response to UV lights ranging from 14.2 to 99.4 mJ/cm^2, but there were no discernible changes to the cider's chemical makeup (such as pH, Brix, and total acids) or organoleptic qualities [74, 309]. Enzymatic activity actually depends on the native structure of the protein, which might conceivably be altered by UV and visible light exposure to promote photo-oxidation.

Even while fresh solid and semi-solid foods may only contain trace levels of vitamins, fruits, and vegetables, some vitamins are thought to be light-sensitive [161]. Vitamins C (ascorbic acid), B12 (cobalamin), B6 (pyridoxine), B2 (riboflavin), and folic acid are among the water-soluble, light-sensitive vitamins.

Among the fat-soluble, light-sensitive vitamins are carotene, A, K, and E (alpha-tocopherol). The majority of research on how light affects vitamins was done in the 290 to 700 nm wavelength range, which covers both UV and visible light. There isn't much information accessible at 253.7 nm, but fluorescent light exposure was a part of them [231]. The amount of vitamin C present also had an impact on the magnitude of the absorption coefficient since vitamin C demonstrates strong UV absorption in the germicidal spectrum, peaking at about 260 nm, but not above 300 nm [306]. The absorption characteristics of juice treated with UV light may change due to the vitamin C being destroyed during exposure. Vitamin C concentration was assessed before and after UV treatment [11].

A 7.11 kJ/m^2 pulsed-UV dosage softened and turned brown tissue on red oak leaves after seven days of holding at 5°C [222]. UV is a non-thermal process; it does not significantly alter the color of cumin seeds or black pepper [85]. According to reports, applying UV to chicken carcass and breast meat has no detrimental effects on the goods' colors [17]. In the investigation, the impact of UV light was seen as a reduction in red color.

Strawberry samples exposed to UV light for longer periods of time had closer-spaced anthocyanins, which intensified the yellow color. According to the findings, the yellow color intensifies as the application duration and UV light exposure increase [269].

UV rays at this wavelength are far more likely to be absorbed when vitamin C is present, since it absorbs at its highest at 260 nm [60, 131, 161]. Therefore, pasteurization of goods supplemented with vitamin C requires the use of higher-energy UV radiation. Another investigation revealed that vitamin C breakdown follows a zero-order reaction and that samples receiving vitamin C supplements had a 2.5 times higher death rate for *E. coli* [60].

First and foremost, UV treatment is used to increase shelf life; throughout this procedure, researchers looked at how UV light affects aroma, color, and nutrient content [286].

Due to ultraviolet light's component-damaging and -changing effects, the use of UV therapy to lengthen shelf life and eliminate hazardous microbes is not universally accepted [60]. The United States Food and Drug Administration asserted in 2000 that ultraviolet light therapy was fully safe and eliminated human harmful bacteria from fruit juices [135]. Additionally, it is known that UV irradiation may result in the production of some substances and the breakdown of others, but neither of these effects pose a threat to human health.

The amount of human pathogenic bacteria must decrease by two orders of magnitude compared to the control sample for UV treatment to be deemed microbiologically safe. When it comes to applications for human use, this level of decrease must always be guaranteed and verified [60].

6.5.3 *SENSORY PROPERTIES*

The food's "organoleptic" qualities, such as its appearance, flavor, aroma, smell, texture, and taste, as well as — including, first and foremost, the vitamins, which may be the substances most vulnerable to UV treatment — can all be used to assess the effects of UV treatment on food quality [60]. Since not all foods are equally vulnerable to UV radiation, our evaluation must also be consistent with the kind of sample we intend to evaluate [29, 60]. Whipped cream, dried vegetable soup powder, butter, margarine, and mayonnaise were the most susceptible to sunshine and fluorescent light, giving off an unpleasant smell after being exposed to UV light for a short time. These foods started to smell awful after just a short period of UV exposure. Visible alterations in sugar, chocolate, cheese, bacon, uncooked sausage, green beans, and salted peanuts occurred after one to three days of irradiation. The third group, which consists of rice and potato chips, did not exhibit significant alterations until after 4–7 days, whereas split peas, almonds, and pastries did so after 10–30 days. The wavelength of UV radiation to which the particular product exhibits sensitivity, the transparency of the container, and the treatment temperature all have an impact on these alterations [166, 168, 248].

6.6 UV TECHNOLOGY WITH OTHER TECHNIQUES: COMBINATION EFFECTS

With the development of food technology, application of UV LEDs [98, 150, 163, 216, 314], pulsed UV light [3, 161, 203, 219], UVC assisted photocatalysis [134, 228, 272], and hurdle technology [152, 242, 320] for food purification and preservation.

For fresh produce, there are several non-thermal pathogen control techniques [55]. Chemical or non-chemical techniques may be used in these alternative treatment methods [225]. Chlorine (Cl), bromine (Br), iodine (I), trisodium phosphate, hydrogen peroxide, or ozone (O_3) are the most used chemical therapy agents. Microwave, ultrasound, ionizing radiation, and ultraviolet radiation are examples of non-chemical treatments (UV) [31]. Because UV is safe and has no aftereffects on the food it treats, it is a potential surface decontamination solution. According to research, UV treatments can boost the nutritional value and shelf life of fruits and vegetables, as well as their antioxidant activity, in addition to being germicidal [107].

Using active packaging, post-packaging decontamination technologies, and food product reformulation together can be a successful strategy for producing high-quality, safe, and nutrient-dense food items with a long shelf life [255]. To determine how well UV LEDs or pulsed-light technologies can alter the food items' nutritive content and sensory attributes, research on microbial inactivation also requires methodological work. In the near future, it is also necessary to solve problems related to worker safety, economic viability, and ozone generation [272].

The use of UV-C radiation in conjunction with modified atmosphere packaging (MAP) may be an effective way to treat freshly cut rocket leaves. The sensory characteristics were maintained for up to 12 days at 5°C for the combined doses of 7.5 and 15 kJ UV-C/m^2. The total polyphenol content and the total antioxidant activity were not affected by the combined treatments and were maintained throughout the storage period. All combination treatments tested delayed microbial development until day 8 at 5°C, compared to a non-treated control under conventional passive MAP. Treatment with both UV-C and MAP postponed the loss of total chlorophyll concentration over the course of the shelf-life [122].

Effective control of dangerous and spoilage microorganisms in food items is necessary to promote food safety and sustain hygienic standards for human health [95]. A cost-effective and ecologically friendly technique, UV light treatment has enormous potential for the food business [69]. It might be a practical non-thermal alternative technique for eliminating dangerous and spoilage germs in foods and beverages. Additionally, there are no hazardous byproducts or chemical residues left behind after UVC-based food disinfection. Furthermore, it is a particularly cost-effective technology for ensuring food safety in developing countries due to its low maintenance, installation, and operation costs with minimal energy consumption [116].

Of late, attention is also being drawn to the use of UV light in addition to other techniques (also known as hurdle technology) [135]. Despite the fact that UVC-based techniques have a number of benefits, some disadvantages may prevent their widespread use in food sanitation. For instance, food's physical characteristics have a significant impact on how effectively UVC light inactivates microbes. For treating food products with vast areas, greater doses are necessary due to UVC's poor penetrating power. The quality of food may be impacted by such prolonged UVC exposure. To effectively inactivate microorganisms in a range of food products, UV dose adjustment and exposure time must be taken into account. Additionally, before being implemented at a wider scale, the regulatory authorities' (FDA) allowed limits must be thoroughly reviewed and tested. Additionally, the right labeling can be used to encourage consumers to accept UV-irradiated food products or substances [272].

There have been claims that inactivation of viruses by UV radiation, whether pulsed or otherwise, is consistent with expectations [258]. UV radiation is a highly effective virucide for transparent liquids and, if turbulent flow is maintained, even opaque liquids. Uneven surfaces on solid foods limit inactivation [109]. We cannot draw any firm conclusions regarding the economic viability of this technique due to the paucity of information and the wide variation in how various types of viruses react to UV radiation.

Meat, berries, and dry herbs are somewhat less effective with UV-C. To get better outcomes, it might be used with other technologies as part of a hurdle strategy. Higher fluencies may produce better results for some virus-food pairings in the first scenario. UV is also helpful for cleaning surfaces that come into contact with food. Learning more about how UV light affects foodborne viruses, and their lab substitutes is still a significant step in improving future food security and our understanding of food safety [189].

6.7 UV LIGHT ADVANCES IN THE ENHANCEMENT OF SHELF-LIFE FOR SOLID AND SEMI-SOLID FOODS

Fruits must be handled properly after being harvested to prevent ripening and senescence during storage because they are highly perishable [4]. Quality loss, discoloration, tissue thinning, weight loss, an increase in respiration rate, and ethylene production are the main signs of deterioration [319]. The traditional method for increasing the fresh food's shelf life involves changing the temperature and environment of the storage space [200]. To prevent negative impacts, these two elements must be optimized. For instance, storing goods at a temperature that is too low can cause chilling damage. Hormetic UV levels can increase the expression of genes involved in the defense response while decreasing the expression of genes involved in photosynthesis, lipid metabolism, and wall disintegration [187, 233].

By 2050, it is predicted that there will be an increase in the global population of up to 9 billion people, necessitating the adoption of sustainable practices that would allow for the satisfaction of existing demands without endangering those of future generations [23]. These include the creation of innovative techniques that reduce resource consumption and the abolition of poverty [270]. They also include better product stewardship, improved energy efficiency, and wiser management of the use of natural resources [73].

A developing non-thermal technology called UV light has a lot to offer for the societal advancement of sustainability [202]. It has been demonstrated that using it to process food results in the production of fresh-tasting fruits and fruit-based goods like juices that are safe and extremely nourishing [291]. Additionally, the loss of fresh food, which is on the order of 20% in wealthy countries and at least 50% in underdeveloped countries, can be considerably reduced by using UV radiation as a post-harvest treatment [217]. Numerous researchers have demonstrated that the use of UV technologies could be an alternative means of postharvest management.

6.8 LIMITATIONS OF UV LIGHT IN SOLID AND SEMI-SOLID FOODS

Insufficient penetrating strength, diminished effectiveness in the presence of additional biological debris, the shadowing effect, and multiple reactivations are only a few of the UV sanitizing agent's acknowledged flaws [109]. Numerous bacteria can remain concealed in weakly penetrating materials like fruit skins and water. For cleaning porous food, UV is ineffective [287]. The shadowing effect of rough or fluctuating surfaces is lessened by rotating lamps with all-around illumination. Bivalve mussels, hog liver, and berry products with cavities are a few dangerous foods. Additionally, it has been shown that superfluous organic matter makes it harder to illuminate food surfaces that might be virus-infected, which reduces the efficacy of photon energy therapy. When possible, mild washing of the matrix would be useful to remove at least some of this organic material. Last but not least, it should be expected that UV exposure will change (mutate) the viral DNA [109]. Genome reassortment may enable the emergence of a new strain if numerous damaged virions reach the same host cell. A method for foodborne viral survival has been identified as several reactivations [184].

Compared to other traditional preservation techniques, UV technology has a number of benefits, but it also has some significant disadvantages [197].

A wide variety of harmful and spoilage bacteria can be inactivated by UV treatments, which are easy to use and have little impact on the food's nutritional and sensory qualities [8, 272]. It does not use water, produce waste, or leave any residues on foods or treated surfaces, in contrast to conventional disinfection techniques [289]. Additionally, it has been approved with few to no restrictions by the majority of other food regulatory authorities, including the European Food Safety Authority (EFSA), the United States Food and Drug Administration (FDA), and others. UV technology has a number of benefits over electron beam or gamma irradiation, including reduced investment requirements, fewer regulations, and the lack of a need for a treatment label on the products [162]. Another advantage is the ease with which UV treatments can be combined with other preservation methods in search of supplemental or even synergistic benefits [65]. In the hurdle technology, which combines various physical, chemical, and/or biological preservation techniques to create safe and stable foods, UV irradiation is regarded as a very valuable tool [1]. In addition to their germicidal effects, in order to stimulate natural defensive mechanisms (such as phytoalexins, free radical scavengers, antimicrobial and antioxidant enzymes), UV treatments have been applied to metabolically active fruit and vegetables [65]. Additionally, cost- and energy-effective compared to many traditional procedures, UV treatments have been used historically for a variety of purposes, and there is validation data available [246].

However, there are a number of drawbacks to UV technology that prohibit its widespread application in the food sector. In solids and turbid media, in particular, it has a poor penetrating power [38]. Furthermore, the lack of individuality may be a disadvantage if long-lasting benefits are needed. Direct and even exposure is necessary to fully express the UV germicide characteristics [65]. Under commercial operations, this could provide difficulties in various situations. Another challenge would be modifying UV treatments to help move processing lines along, especially where long exposure times or low irradiance levels are required [14]. UV therapies are inferior to ionizing irradiation when compared to the times when they could be applied because they can't be employed after packing [239]. Direct exposure to UV radiation could be damaging to operators as well, although this risk can be readily avoided with the use of personal protective equipment and basic safety practices [232]. The following describes the advantages, disadvantages, and limitations of UV light treatment for food protection [47, 52, 235].

6.8.1 ADVANTAGES OF UV RADIATION IN HANDLING OF DAIRY PRODUCTS

- Low maintenance, installation, and service expenses.
- Keeps the food's physicochemical characteristics and nutritional value while preventing undesirable changes in the sensory characteristics.
- No negative environmental effects (no chemical residue, toxins, wastes).
- There isn't any heat production (economical aspects).
- Combining with various non-thermal production techniques is a possibility.
- Having the ability to work continuously.
- A minimal energy footprint.
- Possibility of commercial processing.
- Eradication of harmful and rotting bacteria.

6.8.2 DISADVANTAGES AND LIMITATIONS OF UV RADIATION IN HANDLING OF DAIRY PRODUCTS

- Foods having a high amount of suspended solids perform poorly.
- Just a little amount of UV can pass through opaque liquids.
- Prolonged contact with humans could be harmful (to eyes, causes skin burns and skin cancer).
- Low capability for penetration into dietary solids (restricted to surface decontamination only).
- It is challenging to measure the rate of UV sterilization of food contact surfaces.
- The potential of spores to regenerate following UV exposure.

6.9 ECONOMIC AND COMMERCIAL POSSIBILITY OF UV LIGHT TECHNOLOGY IN SOLID AND SEMI SOLID FOODS

For several years, air, surfaces, and drinking water have all been physically disinfected using ultraviolet radiation. It has been used for a variety of purposes in construction, hospitals, institutions, and businesses. Water used in breweries, industrial effluent, drinking water, and water used for item rinsing are a few examples of liquids that undergo UV-C treatment to lessen the amount of bacteria present in them. UV light can be utilized in the food industry for purposes other than water disinfection because it is a cold disinfection medium. Some of them include: (i) pasteurizing liquid foods rather than thermally pasteurizing them in the wine, cheese, and dairy (raw, low-fat, and skim milk) industries; (ii) washing the surfaces of prepared meals that are packaged for storage or before refrigeration, as well as on fresh produce. Further research on UV radiation is needed to determine the final product's dose, color, and appearance, flavor, microbiological quality, and nutritional value because, as previously stated, different types of liquids respond differently to UV light penetration [119].

Numerous studies have shown that UV technology can be utilized as a substitute approach to treat fungi-caused postharvest diseases. As a result, it may be possible to considerably reduce the use of fungicides and other substances that are dangerous to the environment and human health [115, 190]. The main drawback of UV technology is that its UV sources contain mercury. A health risk is presented by the potential mercury exposure brought on by lamp sleeve fracture [256]. Both when they are in use and when being serviced, lamps may break [33]. The lamp sleeve and lamp envelope that enclose the UV lamp shield its mercury from exposure. The lamp and lamp sleeve must both break in order for the mercury to escape. Typically, a single UV lamp used to cure water contains 0.005 to 0.4 grams of mercury (5–400 mg). LPM lights have less mercury (5–50 mg/lamp) than LPHO (26–150 mg/lamp) and MPM lamps (200–400 mg/lamp). The maximum contamination level (MCL) for mercury was set by the EPA at 0.002 mg/L^{-1}. The EPA found that kidney impairment may result from brief exposures to mercury at levels over the 0.002 $mg\ L^{-1}$ MCL. Concerns about the effects of mercury emissions from food plants on the environment led to the creation and approval of mercury-free special technology lights and LEDs [115].

The UV tumble method created by C&S Equipment Company is an illustration of a commercially available technology to clean food surfaces [48, 161]. To ensure that the product is exposed to the UV source, the business either installed a screw conveyor or a spinning drum that lifts and tumbles the product. The system can handle 95 refrigerated products, including fresh or frozen fruits, vegetables, meats, seafood, bakery goods, etc., cooked, and other items (dairy products, ready-to-cook, etc.). Solutions for these problems are created using Steril-AireTM UV Emitters, a patented product of the C&S Equipment Company. These emitters are protected by plastic sleeves to meet the requirements of food processing facilities for food safety. Rival units fail to function in the chilly conditions of refrigeration or chiller coils, but the 100-unique design enables emitters to do so [163]. The following UV units are examples of those currently in use in commercial processing facilities: (i) a UV tumbling drum for cooked and IQF frozen chicken and beef fajita strips with a capacity of 6,000–7,000 lb/h; (ii) a hooded conveyor with turnover for cooked and IQF frozen hamburger patties with a capacity of 3,000 lb/h; and (iii) a deli meat system (custom conveyor with UVC hood) for formed. Additionally, wrapped partially baked baguettes have undergone UV treatment to lengthen their shelf life and lessen post-baking contamination [75, 161]. On eggshell surfaces, UV light has been demonstrated to drastically lower the TPC (aerobic), *S. Typhimurium* and *E. coli*, *Yersinia enterocolitica*, and the total aerobic plate count [96]. Despite the critical need for UV treatment of eggs to promote egg safety, it has not yet been widely used in commerce.

UV radiation technology is a revolutionary non-thermal method of processing food for sanitizing a variety of foods in various food processing industries. Food that has undergone UV processing gains characteristics that make it appear fresher in terms of colors, flavors, and textures. It also increases the food's shelf life. UV light is used to render bacteria dormant so that they cannot multiply because the majority of bacteria are killed by it. Although UV isn't used all that much or prominently, it has a lot of promise to replace current techniques for both solid and liquid food preservation. However, because radiation harms people, it must be handled with the utmost care. Additionally, it is always advisable to adhere to the FDA's (Food and Drug Administration), FSSAI, and other acceptable limitations, which must first be tested on a small scale before being scaled up and employed on a larger scale. Furthermore, it is always advisable to abide by the FDA's (Food and Drug Administration), FSSAI, and other permissible limitations, which must initially be evaluated on a small scale before being scaled up and used on a bigger scale.

6.10 SUMMARY

A practical non-thermal option for handling fruits and fruit products uses continuous and pulsed UV light technology [109, 256, 318]. Numerous reports that have been examined have successfully used UV light to get rid of or cut back on the microorganisms that live on the surfaces of fresh fruit and fruit products and are pathogenic, non-pathogenic, and spoilage [115]. Finding the ideal UV processing conditions and the appropriate UV source for a particular product is essential for achieving the necessary microbial reduction as well as color, texture, and flavor preservation. Additionally, UV light is a viable option for controlling microbial loads on surfaces that come into contact with food, water, and non-food items in fruit processing facilities. Currently, the most common sources for UV therapy of fruits are LPM lamps and xenon PL, despite the fact that several UV sources are either being developed or are already available since they have the approval of many quality control agencies across many countries. For the pasteurization of a range of fruit juices, numerous thin film devices, static and dynamic mixers, annular laminar and turbulent flow reactors, and other UV-light continuous flow systems were designed and tested. The effectiveness of UV inactivation can be increased by minimizing interference from viscosity, and low UVT is related to specific juices [165]. Further research is needed to create UV systems that can adequately treat all regions of treated liquids with low UVT, including fruit juices. Recent research found that UV radiation may boost the health-promoting properties of substances including antioxidants, polyphenols, and flavonoids [256]. UV therapy has been found to have positive benefits on the preservation of different fruits, both fresh and uncut, in numerous studies that are referenced here. There is significant disagreement over the mechanism underlying the hermetic reaction in fresh vegetables; nevertheless, based on the data in the literature that is currently accessible [165]. Plants respond to UV radiation by activating a variety of peroxidases, reductases, and chitinases, each with unique chemical structures and UV-A, UV-B, and UV-C absorption properties. As a result, plant response differs according to the UV dose and UV emission spectrum used. Additional research is required to keep track of and report on UV treatment circumstances and parameters, including UV fluence levels, lamp parameters, and wavelength of radiation, in order to advance the current state of understanding of how fresh fruit is processed by UV. Further research is also needed to determine how UV radiation affects the fruit and fruit-related product quality. Even though UV is an entirely non-thermal treatment, some UV spectrum and doses may cause damage to vitamins, proteins, and antioxidants, along with variations in color and the development of flavor defects and odors. In order to ascertain whether there are any issues with chemical safety or toxicology related to items that have received UV treatment, additionally, it's critical to evaluate how UV radiation may affect the development of chemical compounds in food that could be harmful to human health [163]. It is also important to look more closely at how UV radiation can remove contaminants or unwanted chemicals. Due to the poor penetration of UV light, processing facilities may find it more advantageous and appealing to combine UV light with other post-harvest technologies (such as ozone, ultrasound, changed packing atmospheres, sanitizing, and anti-browning compounds). Information about UV processing is scarce and is used in conjunction with other therapies, thus more research must be done. Long-term competition for arable land with the food supply and/or food crops.

KEYWORDS

- **food crops**
- **microbial inactivation**
- **non-thermal technology**
- **post-harvest technologies**
- **semisolid foods**
- **solid foods**
- **UV light**

REFERENCES

1. Aaliya, B., Sunooj, K. V., Navaf, M., Akhila, P. P., Sudheesh, C., Mir, S. A., & George, J. (2021). Recent trends in bacterial decontamination of food products by hurdle technology: A synergistic approach using thermal and non-thermal processing techniques. *Food Research International, 147*, 110514.
2. Abdipour, M., Hosseinifarahi, M., & Naseri, N. (2019). Combination method of UV-B and UV-C prevents post-harvest decay and improves organoleptic quality of peach fruit. *Scientia Horticulturae, 256*, 108564.
3. Abida, J., Rayees, B., & Masoodi, F. A. (2014). Pulsed light technology: A novel method for food preservation. *International Food Research Journal, 21*(3), 839.
4. Ahmad, M. S., Siddiqui, M. W., Ahmad, M. S., & Siddiqui, M. W. (2015). Factors affecting postharvest quality of fresh fruits. In *Postharvest Quality Assurance of Fruits: Practical Approaches for Developing Countries* (pp. 7–32), Springer.
5. Ahmed, H., Maunula, L., & Korhonen, J. (2020). Reduction of norovirus in foods by nonthermal treatments: A review. *Journal of Food Protection, 83*(12), 2053–2073.
6. Ailavadi, S. (2017). *Heat Inactivation Kinetics of Tulane Virus and Aichi Virus*. A Thesis Presented for the Master of Science Degree. The University of Tennessee, Knoxville.
7. Ali, A., Yeoh, W. K., Forney, C., & Siddiqui, M. W. (2018). Advances in postharvest technologies to extend the storage life of minimally processed fruits and vegetables. *Critical Reviews in Food Science and Nutrition, 58*(15), 2632–2649.

8. Amit, S. K., Uddin, M., Rahman, R., Islam, S. M., & Khan, M. S. (2017). A review on mechanisms and commercial aspects of food preservation and processing. *Agriculture & Food Security, 6*(1), 1–22.
9. Anderson, J. G., Rowan, N. J., MacGregor, S. J., Fouracre, R. A., & Farish, O. (2000). Inactivation of food-borne enteropathogenic bacteria and spoilage fungi using pulsed-light. *IEEE Transactions on Plasma Science, 28*(1), 83–88.
10. Araque, L. C. O., Rodoni, L. M., Darré, M., Ortiz, C. M., Civello, P. M., & Vicente, A. R. (2018). Cyclic low dose UV-C treatments retain strawberry fruit quality more effectively than conventional pre-storage single high fluence applications. *LWT, 92*, 304–311.
11. Aride, P. H. R., Ferreira, M. S., Duarte, R. M., De Oliveira, A. M., De Freitas, D. V., Dos Santos, A. L. W., & Val, A. L. (2010). Ascorbic acid (vitamin C) and iron concentration in *Tambaqui, Colossoma macropomum*, iron absorption. *Journal of the World Aquaculture Society, 41*, 291–297.
12. Axel, C., Zannini, E., & Arendt, E. K. (2017). Mold spoilage of bread and its biopreservation: A review of current strategies for bread shelf life extension. *Critical Reviews in Food Science and Nutrition, 57*(16), 3528–3542.
13. Baenas, N., Iniesta, C., González-Barrio, R., Nuñez-Gómez, V., Periago, M. J., & García-Alonso, F. J. (2021). Post-harvest use of ultraviolet light (UV) and light emitting diode (LED) to enhance bioactive compounds in refrigerated tomatoes. *Molecules, 26*(7), 1847.
14. Ballari, R. V., & Martin, A. (2013). Assessment of DNA degradation induced by thermal and UV radiation processing: Implications for quantification of genetically modified organisms. *Food Chemistry, 141*(3), 2130–2136.
15. Bambalele, N. L., Mditshwa, A., Magwaza, L. S., & Tesfay, S. Z. (2021). Recent advances on postharvest technologies of mango fruit: A review. *International Journal of Fruit Science, 21*(1), 565–586.
16. Bangel, N. A. (2012). An overview of regulations, guidelines, and intervention strategies for *Listeria monocytogenes* in ready-to-eat meat and poultry products. pub; Kansas State University Master Thesis, pp. 23–35.
17. Baptista, E., Borges, A., Aymerich, T., Alves, S. P., Gama, L. T. D., Fernandes, H., & Fraqueza, M. J. (2022). Pulsed light application for Campylobacter control on poultry meat and its effect on color and volatile profile. *Foods, 11*(18), 2848.
18. Barata, A., Malfeito-Ferreira, M., & Loureiro, V. (2012). The microbial ecology of wine grape berries. *International Journal of Food Microbiology, 153*(3), 243–259.
19. Barbhuiya, R. I., Singha, P., & Singh, S. K. (2021). A comprehensive review on impact of non-thermal processing on the structural changes of food components. *Food Research International, 149*, 110647.
20. Barjasteh, A., Dehghani, Z., Lamichhane, P., Kaushik, N., Choi, E. H., & Kaushik, N. K. (2021). Recent progress in applications of non-thermal plasma for water purification, bio-sterilization, and decontamination. *Applied Sciences, 11*(8), 3372.
21. Barkai-Golan, R., & Paster, N. (2008). Mouldy fruits and vegetables as a source of mycotoxins: Part 1. *World Mycotoxin Journal, 1*(2), 147–159.
22. Begyn, K., Kim, T. D., Heyndrickx, M., Michiels, C., Aertsen, A., Rajkovic, A., & Devlieghere, F. (2020). Directed evolution by UV-C treatment of *Bacillus cereus* spores. *International Journal of Food Microbiology, 317*, 108424.

23. Béné, C., Barange, M., Subasinghe, R., Pinstrup-Andersen, P., Merino, G., Hemre, G. I., & Williams, M. (2015). Feeding 9 billion by 2050–Putting fish back on the menu. *Food Security, 7*, 261–274.
24. Benyam, A. A., Soma, T., & Fraser, E. (2021). Digital agricultural technologies for food loss and waste prevention and reduction: Global trends, adoption opportunities and barriers. *Journal of Cleaner Production, 323*, 129099.
25. Beretta, B., Gaiaschi, A., Galli, C. L., & Restani, P. (2000). Patulin in apple-based foods: Occurrence and safety evaluation. *Food Additives & Contaminants, 17*(5), 399–406.
26. Bhat, R., & Stamminger, R. (2015). Preserving strawberry quality by employing novel food preservation and processing techniques – Recent updates and future scope – An overview. *Journal of Food Process Engineering, 38*(6), 536–554.
27. Bhattacharya, S., & Rattan, S. I. (2019). Primary stress response pathways for preconditioning and physiological hormesis. In S. I. Rattan (Ed.), *The Science of Hormesis in Health and Longevity* (pp. 35–51). Academic Press.
28. Bhilwadikar, T., Pounraj, S., Manivannan, S., Rastogi, N. K., & Negi, P. S. (2019). Decontamination of microorganisms and pesticides from fresh fruits and vegetables: A comprehensive review from common household processes to modern techniques. *Comprehensive Reviews in Food Science and Food Safety, 18*(4), 1003–1038.
29. Bintsis, T., Litopoulou-Tzanetaki, E., & Robinson, R. K. (2000). Existing and potential applications of ultraviolet light in the food industry – A critical review. *Journal of the Science of Food and Agriculture, 80*(6), 637–645.
30. Birch, C. S., & Bonwick, G. A. (2019). Ensuring the future of functional foods. *International Journal of Food Science & Technology, 54*(5), 1467–1485.
31. Bisht, B., Bhatnagar, P., Gururani, P., Kumar, V., Tomar, M. S., Sinhmar, R., & Kumar, S. (2021). Food irradiation: Effect of ionizing and non-ionizing radiations on preservation of fruits and vegetables – A review. *Trends in Food Science & Technology, 114*, 372–385.
32. Bolder, N. M. (1997). Decontamination of meat and poultry carcasses. *Trends in Food Science & Technology, 8*(7), 221–227.
33. Bolton, J. R., & Cotton, C. A. (2011). The Ultraviolet Disinfection Handbook. American Water Works Association. Chapter 2: *Fundamentals of UV Light and Photochemical*, pp. 11–12.
34. Botondi, R., Barone, M., & Grasso, C. (2021). A review into the effectiveness of ozone technology for improving the safety and preserving the quality of fresh-cut fruits and vegetables. *Foods, 10*(4), 748.
35. Bozkurt, H., D'SOUZA, D. H., & Davidson, P. M. (2015). Thermal inactivation of foodborne enteric viruses and their viral surrogates in foods. *Journal of Food Protection, 78*(8), 1597–1617.
36. Bozkurt, H., Phan-Thien, K. Y., van Ogtrop, F., Bell, T., & McConchie, R. (2021). Outbreaks, occurrence, and control of norovirus and hepatitis A virus contamination in berries: A review. *Critical Reviews in Food Science and Nutrition, 61*(1), 116–138.
37. Brasil, I. M., & Siddiqui, M. W. (2018). Postharvest quality of fruits and vegetables: An overview. In *Preharvest Modulation of Postharvest Fruit and Vegetable Quality* (pp. 1–40). Elsevier.
38. Brodowska, A. J., Nowak, A., & Śmigielski, K. (2018). Ozone in the food industry: Principles of ozone treatment, mechanisms of action, and applications: An overview. *Critical Reviews in Food Science and Nutrition, 58*(13), 2176–2201.

39. Calle, A., Fernandez, M., Montoya, B., Schmidt, M., & Thompson, J. (2021). UV-C LED irradiation reduces salmonella on chicken and food contact surfaces. *Foods, 10*(4), 1459.
40. Can, F. O. (2013). *Surface Decontamination of Hard Cheeses by Pulsed UV light.* Master thesis, Penn State University Library, pp. 56–64.
41. Cano-Lamadrid, M., & Artés-Hernández, F. (2022). Thermal and non-thermal treatments to preserve and encourage bioactive compounds in fruit-and vegetable-based products. *Foods, 11*(21), 3400.
42. Cassar, J. R. (2019). *Utilization of Pulsed Ultraviolet Light as a Microbial Reduction Intervention on Raw Chicken.* Penn State University Library. Master Thesis, pp. 57–72.
43. Cassar, J. R., Mills, E. W., & Demirci, A. (2022). Characterization of pulsed light for microbial inactivation. *Journal of Food Engineering, 334*, 111152.
44. Cencic, A., & Chingwaru, W. (2010). The role of functional foods, nutraceuticals, and food supplements in intestinal health. *Nutrients, 2*(6), 611–625.
45. Chandrasekaran, I., Panigrahi, S. S., Ravikanth, L., & Singh, C. B. (2019). Potential of near-infrared (NIR) spectroscopy and hyperspectral imaging for quality and safety assessment of fruits: An overview. *Food Analytical Methods, 12*, 2438–2458.
46. Chang, J. C., Ossoff, S. F., Lobe, D. C., Dorfman, M. H., Dumais, C. M., Qualls, R. G., & Johnson, J. D. (1985). UV inactivation of pathogenic and indicator microorganisms. *Applied and Environmental Microbiology, 49*(6), 1361–1365.
47. Chaplin, L. G., & Lyster, R. L. (1988). Effect of temperature on the pH of skim milk. *Journal of Dairy Research, 55*(2), 277–280.
48. Chapman, S. (2003). New machines use tumbling process to decontaminate food. *Food Chem. News, 10*, 20.
49. Chatziprodromidou, I. P., Bellou, M., Vantarakis, G., & Vantarakis, A. (2018). Viral outbreaks linked to fresh produce consumption: A systematic review. *Journal of Applied Microbiology, 124*(4), 932–942.
50. Chauhan, N., Singh, J., Chandra, S., Chaudhary, V., & Kumar, V. (2018). "Non-thermal techniques: Application in food industries" A review. *Journal of Pharmacognosy and Phytochemistry, 7*(5), 1507–1518.
51. Chavez, C., Knape, K. D., Coufal, C. D., & Carey, J. B. (2002). Reduction of eggshell aerobic plate counts by ultraviolet irradiation. *Poultry Science, 81*(8), 1132–1135.
52. Chawla, A., Lobacz, A., Tarapata, J., & Zulewska, J. (2021). UV light application as a mean for disinfection applied in the dairy industry. *Applied Sciences, 11*(16), 7285.
53. Chevremont, A. C., Boudenne, J. L., Coulomb, B., & Farnet, A. M. (2013). Impact of watering with UV-LED-treated wastewater on microbial and physico-chemical parameters of soil. *Water Research, 47*(6), 1971–1982.
54. Chinchkar, A. V., Singh, A., Singh, S. V., Acharya, A. M., & Kamble, M. G. (2022). Potential sanitizers and disinfectants for fresh fruits and vegetables: A comprehensive review. *Journal of Food Processing and Preservation, 46*(10), e16495.
55. Choudhary, R., & Bandla, S. (2012). Ultraviolet pasteurization for food industry. *International Journal of Food Science and Nutrition Engineering, 2*(1), 12–15.
56. Cohen, D. E. (2019). Occupational dermatoses. In *Handbook of Occupational Safety and Health* (pp. 199–230). Wiley.
57. Corry, J. E., James, S. J., Purnell, G., Barbedo-Pinto, C. S., Chochois, Y., Howell, M., & James, C. (2007). Surface pasteurization of chicken carcasses using hot water. *Journal of Food Engineering, 79*(3), 913–919.

58. Cote, S., Rodoni, L., Miceli, E., Concellón, A., Civello, P. M., & Vicente, A. R. (2013). Effect of radiation intensity on the outcome of postharvest UV-C treatments. *Postharvest Biology and Technology, 83*, 83-89.
59. CS, D., Samreen, F., & Prakash, J. (2015). A review on composition, processed products and medicinal uses of papaya (*Carica papaya* L.). *Int. J. Food Sci. Nutr., 3*, 99–117.
60. Csapó, J., Prokisch, J., Albert, C., & Sipos, P. (2019). Effect of UV light on food quality and safety. *Acta Universitatis Sapientiae, Alimentaria, 12*(1), 21–41.
61. Cubero, S., Aleixos, N., Moltó, E., Gómez-Sanchis, J., & Blasco, J. (2011). Advances in machine vision applications for automatic inspection and quality evaluation of fruits and vegetables. *Food and Bioprocess Technology, 4*, 487–504.
62. D'Souza, D. H., & Joshi, S. S. (2017). Noroviruses: Laboratory surrogates for determining survival and inactivation. In *Laboratory Models for Foodborne Infections* (pp. 75–93). CRC Press.
63. Dar, J. A., Wani, A. A., Ahmed, M., Nazir, R., Zargar, S. M., & Javaid, K. (2019). Peel colour in apple (*Malus× domestica* Borkh.): An economic quality parameter in fruit market. *Scientia Horticulturae, 244*, 50–60.
64. Darré, M., Valerga, L., Araque, L. C. O., Lemoine, M. L., Demkura, P. V., Vicente, A. R., & Concellón, A. (2017). Role of UV-B irradiation dose and intensity on color retention and antioxidant elicitation in broccoli florets (*Brassica oleracea* var. Italica). *Postharvest Biology and Technology, 128*, 76–82.
65. Darré, M., Vicente, A. R., Cisneros-Zevallos, L., & Artés-Hernández, F. (2022). Postharvest ultraviolet radiation in fruit and vegetables: Applications and factors modulating its efficacy on bioactive compounds and microbial growth. *Foods, 11*(5), 653.
66. Darvishi, S., Fatemi, A., & Davari, K. (2012). Keeping quality of use of fresh 'Kurdistan' strawberry by UV-C radiation. *World Applied Sciences Journal, 17*(7), 826-831.
67. De Corato, U. (2020). Improving the shelf-life and quality of fresh and minimally-processed fruits and vegetables for a modern food industry: A comprehensive critical review from the traditional technologies into the most promising advancements. *Critical Reviews in Food Science and Nutrition, 60*(6), 940–975.
68. DeFlorio, W., Liu, S., White, A. R., Taylor, T. M., Cisneros-Zevallos, L., Min, Y., & Scholar, E. M. (2021). Recent developments in antimicrobial and antifouling coatings to reduce or prevent contamination and cross-contamination of food contact surfaces by bacteria. *Comprehensive Reviews in Food Science and Food Safety, 20*(3), 3093–3134.
69. Delorme, M. M., Guimarães, J. T., Coutinho, N. M., Balthazar, C. F., Rocha, R. S., Silva, R., & Cruz, A. G. (2020). Ultraviolet radiation: An interesting technology to preserve quality and safety of milk and dairy foods. *Trends in Food Science & Technology, 102*, 146–154.
70. Deng, L. Z., Mujumdar, A. S., Pan, Z., Vidyarthi, S. K., Xu, J., Zielinska, M., & Xiao, H. W. (2020). Emerging chemical and physical disinfection technologies of fruits and vegetables: A comprehensive review. *Critical Reviews in Food Science and Nutrition, 60*(15), 2481–2508.
71. Dinçer, A. H., & Baysal, T. (2004). Decontamination techniques of pathogen bacteria in meat and poultry. *Critical Reviews in Microbiology, 30*(3), 197–204.
72. Ding, R., Che, X., Shen, Z., & Zhang, Y. (2021). Metabolome and transcriptome profiling provide insights into green apple peel reveals light-and UV-B-responsive pathway in anthocyanins accumulation. *BMC Plant Biology, 21*(1), 1–15.

73. Dobermann, A., & Nelson, R. (2013). Opportunities and solutions for sustainable food production. *Sustainable Development Solutions Network: Paris, France*. Pp. 12–24.
74. Dong, Q., Manns, D. C., Feng, G., Yue, T., Churey, J. J., & Worobo, R. W. (2010). Reduction of patulin in apple cider by UV radiation. *Journal of Food Protection, 73*(1), 69–74.
75. Doulia, D., Katsinis, G., & Mougin, B. (2000). Prolongation of the microbial shelf life of wrapped part baked baguettes. *International Journal of Food Properties, 3*(3), 447–457.
76. Dragland, I. S., Rukke, H. V., Stenhagen, I. S., Lönn-Stensrud, J., & Kopperud, H. M. (2016). Antibacterial and antibiofilm effect of low viscosity chitosan against *Staphylococcus epidermidis. International Journal of Microbiology, 2016*. p.9159761.
77. Duarte-Sierra, A., Charles, M. T., & Arul, J. (2019). UV-C hormesis: A means of controlling diseases and delaying senescence in fresh fruits and vegetables during storage. In *Postharvest Pathology of Fresh Horticultural Produce* (pp. 539–594). CRC Press.
78. Duarte-Sierra, A., Nadeau, F., Angers, P., Michaud, D., & Arul, J. (2019). UV-C hormesis in broccoli florets: Preservation, phyto-compounds and gene expression. *Postharvest Biology and Technology, 157*, 110965.
79. Duarte-Sierra, A., Tiznado-Hernández, M. E., Jha, D. K., Janmeja, N., & Arul, J. (2020). Abiotic stress hormesis: An approach to maintain quality, extend storability, and enhance phytochemicals on fresh produce during postharvest. *Comprehensive Reviews in Food Science and Food Safety, 19*(6), 3659–3682.
80. Dyshlyuk, L., Babich, O., Prosekov, A., Ivanova, S., Pavsky, V., & Chaplygina, T. (2020). The effect of postharvest ultraviolet irradiation on the content of antioxidant compounds and the activity of antioxidant enzymes in tomato. *Heliyon, 6*(1), e03288.
81. El Sohaimy, S. A. (2012). Functional foods and nutraceuticals-modern approach to food science. *World Applied Sciences Journal, 20*(5), 691–708.
82. Elkhatat, A. M. (2022). Recent trends of microbial decontamination for occupational, industrial and domestic applications. *Bulletin of the National Research Centre, 46*(1), 1–21.
83. El-Ramady, H. R., Domokos-Szabolcsy, É., Abdalla, N. A., Taha, H. S., & Fári, M. (2015). Postharvest management of fruits and vegetables storage. *Sustainable Agriculture Reviews: Volume 15*, 65–152.
84. Enwemeka, C. S., Baker, T. L., & Bumah, V. V. (2021). The role of UV and blue light in photo-eradication of microorganisms. *Journal of Photochemistry and Photobiology, 8*, 100064.
85. Erdoğdu, S. B., & Ekiz, H. İ. (2013). Far infrared and ultraviolet radiation as a combined method for surface pasteurization of black pepper seeds. *Journal of Food Engineering, 116*(2), 310–314.
86. Esua, O. J., Chin, N. L., Yusof, Y. A., & Sukor, R. (2019). Effects of simultaneous UV-C radiation and ultrasonic energy postharvest treatment on bioactive compounds and antioxidant activity of tomatoes during storage. *Food Chemistry, 270*, 113–122.
87. Esua, O. J., Chin, N. L., Yusof, Y. A., & Sukor, R. (2020). A review on individual and combination technologies of UV-C radiation and ultrasound in postharvest handling of fruits and vegetables. *Processes, 8*(11), 1433.
88. Ezzatpanah, H., Gómez-López, V. M., Koutchma, T., Lavafpour, F., Moerman, F., Mohammadi, M., & Raheem, D. (2022). Risks and new challenges in the food chain:

Viral contamination and decontamination from a global perspective, guidelines, and cleaning. *Comprehensive Reviews in Food Science and Food Safety, 21*(2), 868-903.

89. Falcó Ferrando, I. L. (2021). Characterization of antiviral activity of green tea extract and applications for improving food safety. Doctoral dissertation, Universitat de València. Facultat de Farmàcia.
90. Falguera, V., Pagán, J., & Ibarz, A. (2011). Effect of UV irradiation on enzymatic activities and physicochemical properties of apple juices from different varieties. *LWT – Food Science and Technology, 44*(1), 115–119.
91. Falguera, V., Pagán, J., Garza, S., Garvín, A., & Ibarz, A. (2011). Ultraviolet processing of liquid food: A review: Part 2: Effects on microorganisms and on food components and properties. *Food Research International, 44*(6), 1580–1588.
92. Fan, L., Muhammad, A. I., Ismail, B. B., & Liu, D. (2021). Sonodynamic antimicrobial chemotherapy: An emerging alternative strategy for microbial inactivation. *Ultrasonics Sonochemistry, 75*, 105591.
93. Fan, X., Huang, R., & Chen, H. (2017). Application of ultraviolet C technology for surface decontamination of fresh produce. *Trends in Food Science & Technology, 70*, 9–19.
94. Farber, J. M., Zwietering, M., Wiedmann, M., Schaffner, D., Hedberg, C. W., Harrison, M. A., & Gummalla, S. (2021). Alternative approaches to the risk management of *Listeria monocytogenes* in low risk foods. *Food Control, 123*, 107601.
95. Farooq, U., Shafi, A., Shahbaz, M., Khan, M. Z., Hayat, K., Baqir, M., & Iqbal, M. (2021). Food quality and food safety: An introduction. In *Sequencing Technologies in Microbial Food Safety and Quality* (pp. 3–24). CRC Press.
96. Favier, G. I., Escudero, M. E., & de GUZMÁN, A. M. (2001). Effect of chlorine, sodium chloride, trisodium phosphate, and ultraviolet radiation on the reduction of *Yersinia enterocolitica* and mesophilic aerobic bacteria from eggshell surface. *Journal of Food Protection, 64*(10), 1621–1623.
97. Fenoglio, D., Ferrario, M., Andreone, A., & Guerrero, S. (2022). Development of an orange-tangerine juice treated by assisted pilot-scale UV-C light and loaded with yerba mate: microbiological, physicochemical, and dynamic sensory studies. *Food and Bioprocess Technology, 15*(4), 915-932.
98. Finardi, S., Hoffmann, T. G., Schmitz, F. R. W., Bertoli, S. L., Khayrullin, M., Neverova, O., & De Souza, C. K. (2021). Comprehensive study of Light-Emitting Diodes (LEDs) and ultraviolet-LED lights application in food quality and safety. *Journal of Pure and Applied Microbiology, 15*(3), 1125–1135.
99. Florea, C., Bratucu, G., & Paunescu, V. S. (2012). Possibilities of using the germicidal effect of UV-C light for disinfecting the water used in food industry. *Bulletin of the Transilvania University of Brasov. Forestry, Wood Industry, Agricultural Food Engineering. Series II, 5*(1), 147.
100. Freeman, S., Eddy, S. L., McDonough, M., Smith, M. K., Okoroafor, N., Jordt, H., & Wenderoth, M. P. (2014). Active learning increases student performance in science, engineering, and mathematics. *Proceedings of the National Academy of Sciences, 111*(23), 8410–8415.
101. Gallo, M., Ferrara, L., Calogero, A., Montesano, D., & Naviglio, D. (2020). Relationships between food and diseases: What to know to ensure food safety. *Food Research International, 137*, 109414.

102. Gamage, G. (2015). Effectiveness of UV-C irradiation on controlling growth of *L. monocytogenes* on fresh cut broccoli: A thesis submitted in partial fulfilment of the requirements for the degree of Doctor of Philosophy in Food Technology at Massey University, New Zealand (Doctoral dissertation, Massey University). pp. 103–107.
103. Gautam, S., Variyar, P. S., & Sharma, A. K. (2015). Quality and safety aspects of vegetables. In Y. H. Hui & E. Ö. Evranuz (Eds.), *Handbook of Vegetable Preservation and Processing* (pp. 925–954). CRX Press.
104. George, D. S., Razali, Z., Santhirasegaram, V., & Somasundram, C. (2015). Effects of ultraviolet light (UV-C) and heat treatment on the quality of fresh-cut Chokanan mango and Josephine pineapple. *Journal of Food Science, 80*(2), S426–S434.
105. Gibson, J., Drake, J., & Karney, B. (2017). UV disinfection of wastewater and combined sewer overflows. In M. D. Farhat (Ed.), *Ultraviolet Light in Human Health, Diseases and Environment* (pp. 267–275). Springer.
106. Gobeil, A., Shankar, S., & Lacroix, M. (2020). Radiosensitivity of feline calicivirus F9 on iceberg lettuce surface after combined treatments with γ-radiation. *Journal of Food Protection, 83*(12), 2134–2146.
107. Gogo, E. O., Opiyo, A. M., Hassenberg, K., Ulrichs, C., & Huyskens-Keil, S. (2017). Postharvest UV-C treatment for extending shelf life and improving nutritional quality of African indigenous leafy vegetables. *Postharvest Biology and Technology, 129*, 107–117.
108. Goldberg, I. (2012). *Functional Foods: Designer Foods, Pharmafoods, Nutraceuticals*. Springer Science & Business Media. Chapter 1; Introduction, pp. 3–18.
109. Gómez-López, V. M., Jubinville, E., Rodríguez-López, M. I., Trudel-Ferland, M., Bouchard, S., & Jean, J. (2021). Inactivation of foodborne viruses by UV light: A review. *Foods, 10*(12), 3141.
110. Gonzales-Aguilar, G. A., Ayala-Zavala, J. F., de La Rosa, L. A., & Ivarez-Parrilla, E. A. (2010). Preserving quality of fresh-cut product using safe technology. *J. Verbrauch. Lebensm, 5*, 65–72.
111. González-Aguilar, G. A., Villegas-Ochoa, M. A., Martínez-Téllez, M. A., Gardea, A. A., & Ayala-Zavala, J. F. (2007). Improving antioxidant capacity of fresh-cut mangoes treated with UV-C. *Journal of Food Science, 72*(3), S197–S202.
112. González-Villagra, J., Reyes-Díaz, M., Alberdi, M., Mora, M. L., Ulloa-Inostroza, E. M., & Ribera-Fonseca, A. E. (2020). Impact of cold-storage and UV-C irradiation postharvest treatments on quality and antioxidant properties of fruits from blueberry cultivars grown in Southern Chile. *Journal of Soil Science and Plant Nutrition, 20*, 1751–1758.
113. Gourama, H. (2020). Foodborne pathogens. In Y. H. Hui & L. W. Heldman (Eds.), *Food Safety Engineering* (pp. 25–49). Taylor and Francis.
114. Graça, A., Santo, D., Pires-Cabral, P., & Quintas, C. (2020). The effect of UV-C and electrolyzed water on yeasts on fresh-cut apple at 4°C. *Journal of Food Engineering, 282*, 110034.
115. Green, A., Popović, V., Pierscianowski, J., Biancaniello, M., Warriner, K., & Koutchma, T. (2018). Inactivation of *Escherichia coli, Listeria* and *Salmonella* by single and multiple wavelength ultraviolet-light emitting diodes. *Innovative Food Science & Emerging Technologies, 47*, 353–361.

116. Guan, T., Xu, Z., Wang, J., Liu, Y., Shen, X., Li, X., & Lei, H. (2022). Multiplex optical bioassays for food safety analysis: Toward on-site detection. *Comprehensive Reviews in Food Science and Food Safety, 21*(2), 1627–1656.
117. Guan, W., Fan, X., & Yan, R. (2012). Effects of UV-C treatment on inactivation of *Escherichia coli* O157: H7, microbial loads, and quality of button mushrooms. *Postharvest Biology and Technology, 64*(1), 119–125.
118. Guan, W., Zhang, J., Yan, R., Shao, S., Zhou, T., Lei, J., & Wang, Z. (2016). Effects of UV-C treatment and cold storage on ergosterol and vitamin D2 contents in different parts of white and brown mushroom (*Agaricus bisporus*). *Food Chemistry, 210*, 129–134.
119. Guerrero-Beltrán, J. A., & Barbosa-Cánovas, G. V. (2004). Advantages and limitations on processing foods by UV light. *Food Science and Technology International, 10*(3), 137–147.
120. Guerrero-Beltrán, J. A., & Barbosa-Cánovas, G. V. (2011). Ultraviolet-C light processing of liquid food products. In H. Q. Zhang, G. V. Barbosa-Cánovas, V. M. Balasubramaniam, C. P. Dunne, D. F. Farkas, & J. T. C. Yuan (Eds.), *Nonthermal Processing Technologies for Food* (pp. 262–270). Wiley Blackwell.
121. Guo, S., Huang, R., & Chen, H. (2019). Evaluating a combined method of UV and washing for sanitizing blueberries, tomatoes, strawberries, baby spinach, and lettuce. *Journal of Food Protection, 82*(11), 1879–1889.
122. Gutierrez, D. R., & Rodriguez, S. D. C. (2017). Combined effect of UV-C and modified atmosphere packaging for keeping antioxidant compounds and extend to shelf-life of fresh-cut rocket leaves. *International Journal of New Technology and Research, 3*(6), 263275.
123. Zhu, Z., Cai, H., & Sun, D. W. (2018). Titanium dioxide (TiO_2) photocatalysis technology for nonthermal inactivation of microorganisms in foods. *Trends in Food Science & Technology, 75*, 23–35.
124. Gutiérrez, D. R., Lemos, L., & Rodríguez, S. D. C. (2018). Effect of UV-C and ozone on the bioactive compounds and antioxidant capacity of minimally processed rocket (*Eruca sativa* Mill.). *International Journal of New Technology and Research, 4*, 23–29.
125. Ha, J. W., Back, K. H., Kim, Y. H., & Kang, D. H. (2016). Efficacy of UV-C irradiation for inactivation of food-borne pathogens on sliced cheese packaged with different types and thicknesses of plastic films. *Food Microbiology, 57*, 172–177.
126. Hakgüder Taze, B., Pelvan Akgün, M., Yıldız, S., Kaya, Z., & Ünlütürk, S. (2021). UV processing and storage of liquid and solid foods: Quality, microbial, enzymatic, nutritional, organoleptic, composition and properties effects. *Trends in Food Science & Technology, 108*, 109–124.
127. Hakguder Taze, B., Unluturk, S., Buzrul, S., & Alpas, H. (2015). The impact of UV-C irradiation on spoilage microorganisms and color of orange juice. *Journal of Food Science and Technology, 52*, 1000–1007.
128. Harguindeguy, M., & Gómez-Camacho, C. E. (2021). Pulsed light (PL) treatments on almond kernels: *Salmonella enteritidis* inactivation kinetics and infrared thermography insights. *Food and Bioprocess Technology, 14*, 2323–2335.
129. Henry-Kirk, R. A., Plunkett, B., Hall, M., McGhie, T., Allan, A. C., Wargent, J. J., & Espley, R. V. (2018). Solar UV light regulates flavonoid metabolism in apple (Malus x domestica). *Plant, Cell & Environment, 41*(3), 675-688.

130. Hijnen, W. A. M., Beerendonk, E. F., & Medema, G. J. (2006). Inactivation credit of UV radiation for viruses, bacteria and protozoan (oo)cysts in water: a review. *Water Research, 40*(1), 3–22.
131. Hinds, L. M., O'Donnell, C. P., Akhter, M., & Tiwari, B. K. (2019). Principles and mechanisms of ultraviolet light emitting diode technology for food industry applications. *Innovative Food Science & Emerging Technologies, 56*, 102153.
132. Holck, A. L., Liland, K. H., Drømtorp, S. M., Carlehög, M., & McLeod, A. (2018). Comparison of UV-C and pulsed UV light treatments for reduction of *Salmonella, Listeria monocytogenes*, and enterohemorrhagic *Escherichia coli* on eggs. *Journal of Food Protection, 81*(1), 6–16.
133. Horton, L., Torres, A. E., Narla, S., Lyons, A. B., Kohli, I., Gelfand, J. M., & Lim, H. W. (2020). Spectrum of virucidal activity from ultraviolet to infrared radiation. *Photochemical & Photobiological Sciences, 19*, 1262–1270.
134. Hu, D., Yang, X., Hu, C., Feng, Z., Chen, W., & Shi, H. (2021). Comparison of ergosterol and vitamin D2 in mushrooms *Agaricus bisporus* and *Cordyceps militaris* using ultraviolet irradiation directly on dry powder or in ethanol suspension. *ACS Omega, 6*(44), 29506–29515.
135. Hui, Y. H. (Ed.). (2006). *Handbook of Food Science, Technology, and Engineering* (Vol. 149). CRC Press. Chapter 102: Unit Operation pp 102 (1–19).
136. Idzwana, M. I. N., Chou, K. S., Shah, R. M., & Soh, N. C. (2020). The effect of ultraviolet light treatment in extending shelf life and preserving the quality of strawberry (*Fragaria x ananassa*) cv. Festival. *International Journal on Food, Agriculture and Natural Resources, 1*(1), 15–18.
137. Jahid, I. K., Han, N. R., Srey, S., & Ha, S. D. (2014). Competitive interactions inside mixed-culture biofilms of *Salmonella typhimurium* and cultivable indigenous microorganisms on lettuce enhance microbial resistance of their sessile cells to ultraviolet C (UV-C) irradiation. *Food Research International, 55*, 445–454.
138. James, J. (2006). Overview of microbial hazards in fresh fruit and vegetables operations. In J. James (Ed.), *Microbial Hazard Identification in Fresh Fruit and Vegetables* (pp. 1–36). Wiley.
139. Jan, A., Sood, M., Sofi, S. A., & Norzom, T. (2017). Non-thermal processing in food applications: A review. *International Journal of Food Science and Nutrition, 2*(6), 171–180.
140. Jasinghe, V. J., & Perera, C. O. (2006). Ultraviolet irradiation: The generator of vitamin D2 in edible mushrooms. *Food Chemistry, 95*(4), 638–643.
141. Jiang, Q., Zhang, M., & Mujumdar, A. S. (2020). UV induced conversion during drying of ergosterol to vitamin D in various mushrooms: Effect of different drying conditions. *Trends in Food Science & Technology, 105*, 200–210.
142. John, D., & Ramaswamy, H. S. (2018). Pulsed light technology to enhance food safety and quality: A mini-review. *Current Opinion in Food Science, 23*, 70–79.
143. Joshi, S. S., & D'Souza, D. H. (2021). Grape seed extract against Aichi virus infectivity in model foods and contact surfaces. *Food Microbiology, 98*, 103784.
144. Joshi, V. K., Lakhanpal, P., & Kumar, V. (2013). Occurrence of patulin its dietary intake through consumption of apple and apple products and methods of its removal. *International Journal of Food and Fermentation Technology, 3*(1), 15–32.

145. Julakanti, S. C. (2016). Mycotoxins detoxification using UV irradiation and advanced oxidation processes (Doctoral dissertation, Tennessee State University). Masters Thesis, Tennessee State University, pp. 78.
146. Kader, A. A. (2002). Quality parameters of fresh-cut fruit and vegetable products. In *Fresh-Cut Fruits and Vegetables: Science, Technology, and Market* (pp. 11–20). CRC Press.
147. Kaewklin, P., Siripatrawan, U., Suwanagul, A., & Lee, Y. S. (2018). Active packaging from chitosan-titanium dioxide nanocomposite film for prolonging storage life of tomato fruit. *International Journal of Biological Macromolecules, 112*, 523–529.
148. Kalaras, M. D., Beelman, R. B., & Elias, R. J. (2012). Effects of postharvest pulsed UV light treatment of white button mushrooms (*Agaricus bisporus*) on vitamin D2 content and quality attributes. *Journal of Agricultural and Food Chemistry, 60*(1), 220–225.
149. Kciuk, M., Marciniak, B., Mojzych, M., & Kontek, R. (2020). Focus on UV-induced DNA damage and repair—Disease relevance and protective strategies. *International Journal of Molecular Sciences, 21*(19), 7264.
150. Kebbi, Y., Muhammad, A. I., Sant'Ana, A. S., do Prado-Silva, L., Liu, D., & Ding, T. (2020). Recent advances on the application of UV-LED technology for microbial inactivation: Progress and mechanism. *Comprehensive Reviews in Food Science and Food Safety, 19*(6), 3501–3527.
151. Keklik, N. M., Elik, A., Salgin, U., Demirci, A., & Koçer, G. (2019). Inactivation of *Staphylococcus aureus* and *Escherichia coli* O157 on fresh kosher cheese with pulsed ultraviolet light. *Food Science and Technology International, 25*(8), 680–691.
152. Khan, I., Tango, C. N., Miskeen, S., Lee, B. H., & Oh, D. H. (2017). Hurdle technology: A novel approach for enhanced food quality and safety–A review. *Food Control, 73*, 1426–1444.
153. Khan, S., Dar, A. H., Shams, R., Aga, M. B., Siddiqui, M. W., Mir, S. A., & Altaf, A. (2022). Applications of ultraviolet light–emitting diode technology in horticultural produce: A systematic review and meta-analysis. *Food and Bioprocess Technology*, 1–11.
154. Khandpur, P., & Gogate, P. R. (2015). Effect of novel ultrasound-based processing on the nutrition quality of different fruit and vegetable juices. *Ultrasonics Sonochemistry, 27*, 125–136.
155. Koca, N., & Öztürk, M. U. (2020). Application of batch system ultraviolet light on the surface of Kasher cheese, a kind of pasta-filata cheese: Effects on mold inactivation, lipid oxidation, color, hardness and sensory properties. *Journal of Dairy Research, 87*(4), 493–497.
156. Koca, N., Urgu, M., & Saatli, T. E. (2018). Ultraviolet light applications in dairy processing. In *Technological Approaches for Novel Applications in Dairy Processing* (pp. 1–16).
157. Koch, F., Wiacek, C., & Braun, P. G. (2019). Pulsed light treatment for the reduction of *Salmonella Typhimurium* and *Yersinia enterocolitica* on pork skin and pork loin. *International Journal of Food Microbiology, 292*, 64–71.
158. Kokalj, D., Zlatić, E., Cigić, B., Kobav, M. B., & Vidrih, R. (2019). Postharvest flavanol and anthocyanin accumulation in three apple cultivars in response to blue-light-emitting diode light. *Scientia Horticulturae, 257*, 108711.

159. Kolb, A. Y., & Kolb, D. A. (2005). Learning styles and learning spaces: Enhancing experiential learning in higher education. *Academy of Management Learning & Education, 4*(2), 193–212.
160. Kong, J., Lu, Y., Ren, Y., Chen, Z., & Chen, M. (2021). The virus removal in UV irradiation, ozonation and chlorination. *Water Cycle, 2*, 23–31.
161. Koutchma, T. (2008). UV light for processing foods. *Ozone: Science and Engineering, 30*(1), 93–98.
162. Koutchma, T. (2018). Status of international regulations for ultraviolet treatment of foods. *IUVA News, 20*, 13–16.
163. Koutchma, T. (2019). *Ultraviolet Light in Food Technology: Principles and Applications*. IntechOpen/CRC Press.
164. Koutchma, T., & Stewart, C. (2005). Applications and efficacy of UV light for foods. *IUVA News, 7*(4), pp. 1–48.
165. Koutchma, T., & Orlowska, M. (2012). Ultraviolet light for processing fruits and fruit products. In *Advances in Fruit Processing Technologies* (pp. 2–27).
166. Koutchma, T., Englert, R., & Adhikari, C. (2002). The effect of browning and suspended particles on the UV treatment of apple juice/cider. In *IFT Annual Meeting*, Anaheim, CA, Springer, New York, NY.
167. Koutchma, T., Orlowska, M., & Zhu, Y. (2018). Fruits and fruit products treated by UV light. In *Fruit Preservation* (pp. 457–504). Springer, New York, NY.
168. Koutchma, T., Popović, V., & Green, A. (2019). Overview of Ultraviolet (UV) LEDs technology for applications in food production. In *Ultraviolet LED Technology for Food Applications* (pp. 1–23). Academic Press.
169. Kowalski, W. (2010). *Ultraviolet Germicidal Irradiation Handbook: UVGI for Air and Surface Disinfection*. Springer Science & Business Media. pp. 427–430.
170. Kramer, B. (2022). Pulsed light for food surface disinfection—Mode of microbial inactivation and application for fresh produce (Doctoral dissertation, Technische Universität München, pp, 59–76).
171. Kumar, M., Tomar, M., Amarowicz, R., Saurabh, V., Nair, M. S., Maheshwari, C., & Satankar, V. (2021). Guava (*Psidium guajava* L.) leaves: Nutritional composition, phytochemical profile, and health-promoting bioactivities. *Foods, 10*(4), 752.
172. Kumari, B., Devi, E. P., Sharma, G., Rawat, S., & Jaiswal, J. P. (2017). Irradiation as an alternative method for post-harvest disease management: an overview. *International Journal of Agriculture, Environment and Biotechnology, 10*(5), 625–633.
173. Kyriazis, M. (2016). "Hormesis and Adaption." In: *Challenging Aging: The Anti-Senescence Effects of Hormesis*, Environmental Enrichment, and Information Exposure (Vol. 1). Bentham Science Publishers. pp. 3–28.
174. Lacivita, V., Conte, A., Manzocco, L., Plazzotta, S., Zambrini, V. A., Del Nobile, M. A., & Nicoli, M. C. (2016). Surface UV-C light treatments to prolong the shelf-life of Fiordilatte cheese. *Innovative Food Science & Emerging Technologies, 36*, 150–155.
175. Lammertyn, J., De Ketelaere, B., Marquenie, D., Molenberghs, G., & Nicolaï, B. M. (2003). Mixed models for multicategorical repeated response: modelling the time effect of physical treatments on strawberry sepal quality. *Postharvest Biology and Technology, 30*(2), 195–207.
176. Latief, M., & Tarigan, I. L. (2022). Preliminary data on the antibacterial activity of Coffea Arabica, *Coffea canephora*, and *Coffea liberica*. *Pharmacognosy Journal, 14*(2), 413–424.

177. Leggott, N. L., Shephard, G. S., Stockenström, S., Staal, E., & Van Schalkwyk, D. J. (2001). The reduction of patulin in apple juice by three different types of activated carbon. *Food Additives & Contaminants, 18*(9), 825–829.
178. Lemessa, A., Popardowski, E., Hebda, T., & Jakubowski, T. (2022). The effect of UV-C irradiation on the mechanical and physiological properties of potato tuber and different products. *Applied Sciences, 12*(12), 5907.
179. Leneveu-Jenvrin, C., Charles, F., Barba, F. J., & Remize, F. (2020). Role of biological control agents and physical treatments in maintaining the quality of fresh and minimally-processed fruit and vegetables. *Critical Reviews in Food Science and Nutrition, 60*(17), 2837–2855.
180. Li, B., Wang, L., Bai, W., Chen, W., Chen, F., Shu, C., & Shu, C. (2021). Unveiling the metabolic modulatory effect of anthocyanin and gut microbiota involvement. In *Anthocyanins: Chemistry, Processing & Bioactivity* (pp. 339–384).
181. Li, J., Luo, W., Han, L., Cai, Z., & Guo, Z. (2022). Two-wavelength image detection of early decayed oranges by coupling spectral classification with image processing. *Journal of Food Composition and Analysis, 111*, 104642.
182. Li, T., Yamane, H., & Tao, R. (2021). Preharvest long-term exposure to UV-B radiation promotes fruit ripening and modifies stage-specific anthocyanin metabolism in highbush blueberry. *Horticulture Research, 8*, 67.
183. Lim, W., & Harrison, M. A. (2016). Effectiveness of UV light as a means to reduce *Salmonella* contamination on tomatoes and food contact surfaces. *Food Control, 66*, 166–173.
184. Liu, C., Zheng, H., Sheng, K., Liu, W., & Zheng, L. (2018). Effects of postharvest UV-C irradiation on phenolic acids, flavonoids, and key phenylpropanoid pathway genes in tomato fruit. *Scientia Horticulturae, 241*, 107–114.
185. Liu, D. (Ed.). (2018). *Handbook of Foodborne Diseases*. CRC Press.
186. Liu, H., Cao, X., Liu, X., Xin, R., Wang, J., Gao, J., & Chen, K. (2017). UV-B irradiation differentially regulates terpene synthases and terpene content of peach. *Plant, Cell & Environment, 40*(10), 2261–2275.
187. Liu, X., Garoma, T., Chen, Z., Wang, L., & Wu, Y. (2012). SMX degradation by ozonation and UV radiation: a kinetic study. *Chemosphere, 87*(10), 1134–1140.
188. Loconsole, D., & Santamaria, P. (2021). UV lighting in horticulture: A sustainable tool for improving production quality and food safety. *Horticulturae, 7*(1), 9.
189. Londhe, D. H., Nalawade, S. M., Pawar, G. S., Atkari, V. T., & Wandkar, S. V. (2013). Grader: A review of different methods of grading for fruits and vegetables. *Agricultural Engineering International: CIGR Journal, 15*(3), 217–230.
190. Lu, J. Y., Stevens, C., Khan, V. A., Kabwe, M., & Wilson, C. L. (1991). The effect of ultraviolet irradiation on shelf-life and ripening of peaches and apples. *Journal of Food Quality, 14*(4), 299–305.
191. Luo, W., Fan, G., Tian, P., Dong, W., Zhang, H., & Zhan, B. (2022). Spectrum classification of citrus tissues infected by fungi and multispectral image identification of early rotten oranges. *Spectrochimica Acta Part A: Molecular and Biomolecular Spectroscopy, 279*, 121412.
192. Ma, L., Wang, Q., Li, L., Grierson, D., Yuan, S., Zheng, S., & Zuo, J. (2021). UV-C irradiation delays the physiological changes of bell pepper fruit during storage. *Postharvest Biology and Technology, 180*, 111506.

193. Maharaj, R. (2015). Effects of abiotic stress (UV-C) induced activation of phytochemicals on the postharvest quality of horticultural crops. In *Phytochemicals: Isolation, Characterization and Role in Human Health* (pp. 221–244).
194. Maharaj, R. (2018). Improving quality of ready-to-eat, minimally processed produce using non-thermal postharvest technologies. In: *Innovative Food Science and Emerging Technologies*. pp. 290–313.
195. Maharaj, R., Arul, J., & Nadeau, P. (2014). UV-C irradiation effects on levels of enzymic and non-enzymic phytochemicals in tomato. *Innovative Food Science & Emerging Technologies, 21*, 99–106.
196. Mahendran, R., Jayashree, G. C., & Alagusundaram, K. (2012). Application of computer vision technique on sorting and grading of fruits and vegetables. *J. Food Process. Technol, 10*, 2157–7110.
197. Mahmoud, M. Z., Fagiry, M. A., Davidson, R., & Abdelbasset, W. K. (2022). The benefits, drawbacks, and potential future challenges of the most commonly used ultrasound-based hurdle combinations technologies in food preservation. *Journal of Radiation Research and Applied Sciences, 15*(1), 206–212.
198. Malek, N. I., Abdullah, W. Z. W., & Sembok, W. Z. W. (2021). Effects of UV-C radiation in delaying ripening of Berangan banana (*Musa* sp. AAA Berangan). *Universiti Malaysia Terengganu Journal of Undergraduate Research, 3*(4), 173–182.
199. Mandal, R., Mohammadi, X., Wiktor, A., Singh, A., & Pratap Singh, A. (2020). Applications of pulsed light decontamination technology in food processing: An overview. *Applied Sciences, 10*(10), 3606.
200. Mangaraj, S., & Goswami, T. K. (2009). Modified atmosphere packaging of fruits and vegetables for extending shelf-life: A review. *Fresh Produce, 3*(1), 1–31.
201. Mansourbahmani, S., Ghareyazie, B., Zarinnia, V., Kalatejari, S., & Mohammadi, R. S. (2018). Study on the efficiency of ethylene scavengers on the maintenance of postharvest quality of tomato fruit. *Journal of Food Measurement and Characterization, 12*, 691–701.
202. María, L. P., Díaz-Reinoso, B., Giancarlo, C., Barba, F. J., Moure, A., Domínguez, H., & Daniel, F. (2019). Green technologies for food processing: principal considerations. In *Innovative Thermal and Non-Thermal Processing, Bioaccessibility and Bioavailability of Nutrients and Bioactive Compounds* (pp. 55–103). Woodhead Publishing.
203. Marquenie, D., Geeraerd, A. H., Lammertyn, J., Soontjens, C., Van Impe, J. F., Michiels, C. W., & Nicolaı, B. M. (2003). Combinations of pulsed white light and UV-C or mild heat treatment to inactivate conidia of *Botrytis cinerea* and *Monilia fructigena. International Journal of Food Microbiology, 85*(1–2), 185–196.
204. Maya-Cano, D. A., Arango-Varela, S., & Santa-Gonzalez, G. A. (2021). Phenolic compounds of blueberries (*Vaccinium* spp) as a protective strategy against skin cell damage induced by ROS: A review of antioxidant potential and antiproliferative capacity. *Heliyon, 7*(2), e06297.
205. Mazaheri, T., Cervantes-Huamán, B. R., Bermúdez-Capdevila, M., Ripolles-Avila, C., & Rodríguez-Jerez, J. J. (2021). *Listeria monocytogenes* biofilms in the food industry: Is the current hygiene program sufficient to combat the persistence of the pathogen? *Microorganisms, 9*(1), 181.
206. Mendoza, I. C., Luna, E. O., Pozo, M. D., Vásquez, M. V., Montoya, D. C., Moran, G. C., & León, J. C. (2022). Conventional and non-conventional disinfection methods to

prevent microbial contamination in minimally processed fruits and vegetables. *LWT, 165*, 113714.
207. Michel, C., Samtlebe, M., Wagner, N., Neve, H., Franz, C. M., Hinrichs, J., & Atamer, Z. (2021). Orthogonal processing strategies to create "phage-free" whey – Membrane filtration followed by thermal or ultraviolet C treatment for the reduction of *Lactococcus lactis* bacteriophages. *International Dairy Journal, 122*, 105149.
208. Moazzami, M., Fernström, L. L., & Hansson, I. (2021). Reducing *Campylobacter jejuni*, Enterobacteriaceae and total aerobic bacteria on transport crates for chickens by irradiation with 265-nm ultraviolet light (UV–C LED). *Food Control, 119*, 107424.
209. Mostafidi, M., Sanjabi, M. R., Shirkhan, F., & Zahedi, M. T. (2020). A review of recent trends in the development of the microbial safety of fruits and vegetables. *Trends in Food Science & Technology, 103*, 321–332.
210. Munhuweyi, K., Mpai, S., & Sivakumar, D. (2020). Extension of avocado fruit postharvest quality using non-chemical treatments. *Agronomy, 10*(2), 212.
211. Nasheri, N., Vester, A., & Petronella, N. (2019). Foodborne viral outbreaks associated with frozen produce. *Epidemiology & Infection, 147*, e291.
212. Nassarawa, S. S., Abdelshafy, A. M., Xu, Y., Li, L., & Luo, Z. (2021). Effect of Light-Emitting Diodes (LEDs) on the quality of fruits and vegetables during postharvest period: A review. *Food and Bioprocess Technology, 14*, 388–414.
213. Nawkar, G. M., Maibam, P., Park, J. H., Sahi, V. P., Lee, S. Y., & Kang, C. H. (2013). UV-induced cell death in plants. *International Journal of Molecular Sciences, 14*(1), 1608–1628.
214. Nazir, M., Asad Ullah, M., Mumtaz, S., Siddiquah, A., Shah, M., Drouet, S., & Abbasi, B. H. (2020). Interactive effect of melatonin and UV-C on phenylpropanoid metabolite production and antioxidant potential in callus cultures of purple basil (*Ocimum basilicum* L. var purpurascens). *Molecules, 25*(5), 1072.
215. Nerandzic, M. M., Thota, P., Sankar, T., Jencson, A., Cadnum, J. L., Ray, A. J., & Donskey, C. J. (2015). Evaluation of a pulsed xenon ultraviolet disinfection system for reduction of healthcare-associated pathogens in hospital rooms. *Infection Control & Hospital Epidemiology, 36*(2), 192–197.
216. Nyhan, L., Przyjalgowski, M., Lewis, L., Begley, M., & Callanan, M. (2021). Investigating the use of ultraviolet light emitting diodes (UV-LEDs) for the inactivation of bacteria in powdered food ingredients. *Foods, 10*(4), 797.
217. Obande, M. A., & Shama, G. (2011). The use of biodosimetry to measure the UV-C dose delivered to a sphere, and implications for the commercial treatment of fruit. *Journal of Food Engineering, 104*(1), 1–5.
218. Odeyemi, O. A., Alegbeleye, O. O., Strateva, M., & Stratev, D. (2020). Understanding spoilage microbial community and spoilage mechanisms in foods of animal origin. *Comprehensive Reviews in Food Science and Food Safety, 19*(2), 311–331.
219. Oms-Oliu, G., Martín-Belloso, O., & Soliva-Fortuny, R. (2010). Pulsed light treatments for food preservation. A review. *Food and Bioprocess Technology, 3*, 13–23.
220. Onwude, D. I., & Chen, G. (2021). Special Issue "Advances in Postharvest Process Systems." Processes 2021, 9, 1426. https:// doi.org/10.3390/pr9081426.
221. Ortiz-Solà, J., Abadias, I., Colàs-Medà, P., Anguera, M., & Viñas, I. (2021). Inactivation of *Salmonella enterica, Listeria monocytogenes* and murine norovirus (MNV-1) on fresh strawberries by conventional and water-assisted ultraviolet light (UV-C). *Postharvest Biology and Technology, 174*, 111447.

222. Pandith, J. A., Bansal, S., Katiyar, S. K., Manzoor, A., Junaid, P. M., & Ahmad, S. (2023). Scope of novel pulsed light technology: Safety and shelf-life of minimally processed fruits and vegetables. In *Quality Control in Fruit and Vegetable Processing: Methods and Strategies* (pp. 77–98).
223. Pandohee, J., Wani, S. A., Rafiq, S., & Dar, B. N. (2022). Effect of preharvest and postharvest technologies on bioactive compounds. In *Apples* (pp. 245–264). CRC Press.
224. Panjai, L., Noga, G., Fiebig, A., & Hunsche, M. (2017). Effects of continuous red light and short daily UV exposure during postharvest on carotenoid concentration and antioxidant capacity in stored tomatoes. *Scientia Horticulturae, 226*, 97-103.
225. Papoutsis, K., Mathioudakis, M. M., Hasperué, J. H., & Ziogas, V. (2019). Non-chemical treatments for preventing the postharvest fungal rotting of citrus caused by *Penicillium digitatum* (green mold) and *Penicillium italicum* (blue mold). *Trends in Food Science & Technology, 86*, 479–491.
226. Pataro, G., Sinik, M., Capitoli, M. M., Donsì, G., & Ferrari, G. (2015). The influence of post-harvest UV-C and pulsed light treatments on quality and antioxidant properties of tomato fruits during storage. *Innovative Food Science & Emerging Technologies, 30*, 103–111.
227. Patel, K. K., Kar, A., Jha, S. N., & Khan, M. A. (2012). Machine vision system: a tool for quality inspection of food and agricultural products. *Journal of Food Science and Technology, 49*, 123–141.
228. Pathak, N., Caleb, O. J., Geyer, M., Herppich, W. B., Rauh, C., & Mahajan, P. V. (2017). Photocatalytic and photochemical oxidation of ethylene: Potential for storage of fresh produce—A review. *Food and Bioprocess Technology, 10*, 982–1001.
229. Pichel, N., Vivar, M., & Fuentes, M. (2019). The problem of drinking water access: A review of disinfection technologies with an emphasis on solar treatment methods. *Chemosphere, 218*, 1014–1030.
230. Pillai, S., Oresajo, C., & Hayward, J. (2005). Ultraviolet radiation and skin aging: roles of reactive oxygen species, inflammation and protease activation, and strategies for prevention of inflammation-induced matrix degradation–A review. *International Journal of Cosmetic Science, 27*(1), 17–34.
231. Pokhrel, R. (2020). An analysis of UVGI technology in disinfection of *Escherichia coli* (gram negative bacteria). A Term Paper Submitted to Birendra Multiple Campus, Chitwan Tribhuvan University, Kirtipur, In Partial Fulfillment for the Requirement of Master's Degree of Science in Physics, pp.24.
232. Poller, B., Hall, S., Bailey, C., Gregory, S., Clark, R., Roberts, P., & Evans, C. (2018). 'VIOLET': a fluorescence-based simulation exercise for training healthcare workers in the use of personal protective equipment. *Journal of Hospital Infection, 99*(2), 229–235.
233. Pombo, M. A., Dotto, M. C., Martínez, G. A., & Civello, P. M. (2009). UV-C irradiation delays strawberry fruit softening and modifies the expression of genes involved in cell wall degradation. *Postharvest Biology and Technology, 51*(2), 141–148.
234. Portero, L. R., Alonso-Reyes, D. G., Zannier, F., Vazquez, M. P., Farías, M. E., Gärtner, W., & Albarracín, V. H. (2019). Photolyases and Cryptochromes in UV-resistant Bacteria from High-altitude Andean Lakes. *Photochemistry and Photobiology, 95*(1), 315–330.
235. Pouliot, Y., Boulet, M., & Paquin, P. (1989). Observations on the heat-induced salt balance changes in milk I. Effect of heating time between 4 and 90°C. *Journal of Dairy Research, 56*(2), 185–192.

236. Prajapati, U., Asrey, R., Varghese, E., Singh, A. K., & Singh, M. P. (2021). Effects of postharvest ultraviolet-C treatment on shelf-life and quality of bitter gourd fruit during storage. *Food Packaging and Shelf Life, 28*, 100665.
237. Pristijono, P., Golding, J. B., & Bowyer, M. C. (2018). Postharvest UV-C treatment, followed by storage in a continuous low-level ethylene atmosphere, maintains the quality of 'Kensington pride' mango fruit stored at 20°C. *Horticulturae, 5*(1), 1.
238. Pristijono, P., Papoutsis, K., Scarlett, C. J., Bowyer, M. C., Vuong, Q. V., Stathopoulos, C. E., & Golding, J. B. (2017). Postharvest UV-C treatment combined with 1-methylcyclopropene (1-MCP), followed by storage in continuous low-level ethylene atmosphere, improves the quality of tomatoes. *The Journal of Horticultural Science and Biotechnology, 92*(5), 521–529.
239. Priyadarshi, R., & Rhim, J. W. (2020). Chitosan-based biodegradable functional films for food packaging applications. *Innovative Food Science & Emerging Technologies, 62*, 102346.
240. Proulx, J., Sullivan, G., Marostegan, L. F., VanWees, S., Hsu, L. C., & Moraru, C. I. (2017). Pulsed light and antimicrobial combination treatments for surface decontamination of cheese: Favorable and antagonistic effects. *Journal of Dairy Science, 100*(3), 1664–1673.
241. Prusty, P. (2020). Effect of pulsed light (PL) and UV-C light treatments on the shelf-life and quality enhancement of fresh carrot and radish. McGill University (Canada). Masters Thesis, McGill University, Canada. pp. 74.
242. Putnik, P., Pavlić, B., Šojić, B., Zavadlav, S., Žuntar, I., Kao, L., & Kovačević, D. B. (2020). Innovative hurdle technologies for the preservation of functional fruit juices. *Foods, 9*(6), 699.
243. Qadri, O. S., Yousuf, B., & Srivastava, A. K. (2015). Fresh-cut fruits and vegetables: Critical factors influencing microbiology and novel approaches to prevent microbial risks—A review. *Cogent Food & Agriculture, 1*(1), 1121606.
244. Quarshi, H. Q., Ahmed, W., Azmant, R., Chendouh-Brahmi, N., Quyyum, A., & Abbas, A. (2023). Post-harvest problems of strawberry and their solutions. In *Recent Studies on Strawberries*. IntechOpen. pp. 1–14.
245. Rajkovic, A., Tomasevic, I., De Meulenaer, B., & Devlieghere, F. (2017). The effect of pulsed UV light on *Escherichia coli* O157: H7, *Listeria monocytogenes*, *Salmonella typhimurium*, *Staphylococcus aureus* and *staphylococcal enterotoxin* A on sliced fermented salami and its chemical quality. *Food Control, 73*, 829–837.
246. Ramos, B., Miller, F. A., Brandão, T. R. S., Teixeira, P., & Silva, C. L. M. (2013). Fresh fruits and vegetables—An overview on applied methodologies to improve its quality and safety. *Innovative Food Science & Emerging Technologies, 20*, 1–15.
247. Ravishankar, S., Zhu, L., & Jaroni, D. (2010). Assessing the cross contamination and transfer rates of *Salmonella enterica* from chicken to lettuce under different food-handling scenarios. *Food Microbiology, 27*(6), 791–794.
248. Rawson, A., Patras, A., Tiwari, B. K., Noci, F., Koutchma, T., & Brunton, N. (2011). Effect of thermal and nonthermal processing technologies on the bioactive content of exotic fruits and their products: Review of recent advances. *Food Research International, 44*(7), 1875–1887.
249. Razali, Z., Somasundram, C., Nurulain, S. Z., Kunasekaran, W., & Alias, M. R. (2021). Postharvest quality of cherry tomatoes coated with mucilage from dragon fruit and irradiated with UV-C. *Polymers, 13*, 2919.

250. Ribeiro, C., & Alvarenga, B. (2012). Ribeiro, C., & Alvarenga, B. (2012). Prospects of UV radiation for application in postharvest technology. *Emirates Journal of Food and Agriculture, 24*(6), 586–597.
251. Rodoni, L. M., Hasperué, J. H., Ortiz, C. M., Lemoine, M. L., Concellón, A., & Vicente, A. R. (2016). Combined use of mild heat treatment and refrigeration to extend the postharvest life of organic pepper sticks, as affected by fruit maturity stage. *Postharvest Biology and Technology, 117*, 168–176.
252. Rodoni, L. M., Zaro, M. J., Hasperué, J. H., Concellón, A., & Vicente, A. R. (2015). UV-C treatments extend the shelf life of fresh-cut peppers by delaying pectin solubilization and inducing local accumulation of phenolics. *LWT – Food Science and Technology, 63*(1), 408–414.
253. Rodríguez-Ramírez, J., Méndez-Lagunas, L. L., López-Ortiz, A., Muñiz-Becerá, S., & Nair, K. (2021). Solar drying of strawberry using polycarbonate with UV protection and polyethylene covers: Influence on anthocyanin and total phenolic content. *Solar Energy, 221*, 120–130.
254. Roh, S. H., Oh, Y. J., Lee, S. Y., Kang, J. H., & Min, S. C. (2020). Inactivation of *Escherichia coli* O157: H7, *Salmonella, Listeria monocytogenes*, and Tulane virus in processed chicken breast via atmospheric in-package cold plasma treatment. *LWT, 127*, 109429.
255. Roobab, U., Afzal, R., Ranjha, M. M. A. N., Zeng, X. A., Ahmed, Z., & Aadil, R. M. (2022). High pressure-based hurdle interventions for raw and processed meat: A clean-label prospective. *International Journal of Food Science & Technology, 57*(2), 816–826.
256. Rosenthal, A., Deliza, R., Welti-Chanes, J., & Barbosa-Cánovas, G. V. (Eds.). (2018). *Fruit Preservation: Novel and Conventional Technologies.* Springer. https://doi.org/10.1007/978-1-4939-3311-2.
257. Ruiz-García, Y., & Gómez-Plaza, E. (2013). Elicitors: a tool for improving fruit phenolic content. *Agriculture, 3*(1), 33–52.
258. Sadraeian, M., Zhang, L., Aavani, F., Biazar, E., & Jin, D. (2022). Viral inactivation by light. *Elight, 2*(1), 1–18.
259. Sahu, M., & Bala, S. (2017). Food processing, food spoilage and their prevention: An overview. *International Journal of Life Sciences Scientific Research, 3*(1), 753–759.
260. Sajid, M., Mehmood, S., Yuan, Y., & Yue, T. (2019). Mycotoxin patulin in food matrices: Occurrence and its biological degradation strategies. *Drug Metabolism Reviews, 51*(1), 105–120.
261. Salemi, S., Saedisomeolia, A., Azimi, F., Zolfigol, S., Mohajerani, E., Mohammadi, M., & Yaseri, M. (2021). Optimizing the production of vitamin D in white button mushrooms (*Agaricus bisporus*) using ultraviolet radiation and measurement of its stability. *LWT, 137*, 110401.
262. Sánchez, G. J., Contigiani, E. V., Coronel, M. B., Alzamora, S. M., García-Loredo, A., & Nieto, A. B. (2021). Study of UV-C treatments on postharvest life of blueberries 'O'Neal'and correlation between structure and quality parameters. *Heliyon, 7*(6), e07190.
263. Sanchez-Moreno, C., De Ancos, B., Plaza, L., Elez-Martinez, P., & Cano, M. P. (2009). Nutritional approaches and health-related properties of plant foods processed by high pressure and pulsed electric fields. *Critical Reviews in Food Science and Nutrition, 49*(6), 552–576.

264. Setlow, B., Atluri, S., Kitchel, R., Koziol-Dube, K., & Setlow, P. (2006). Role of dipicolinic acid in resistance and stability of spores of *Bacillus subtilis* with or without DNA-protective α/β-type small acid-soluble proteins. *Journal of Bacteriology, 188*(11), 3740–3747.
265. Setlow, P. (2001). Resistance of spores of Bacillus species to ultraviolet light. *Environmental and Molecular Mutagenesis, 38*(2–3), 97–105.
266. Shama, G., & Alderson, P. (2005). UV hormesis in fruits: a concept ripe for commercialization. *Trends in Food Science & Technology, 16*(4), 128–136.
267. Shehata, S. A., Abdeldaym, E. A., Ali, M. R., Mohamed, R. M., Bob, R. I., & Abdelgawad, K. F. (2020). Effect of some citrus essential oils on post-harvest shelf life and physicochemical quality of strawberries during cold storage. *Agronomy, 10*(10), 1466.
268. Sheng, L., Zhang, Z., Sun, G., & Wang, L. (2020). Light-driven antimicrobial activities of vitamin K_3 against *Listeria monocytogenes*, *Escherichia coli* O157: H7 and *Salmonella enteritidis*. *Food Control, 114*, 107235.
269. Shipp, J., & Abdel-Aal, E. S. M. (2010). Food applications and physiological effects of anthocyanins as functional food ingredients. *The Open Food Science Journal, 4*(1), 7–22.
270. Si, S., Ahlstrom, D., Wei, J., & Cullen, J. (2021). Introduction: Business, entrepreneurship and innovation toward poverty reduction. In *Business, Entrepreneurship and Innovation Toward Poverty Reduction* (pp. 1–20). Routledge.
271. Silberbauer, A., & Schmid, M. (2017). Packaging concepts for ready-to-eat food: Recent progress. *Journal of Packaging Technology and Research, 1*, 113–126.
272. Singh, H., Bhardwaj, S. K., Khatri, M., Kim, K. H., & Bhardwaj, N. (2021). UVC radiation for food safety: An emerging technology for the microbial disinfection of food products. *Chemical Engineering Journal, 417*, 128084.
273. Singhal, R. S., Kulkarni, P. R., & Reg, D. V. (1997). *Handbook of Indices of Food Quality and Authenticity*. Elsevier.
274. Sinha, R. P., & Häder, D. P. (2015). Physiological aspects of UV-excitation of DNA. In *Photoinduced Phenomena in Nucleic Acids II: DNA Fragments and Phenomenological Aspects* (pp. 203–248).
275. Skowron, K., Bauza-Kaszewska, J., Dobrzański, Z., Paluszak, Z., & Skowron, K. J. (2014). UV-C radiation as a factor reducing microbiological contamination of fish meal. *The Scientific World Journal, 1*, 928094.
276. Skudra, A., Mezule, L., Spunde, K., Revalde, G., Zajakina, A., Ziņģe, M., & Juhna, T. (2021). Alternative UV light sources for surface disinfection. In *environment. Technologies. Resources. Proceedings of the International Scientific and Practical Conference* (Vol. 1, pp. 218–222).
277. Sofos, J. N. (2008). Challenges to meat safety in the 21st century. *Meat Science, 78*(1–2), 3–13.
278. Sofos, J. N. (2014). Meat and meat products. In *Food Safety Management* (pp. 119–162). Academic Press.
279. Sripong, K., Jitareerat, P., & Uthairatanakij, A. (2019). UV irradiation induces resistance against fruit rot disease and improves the quality of harvested mangosteen. *Postharvest Biology and Technology, 149*, 187–194.

280. Stephen, J., & Radhakrishnan, M. (2022). Avocado (*Persea americana* Mill.) fruit: Nutritional value, handling and processing techniques, and health benefits. *Journal of Food Processing and Preservation, 46*(12), e17207.
281. Stevens, C., Khan, V. A., Lu, J. Y., Wilson, C. L., Pusey, P. L., Igwegbe, E. C. K., & Droby, S. (1997). Integration of ultraviolet (UV-C) light with yeast treatment for control of postharvest storage rots of fruits and vegetables. *Biological Control, 10*(2), 98–103.
282. Stevens, C., Khan, V. A., Wilson, C. L., Lu, J. Y., Chalutz, E., & Droby, S. (2005). The effect of fruit orientation of postharvest commodities following low dose ultraviolet light-C treatment on host induced resistance to decay. *Crop Protection, 24*(8), 756–759.
283. Sun, Y., Qian, M., Wu, R., Niu, Q., Teng, Y., & Zhang, D. (2014). Postharvest pigmentation in red Chinese sand pears (*Pyrus pyrifolia* Nakai) in response to optimum light and temperature. *Postharvest Biology and Technology, 91*, 64–71.
284. Sun-Waterhouse, D. (2011). The development of fruit-based functional foods targeting the health and wellness market: A review. *International Journal of Food Science & Technology, 46*(5), 899–920.
285. Surjadinata, B. B., Jacobo-Velázquez, D. A., & Cisneros-Zevallos, L. (2021). Physiological role of reactive oxygen species, ethylene, and jasmonic acid on UV light induced phenolic biosynthesis in wounded carrot tissue. *Postharvest Biology and Technology, 172*, 111388.
286. Tandon, K., Worobo, R. W., Churey, J. J., & Padilla-Zakour, O. I. (2003). Storage quality of pasteurized and UV treated apple cider. *Journal of Food Processing and Preservation, 27*(1), 21–35.
287. Tang, T., Zhang, M., & Mujumdar, A. S. (2022). Intelligent detection for fresh-cut fruit and vegetable processing: Imaging technology. *Comprehensive Reviews in Food Science and Food Safety, 21*(6), 5171–5198.
288. Terao, D., de Carvalho Campos, J. S., Benato, E. A., & Hashimoto, J. M. (2015). Alternative strategy on control of postharvest diseases of mango (*Mangifera indica* L.) by use of low dose of ultraviolet-C irradiation. *Food Engineering Reviews, 7*, 171–175.
289. Thirumdas, R., Sarangapani, C., & Annapure, U. S. (2015). Cold plasma: a novel non-thermal technology for food processing. *Food Biophysics, 10*, 1–11.
290. Tian, X., Fan, S., Huang, W., Wang, Z., & Li, J. (2020). Detection of early decay on citrus using hyperspectral transmittance imaging technology coupled with principal component analysis and improved watershed segmentation algorithms. *Postharvest Biology and Technology, 161*, 111071.
291. Turatti, A. (2016). Process design and equipment for fresh-cut fruit and vegetable industry. In *Fresh-Cut Fruits and Vegetables* (pp. 169–220). CRC Press.
292. Turtoi, M. (2017). Ultraviolet light treatments. In *Novel Postharvest Treatments of Fresh Produce* (pp. 341–402). CRC Press.
293. Turtoi, M., & Borda, D. (2014). Decontamination of egg shells using ultraviolet light treatment. *World's Poultry Science Journal, 70*(2), 265–278.
294. Urban, L., Charles, F., de Miranda, M. R. A., & Aarrouf, J. (2016). Understanding the physiological effects of UV-C light and exploiting its agronomic potential before and after harvest. *Plant Physiology and Biochemistry, 105*, 1–11.
295. Urban, L., Sari, D. C., Orsal, B., Lopes, M. M. D. A., Miranda, R., & Aarrouf, J. (2018). UV-C light and pulsed light as alternatives to chemical and biological elicitors for stimulating plant natural defenses against fungal diseases. *Scientia Horticulturae, 235*, 452–459.

296. Üstün, H., & Doğan, A. (2022). The effect of different UV-C illumination doses on postharvest quality of fresh fig. *Gıda, 47*(5), 744–753.
297. Van Kleef, E., Van Trijp, H. C., & Luning, P. (2005). Functional foods: Health claim-food product compatibility and the impact of health claim framing on consumer evaluation. *Appetite, 44*(3), 299–308.
298. Vanhaelewyn, L., Van Der Straeten, D., De Coninck, B., & Vandenbussche, F. (2020). Ultraviolet radiation from a plant perspective: The plant-microorganism context. *Frontiers in Plant Science, 11*, 597642.
299. Vollmer, K., Chakraborty, S., Bhalerao, P. P., Carle, R., Frank, J., & Steingass, C. B. (2020). Effect of pulsed light treatment on natural microbiota, enzyme activity, and phytochemical composition of pineapple (*Ananas comosus* [L.] Merr.) juice. *Food and Bioprocess Technology, 13*, 1095–1109.
300. Waldrop, M. M. (2015). Why we are teaching science wrong, and how to make it right. *Nature, 523*(7560), 272–274.
301. Walters, E., Graml, M., Behle, C., Müller, E., & Horn, H. (2014). Influence of particle association and suspended solids on UV inactivation of fecal indicator bacteria in an urban river. *Water, Air, & Soil Pollution, 225*, 1–9.
302. Wang, Q., Chu, L., & Kou, L. (2017). UV-C treatment maintains quality and delays senescence of oyster mushroom (*Pleurotus ostreatus*). *Scientia Horticulturae, 225*, 380–385.
303. Wang, X., Ouyang, Y., Liu, J., Zhu, M., Zhao, G., Bao, W., & Hu, F. B. (2014). Fruit and vegetable consumption and mortality from all causes, cardiovascular disease, and cancer: Systematic review and dose-response meta-analysis of prospective cohort studies. *BMJ, 349*, g4490.
304. Ward, D. M. G. (2018). *Treatment of Ultra-Low UV Transmittance Fluids: Fundamentals and Applications*. Doctoral dissertation, Tennessee State University.
305. Wituschek, A. (2020). *Phase-Modulated Coherent Time-Domain Spectroscopy in the Extreme Ultraviolet Regime*. Doctoral dissertation, Universität Freiburg.
306. Woo, H., Beck, S. E., Boczek, L. A., Carlson, K. M., Brinkman, N. E., Linden, K. G., & Ryu, H. (2019). Efficacy of inactivation of human enteroviruses by dual-wavelength germicidal ultraviolet (UV-C) light emitting diodes (LEDs). *Water, 11*(6), 1131.
307. World Health Organization. (2015). *Food Safety: What You Should Know (No. SEA-NUT-196)*. World Health Organization.
308. Wu, X., Guan, W., Yan, R., Lei, J., Xu, L., & Wang, Z. (2016). Effects of UV-C on antioxidant activity, total phenolics and main phenolic compounds of the melanin biosynthesis pathway in different tissues of button mushroom. *Postharvest Biology and Technology, 118*, 51–58.
309. Zhu, Y. (2014). *Modeling of Patulin Degradation in Apple-based Beverages by Multiple Wavelengths of Ultraviolet-C Irradiation*. Doctoral Dissertation, University of Guelph.
310. Yahia, E. M., Gardea-Béjar, A., de Jesús Ornelas-Paz, J., Maya-Meraz, I. O., Rodríguez-Roque, M. J., Rios-Velasco, C., & Salas-Marina, M. A. (2019). Preharvest factors affecting postharvest quality. In *Postharvest Technology of Perishable Horticultural Commodities* (pp. 99–128). Woodhead Publishing.
311. Yan, R., Yun, J., Gurtler, J., & Fan, X. (2017). Radiochromic film dosimetry for UV-C treatments of apple fruit. *Postharvest Biology and Technology, 127*, 14–20.

312. Yang, B., Lee, C., Xiang, W., Xie, J., He, J. H., Kotlanka, R. K., & Feng, H. (2009). Electromagnetic energy harvesting from vibrations of multiple frequencies. *Journal of Micromechanics and Microengineering, 19*(3), 035001.
313. Yang, J., Li, B., Shi, W., Gong, Z., Chen, L., & Hou, Z. (2018). Transcriptional activation of anthocyanin biosynthesis in developing fruit of blueberries (*Vaccinium corymbosum* L.) by preharvest and postharvest UV irradiation. *Journal of Agricultural and Food Chemistry, 66*(42), 10931–10942.
314. Yang, J., Shi, W., Li, B., Bai, Y., & Hou, Z. (2019). Preharvest and postharvest UV radiation affected flavonoid metabolism and antioxidant capacity differently in developing blueberries (*Vaccinium corymbosum* L.). *Food Chemistry, 301*, 125248.
315. Yang, S., Sadekuzzaman, M., & Ha, S. D. (2017). Reduction of *Listeria monocytogenes* on chicken breasts by combined treatment with UV-C light and bacteriophage ListShield. *LWT, 86*, 193–200.
316. Yang, X., Kalchayanand, N., Belk, K. E., & Wheeler, T. L. (2019). Photohydroionization reduces Shiga toxin-producing *Escherichia coli* and *Salmonella* on fresh beef with minimal effects on meat quality. *Meat and Muscle Biology, 3*(1), 105–115.
317. Yeh, Y., De Moura, F. H., Van Den Broek, K., & De Mello, A. S. (2018). Effect of ultraviolet light, organic acids, and bacteriophage on Salmonella populations in ground beef. *Meat Science, 139*, 44–48.
318. Yildiz, S., Pokhrel, P. R., Unluturk, S., & Barbosa-Cánovas, G. V. (2019). Identification of equivalent processing conditions for pasteurization of strawberry juice by high pressure, ultrasound, and pulsed electric fields processing. *Innovative Food Science & Emerging Technologies, 57*, 102195.
319. Yousuf, B., Qadri, O. S., & Srivastava, A. K. (2018). Recent developments in shelf-life extension of fresh-cut fruits and vegetables by application of different edible coatings: A review. *LWT, 89*, 198–209.
320. Yuan, L., Sadiq, F. A., Wang, N., Yang, Z., & He, G. (2021). Recent advances in understanding the control of disinfectant-resistant biofilms by hurdle technology in the food industry. *Critical Reviews in Food Science and Nutrition, 61*(22), 3876–3891.
321. Yun, J., Yan, R., Fan, X., Gurtler, J., & Phillips, J. (2013). Fate of *E. coli* O157: H7, *Salmonella* spp. and potential surrogate bacteria on apricot fruit, following exposure to UV-C light. *International Journal of Food Microbiology, 166*(3), 356–363.
322. Zhang, D., Yu, B., Bai, J., Qian, M., Shu, Q., Su, J., & Teng, Y. (2012). Effects of high temperatures on UV-B/visible irradiation induced postharvest anthocyanin accumulation in 'Yunhongli No. 1'(*Pyrus pyrifolia* Nakai) pears. *Scientia Horticulturae, 134*, 53–59.
323. Zhang, W., & Jiang, W. (2019). UV treatment improved the quality of postharvest fruits and vegetables by inducing resistance. *Trends in Food Science & Technology, 92*, 71–80.
324. Zhang, X., Zhang, M., Xu, B., Mujumdar, A. S., & Guo, Z. (2022). Light-Emitting Diodes (below 700 nm): Improving the preservation of fresh foods during postharvest handling, storage, and transportation. *Comprehensive Reviews in Food Science and Food Safety, 21*(1), 106–126.
325. Zhang, Z. H., Wang, L. H., Zeng, X. A., Han, Z., & Brennan, C. S. (2019). Non-thermal technologies and its current and future application in the food industry: a review. *International Journal of Food Science & Technology, 54*(1), 1–13.
326. Zhao, W., Dai, J., Liu, F., Bao, J., Wang, Y., Yang, Y., & Zhao, D. (2012). Photocatalytic oxidation of indoor toluene: Process risk analysis and influence of relative humidity, photocatalysts, and VUV irradiation. *Science of the Total Environment, 438*, 201–209.

327. Zhao, W., Yang, Y., Dai, J., Liu, F., & Wang, Y. (2013). VUV photolysis of naphthalene in indoor air: Intermediates, pathways, and health risk. *Chemosphere, 91*(7), 1002–1008.
328. Zhao, Y., Zhu, X., Hou, Y., Wang, X., & Li, X. (2019). Effects of nitric oxide fumigation treatment on retarding cell wall degradation and delaying softening of winter jujube (*Ziziphus jujuba* Mill. cv. Dongzao) fruit during storage. *Postharvest Biology and Technology, 156*, 110954.
329. Zhong, L., Carere, J., Lu, Z., Lu, F., & Zhou, T. (2018). Patulin in apples and apple-based food products: The burdens and the mitigation strategies. *Toxins, 10*(11), 475.
330. Zhou, D., Sun, Y., Li, M., Zhu, T., & Tu, K. (2019). Postharvest hot air and UV-C treatments enhance aroma-related volatiles by simulating the lipoxygenase pathway in peaches during cold storage. *Food Chemistry, 292*, 294–303.
331. Zhou, D., Wang, Z., Tu, S., Chen, S., Peng, J., & Tu, K. (2019). Effects of cold plasma, UV-C or aqueous ozone treatment on *Botrytis cinerea* and their potential application in preserving blueberry. *Journal of Applied Microbiology, 127*(1), 175–185.

PART III

PULSED LIGHT (PL) TECHNOLOGY: ADVANCES IN FOOD PROCESSING

CHAPTER 7

INTRODUCTION TO PULSED LIGHT: PRINCIPLES, MECHANISM OF ACTION, ADVANTAGES, LIMITATIONS, AND APPLICATIONS

N. U. SRUTHI, PAVULURI SRINIVASA RAO, and
REWATI RAMAN BHATTARAI

ABSTRACT

Pulsed light (PL) is an emerging non-thermal processing method for decontaminating food products (such as fresh fruits, vegetables, meat items, fruit juices, etc.) using high-intensity light pulses with an intense broad spectrum for a short duration of time. The technique uses high-power pulses of radiation with wavelengths ranging from 200–1,100 nm, which includes ultraviolet, visible, and near-infrared regions. The current chapter reviews the pulsed light technology in the context of the critical fundamentals, the theories and principles that govern its action, associated food safety aspects, and its various applications in food. The chapter also provides a brief outline of the effect of pulsed light processing on food quality attributes, advantages and accompanying limitations, safety concerns, and future perspectives.

7.1 INTRODUCTION

With the increasing interest in processed foods and rising knowledge and awareness among consumers regarding the quality and safety of food, novel non-thermal technologies are being adopted in the market to minimally process the food and thereby preserve the nutritional and organoleptic

Nonthermal Light-Based Technologies in Food Processing.
C. K. Sunil, Preeti Birwal, Megh R. Goyal, and R. Mahendran (Eds.)

attributes. Pulsed Light (PL) processing is one such nonthermal technique gaining popularity in the food industry where cycles of high energy (0.01 to 50 J/cm^2) short time (1 μs to 0.1 s) pulses in random sequence are used as mild treatment conditions for achieving microbial decontamination and other functional changes in food products [42].

The US Food and Drug Administration (FDA) has sanctioned using PL in the processing, packaging, and handling stages of foods since 1996 till an aggregate dosage of 12 J/cm^2 and pulse duration of less than 2 ms. The scientific principle behind pulsed light processing is grounded on the build-up of high-discharge voltage in a capacitor and subsequent delivery of the stored energy in rapid and prompt pulses through a radiance source sealed with xenon gas. High-power electric pulses are converted to high-power radiation pulses using inert gas flash lamps, which result in a wavelength range that corresponds to a spectrum that covers ultraviolet (UV) (200–400 nm), visible light (VIS) (400–700 nm), and near-infrared region (IR) (700–1,100 nm) regions, hence being given several nomenclatures such as pulsed UV light, pulsed white light, intense light pulses, high-intensity broad-spectrum pulsed light, etc. [65].

The primary mechanism behind microbial inactivation is the UV-C induced dimerization of nucleic acids present in the microbes, which inhibits transcription and replication, causing cellular inactivation [25]. With better and much more promising results than continuous UV treatment in decontaminating microorganisms, PL has found other significant applications in the food industry, which include enzyme inactivation [28], food package decontamination [11], bioactive extraction [31], etc. Furthermore, being a gentle yet targeted processing technique, PL has also found application in modifying the functional properties and structure of proteins, thereby improving certain surface and foaming properties [60]. With such potential in the food industry in producing high-value-added ingredients, the present chapter aims to corroborate the recent advances in pulsed light processing, its mechanism of action, impact on food quality, and associated limitations. The chapter details potential applications of PL and associated health risks and regulatory concerns.

7.2 PRINCIPLE OF PULSED LIGHT

The use of continuous UV in food processing and medicinal applications, for instance, microbial inactivation and surface decontamination, dates back to the early 1900s. However, investigations on light pulses began in the 1970s.

Nevertheless, it was only in 1996 that the technique was adopted in the food industry following its approval by the Food and Drug Administration authority [63]. As mentioned, PL processing entails continuous repetition of short-duration pulses of high-power wideband emission light (200–1,100 nm) with a significant volume of energy in the short-wave UV region (Figure 7.1). Nonetheless, the peak power generated using pulses is much superior to that provided in equivalent continuous UV energy input, giving PL a higher penetrating capability and energetic efficiency [55].

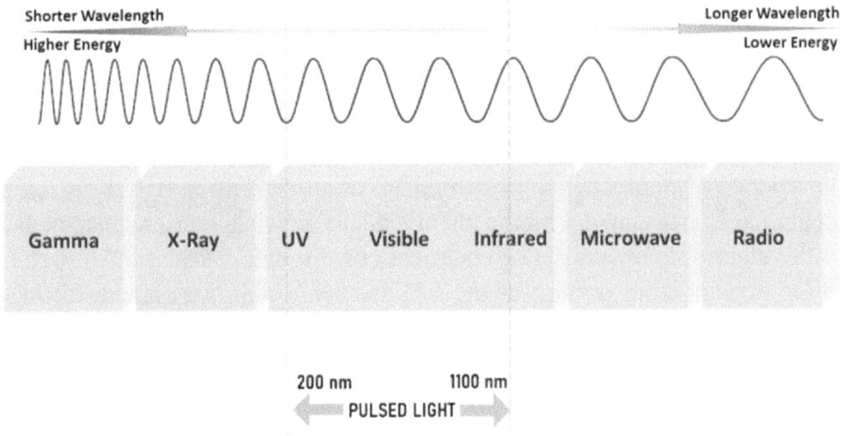

FIGURE 7.1 Spectrum of electromagnetic radiation showing the pulsed light region.

To understand the principle behind the impact of PL on different quality aspects and the safety of food, a broad discussion on electromagnetic (EM) emission and illumination sources is a requisite, enabling a more perceptive interpretation of the outcomes of pulsed light processing. Light is a form of electromagnetic radiation that includes a wide range of wavelengths. The EM spectrum comprises frequencies and corresponding energies ranging from radio waves to gamma rays, the source being piezoelectric or magnetostrictive transducers to radioactive decay, respectively. There exists no single universal source that practically generates all EM radiations [54]. Additionally, each mode of radiation has a distinct way of expression in terms of energy, wavelength, and frequency, where the ionizing radiation is expressed in energy terms, wavelength units are used to express the range from UV to IR, and microwaves in frequency units.

The emission of radiant energy from any radiation source is the result of the descending transitions between two energy states when the valence electrons or nucleons experience a change in configuration. For transitioning from a stable ground level to a temporary excited state, the system requires additional energy, which is electrical energy in the case of inert gas flash lamps. However, these molecules prefer to remain in the stable low energy, and in the process of this transition to the ground state, a discrete amount of energy is released in the form of photon annihilation, and this phenomenon is designated as the absorption of radiation which is also accompanied by heat and induction of certain chemical changes in the target matrix.

The principle behind the lethal effect of PL can be attributed to this basic phenomenon of energy dissipation which can be described by three pathways, namely, thermal relaxation as a result of molecular motion, luminescence due to emission of radiation, and chemical modifications induced by reorganization of the molecule's architecture. In other words, it can be summarized that the science of photochemistry behind PL or any radiation-based processing technique is associated with the chemical and physical consequences related to the electronic excitation of chemical constituents. However, PL belongs to the non-ionizing portion of the EM spectrum that does not result in the ionization of small molecules and exhibits a peak of normalized intensity between 400 and 500 nm [48]. In theory, the spectrum of PL can be seen as similar to sunlight but also includes fractions of UV light blocked by the ozone layer of the Earth's atmosphere. Specifically, unlike sunlight, which is abundant in visible regions of the spectrum, PL is rich in UV, contributing to about 54% of the energy. With a comparable peak emission to sunlight, each flash of PL is estimated to deliver an intensity per unit time in the order of 20,000 times more pronounced than that of sunlight reaching the Earth's surface [12].

When the light of such high intensity is incident on any food substance, the energy absorption by the matter occurs according to the Beer-Lambert law. It is stated that the transmitted intensity depends on the initial intensity, the thickness of the material, and the optical properties of the substance, as expressed by the following equation:

$$I(x) = (1 - r)I_o e^{-aX} \tag{1}$$

where; I(X) is the radiation intensity at a depth X beneath the material surface; I_0 is the incident intensity hitting the product surface; *r* is the product reflection coefficient; and α describes the extinction coefficient of the material at a given wavelength. The α value is 1 for most solids that are opaque transmitting no incident radiation, while it is 0 for many liquids and gases that do not absorb any part of the energy.

From Eqn. (1), it is apprehended that the intensity of light decreases exponentially when light is penetrated into the food matrix. Since most food products are opaque, PL has a short penetration confining its applications to surface disinfection of fresh produce and treatment of transparent fluids. Generally, PL shows a penetration depth of about 10 mm and the food topography, opacity, and flow characteristics play a key aspect [43]. Also, for food applications, the light pulses discharge 1–20 flashes in a second with fluencies between 0.01 and 50 J cm^{-2} at the surface.

7.3 ROLE OF PULSED LIGHT TECHNIQUE IN FOOD SAFETY

With the rising popularity of fresh-cut produce and minimal processed foods, novel strategies have been investigated to combat the microbial safety hazards and shelf-life constraints of food and food contact surfaces at the handling and processing locations. Traditionally, chemical or synthetic preservatives and thermal processing methods have been utilized in the food industry to ensure food safety. Processing food using pulsed light has earned relevance in recent years as an alternative preservation method due to its effective inactivation kinetics on several pathogenic organisms like bacteria, yeasts, molds, and even viruses. The past decade has shown a tremendous rise in research and investigations aiming to document and demonstrate the efficacy of PL in decontaminating several food products and enhancing microbiological safety. PL also serves as a viable alternative to chemical sterilizers for decontaminating food packaging materials and contact surfaces and congruently improving food safety.

Table 7.1 shows a recent summary of published studies on Pulsed Light treatment of different food products with respect to attained microbial reductions and critical processing parameters. It is to be noted that in most of the research, the food product was artificially inoculated with the test bacteria. Some studies also focused on investigating the inactivation efficiency of the native microflora. It is to be remarked that the native microbial populations typically demonstrate different behavior in comparison to inoculated microorganisms on produce surfaces following Pulsed Light exposure in the course of storage due to their adaptation. Nevertheless, the differences between product properties and the operating conditions for different studies render any comparison difficult owing to the variability in the obtained results.

TABLE 7.1 Selected Studies Reporting the Impact of Pulsed Light Processing on Different Food Commodities and Microbial Reduction Levels

Food Product	Opt. Treatment Conditions	Pathogens Studied	Microbial Reduction	Food Quality Changes	References
Cheese	D = 5.8 cm n = 1–12 F = 1.02–12.29 J/cm^2	*Pseudomonas fluorescens* *Escherichia coli* *Listeria innocua*	2–3.73 log CFU/g	NA	[52]
Endive salad	V = 3 kV F = 1 J/cm^2 D = 10 cm	*Escherichia coli* *Listeria innocua*	2 log CFU/g	Pronounced discoloration, increased respiration activity.	[35]
Fresh cut tomato	D = 8.5 cm F = 4, 6, 8 J/cm^2	*Listeria innocua* *Escherichia coli*	0.9 log CFU/g 1.4 log CFU/g	Accelerated O_2 consumption during cold storage.	[64]
Fresh-cut apple	D = 8.5 cm n = 20 F = 8 J/cm^2	Mesophilic and psychrophilic aerobic counts.	1.55 log CFU/g	Increased dosage affected firmness and color.	[39]
Fresh-cut cucumber	D = 10.8 cm t = 12.4 s F = 12 J/cm^2	*Escherichia coli*	2.8 log CFU/g	NA	[61]
Fresh-cut red bell pepper	D = 7.58 cm V = 2.1 kV F = 16 J/cm^2	*Listeria monocytogenes* *Salmonella enterica*	1.59–2.13 log CFU/g	Higher fluencies caused changes in bioactive compounds during storage.	[56]
Goat meat Beef	D = 4.47 cm t = 60 s F = 1.27 J/cm^2	*Escherichia coli*	1.66 log CFU/mL 1.74 log CFU/mL	NA	[7]
Ham slice	t = 1 mm V = 2.5 kV F = 8 J/cm^2	*Listeria innocua*	1–2 log CFU/cm^2	No change in peroxide and rancid notes. Increased conc. of volatile compounds.	[16]
Mung bean sprout	V = 3 kV F = 1 J/cm^2 D = 10 cm	*Escherichia coli* *Listeria innocua*	1.6 log CFU/g	Nil	[35]
Pork skin Pork loin	D = 8.3, 10.8, 13.4 cm. t = 1–30 s F = 0.52–19.11 J/cm^2	*Salmonella typhimurium* *Yersinia enterocolitica*	1.73 to 3.16 log CFU/g 1.48 to 4.37 log CFU/g	Higher fluencies provoked color changes and fatty smell.	[33]
Raspberry	D = 13 cm t = 5 s F = 5 J/cm^2	*Salmonella* *Escherichia coli*	3.4–3.6 log CFU/g	Higher exposure time affected color and texture.	[69]

TABLE 7.1 *(Continued)*

Food Product	Opt. Treatment Conditions	Pathogens Studied	Microbial Reduction	Food Quality Changes	References
Raw salmon	D = 3.5 cm t = 9 s F = 152.6 mJ/cm^2	*Pseudomonas* spp. *Brochothrix* spp. *Listeria* spp. *Photobacterium* spp.	0.6–1.6 log CFU/g	Higher dosage caused cooked appearance.	[49]
Rough rice	D = 13 cm t = 21 s V = 3.8 kV I = 1.08 W/cm^2	*Aspergillus flavus*	0.29 log CFU/g	Improved rice drying efficiency and milling quality.	[67]
Sesame seed	t = 120 s V = 2.4 kV F = 44.46 J/cm^2	Overall bacteria, coliform group, and yeast/mold.	1.46 log CFU/g	NA	[26]
Shelled walnut	D = 8 cm t = 45 s F = 5.6 J/cm^2	*Salmonella Enteritidis*	3.18 log CFU/g	Insignificant changes in the color and lipid oxidation.	[27]
Spinach	D = 8.5 cm n = 5 F = 8 J/cm^2	*Listeria innocua* *Escherichia coli*	1.85 log CFU/g 1.72 log CFU/g	Increase in total phenolic content and antioxidant capacity.	[3]
Whey	t = 2.15 mm f = 5 L/min V = 1–3 kV F = 11 J/cm^2	*Listeria innocua*	5 log CFU/g	Shelf-life extension up to 7 days.	[5]
Whole egg	D = 15 cm n = 2 F = 1.32 J/cm^2	*Escherichia coli*	3.77 log CFU/egg.	Decomposition of cuticle layer, decrease in yolk index, increase in albumen pH, and weight loss with storage.	[68]

Legend: D: distance between sample and lamp; n: number of pulses; F: fluence; V: applied voltage; t: exposure time; I: intensity; CFU: colony-forming units; and NA: not available.

The application of pulsed light with fluencies ranging from 0.01–50 J/cm^2 is capable of causing 9 logarithmic reductions of the microbial population for vegetative microorganisms and over 7 log cycle reduction of bacterial spores on even and smooth surfaces such as those of packaging materials [48]. Such higher microbial inactivation efficacy of PL on smooth surfaces can be attributed to the lower penetration of PL, which mostly limits it to surface inactivation where the microbial population will be entirely exposed to flashes of light. However, the surface of most food products is not always smooth and even. It typically contains minute cracks, crevices, fissures, folds, and irregularities, rendering space for microbes to hide, which restricts light exposure. Such surface asymmetries and stacking up microbial cells on top of one another shield the sub-surface layers from highly directional coherent PL beam resulting in a shadow effect. The occurrence of shadowing has been widely documented in many studies involving the treatment of fresh horticultural produce, cereals, oilseeds, and meat products, including cabbage, sesame seeds, pork skin, sliced boiled ham, etc. [26, 32, 33, 36].

Additionally, impurities or large particles on food surfaces reflect and scatter the incident photons, consequently minimizing the energy available for microbial inactivation. In such situations, the achievable microbial load reduction is only of the order of 2–3 log cycles regardless of the fluence applied. Addition of devices in the treatment chamber of the PL system that can create multidirectional pulsed light or system configurations that can allow random movement or rotation of the treated product can, to a certain extent, increase the uniformity of surface exposure and eradicate any sheltered microorganisms [18].

The United States (US) Food and Drug Administration (FDA) recommended safety regulation for fruit and vegetable juices implies a 5-log cycle reduction in microbes and considerable microbial reduction in a very short exposure time with better process flexibility is another advantage of PL as far as food safety of beverages and liquid products are concerned. In such products, much deeper penetration can be achieved depending on the optical properties and flow characteristics. It is seen that turbid liquid food with a large amount of suspended matter significantly scatters the light, limiting the process efficiency [6]. The depth of penetration is decided by the turbidity, transmittance, and total soluble solids of liquids, as was observed in a study where fruit juices with different matrix properties were treated with PL to establish the inactivation efficacy against *Escherichia coli* [51].

The highest microbial reduction was achieved in coconut water, having maximum transmittance to light, lower turbidity, and fewer soluble solids than orange and pineapple juice. Similarly, rheological and flow properties like viscosity, density, and film thickness also affect microbial inactivation efficacy of PL when designing a continuous flow treatment system. Also, with the microbes dwelling in the whole volume of liquid, it is necessary to ensure uniform exposure of the entire liquid bulk to PL for maximum decontamination.

The status of food safety by any decontamination technique can only be authenticated when the attained microbial reduction through initial processing is maintained for the entire period of storage, depending on the specific food characteristics and shelf life. Thus, microbiological trials under specific storage conditions, including the assessment of quality attributes, are recommended since significant changes are mostly not found straightaway after PL exposure but in the course of storage [37].

Furthermore, such trials should be able to track the reoccurrence of microbial load concerning lag phases and growth rates. Customarily, with an increased degree of robustness of the remaining microbes that persist after the PL treatment even after a significant reduction in the number of competitors, an enhanced rate of microbial proliferation is normally observed for the duration of storage [37, 69]. Nonetheless, it should also be mentioned that the failure to prolong the microbial safety throughout the entire storage period does not essentially indicate the inadequacy of PL but can also be a corollary of the storage environments. PL can be a potential technique among other minimal processing technologies for effectively tackling food safety issues.

7.4 MECHANISM OF ACTION

There has been considerable research on the mechanism of action of pulsed light, especially its lethal effects, for exploiting the use in food decontamination. The pulsed light processing method involves using engineering technologies to multiply power by many folds by releasing the accumulated energy in a storage capacitor in a much shorter time, thus magnifying the applied power. The efficacy of PL in any food application can be attributed to this instantaneous high power duty cycle. With the effects varying concerning different wavelengths, a broad spectrum of high-intensity wavelengths is applied for food treatment.

As depicted in Figure 7.1, much higher energies are achieved at shorter wavelengths, typically in the UV region of the PL spectrum; thus, the lethal outcomes of PL can be accredited predominantly to the rich broad-spectrum UV content [57]. Moreover, the short-duration high peak power delivery and the adeptness to control both the pulse length and the output frequency of xenon-based lamps play a major part in heightening the lethal effect. Theoretically, the damages or changes induced by PL in the food or microbial load can be categorized into three coexisting mechanisms: photochemical, photothermal, and photophysical effects, making it a multi-target process [6].

7.4.1 PHOTOCHEMICAL EFFECT

Although xenon flash lamps emit a broad spectrum of radiation stretching from UV to IR, the UV-C (200–280 nm) part of the spectrum with a theoretical inactivation peak at 253.7 nm is the most important for microbial destruction due to the absorption of photons of light at this peak wavelength by the microbial DNA or RNA [65]. This was confirmed in a study where light filters were placed to block out UV and transmit only the visible and infrared portion of the light to understand the microbial inactivation kinetics in the absence of UV spectrum [8]. It was observed that visible (VIS) and near-IR radiation contributed significantly less to total microbial destruction and caused negligible reductions after 15 pulses compared to the total inactivation observed for the unfiltered broad-spectrum PL.

The germicidal effect of the UV spectrum of PL is predominantly due to the light-induced chemical transformation of pyrimidine bases in the microbial DNA to form dimers which inhibit cell multiplication and eventually result in cell death. This was elucidated in a study where the possible photochemical effects of PL that caused loss of cultivability of *Listeria innocua* were investigated by detecting the formation of reactive oxygen species (ROS) and DNA damage by ROS-sensitive 2′7′-dichlorofluorescin diacetate probe and random amplified polymorphic DNA (RAPD) polymerase chain reaction (PCR) marker, respectively [34]. It was found that alterations were observed in the banding patterns of isolated genomic DNA of PL-treated strain, and the magnitude intensified with increasing fluence. Also, the change in band patterns was attributed to double-strand breaks, single-strand breaks, and the formation of different photo products like cyclobutane-pyrimidine dimers (CPDs), 6–4 photoproducts, and Dewar valence isomers. Such alterations in the bond structure prevent the unzipping of DNA for replication rendering the organism incompetent for reproduction, causing mutations and impaired gene transcription, and eventually causing its death.

The UV portion is also absorbed by the highly conjugated carbon-carbon double bonds (C=C) of DNA, protein molecules, and nucleic acids, which also results in structural alterations [50]. These mutations are often proven irreversible under different experimental conditions due to the severe damage caused by UV, making the repair mechanisms challenging to operate, unlike the changes that occur during conventional UV processing. The photochemical effects were also proven detrimental to different fungal contamination in food products. In many foods, secondary metabolites of fungi such as aflatoxins are potent toxigenic, carcinogenic, and immunosuppressive compounds. It was established that the photochemical effects of photons increase with an increase in water content of the matrix owing to the opening of the lactone ring in aflatoxins that form a carboxylic acid in the terminal portion of the ring [2]. The chemical changes induced include fragmentation of the mycotoxin molecules, mutagenic activity, detoxification of aflatoxins, and disruption of the terminal furan double bond.

The application of light flashes and the administration of certain photoactive compounds called photosensitizers have been reported to amplify the photochemical effect of PL [42]. These compounds are positively charged and can attach or accumulate in the targeted microorganisms. Upon absorption of light, certain ROS such as singlet oxygen are generated, which is accompanied by several cytotoxic reactions in the cell leading to disruption of the cellular membrane, enzyme inactivation, DNA damage, and eventually cell death. Furthermore, the occurrence of multiple positive charges allows the photosensitizer compound to interact with the negatively charged outer cell wall parts of bacteria, particularly with the lipopolysaccharides of the Gram-negative bacteria inducing irreversible oxidative damage to the pathogen [41].

7.4.2 PHOTOTHERMAL EFFECT

The lethal action of PL can also be attributed to the temperature rise of the exposed surface upon heat dissipation when higher flux densities are achieved. This ultimately triggers the photothermal mechanism to intercede with longer exposure times, instigating some heat to leak out from the xenon lamp to the surroundings and the treated sample. The sample heating upon prolonged exposure to PL can also be attributed to the presence of the near-infrared domain in the broad spectrum of PL. The observed temperature rise can create a temperature gradient (Eqn. (2)) between the surface of the product and its internal layers, setting up conductive heat transfer within the material.

$$\frac{\mathrm{d}T}{\mathrm{d}d} = \frac{I_d}{kA}$$

$$I_d = I(x)(1 - e^{-[a]d}) \tag{2}$$

where; I_d is the energy absorbed by the sample from the transmitted energy beneath the distance x from the surface till a depth d; α is the extinction coefficient; A is the product surface area; and k is the thermal conductivity of the treated product [42].

With a rise in temperature, considerable damage to the cell membrane and microbial proteins occurs, accompanied by the expansion of vacuoles [33]. The significant differences in the absorption of the cell and the surrounding matrix are evident to be the cause of the cell rupture. Also, the possibility of overheating is certain as the total rate of cooling is lower than the overall energy that is discharged into the cell [23]. This causes spontaneous evaporation of the water present in the cell in the form of micro-bubbles causing a rupture of the functional membranes. It will eventually lead to the leakage and emission of cell constituents causing possible disintegration of the organisms into the matrix on attaining a particular surface temperature. Concurrently, in the absence of UV light, the primary cause of destruction is the photothermal effect. The sudden increase in temperature of the intracellular fluid of bacterial cells and the associated loss in membrane integrity leading to flattening out of the cells from the edges has been reported in a study on the inactivation efficacy of PL on *Salmonella* and *Escherichia coli* in fresh raspberries [69]. They attributed these changes to the intercellular water vaporization in a relatively short time induced by the photothermal stress.

Despite the fact that a photothermal effect occurs locally on biological structures, with the temperature escalating up to 130C causing prompt heating of microbial membranes, the overall impact of the thermal effect on specific food matrices can be considered insignificant [59]. These effects increase the efficiency of pulsed light (PL) compared to continuous UV due to the combined chemical and thermal effects resulting in a "multi-hit target" approach. It is presumed that the shoulders of survival curves are associated with DNA damage and repair events. Repair mechanisms can mend the damage encountered up to specific UV dosages resulting in shoulders [29]. With much higher dosages, the DNA repair capability reaches its threshold, leading to an exponential decline in survivors and, ultimately, death. However, different studies have shown that temperature rise due to light pulses is considered to be a lesser causative factor in microbial destruction compared to the photochemical effects. Due to the short duration of action of PL and relatively limited exposure with a dosage restricted to 12 J/cm^2, there is no possibility of a steady-state thermal conduction. This is because only a thin superficial surface layer becomes heated rapidly, leaving the inner layers of the treated product unaffected, in contrast to the overall high temperature achieved with continuous approaches. Therefore, it can be asserted that PL treatment is a non-thermal processing technique as the inner layers of the product matrix do not experience any rise in temperature.

However, prolonged exposure to PL may materialize as a drawback of this technique. As a result, it is suggested to cool the exposed sample for the duration of PL processing or immediately after the treatment to prevent the temperature-induced quality deterioration of the product [13]. The textural and colorimetric properties were evident in fresh-cut mushrooms treated with PL at high fluencies (28 J/cm^2) [46]. Textural changes can be attributed to the structural and physicochemical changes resulting from dehydration, making the tissue more flattened and causing it to collapse. The senescent breakdown of intracellular membranes at a high temperature can also increase the incidence of browning, affecting the product's color. Similar changes in total profile analysis, dynamic-creep behavior, and the browning index were witnessed in fresh-cut apples treated by PL at higher fluences (100 s) that can dramatically affect the product quality [21].

In contrast, an increase in fruit firmness was noticed in fresh-cut mangoes treated with PL at lower fluence (8 J/cm^2) owing to the consecutive moderate heating which induces the formation of a delicate dried film over the product surface preventing any physiological degradation [10]. While considering the photothermal effects of PL, it is also important to examine the heat-resistant properties of the packaging materials in addition to the light transmission properties, as PL is also a potent sterilizing technique for food packaging materials. PL being pronounced as a non-thermal technique, a high melting point is not considered a requirement for packaging systems; however, the surface heating phenomenon accompanied by PL exposure can cause changes in the polymer structure of the packaging material [43]. Moreover, most foods processed by PL are commercialized at colder processing temperatures which mandate the packaging material to withstand the cooling conditions for the period of storage, transport, and marketing.

7.4.3 PHOTOPHYSICAL EFFECT

Pulsed light is also anticipated to result in certain physical disruption of the cellular structures of the microbes due to the intermittent but high-energy pulses. These physical alterations include single-strand breakage of RNA, damage of virion structure, damage to the cell membrane, protein elution, breakage of viral proteins, expansion of vacuole, and eventually the deformation and death of cells [13, 66]. Such structural changes can be induced due to the photochemical and photothermal consequences of PL.

The photophysical effect can also be seen as the water vapor-induced damages in the cell, such as cell wall disruption, shrinkage of the cytoplasmic membrane, mesosome rupture, and leakage of cellular contents and genetic materials. The concentration of eluted proteins quantifies the severity or extent of cell damage. The transmission electron microscopy micrographs of *Escherichia coli* and *Listeria innocua* of PL-treated fresh-cut mushrooms elucidated considerable damage to cell cytoplasm with agglutination of cytoplasmic content observed when exposure was done at 6 J/cm^2 while more changes in cytoplasm membrane being evident at 12 J/cm^2 [53].

The photophysical effect of PL can also be a contributing factor behind allergen and antibody binding reduction in tree nuts like almonds, where cross-linking and/or fragmentation were observed in the protein extracts upon PL exposure [38]. In addition, major changes in allergen conformation could be observed, causing protein aggregation resulting in a reduced antibody-binding capacity as a result of the altered epitopes. This is also accompanied by molecular ionization as a result of the high-energy burst, which leads to modification of epitopes, thus decreasing the immunoreactivity [44]. These structural damages are also prompted by the regular disturbance in the food matrix caused by the high-energy pulses.

It is also worth noting that these physical effects or damages caused by PL are much more intense and higher than those caused by continuous UV [19]. However, few incidences of nutritional disparities have been found because of the photophysical effect where off-odor was observed in the PL-treated fresh-cut apple samples when kept for storage for up to 14 days [45].

These major mechanisms are accountable for the destruction of microorganisms and other disinfection aspects independently or in combination, as influenced by the processing conditions, the target organisms, and the food matrix. Although photothermal and photophysical effects might influence inactivation, a majority of works point out the photochemical impact of PL to be the major contributing factor that causes microbial destruction.

7.5 HIGH-INTENSITY LIGHT PULSES FOR FOOD APPLICATIONS

With the first patent (Patent No. US4464336 A, 1984) for successful application in microbial inactivation dating back to the 1980s [65] and attaining US FDA approval only in the late 1990s [22], pulsed light has found a myriad of applications in the food industry for various food processing operations. This gaining momentum over the past years has encouraged researchers and industrialists to execute different experimental trials to investigate every practicable application of PL mainly owing to its non-thermal nature and cost-effectiveness. The following sections briefly outline different evidenced and established applications of PL at commercial as well as laboratory levels.

The central and most explored field of PL application is microbial decontamination and rapid disinfection of foods, both solids and liquids. The mechanism is similar to UV processing, with additional benefits from the broad spectrum of PL. The injury to the microbial population can be lethal or sub-lethal through multiple damages to redundant vital structures in the microbe [18]. This adeptness of PL makes it helpful in disinfecting foods that are subjected to a wide class of bacteria, yeast, mold, and viruses. PL can also be employed as a subaltern treatment providing supplementary stresses to the microbes in multiple targets, which enhances the efficiency of other antimicrobial treatments.

The process efficacy could depend on the type and initial population of the microorganisms, the product characteristics, and other processing parameters. With the limitation of low penetrating power, Pulsed Light finds major application in the primary surface sterilization of fresh produce, which can later be treated with other sterilization techniques using a hurdle approach. Food applications of Pulsed Light are not just confined to pathogenic microorganisms but are also found to be efficient against food allergens, toxins, enzymes, and antinutrients. Pulsed Light can alter the protein configurations of allergens like β-lactoglobulins, reducing their immunoreactivity and making them susceptible to being hydrolyzed by digestive enzymes [47]. Similarly, photooxidation could change the protein structure through an aggregation effect that can alter the biological activity. In this way, Pulsed Light could modify the conformation of enzyme active sites, leading to increased activity levels depending on the intensity of treatments. With the advantage of such a surface enzyme inactivation process, Pulsed Light can help prevent the enzymatic browning of product surfaces without affecting the interior tissues.

Apart from the concerns associated with the consumption of safe and fresh horticultural produce and meat products, significant apprehensions are also related to the food safety associated with food contact surfaces and packaging materials. The low penetration property of PL and the requirement of surface smoothness and transparency of the treated product can be successfully utilized for bulk processing of completely transparent materials or surface disinfection of less transparent ones. Materials such as polyethylene, polypropylene, ethyl vinyl alcohol, and ethyl vinyl acetate are entirely transparent to UV light, making them suitable for PL treatment [42].

The technique can thus be seen as an efficient alternative to continuous UV sanitization or hydrogen peroxide treatments that are currently employed for sterilizing form fill seal-type or aseptic packaging films. Similarly, the same superficial treatment characteristics can be used to sterilize equipment surfaces and food handling areas in processing plants where higher fluencies and exposure time can be necessary for fouled surfaces [66]. The surface topography plays a major role in this case, too, where the PL efficiency was reduced in highly smooth and highly rough stainless-steel surfaces because of surface reflectivity and shrouding of cells within the rough surfaces, respectively.

7.6 EFFECTS OF PULSED LIGHT ON FOOD QUALITY PARAMETERS

With the substantial potential for destroying a wide range of microbes in food, the major limitation of the conventionally employed thermal treatments is their negative impact on food quality attributes like nutritional and bioactive composition. With complex interactions between different senses and the diverse sensory perception of different individuals, the consumer preference or acceptance of any processed or unprocessed food item depends majorly on the sensory attributes, which include appearance, taste, flavor, odor, and other textural qualities. Under such circumstances, attempts to popularize non-thermal processing techniques heightened where alternative treatment methods were explored that can ensure microbial stability but do not compromise food quality. The present section discusses the effect of PL processing on different quality attributes of food to ascertain its potency.

Inclusively, PL is reported to maintain the organoleptic properties of food under optimum processing conditions. As can be seen from Table 7.1, higher dosage and longer exposure time have led to considerable changes in the color properties of different foods. In this case, the thermal effect might have dominated the UV effect causing overheating of the sample and resulting in undesirable color alterations even before the decontamination threshold is reached. With dark-colored products absorbing more energy than pale ones, color degradation due to PL could be more significant in the former [17]. The color modification could also be accredited to oxidation and photodegradation of color pigments.

Light can also induce enzymatic and non-enzymatic browning reactions in foods altering the color parameters and naturally affecting the sensory properties. The degradation of color pigments can largely depend on the product matrix as PL is reported to decrease carotenoids in watermelon while causing the conversion of chlorophyll to carotenoids in mango. Similar observations can be made with other properties such as flavor and odor, where lipid oxidation induced by light is a major contributing factor. In meat products, this largely relies on the fat content of the product, whereas meats with higher fat content could show more sensory changes [62].

The presence of fat could also increase the resistance of the bacteria to PL, thus requiring severe treatment conditions that adversely affect the sensory quality. Conversely, PL is also known to inactivate the lipoxygenase enzyme, which catalyzes the lipid oxidation process. This is evident in a study of soymilk where no negative sensory effect was observed in the PL-treated samples and even improved the flavor [4].

Fruit and vegetable firmness is an important quality parameter used to grade fresh produce and is highly related to weight loss. Reduced fresh weight and associated decrease in product firmness usually cause direct economic loss for producers and retailers. The occasional sample heating during the PL process could cause the removal of moisture and thus promote weight loss. Change in firmness can also be a consequence of the modulation of pectinolytic enzyme activities in fruits triggered by the radiation treatment [39]. Fruits tend to soften when pectic acids get hydrolyzed in the cell wall. Any stress applied to plant sources has been reported to result in increased activities of pectin enzymes such as pectin methyl esterase and polygalacturonase at specific temperatures, affecting the firmness. With PL already increasing the permeabilization of the cell wall, combining the treatment with an appropriate coating method or dipping process can minimize the susceptibility to rupture. The presence of UV-C in the PL spectrum can also cause the formation of high-level polyamines which function similarly to calcium forming cation cross-links with pectic acid, thus restricting the accessibility of cell wall degrading enzymes [10]. This could help the product retain its firmness, but the extent of this retention could be product and dose-dependent.

Other physicochemical properties like TSS, pH, and titratable acidity have shown different results with respect to the treated product and the processing conditions. In some cases, no significant changes were obtained in these quality parameters, which can be explained by the energy associated with the applied PL intensity, which could be insufficient to induce ionic dissociation. The product acidity thus remains unaffected as there occurs no release of hydronium ions into the medium. Similarly, the energy could be even inadequate to increase the dissociation enthalpy of high molecular weight sugars to be converted into their monomeric unit keeping the TSS unaffected [9]. On the other hand, any decrease observed in the TSS of the treated product could be a result of the condensation of soluble tannins and the subsequent formation of insoluble compounds that are non-astringents in nature. Yet, no proper concluding findings can be drawn from the studies, and it could be stated that the effect of PL on physicochemical properties is driven by the product matrix, and further investigations are essential to confirm any hypothesis.

As PL treatment only tends to react with the surface layers of the foods, it does not seem to impede the inner layers from maintaining the nutrient and bioactive constituents unharmed. The inherent phenolic compounds in foods are highly important to food quality as they have potential therapeutic effects that could prevent various diseases. Any exposure to unfavorable environments could increase the concentration of these components as these are products of the plant's natural defense mechanism. The photochemical mechanism of PL as a result of the UV content in the spectrum has been reported to promote the biosynthesis of secondary metabolites through a process called phenylpropanoid metabolism [65]. Additionally, other biological effects also indicate the presence of different perception mechanisms in the form of photoreceptors and signaling pathways induced by oxidative stress, which is involved in the control of defense metabolism and consequently accumulate secondary metabolites (phenolic compounds).

The concrete application of PL is food decontamination, and more experimental investigations are necessary to draw further concluding remarks on the effect of PL on food quality. However, any adverse consequences of PL on the sensory and nutritional attributes of food could be minimized by proper optimization of the treatment parameters. Likewise, using chemicals like anti-browning substances or textural stabilizers that can counter the remarked changes is also possible, thereby improving the shelf life and sustaining the product quality.

7.7 ADVANTAGES AND LIMITATIONS OF PL

The multi-target process of employing intense pulsed light as a nonthermal disinfection technique has evolved to be a promising minimal process for boosting the safety and quality of food products. With considerable progress being made in the research advancements related to the application of PL in the food industry and improving the process efficiency, it is also important to scrutinize the significant advantages and limitations of the process for exploiting it to the fullest potential.

A notable advantage of Pulsed Light (PL) processing of food will be its rapid disinfection propensity against a wide range of contaminating pathogenic organisms where surface contamination is a concern, as in fresh horticultural products, hard cheeses, meat slices, and many more. This can be attributed to the short pulse duration and high dosage of broad-spectrum radiation emitted from the source, resulting in a high peak power that can act against the contaminating agents at very short exposure times [40]. This pulsed nature of the emitted radiation makes it superior to continuous UV, where dispensing a similar amount of energy takes a long time, rendering the treated product vulnerable to other unwanted light-induced changes. Nonetheless, the major causative factor prompting microbial inactivation is the dominance of the UV region in the PL spectrum, which kindles photochemical reactions in the food matrix that are proven lethal to pathogens. The shorter energy delivery time in PL also allows the product bulk to remain near ambient temperatures, leading to no damage to the internal constituents of the product. Compared to other non-thermal processing methods like high-pressure processing (HPP) and pulsed electric field (PEF), PL stands preeminent in its ability to inactivate spores and cope with large sample volumes. In HPP, the food to be treated usually needs to be packed, and the process is often reported to promote germination, which is not a concern with PL [42].

PL allows treatment of packed and unpacked foodstuffs and can decontaminate food contact surfaces without generating toxic chemical residues or solid wastes. This can be credited to the use of xenon-gas flash lamps that are more environmentally friendly because of their mercury-free characteristics and do not involve chemical sterilizers or synthetic preservatives. Furthermore, the better economics of processing, lower operating costs, greater flexibility, and high throughput add to the advantages of using PL in food processing. Also, the light source does not require any initial warming phase allowing instant action, and the radiation, being in pulse nature, allows natural cooling of the lamps [23].

PL has added benefits to the water industry as it generates a broad-spectrum germicidal pulse which is effective against chlorine-resistant water-borne enteroparasites such as *Cryptosporidium parvum* or *Giardia lamblia* [47]. Moreover, PL has been related to enhanced decontamination efficiency compared to other common approaches such as chlorine treatment, organic acid dips, and hydrogen peroxide treatment. This has been documented in studies on shredded lettuce and mango slices, where higher levels of microbial reduction were achieved with PL compared to chlorine and malic acid, respectively, for *Salmonella enterica* and *Listeria innocua* [24, 58].

Similarly, PL has been shown to be more expedient in inactivating *Bacillus subtilis, B. atrophaeus, B. cereus, Geobacillus stearothermophilus,* and *Aspergillus niger* spores in such a way that the exposure time necessary to obtain a 4-log reduction was almost 20,000 times quicker than continuous UV [31]. Being a minimal processing method, the foods treated have good consumer acceptance, which increases the prospects of this technique to be commercialized as it also allows operation in both batch and continuous modes, fitting it into existing processes and consuming minimum space. PL technology also serves as a strong candidate to be used in combination with different sub-lethal hurdle processes like ultrasound, HPP, PEF, irradiation, and chemical and biochemical hurdles to lessen the threat of food becoming a vector for microbial contagion and toxin production [23].

As previously discussed, the major drawback of this technique is its low penetration power, restricting its use to surface decontamination for opaque solid products due to the decreasing light intensity when the light travels through the samples as absorption and scattering are reduced [1]. Thus, product characteristics such as opacity and matrix composition greatly influence the process's effectiveness. To achieve maximum decontamination in foodstuffs with rough and uneven surfaces that can create a shadow effect reducing the radiation dose, the complete superficial surface of the food needs to be exposed to flashes, which makes the process complex [18]. This makes PL an unfitting option for cereals, grains, and spices disinfection attributable to their opaque characteristics with minor openings, crevices, or pores that can harbor microorganisms [6].

For liquid foods, the microbes will be suspended in the entire bulk, which mandates ensuring proper flow of the product through the treatment chamber so that the microbes are closer to the light source for uniform exposure, requiring appropriate engineering solutions. Even so, photoreactivation of microbes after processing is still a potential limitation to PL processing, which occurs as a result of the DNA repair mechanism by DNA photolyases in certain microorganisms that get activated on flashing at wavelengths in the range of 310–700 nm [13]. This allows the injured microbial DNA to heal, building possible recontamination chances during the storage phase. Also, there hasn't been much investigation on enzyme inactivation efficacy of PL as far as fruit juices are concerned, making it difficult to comment on the susceptibility of enzymes to PL processing, creating a gap in the field.

To boot, Pulsed Light (PL) technology involves high investment costs for the PL instrument depending on the design and lamp properties, which can discourage its adoption in regular food sanitization applications and make it mostly interesting for high value-added foods or particular market conditions [30].

It is also essential to highlight that the practical applications of PL are limited by the possible overheating of the sample due to high fluencies, a higher number of pulses, and accompanying lamp heating. Studies have shown a rise in temperature of around 20°C to even beyond 100°C when the treatment time was around 150 s [4, 40]. Such high temperatures can have a negative impact on some food constituents, which could lead to changes in sensory properties that can affect consumer acceptance. However, since the PL-induced temperature increase is mainly limited to the product surface, fewer photoproducts could result from this phenomenon, thus minimizing product degradation. Overheating could be prevented by providing water circulation configurations around the lamp through a quartz jacket or other cooling modules, including filtered air streams. Additionally, scaling-up and commercialization would be much easier if the important process parameters of processing to achieve the desired microbial reduction are optimized so that the treatment would not jeopardize the food quality.

7.8 REMARKS ON HEALTH AND SAFETY CONCERNS

The technology of pulsed light processing was permitted by the FDA according to the code 21CFR179.41 in 1996, where the agency considered potential variations in the chemical and nutrient composition of the treated food on account of the applied radiation in assessing the safety of the product [15].

The FDA recommended processing using PL proposes the use of xenon gas lamps covering surface emission of wavelengths ranging from 200–1,100 nm, with the aggregate treatment in terms of entire fluence not exceeding 12 J/cm^2 and the pulse length being less than 2 ms with typical pulse frequencies in the range of 1–20 pulses per second. However, the legal status of employing PL for food applications is still unclear in the European Union (EU). This may be because the legal status has a different approach with more food and ingredient-oriented than technology-oriented and is covered by regulation 258/97 on "novel foods and novel food ingredients," Article 1, item f [14]. Under this regulation, the safety of food is to be assessed based on likely changes in its nutritional value, metabolism, and formation of unfavorable substances upon treatment with any process currently not in use. In this picture, there is currently no approval for the PL technology, but distinct foods and ingredients processed with PL could be approved.

Besides, industrial-scale expansion of PL should also contemplate the 5-log reduction of pathogens as necessitated by the hazard analysis critical control point (HACCP) regulation aimed at the 'pertinent pathogen' that is the most resistant microbe present in food. To achieve such a significant level of inactivation, many researchers have even crossed the recommended limits of PL doses (Table 7.1). Determining the actual fluence absorbed by the sample being treated is the most critical parameter that characterizes PL processing, which is often neglected or inappropriately reported [20]. The energy received by the sample is noticeably different from the energy emitted by the light source, and the measurement of which can be a complex procedure and can require sufficient physics background with good knowledge of light properties.

7.9 SUMMARY

Pulsed light processing is a budding non-thermal technique for various food applications, mainly aimed at improving the safety of fresh produce by combining photochemical, photothermal, and photophysical mechanisms. The technique is rapid and environmentally friendly, opening up a new array of possibilities in the food industry either individually or in combination with other promising techniques. Currently available data indicate the suitability of this technique for the treatment of transparent beverages or products with smooth surfaces that do not present complex microstructures which could embed microorganisms, protecting them from irradiation. However, there is a need to establish standardized treatment procedures and protocols in agreement with regulatory requirements to achieve the target inactivation levels in specific food products without affecting quality. Additional research is needed to address the issues of overheating and treating the food homogeneously, considering the associated nutritional consequences. Until now, the majority of studies reported are laboratory-based on *in-vitro* tests, which lack the practical implications of the process. Thus, future studies should also focus on attempts to bridge the gap between basic and practical research.

KEYWORDS

- **food quality**
- **in-vitro tests**
- **microbial inactivation**
- **non-thermal processing**
- **nutritional quality**
- **pulsed light**

REFERENCES

1. Abida, J., Rayees, B., & Masoodi, F. A. (2014). Pulsed light technology: A novel method for food preservation. *International Food Research Journal, 21*(3), 839–848.
2. Abuagela, M. O., Iqdiam, B. M., Mostafa, H., Gu, L., Smith, M. E., & Sarnoski, P. J. (2018). Assessing pulsed light treatment on the reduction of aflatoxins in peanuts with and without skin. *International Journal of Food Science & Technology, 53*(11), 2567–2575.
3. Agüero, M. V., Jagus, R. J., Martín-Belloso, O., & Soliva-Fortuny, R. (2016). Surface decontamination of spinach by intense pulsed light treatments: impact on quality attributes. *Postharvest Biology and Technology, 121*, 118–125.
4. Alhendi, A., Yang, W., Goodrich-Schneider, R., & Sarnoski, P. J. (2018). Total inactivation of lipoxygenase in whole soya bean by pulsed light and the effect of pulsed light on the chemical properties of soya milk produced from the treated soya beans. *International Journal of Food Science & Technology, 53*(11), 2567–2575.
5. Artíguez, M. L., & de Marañón, I. M. (2015). Improved process for decontamination of whey by a continuous flow-through pulsed light system. *Food Control, 47*, 599–605.
6. Bhavya, M. L., & Umesh Hebbar, H. (2017). Pulsed light processing of foods for microbial safety. *Food Quality and Safety, 1*(3), 187–202.
7. Bryant, M. T., Degala, H. L., Mahapatra, A. K., Gosukonda, R. M., & Kannan, G. (2021). Inactivation of *Escherichia coli* K12 by pulsed UV light on goat meat and beef: Microbial responses and modelling. *International Journal of Food Science & Technology, 56*(2), 563–572.
8. Cassar, J. R., Mills, E. W., & Demirci, A. (2022). Characterization of pulsed light for microbial inactivation. *Journal of Food Engineering, 334,* 111152.
9. Chakraborty, S., Mahale, S., Dhar, R., & Basak, S. (2022). Development of a mixed fruit beverage and pulsed light treatment thereof to obtain a microbially safe and enzymatically stable product. *Food Bioscience, 45*, 101508.
10. Charles, F., Vidal, V., Olive, F., Filgueiras, H., & Sallanon, H. (2013). Pulsed light treatment as a new method to maintain the physical and nutritional quality of fresh-cut mangoes. *Innovative Food Science & Emerging Technologies, 18*, 190–195.

11. Chen, B. Y., Lung, H. M., Yang, B. B., & Wang, C. Y. (2015). Pulsed light sterilization of packaging materials. *Food Packaging and Shelf Life, 5*, 1–9.
12. Choi, M. S., Cheigh, C. I., Jeong, E. A., Shin, J. K., & Chung, M. S. (2010). Nonthermal sterilization of *Listeria monocytogenes* in infant foods by intense pulsed-light treatment. *Journal of Food Engineering, 97*(4), 504–509.
13. Dhar, R., Basak, S., & Chakraborty, S. (2022). Pasteurization of fruit juices by pulsed light treatment: a review on the microbial safety, enzymatic stability, and kinetic approach to process design. *Comprehensive Reviews in Food Science and Food Safety, 21*(1), 499–540.
14. European Union. (1997). Regulation (EC) No 258/97 of the European parliament and of the council of 27 January 1997 concerning novel foods and novel food ingredients. *Official Journal of the European Communities, 40*, 1–7.
15. FDA. (2005). Code of Federal Regulations 21CFR179.41; Sec. 179.41 ; Title 21, Volume 3], https://www.accessdata.fda.gov/scripts/cdrh/cfdocs/cfcfr/CFRSearch.cfm?fr=179.41.
16. Fernández, M., Hospital, X. F., Cabellos, C., & Hierro, E. (2020). Effect of pulsed light treatment on Listeria inactivation, sensory quality and oxidation in two varieties of Spanish dry-cured ham. *Food Chemistry, 316*, 126294.
17. Fine, F., & Gervais, P. (2004). Efficiency of pulsed UV light for microbial decontamination of food powders. *Journal of Food Protection, 67*(4), 787–792.
18. Franco-Vega, A., Reyes-Jurado, F., González-Albarrán, D., Ramírez-Corona, N., Palou, E., & López-Malo, A. (2021). Developments and advances of high intensity pulsed light and its combination with other treatments for microbial inactivation in food products. *Food Engineering Reviews, 13*(4), 741–768.
19. Gómez-López, V. M., Koutchma, T., & Linden, K. (2012). Ultraviolet and pulsed light processing of fluid foods. In *Novel Thermal and Non-Thermal Technologies for Fluid Foods* (pp 185–223). Elsevier.
20. Gomez-Lopez, V. M., Ragaert, P., Debevere, J., & Devlieghere, F. (2007). Pulsed light for food decontamination: A review. *Trends in Food Science & Technology, 18*(9), 464–473.
21. Gómez, P. L., García-Loredo, A., Nieto, A., Salvatori, D. M., Guerrero, S., & Alzamora, S. M. (2012). Effect of pulsed light combined with an antibranding pretreatment on the quality of fresh-cut apple. *Innovative Food Science & Emerging Technologies, 16*, 102–112.
22. Gunjal, M., Tazeen, H., Gundu, P. M., Birwal, P., & Krishna, A. (2021). Pulsed light technology in food processing and preservation. In *Handbook of Research on Food Processing and Preservation Technologies: Volume 5: Emerging Techniques for Food Processing, Quality, and Safety Assurance* (p 115). CRC Press.
23. Heinrich, V., Zunabovic, M., Varzakas, T., Bergmair, J., & Kneifel, W. (2016). Pulsed light treatment of different food types with a special focus on meat: a critical review. *Critical Reviews in Food Science and Nutrition, 56*(4), 591–613.
24. Huang, R., & Chen, H. (2018). Evaluation of inactivating *Salmonella* on iceberg lettuce shreds with washing process in combination with pulsed light, ultrasound, and chlorine. *International Journal of Food Microbiology, 285*, 144–151.
25. Huang, Y., & Chen, H. (2014). A novel water-assisted pulsed light processing for decontamination of blueberries. *Food Microbiology, 40*, 1–8.

26. Hwang, H. J., Cheigh, C. I., & Chung, M. S. (2017). Construction of a pilot-scale continuous-flow intense pulsed light system and its efficacy in sterilizing sesame seeds. *Innovative Food Science & Emerging Technologies, 39*, 1–6.
27. Izmirlioglu, G., Ouyang, B., & Demirci, A. (2020). Utilization of pulsed UV light for inactivation of *Salmonella enteritidis* on shelled walnuts. *LWT, 134*, 110023.
28. Jeon, M. S., Park, K. M., Yu, H., Park, J. Y., & Chang, P. S. (2019). Effect of intense pulsed light on the deactivation of lipase: enzyme-deactivation kinetics and tertiary structural changes by fragmentation. *Enzyme and Microbial Technology, 124*, 63–69.
29. John, D., & Ramaswamy, H. S. (2020). Comparison of pulsed light inactivation kinetics and modeling of *Escherichia coli* (ATCC-29055), *Clostridium sporogenes* (ATCC-7955) and *Geobacillus stearothermophilus* (ATCC-10149). *Current Research in Food Science, 3*, 82–91.
30. John, D., & Ramaswamy, H. S. (2018). Pulsed light technology to enhance food safety and quality: A mini-review. *Current Opinion in Food Science, 23*, 70–79.
31. Kim, S. W., Ko, M. J., & Chung, M. S. (2019). Extraction of the flavanol quercetin from onion waste by combined treatment with intense pulsed light and subcritical water extraction. *Journal of Cleaner Production, 231*, 1192–1199. https://doi.org/10.1016/j.jclepro.2019.05.280.
32. Kim, Y. E., & Min, S. C. (2022). Consecutive treatments of cold plasma and intense pulsed light for microbial decontamination of fresh cabbage slices in plastic containers. *International Journal of Food Microbiology, 369*, 109626.
33. Koch, F., Wiacek, C., & Braun, P. G. (2019). Pulsed light treatment for the reduction of *Salmonella typhimurium* and *Yersinia enterocolitica* on pork skin and pork loin. *International Journal of Food Microbiology, 292*, 64–71.
34. Kramer, B., & Muranyi, P. (2014). Effect of pulsed light on structural and physiological properties of *Listeria innocua* and *Escherichia coli*. *Journal of Applied Microbiology, 116*(3), 596–611.
35. Kramer, B., Wunderlich, J., & Muranyi, P. (2015). Pulsed light decontamination of endive salad and mung bean sprouts and impact on color and respiration activity. *Journal of Food Protection, 78*(2), 340–348.
36. Kramer, B., Wunderlich, J., & Muranyi, P. (2019). Inactivation of *Listeria innocua* on packaged meat products by pulsed light. *Food Packaging and Shelf Life, 21*, 100353.
37. Kramer, B., Wunderlich, J., & Muranyi, P. (2017). Recent findings in pulsed light disinfection. *Journal of Applied Microbiology, 122*(4), 830–856.
38. Li, Y., Yang, W., Chung, S. Y., Chen, H., Ye, M., Teixeira, A. A., Gregory, J. F., Welt, B. A., & Shriver, S. (2013). Effect of pulsed ultraviolet light and high hydrostatic pressure on the antigenicity of almond protein extracts. *Food and Bioprocess Technology, 6*(2), 431–440.
39. Llano, K. R. A., Marsellés-Fontanet, A. R., Martín-Belloso, O., & Soliva-Fortuny, R. (2016). Impact of pulsed light treatments on antioxidant characteristics and quality attributes of fresh-cut apples. *Innovative Food Science & Emerging Technologies, 33*, 206–215.
40. Mahendran, R., Ramanan, K. R., Barba, F. J., Lorenzo, J. M., López-Fernández, O., Munekata, P. E. S., Roohinejad, S., Sant'Ana, A. S., & Tiwari, B. K. (2019). Recent advances in the application of pulsed light processing for improving food safety and increasing shelf life. *Trends in Food Science & Technology, 88*, 67–79.

41. Maisch, T., Spannberger, F., Regensburger, J., Felgenträger, A., & Bäumler, W. (2012). Fast and effective: Intense pulse light photodynamic inactivation of bacteria. *Journal of Industrial Microbiology and Biotechnology, 39*(7), 1013–1021.
42. Mandal, R., Mohammadi, X., Wiktor, A., Singh, A., & Pratap Singh, A. (2020). Applications of pulsed light decontamination technology in food processing: an overview. *Applied Sciences, 10*(10), 3606.
43. Marangoni Jr., L., Cristianini, M., & Anjos, C. A. R. (2020). Packaging aspects for processing and quality of foods treated by pulsed light. *Journal of Food Processing and Preservation, 44*(11), e14902.
44. Meinlschmidt, P., Ueberham, E., Lehmann, J., Reineke, K., Schlüter, O., Schweiggert-Weisz, U., & Eisner, P. (2016). The effects of pulsed ultraviolet light, cold atmospheric pressure plasma, and gamma-irradiation on the immunoreactivity of soy protein isolate. *Innovative Food Science & Emerging Technologies, 38*, 374–383.
45. Moreira, M. R., Álvarez, M. V., Martín-Belloso, O., & Soliva-Fortuny, R. (2017). Effects of pulsed light treatments and pectin edible coatings on the quality of fresh-cut apples: a hurdle technology approach. *Journal of the Science of Food and Agriculture, 97*(1), 261–268.
46. Oms-Oliu, G., Aguiló-Aguayo, I., Martín-Belloso, O., & Soliva-Fortuny, R. (2010). Effects of pulsed light treatments on quality and antioxidant properties of fresh-cut mushrooms (*agaricus bisporus*). *Postharvest Biology and Technology, 56*(3), 216–222.
47. Orcajo, J., Lavilla, M., & Martinez-de-Maranon, I. (2019). Effect of pulsed light treatment on β-lactoglobulin immunoreactivity. *LWT, 112*, 108231.
48. Ortega-Rivas, E. (2012). Pulsed light technology. In *Non-Thermal Food Engineering Operations* (pp 263–273). Springer.
49. Pedrós-Garrido, S., Condón-Abanto, S., Clemente, I., Beltrán, J. A., Lyng, J. G., Bolton, D., Brunton, N., & Whyte, P. (2018). Efficacy of ultraviolet light (UV-C) and pulsed light (PL) for the microbiological decontamination of raw salmon (*Salmo salar*) and food contact surface materials. *Innovative Food Science & Emerging Technologies, 50*, 124–131.
50. Pollock, A. M., Singh, A. P., Ramaswamy, H. S., & Ngadi, M. O. (2017). Pulsed light destruction kinetics of *L. Monocytogenes*. *LWT, 84*, 114–121.
51. Preetha, P., Pandiselvam, R., Varadharaju, N., Kennedy, Z. J., Balakrishnan, M., & Kothakota, A. (2021). Effect of pulsed light treatment on inactivation kinetics of *Escherichia coli* (MTCC 433) in fruit juices. *Food Control, 121*, 107547.
52. Proulx, J., Sullivan, G., Marostegan, L. F., VanWees, S., Hsu, L. C., & Moraru, C. I. (2017). Pulsed light and antimicrobial combination treatments for surface decontamination of cheese: Favorable and antagonistic effects. *Journal of Dairy Science, 100*(3), 1664–1673.
53. Ramos-Villarroel, A. Y., Aron-Maftei, N., Martín-Belloso, O., & Soliva-Fortuny, R. (2012). The role of pulsed light spectral distribution in the inactivation of *Escherichia coli* and *Listeria innocua* on fresh-cut mushrooms. *Food Control, 24*(1–2), 206–213.
54. Rosenthal, I. (2012). *Electromagnetic Radiations in Food Sciences* Springer Science & Business Media. Pp. 64–104. Chapter 2 Ultraviolet-Visible Radiation.
55. Rowan, N. J. (2019). Pulsed light as an emerging technology to cause disruption for food and adjacent industries – Quo Vadis? *Trends in Food Science and Technology, 88*, 316–332. https://doi.org/10.1016/j.tifs.2019.03.027.

56. Rybak, K., Wiktor, A., Pobiega, K., Witrowa-Rajchert, D., & Nowacka, M. (2021). Impact of pulsed light treatment on the quality properties and microbiological aspects of red bell pepper fresh-cuts. *LWT, 149*, 111906.
57. Salehi, F. (2022). Application of pulsed light technology for fruits and vegetables disinfection: a review. *Journal of Applied Microbiology, 132*(4), 2521–2530.
58. Salinas-Roca, B., Soliva-Fortuny, R., Welti-Chanes, J., & Martín-Belloso, O. (2016). Combined effect of pulsed light, edible coating, and malic acid dipping to improve fresh-cut mango safety and quality. *Food Control, 66*, 190–197.
59. Santamera, A., Escott, C., Loira, I., del Fresno, J. M., González, C., & Morata, A. (2020). Pulsed light: Challenges of a non-thermal sanitation technology in the winemaking industry. *Beverages, 6*(3), 45.
60. Siddique, M. A. B., Maresca, P., Pataro, G., & Ferrari, G. (2016). Effect of pulsed light treatment on structural and functional properties of whey protein isolate. *Food Research International, 87*, 189–196.
61. Taştan, Ö., Pataro, G., Donsì, F., Ferrari, G., & Baysal, T. (2017). Decontamination of fresh-cut cucumber slices by a combination of a modified chitosan coating containing carvacrol nanoemulsions and pulsed light. *International Journal of Food Microbiology, 260*, 75–80.
62. Tomašević, I. (2015). Intense light pulses upset the sensory quality of meat products. *Tehnologija Mesa, 56*(1), 1–7.
63. Unni, L. E., & Chauhan, O. P. (2019). Use of pulsed light in food processing. In O. P. Chauhan (Ed.), *Non-Thermal Processing of Foods* (pp. 173–185). CRC Press.
64. Valdivia-Nájar, C. G., Martín-Belloso, O., Giner-Seguí, J., & Soliva-Fortuny, R. (2017). Modeling the inactivation of *Listeria innocua* and *Escherichia coli* in fresh-cut tomato treated with pulsed light. *Food and Bioprocess Technology, 10*(2), 266–274.
65. Vargas-Ramella, M., Pateiro, M., Gavahian, M., Franco, D., Zhang, W., Khaneghah, A. M., Guerrero-Sánchez, Y., & Lorenzo, J. M. (2021). Impact of pulsed light processing technology on phenolic compounds of fruits and vegetables. *Trends in Food Science & Technology, 115*, 1–11.
66. Vimont, A., Fliss, I., & Jean, J. (2015). Efficacy and mechanisms of murine norovirus inhibition by pulsed-light technology. *Applied and Environmental Microbiology, 81*(8), 2950–2957.
67. Wang, B., Khir, R., Pan, Z., Wood, D., Mahoney, N. E., El-Mashad, H., Wu, B., Ma, H., & Liu, X. (2016). Simultaneous decontamination and drying of rough rice using combined pulsed light and holding treatment. *Journal of the Science of Food and Agriculture, 96*(8), 2874–2881.
68. Wang, B., Wei, W., Aputexiakere, J., Li, Y., & Ma, H. (2022). Surface decontamination of whole eggs using pulsed light technology and shelf-life study of combined pulsed light and Vaseline coating during room temperature storage. *Food Control, 137*, 108411.
69. Xu, W., & Wu, C. (2016). The Impact of pulsed light on decontamination, quality, and bacterial attachment of fresh raspberries. *Food Microbiology, 57*, 135–143.

CHAPTER 8

DESIGN CONSIDERATIONS AND FACTORS AFFECTING THE PULSED LIGHT EFFICACY

PALANISAMY PREETHA, VEERAPANDIAN CHANDRASEKAR, and SULTAN SHAHIR

ABSTRACT

Pulsed light technology is a non-thermal technology used for preservation of liquid foods. The generation of pulsed light is vital for pulsed light processing. The design and construction of pulsed light equipment are required for commercial-scale processing. This chapter contains information on the history and origin of pulsed light technology, various pulsed light sources, construction of pulsed light equipment, considerations for designing equipment and design calculations, factors affecting pulsed light efficiency, and optimization techniques.

8.1 INTRODUCTION

Pulsed light (PL) technology has evolved in prominence over the past decade as a non-thermal processing method to preserve food products by reducing microorganisms, improving microbial protection, extending shelf life, and maintaining food quality [65]. Pulsed light is produced as short-duration, high-intensity light pulses that are approximately 20,000 times stronger than sunlight at atmosphere. Pulsed light technology is a low-cost, low-temperature treatment that leaves no residue on food surfaces [31]. Other terms used to describe pulsed light technology include pulsed white light,

Nonthermal Light-Based Technologies in Food Processing.
C. K. Sunil, Preeti Birwal, Megh R. Goyal, and R. Mahendran (Eds.)

intense pulsed light, high-intensity pulsed ultraviolet light, pulsed ultraviolet light, and high-intensity broad-spectrum ultraviolet light [65].

The FDA has recognized the use of pulsed light technology for processing, preservation, and handling of food since 1996. The fluence, emission spectra, and duration of pulsed light used for decontamination are 12 $J.cm^{-2}$ and a wavelength range of 200 to 1,100 nm for less than 2 meters [1]. Pulsed light technology works by exposing food material to ultra-short bursts of energy stored in a capacitor from the xenon-filled light source. PL technology operates by storing and releasing high discharge voltages in ultra-short bursts.

Xenon light source emits bright flashes in the range between 200 and 1,100 nm in which 25% of the light is in the ultraviolet range [36]. High disinfection efficiency was observed in pulsed light technology compared to continuous-wave low-pressure UV irradiation. Pulsed light technology has high peak power and is able to deliver the stored pulses in the form of energy in a short time, about 1 to 10 pulses per second. The factors affecting the efficiency of pulsed light technology are fluence, exposure time, number of applied pulses, frequency of pulse, pulse width, and peak power. The inactivation efficiency of pulsed light is based on intensity ($Joule/cm^2$) and the number of pulses delivered.

8.1.1 ORIGIN OF THE PULSED LIGHT TECHNOLOGY

Pulsed light technology utilizes short duration pulses containing a broad-spectrum range similar to UV-C light to decontaminate microorganisms on surfaces. UV-C refers to the wavelength range between 200 and 280 nm in the electromagnetic spectrum. PL is made with technologies that multiply the power by a factor of ten. The pulse is generated by saving electricity in an energy storage capacitor for longer seconds and released in a short time (less than the stored time), so the power is multiplied to millionths to thousandths of a second. The released pulse has ultimate strength with a wavelength range of 200–1,100 nm [19].

The pulsed light technology produces light flashes with high peak power and shorter bactericidal wavelengths [62]. The first pulsed light technology using flash lamps for disinfection was accomplished in Japan in the 1970s, and the first patent was granted in 1984 [73]. The results of the first pulsed light technology inactivation of microorganisms were published in the scientific literature by Bank et al. [7].

Pulsed light technology uses a xenon lamp to produce flashes several times per second. The fluence rate is the amount of energy received by the unit area of the sample per second, W/m^2; and the fluence is the amount of energy received by the unit area of the sample from the lamp during the treatment, J/m^2. Fluence and dose are often used interchangeably. The treatment's exposure period is measured in seconds. Pulse width is the interval of energy delivering time during pulse treatment. PRR is the number of pulses per second (Hz). Peak power is defined as pulse energy divided by pulse duration in watts (W).

Pulsed light technology has shown to be more effective than continuous UV light irradiation in diminishing pathogens such as *E. coli* O157:H7 and *Listeria monocytogenes* [9]. The pulsed ultraviolet light combined with LED treatment is more effective than continuous treatment in inactivating *E. coli* [79]. As the percentage of on-time of the circuit (duty cycle) is reduced from 100% to 5%, the inactivation rate of *E. coli* increases dramatically by administering the same dose [79]. The penetration power of high-intensity pulsed light in food material is greater than that of ultraviolet light despite the same penetration depth [33, 38]. There is a debate about whether pulsed light or UV has a greater penetration depth. The light penetration depends on wavelength. The penetration depth of wavelengths around 200 nm is higher than continuous wave ultraviolet light because pulsed light includes wavelengths from 200 to 1,100 nm (considering 253.7 nm) [47]. Pulsed light can reach all depths, volumes, and surfaces, so it is used for sterilizing packaging materials, air, water, and many solutions, as well as a wide range of medications and medical products. The safety and effectiveness of pulsed light-treated products and processes were approved by the Food and Drug Administration (FDA) in 1999 [61]. Pulsed UV light is created by pulsing a high-powered xenon lamp for 100 to 10 microseconds, depending on other sources.

If plant material is treated with a few pulses of high-intensity light that transmit huge quantities of thermal energy, raising the temperature of a very thin surface layer of the food where adherent microorganisms occur, sterilization up to a 6-log reduction appears to be possible [17].

8.2 PRINCIPLES

The pulsed light is generated and released in the form of highly concentrated energy. The highly concentrated energy is stored in the capacitor and discharged as short bursts in the form of a light flash in 1 billionth of a second. The short bursts result in power amplification and less energy consumption, and it is exposed to food to control and ensure microbial decontamination. The energy is augmented to many folds in PL technology. Since the augmented electrical energy is stored in a capacitor for a fraction of a second and released in millionths or thousandths of a second to do the work, that results in very high energy output [19]. The released energy has a spectrum 20,000 times more intense than the spectrum released by the sun on Earth.

The pulsed light flashes duration is less than 1 millisecond. The micro-biocidal effect of pulsed light technology is due to light energy absorbed by the protein's conjugated carbon double bond and nucleic acids that break the cellular metabolism. The microbial spores exposed to pulsed light are broken and internally exploded, and the inside content of the spore is evacuated. The spores become empty and collapse due to UV A and B light only. It was observed that visible light did not significantly disintegrate spores [73]. The antimicrobial effects of UV light may be due to the flow of ions to food material, etching effect to increase the cell membrane permeability, and cell membrane depolarization [77]. The mode of action of pulsed light technology is ascribed to selected effects of high peak power and broad-spectrum of light. Nucleic acids are a fundamental biological target. Chemical changes and DNA cleavage are among the methods that cause inactivation.

Pulsed light's effect on proteins, membranes, and other biological components most likely occurs at the same time as nucleic acid destruction [4]. The inactivation modes of pulsed light as proposed by Demirci and Krishnamurthy [18] are – (i) photochemical effect in which the ultraviolet light component of pulsed UV-light causes cell death or spore inactivation by creating thymine dimers in bacterial DNA; (ii) photothermal effect in which the visible and infrared portions of pulsed UV-light generate restricted heating on the bacterial cell, eventually inactivating the enzymes and other cellular contents; and (iii) pulse effect of UV light in which the continuous disturbances caused by intermittent pulses may damage the cell wall and other cellular constituents. Microbes are inactivated by PL through a variety of complex physicochemical processes. Photochemical, photothermal, and photophysical processes play a role in microbial destruction. Depending on the treatment settings and target organisms, these systems are responsible for inactivating microorganisms individually or in combination [28].

Nonetheless, the photo-thermal, photo-chemical, and photo-physical effects inactivate the microorganism; however, the photo-chemical effect is the primary driver of microbial eradication. Microbial destruction by PL is influenced by spectrum range and microbial shape, in addition to physicochemical causes. The influence of light pulse spectral range has also been thoroughly explored. The UV spectral range was discovered to be the most effective in microorganism killing [9]. UV-C (200–280 nm) has the most important effect, and the inactivation peak is around 253.7 nm [28]. The highest *E. coli* reduction by UV-C was found at 270 nm [72]. Using the UV-C range spectrum of pulsed light resulted in more deactivation on fresh-cut avocados than the Vis-NIR spectrum [60]. The visible and infrared areas of light radiation, on the other hand, were ineffective in the PL-based inactivation process [9]. Only at greater levels did the IR and visible areas have an antimicrobial impact [21]. The order of microorganism susceptibility to pulsed light treatment is gram-negative bacteria followed by gram-positive bacterial and fungal spores [3, 62]. On the other hand, they did not find a comparable pattern of vulnerability in their research [28]. Spores are less susceptible than vegetative bacterial cells, which is thought to be due to the cell's own DNA repair systems [3]. However, pulsed light application has a greater impact on DNA repair and is more effective than continuous UV treatment [29]. The color of the microbial pigments plays a role in determining the efficacy of pulsed light treatment. The spore color may influence pulsed light activity [3]. Dark-colored spores of *Aspergillus niger* were found to be more resistant to pulsed light technology than pink-colored spores of *Fusarium culmorum* because of more UV-C absorption by dark-colored spores of *Aspergillus niger*.

8.2.1 EFFECT OF PHOTO-CHEMICAL MECHANISM

The photochemical effect of pulsed light is the most critical factor in microbial inactivation due to the absorption of UV light by DNA. UV radiation has antimicrobial properties because it causes pyrimidine dimers to form, most commonly thymine [28, 52]. The threshold for dimer production in microbial DNA corresponds to UV radiation, specifically 253.7 nm. The light energy with a wavelength of 253.7 nm or less has enough energy to cause DNA dimerization [29]. Viruses have been shown to be destroyed by wavelengths greater than 295 nm. In another investigation, wavelengths greater than 400 nm were found to kill viruses by rupturing phage capsids [9, 52].

The photochemical effect also includes photosensitization, which introduces a photoactive substance (photosensitizer) into the food that accumulates in the target microorganism cells exposed to light [43]. Lukšiene et al. [44] used 5-aminolevulinic acid (ALA) to photosensitize Bacillus cereus *in vitro* as well as on the surface of packaging material. Several 'cytotoxic responses' were identified in the cells as a result of the photosensitization. Consequently, the cellular membrane was disrupted, enzymes were inactivated, and DNA was damaged [45]. The inactivation of microorganisms by photosensitization is another principle for using pulsed light inactivation technology. The emitted light spectrum is also known to be affected by the discharge voltage used to generate the pulse.

8.2.2 EFFECT OF PHOTO-THERMAL MECHANISM

Light wavelengths that correspond to IR areas demonstrate photo-thermal effects. If the final temperature reaches pasteurization temperature, then the microorganism gets inactivated. Although a greater temperature caused by intensive pulsed light treatment may contribute to inactivation, food sensory qualities have been proven to be negatively impacted. The temperature of the surface layer rises, destroying harmful organisms on the surface, then the heat infiltrates into the food by conduction. Subsequently, the heat dissipates before it reaches the core.

Absorption of light pulses into *A. niger* cells made the mold become hot and led to cell death. Dunn et al. [19] suggested that light pulses may be used to heat food surfaces and kill microorganisms. Cell death due to photo-thermal stress caused by altering the structure and physiology of the cells. The local heating occurred during pulsed light treatment on bacteria caused cell damage and rupture resulting in cellular 'explosion' and 'cell bursting' and demonstrated the release of hot contents of spores of PL-treated *A. niger* using scanning electron microscopy (SEM) images [75]. When *Bacillus subtilis* was treated with PL, the cell wall was destroyed due to photo-thermal stress [48]. Hotness on the cell membrane, intercellular fluid vaporization, and subsequent membrane breakdown perhaps the reason for structural alterations in *E. coli* after PL treatment [78]. However, heat generated by pulsed light meagerly contributes to microbial inactivation when compared to photochemical effects, according to the study.

8.2.3 EFFECT OF PHOTO-PHYSICAL MECHANISM

The photophysical mechanism of pulsed light inactivates the microbes. The high-energy pulses have a tendency to damage cell structure [37], and researchers found that *Staphylococcus aureus* experienced degradation of cell partition, reduction of plasma membrane, and leakage of cell organelles.

Pulsed light breaks the cell structure and membrane, allowing intracellular components and proteins to escape from *S. cerevisiae* [68], and causes cytoplasmic damage in *L. innocua* and *E. coli* [58]. Structural damage in *E. coli* cells is caused by central depression due to the photo-physical effect [46]. Pulsed light has a stronger tendency to cause structural damage than continuous UV light [29]. All of these experiments reveal a photophysical mechanism for microbial inactivation by pulsed light.

8.3 DESIGN CONSIDERATIONS AND CALCULATIONS OF PULSED LIGHT EQUIPMENT

8.3.1 CALCULATION FOR FINDING THE CAPACITANCE AND SUPPLY VOLTAGE

Lamps produce flashes; generally, Xenon lamps are used as a source for pulsed light generation (Table 8.1). The lamp's output is determined by the impedance constant (Ko), which depends on the dimensions of the lamp and the pressure of gas filled in (Eqn. (1)). The energy required for generating one pulse with a pulse width of 360 μs is calculated based on the inductance-capacitance network using Eqn. (2). The duration of the pulse (3t) is calculated using the critically damped circuit value (α) (Eqn. (3)). The voltage required for generating the desired pulse energy and the lamp current required to produce the pulse are calculated using Eqns. (4) and (5). The pulse power is the power expended over the substrate material by the lamp during the pulse, and the overall power consumed by the circuit for producing the high pulse power are calculated using Eqns. (6) and (7).

$$K_O = 1.28 \times \frac{L_A}{d} \times \left(\frac{P}{N_x}\right)^{-1/5} \tag{1}$$

$$C = \left(\frac{2E_O \alpha^4 T^2}{K_O^4}\right)^{1/3} \tag{2}$$

$$T_d = 3t \tag{3}$$

$$V_0 = \left(\frac{2E_0}{C}\right)^{0.5} \tag{4}$$

$$I = \left(\frac{E_0}{K_0 T} \right)^{2/3} \tag{5}$$

$$P_P = K_0 I^{2/3} \tag{6}$$

$$P = P_P \times T \times f \tag{7}$$

where; 'K_0' is the Impedance constant of the lamp; 'C' is the inductance-capacitance 'C' (Farads); 'L_A' is the arc length of the lamp, mm; 'd' is the bore diameter of the lamp 'mm;' P is the gas fill pressure in the lamp, Torr; 'N_x' is the constant specific to the filled gas, for Xenon, 450 Torr (59 kPa); E_0 is the energy stored in capacitor (J); 'α' is the damping coefficient; 'T_d' is the duration of pulse (s); 't' is the time constant of circuit (s); 'I' is the lamp current 'A;' 'E_0' is the energy stored in the capacitor C (J); 'K_0' is the impedance constant of the lamp '$\Omega\, A^{1/2}$;' 'T' is the pulse width (s); 'P_p' is the pulse power; and 'f' is the frequency of flashing.

8.3.2 ELECTRONICS COMPONENTS AND ARRANGEMENT FOR PULSED LIGHT APPARATUS

Xenon and krypton gases are normally used to fill DC and pulsed flash lamps. Xenon flash lamps are used because they have no mercury, are widely used for disinfection, have the ability to increase the inactivation of microbes, and reduce rates of organic chemicals from the pulsed light [32, 66, 74]. Pulsed light is produced using the xenon flash lamp, which produces a higher total radiative output for a given electrical input than do other gases like krypton. In order to produce pulsed light, a circuit with components and parameters such as arc length (mm), arc diameter (mm), arc impedance ($\Omega A^{1/2}$), gas fill pressure in the lamp (kPa), damping coefficient, pulse width (μs), capacitance (μ Farads), current (A), frequency (Hertz), power consumption, output energy (J), and output voltage is required.

8.3.2.1 XENON FLASH LAMP

The Xenon flash lamp comprises a rod-sealed tube with a thickness of 3 mm and made of clear fused quartz (CFQ) filled with xenon gas, electrodes to transfer the current to the filled gas, and a trigger electrode. CFQ transmits short-wave ultraviolet (UV) light, while other forms of quartz restrict

TABLE 8.1 Light Sources and Parameters of PL Treatments of Foods

Food Sample, Matrix	Source	Distance between Lamp and Food	Inactivation of Microorganism	Dose/ Fluence, Pulse Energy	Treatment Time (s)	Log10 or Percent Reductions	References
Eggs	Xenon	15 cm	*Escherichia coli* ATCC 8879	1.32 J/cm^2	–	3.77 log CFU/ egg	[71]
Water decontamination	Xenon	5.5 cm	Atrazine, malathion, chlorpyriphos-methyl, and bromopropylate.	10 J/cm^2	–	93%	[8]
Iceberg lettuce	–	–	*Salmonella*	6 J/cm^2	–	< 2 log	[34]
Almonds	Xenon	19 cm	*Salmonella*	–	3 min	>5 log	[42]
Cilantro, mesclun lettuce, spinach, and tomatoes.	Xenon	40 cm.	*Cryptosporidium parvum*	0.0675–0.0680 J/cm^2	0–90 s	2.4, 4.3, 2.5, and 2.2, r	[16]
Lettuce	Xenon	9 cm	*Staphylococcus aureus*	0.33 J/cm^2	6, 13, 19, and 25 s	7.0	[69]
Apple juice	Xenon	150 mm	*Alicyclobacillus*, *Saccharomyces cerevisiae* KE162	2.4–1.6 J/cm^2	–	6.4 and 5.8 log	[26]
Tender coconut water, orange, and pineapple juice.	Xenon	–	*Escherichia coli* (MTCC 433)	95.2 J/cm^2	0 and 15 s	4.0, 4.5, and 5.33 log CFU	[56]
Red bell pepper fresh-cuts	Xenon	7.58 cm	*Listeria monocytogenes*, *Salmonella enterica serovar*, and *Typhimurium*	0.6 J/cm^2	3.5 and 26.5 ms	1.59–2.13 log cycles	[63]
Mesquite pod flour	Xenon	–	*Bacillus cereus* spore	700 J/pulse	0 to 28s	3.51 log10 CFU/g	[14]

wavelengths in the UV region. Along with this, a power source with high voltage is required to electrify the gas. This high voltage (500, 1,000, and 1,500 V) is usually stored on a capacitor to allow accelerated discharge of high voltage when the lamp is triggered. The electrodes protrude at both ends of the tube and are connected to a capacitor that is charged to a relatively high voltage. The Xenon flash lamp (Make: Heraeus Noblelight Ltd., U.K.) had a bore diameter of 8 mm and an arc length of 300 mm, with a Clear Fused Quartz envelope used to produce the broad-spectrum flash from 100 to 1,100 nm. The tube was filled with xenon at a pressure of 450 Torricelli (59 kPa). The flash lamp does not flash when the electrodes are connected to a charged capacitor. These tubes are flashed through the use of a trigger electrode, which is a thin wire made of nickel wound around the length of the lamp. A high voltage of 30 KV (as per the lamp manufacturer's instruction) applied as a pulse to the trigger electrode ionizes the xenon gas and initiates conduction. A current (16 A) flowing through the lamp at this stage produces more ionization and hence conduction of electrons occurs inside the tube.

8.3.2.2 MINIATURE CIRCUIT BREAKER

Miniature Circuit Breaker (MCB) is provided for the protection of the entire system. It gets switched off automatically during abnormal conditions as well as faulty conditions. The function of MCB is to interrupt the flow of electricity in a circuit when a fault is detected. It is a switch that automatically turns off when the current flowing through it exceeds the maximum allowable limit. MCB1 and MCB2 are provided in the circuit with a capacity of 32A, 4Pole for the main circuit and 32A, 3Pole for the triggering circuit in the main electrical panel board, respectively.

8.3.2.3 CONTACTOR

It is used to operate the high voltage 3-phase 0.25 Hp motor with 1,440 rpm. Contactors provided in the circuit are MNX three-pole contactors with a capacity range from 9 A to 650 A.

8.3.2.4 RELAY

A relay is a switch that performs an ON/OFF function. In the circuit, the time gap between the triggering (pulse width of 360 μs) is controlled using the relay. The contactor relay is provided with a capacity of 32 A.

8.3.2.5 TRANSFORMER

Transformers are electrical components that are able to increase or decrease the voltage and current levels without modifying their frequency. The high voltage is needed for flashing the xenon lamp; hence, a transformer is used to step up the voltage from 220 V to 500, 1,000, and 1,500 V. Two transformers were provided in the circuit. One is used for stepping up the current from 220 V to 500, 1,000, and 1,500 V in the flashing supply, and another is provided in the triggering circuit to step down the 220 V current to 80 V as needed for the input of the trigger transformer (as per the manufacturer's design).

8.3.2.6 RESISTORS

In order to protect the rectifiers, transformers, and other components in the circuit, current must be limited. Metal film resistors act as current limiting components in the circuit. Two resistors were provided with a capacity of 500 W, 150 E.

8.3.2.7 RECTIFIER

Pulsed light treatment works with DC current; hence, a full-wave rectifier was used to convert the incoming 500 V alternating current from the transformer into direct current. Two rectifiers with a capacity of 70 A were provided in the flashing circuit and another in the triggering circuit.

8.3.2.8 CAPACITOR

A capacitor is an electrical component with passive two terminals used to store electrical energy temporarily in an electric field. The capacitor is the main component of the xenon lamp since the lamp flashes at a higher voltage of 500 V. Three capacitors were provided in the circuit with capacities of 18 μF 500 V, 33 μF 1,000 V, and 50 μF 1,500 V.

8.3.2.9 TRIGGER PULSE TRANSFORMER

The trigger pulse transformer is used to supply the high voltage 30 kV required for ionization of the xenon gas as a very sharp pulse in a very short duration of time. By using the transformer, 220 V AC current was stepped down to 80 V AC. A rectifier was provided to convert AC into DC current. The trigger transformer was designed in such a way to step up 80 V DC current to 30 kV DC and to produce a sharp pulse. This results in a very high impulse voltage across the secondary coil of the trigger pulse transformer which is conducted into the trigger electrode wire that surrounds the lamp.

8.3.2.10 GENERATION OF PULSED LIGHT

A flash is started by first ionized gas subsequently creating very strong current pulses. Ionization is required to reduce the gas's electrical resistance to make the pulse with thousands of amperes flow down the tube. When this current pulse travels through the pulse light tube, it stimulates electrons surrounding the xenon atoms and makes them hurdle to higher energy levels followed by immediate come back to a lower orbit to produce photons. Thus, conversion of electrical energy into light energy is achieved.

8.3.2.11 DESIGN OF PULSED LIGHT CIRCUIT

The pulsed light treatment circuit was designed based on the calculated values from Table 8.1. The circuit design was to produce flashes of 3 pulses per second with a pulse width of 360 μs. The input energy supplied to the lamp, 2 joules (500 V), 16 joules (1,000 V), and 56 joules (1,500 V), were selected based on earlier research [51, 64]. The circuit can be divided into two major parts, namely: (i) flashing supply; and (ii) triggering supply. The electrical circuit was completely earthed.

8.3.2.12 FLASHING SUPPLY

The power line of 220 V was connected to the transformer to step up the voltage to 500, 1,000, and 1,500 V. Two rectifiers of 100 Amps provided in the circuit rectify the AC voltage into DC voltage. The rectified voltage charges the capacitors C (16 μFd 500 V, 33 μFd 1,000 V, 50 μFd 1,500 V) through two limiting resistors. The storage capacitors dump the energy stored in them into the flash lamp (positive supply to the anode), giving an intense flash of light when the trigger voltage fires the flash lamp. The storage capacitor recharges again, and the cycle continues. The selector switch provided on the main control board was used to vary the input voltage to the lamp (500, 1,000, and 1,500 V).

8.3.2.13 TRIGGERING SUPPLY

Trigger transformer is designed with an input of 80 V DC, so the power from the line 220 V has been stepped down using the Transformer TR2. The rectifier provided in the circuit (Figure 8.1) rectifies the input AC voltage (80 V) into DC voltage. This DC voltage is stored in the capacitor C1 (50 μFd, 500 V capacity) which is provided as an input to the trigger pulse transformer. Trigger pulse transformer was stepped up from 80 V to 30 kV producing a sharp pulse which has been conducted into the trigger electrode wire that surrounds the lamp (external triggering). Thus, the lamp is triggered. The contractor relay was provided to give a gap of 2 seconds between charging and discharging of the voltage from the capacitor in order to build up again and continue the process.

8.3.3 DESIGN OF CONTINUOUS PULSED LIGHT TREATMENT SYSTEM

A pulsed light continuous system for the preservation of liquid foods is discussed. The basic properties of the selected liquid foods, namely density and viscosity using DV-III ultra-Programmable Rheometer (Brookfield's), were determined.

8.3.3.1 FLOW CHARACTERISTICS

The capacity for the developed unit was assumed as 100 ml of liquid food per minute. Based on the assumption, all the other parameters have been designed. Velocity is the important parameter that reflects the residence time of the liquid foods in the treatment chamber. In order to have more residence of the liquid in the treatment chamber, the diameter of the tube (quartz) was selected as 5 mm. The velocity of the liquid foods was determined using the formula (Eqn. (8)) [35]. Reynolds number (Eqn. (9)) specifies the flow pattern such as laminar or turbulent flow in a system.

$$q = A \times v \tag{8}$$

$$N_{Re} = \frac{ñvD}{\mu} \tag{9}$$

where; 'q' is the flow rate (ml/min); 'A' is the cross sectional area of the tube (cm^2); 'v' is the velocity (m/s); 'N_{Re}' is the Reynolds number; 'ρ' is the density (g/cc); 'μ' is the viscosity (g cm/s); and 'D' is the diameter of the treatment tube (mm).

8.3.3.2 RESIDENCE TIME

Residence time is one of the prime factors that influence the penetration of the pulsed light in liquid foods. As the liquid passes through the quartz tube, the incidence of the pulsed light has to fall throughout the tube. The length of the selected xenon lamp was 300 mm, and hence the length of the treated tube was also taken as 300 mm. The residence time of the liquid food inside the treatment tube was calculated using [35].

$$\theta = \frac{\mathrm{v}}{\mathrm{q}} \tag{10}$$

where; 'θ' is the residence time (s); 'v' is the volume of the liquid food passing through the treatment chamber (ml); and 'q' is the flow rate (ml/s).

8.3.3.3 NUMBER OF PASSES

The residence time in a single pass through the tube was much less than the selected exposure time; hence, the required processing time of the liquid inside the treatment chamber was attained by recirculating the liquid foods. The number of passes that the liquid food has to undergo was calculated [35].

$$TP = \frac{t}{\theta} \tag{11}$$

where; 'TP' is the number of times the liquid passes through the treatment system; 't' is the processing time (s); and 'θ' is the residence time (s).

8.3.3.4 PULSED LIGHT RADIATION DOSE

The Bunsen-Roscoe reciprocity law for photochemical processes states that the magnitude of a photochemical reaction is directly related to the total dosage of radiant energy that hits the target. The intensity of the pulsed light treatment is expressed as pulsed light dose. The irradiance (I) of the lamp was measured using a UV-C radiometer (Delta OHM, Italy) placed below the quartz tube where the liquids flow through it. The overall effect of the radiation depends on a combination of applied intensity and exposure time. In order to find the pulsed light dose resulting in the maximum log reduction of the microbial population in liquid foods, preliminary experiments were done at different input voltages to the lamp, distances from the lamp, and exposure times. The applied dosage was calculated using the following formula [67].

$$\text{DpI} = \text{I} \times t \tag{12}$$

where; 'D_{pl}' is the pulsed light dose (J/cm^2); 'I' is the intensity (W/cm^2); and 't' is the exposure time (s).

8.3.4 LABORATORY MODEL PULSED LIGHT CONTINUOUS TREATMENT APPARATUS

A lab scale pulsed light technology apparatus is shown in Figure 8.1. This experimental setup consists of the following components:

- Feed tank;
- Pulsed light treatment chamber;
- Radiometer;
- Circulation system;
- Collection chamber;
- Main control board;
- Electrical panel board.

8.3.5 ESTIMATION OF EFFICIENCY OF THE EQUIPMENT

The inactivation efficacy of PL treatment depends on the delivered fluence. The efficacy is calculated by applying the fundamentals of photochemistry and biodosimetry. Further indicators like fluence rate, spectral irradiance, and radiance exposure are used for determining the inactivation efficacy. The fluence is calculated using the formula (Eqn. (13)). The efficiency of the pulsed light system is calculated using Eqn. (14).

$$\text{F} = e \times t \times \text{f} \tag{13}$$

$$\eta = \frac{F}{\text{E}} \times \frac{t}{3600} \tag{14}$$

where; 'F' is the fluence, j/cm^2; 'e' is the energy of one pulse j/cm^2 per pulse; 't' is the time of the treatment, s; 'f' is the pulse frequency, 1/s; and 'E' is the electrical input energy in measured by an induction energy meter, Watt-hours.

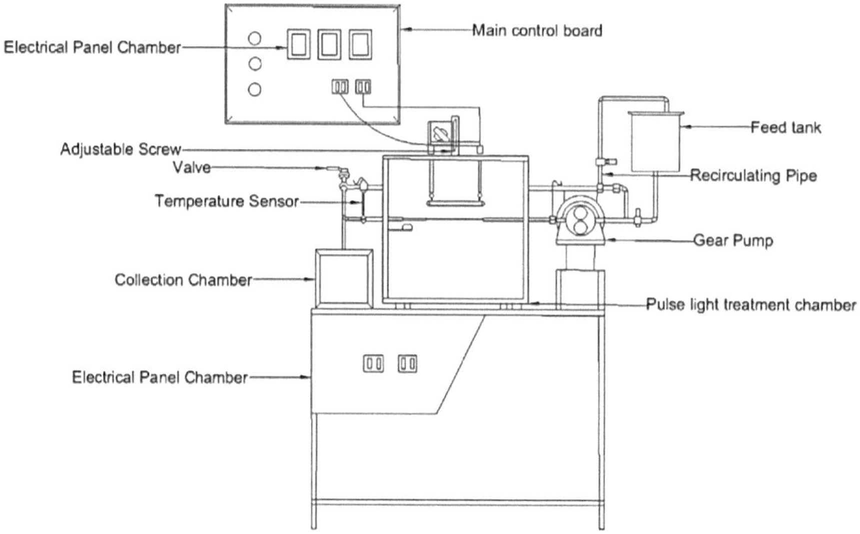

FIGURE 8.1 Lab scale pulsed light technology apparatus.

8.4 FACTORS AFFECTING THE DESIGN OF THE EQUIPMENT

Three major variables that affect the efficacy of PL treatment are: the type of foods and food matrices to be treated, the solid-liquid ratio of the food; the amount of contamination on food and the degree of microbial contamination; and the process parameters such as the shape and size of the treatment chamber, the position of the light and the position of the food, and the intensity of light. As a result, it appears that PL treatments and their effects differ greatly due to challenges in standardizing the parametrical configuration to allow interpretable, comparable, and repeatable decontamination results. Furthermore, users of this technology should be concerned not only with the cleaning rate, but also with the prevention of matrix damage [28].

In microbial Pulsed Light inactivation trials, several investigations have been focused on elucidating significant affecting elements. Food-related microorganisms and, to a lesser extent, medically important microorganisms have data available. The overall fluence impacting the cell surface was shown to be the most important variable for microbial death. Higher input voltage, less distance between flash lamp and sample, and more light flashes result in increased striking energy and, as a result, improved microbial death [6, 23, 39]. The instrumental set-up, experimental set-up, and data reporting methods vary greatly amongst investigations. Fluence ($J.cm^{-2}$) is a good way to describe the applied treatment intensity of PL [6, 39]. However, in many research, the energy input is merely specified as treatment times, the number of light flashes, the voltage, or the total emitted Joules.

8.4.1 EFFECT OF LIGHT ABSORPTION

The position of the microorganism and the media in which the microorganism is present, arrangement of proteins and intercellular liquids in the microorganism are critical factors to absorb the light. Sometimes the particles inside the food scatter the incident light, which diminishes the quantity of photons that impact cells and hence reduce the lethal effect [2]. The inactivation of *L. innocua* [5] and its efficiency depended on the light (UV-C) absorption in the 230–290 nm region that cleaved the lactoglobulin and -casein protein [28].

It has also been noted that a high protein or oil content reduces the microbicidal action of PL. Light absorption caused by microorganism pigmentation has also been demonstrated to affect PL efficiency. Melanin is well-known for its ability to protect microbial cells against irradiation by absorbing large amounts of energy. Some of the resistant developed microorganisms are dark pigmented *Aspergillus* [22] compared to *A. niger* and *F. culmorum* [3]. Furthermore, Farrell et al. [24] found that *P. aeruginosa* produces the pigment pyocyanin, as well as alginate slime, which protects against PL. Similarly, *B. subtilis* and *Geobacillus stearothermophilus* endospores are actually more resistant than their vegetative counterparts. This is due to vegetative cells' stronger light absorption, which provides an elevated amount of protection.

8.4.2 EFFECT OF SPECTRAL RANGE

The UV-C fraction of the PL spectrum (200–280 nm) has demonstrated microbial inactivation. The inactivation of *E. coli* by pulsed light treatment was observed around 270 nm, whereas no inactivation was observed above 300 nm [72]. Wavelengths below 300 nm are very critical to kill microorganisms, whereas the wavelengths between 300 and 400 nm result in a slight reduction in *L. innocua*, and complete removal of UV light results in a complete loss of microbicidal action [76]. The spectral range between 225 and 280 nm is significant for the inactivation of *B. subtilis* spores, and the spectrum above 280 nm did not show a significant effect [41]. Visible light and near-infrared irradiation alone did not show any significant bactericidal effect on foods [58, 59]. These findings are consistent with the absorption spectra by DNA and proteins having a peak in the 250–280 nm region.

Larger voltages result in a shift toward shorter wavelengths and, as a result, higher UV-C proportions, which boost the germicidal action [27, 29]. Depending on the applied voltage, the UV fraction is usually around 20% of the overall fluence [36]. Pulsed light treatments through the packaging material with high UV transmission have been found to be efficient in reducing microbes on food surfaces [13].

As substantial quantities of UV-C are transmitted, multilayer films made of polyamide and polyethylene, polypropylene, or low-density polyethylene appear to be acceptable. The pulsed light treatment through different plastic copolymer films made up of polyamide, vinyl acetate, and polyethylene had no effect on the inactivation of *L. monocytogenes* [25]. However, the impact of pulsed light treatment on the packaged food's physical and chemical qualities needs to be investigated.

8.4.3 EFFECT OF APPLICATION ENERGY

The total number of photons defines the degree of inactivation of microbes by phytochemicals rather than the energy dosage, as stated by the Bunsen–Roscoe principle [27]. So, the effect of one light flash with a high energy dose is on par with the effect of multiple light flashes with a low energy dose. Similar results for *B. subtilis* and *B. cereus* endospores, as well as vegetative cells, were demonstrated [41]. In contrast, the inactivation of *A. niger* conidiospores on polystyrene was highly influenced by the energy input, which was much lower when numerous low-energy light pulses were used instead of a single high-energy light flash.

The applied pulse frequency range of 1–5 Hz did not affect the inactivation of *Salmonella* typhimurium [45]. The total fluence attained in the sample is the most important factor for microbial inactivation, not the input voltage [39]. Pulsed light inactivates *L. monocytogenes*, phages, and *E. coli* more quickly and effectively than UV-C [10, 12]. The germicidal efficiency of continuous and pulsed UV sources is identical and better when compared to a low-pressure UV lamp emitting 254 nm and a broad-spectrum flash lamp [50, 72].

8.5 OPTIMIZATION OF EQUIPMENT DESIGN

Pulse light treatment apparatus must be tested for its efficiency and safety for optimized design [40].

8.5.1 TEST FOR ELECTRICAL AND MECHANICAL SAFETY

The assessment of the reciprocity law and probable peak power dependence is the key element in pulse light treatment apparatus. The efficiency of the pulsed light apparatus depends on multiple parameters; hence, to synergistically assess the efficiency of the pulsed light apparatus, photosensitizers are required additionally. The electromagnetic wave safety test for pulsed light apparatus is prescribed by the IEC 60601-1 [15].

8.5.2 TEST FOR OUTPUT ACCURACY AND CONSISTENCY

Pulsed light technology output is assessed to identify the error tolerance range. A light output measurer was used to verify output radiation and test the protection of harmful radiation from the light. The output accuracy was also examined for output consistency.

8.5.3 WAVELENGTH OF OUTPUT TEST

This test looks at the spectrum of wavelengths emitted by IPL. The wavelength sensor was irradiated and positioned close to the illumination region during the experiment. A spectrometer was then used to determine the wavelength of the radiated light. All the tests recommended by the makers to determine the accuracy of the output wavelength for the filter and handpiece were repeated.

The measurement values must be within the criterion set by manufacturers, based on the performance testing of certified products. However, in accordance with IEC 60601-2-57, the error tolerance range is set at less than 5% when the output wavelength is on the short wavelength boundary (360 nm–480 nm) [15].

8.6 ADVANTAGES AND DISADVANTAGES

Despite their decontamination abilities, pulsed light technology may not be employed in a food-processing facility when one or more of the following requirements are significantly different: (i) increased food safety and shelf life; (ii) preservation of organoleptic and nutritional attributes; (iii) no residues; (iv) convenience; (v) economic efficiency; and (vi) no consumer or regulatory concerns.

If sufficient measures are taken, PL is particularly safe to use for the operator and has no impact on the environment [24, 28]. Furthermore, under typical conditions, needed chemicals, disinfectants, and preservatives can be eliminated, allowing for the production of high-quality meals with no remaining contaminants from the decontamination process [19]. Xenon lamps are a good alternative to the often-used CW UV lamps, which contain mercury, damaging to the environment [49]. Finally, the low energy input completes the ecological benefits [20]. Although the creation of ozone is undesirable, it may be regulated. When it comes to the technique's ease, the key advantages are the fast-processing durations, the ability to operate in batch or continuous mode, the ability to integrate PL treatment into current operations, and the small footprint [20, 49, 57].

Pulsed light technology can be applied using UV lamps. The ability to create tailored high-efficiency plants with high throughput is enabled by the instant action of the light source, natural cooling of the lamps, matrix between the pulses, and changeable light spectrum (via the pulse generating network or applied filters) with high UV output [19, 73, 75]. However, one disadvantage is that geometric configurations and process parameters can be difficult to standardize at times [28]. Currently, the market environment, which includes only a few global players and a few minor market competitors, may limit food companies' ability to be as flexible as they would like.

The hefty investment requirements of industrial PL equipment may deter buyers, making the technology more appealing to high-value foods or niche markets [11]. The highly complicated and expensive driving circuits (10 to 100 times more expensive than CW UV) that are required to ensure preferably high UV outputs and long bulb lifetimes are cited as justifications for the expense.

The lifespan of equipment varies based on the operating factors and can range from 6 to 12 months. However, the costs are several times that of CW UV lamps. This contrasts with the described emerging technology's low operating expenses and thus increased cost efficiency.

8.7 SUMMARY

Pulsed light is a non-thermal decortication technique used to sterilize pathogenic and spoilage bacteria. It brings new opportunities in food decontamination, especially for meat and meat products. Although its efficiency is limited to the surface, reducing vegetative cells and spores before or after packaging improves product safety without the need for traditional thermal or chemical decontamination methods. Factors affecting the efficacy of the machine include the light source, position, input energy, and type of food materials. Designing a pulsed light technology apparatus poses a major challenge in implementing this technology on a commercial scale. Further standard procedures for evaluating the apparatus are required to develop efficient pulsed light technology.

ACKNOWLEDGMENT

The authors acknowledge the Department of Food and Agricultural Process Engineering, Tamil Nadu Agricultural University, Coimbatore for providing the facilities.

KEYWORDS

- **chemical decontamination**
- **circuit design**
- **efficiency**
- **food materials**
- **non-thermal decortication technique**
- **pulsed light**
- **spoilage bacteria**
- **ultraviolet**

REFERENCES

1. Administration, Food & Drug (FDA). (2015). Code of federal regulations title 21, sec. 179.41 pulsed light for the treatment of food: 21CFR179. 41. Washington DC: US–FDA. https://www.accessdata.fda.gov/scripts/cdrh/cfdocs/cfcfr/CFRSearch.cfm?fr=179.41.
2. Aguirre, J. S., Hierro, E., Fernández, M., & de Fernando, G. D. G. (2014). Modelling the effect of light penetration and matrix color on the inactivation of *Listeria innocua* by pulsed light. *Innovative Food Science & Emerging Technologies, 26*, 505–510.
3. Anderson, J. G., Rowan, N. J., MacGregor, S. J., Fouracre, R. A., & Farish, O. (2000). Inactivation of food-borne enteropathogenic bacteria and spoilage fungi using pulsed-light. *IEEE Transactions on Plasma Science, 28*(1), 83–88.
4. Ansari, I. A., & Datta, A. K. (2003). An overview of sterilization methods for packaging materials used in aseptic packaging systems. *Food and Bioproducts Processing, 81*(1), 57–65.

5. Artíguez, M. L., Arboleya, J.-C., & De Maranon, I. M. (2012). Influence of β-lactoglobulin and β-casein on *Listeria innocua* inactivation by pulsed light. *International Journal of Food Microbiology, 153*(1–2), 223–228.
6. Artíguez, M. L., & Martínez de Marañón, I. (2014). Process parameters affecting *Listeria innocua* inactivation by pulsed light. *Food and Bioprocess Technology, 7*(9), 2759–2765.
7. Bank, H. L., John, J., Schmehl, M. K., & Dratch, R. J. (1990). Bactericidal effectiveness of modulated UV light. *Applied and Environmental Microbiology, 56*(12), 3888–3889.
8. Baranda, A. B., Lasagabaster, A., & de Marañón, I. M. (2017). Static and continuous flow-through pulsed light technology for pesticide abatement in water. *Journal of Hazardous Materials, 340*, 140–151.
9. Bhavya, M. L., & Umesh Hebbar, H. (2017). Pulsed light processing of foods for microbial safety. *Food Quality and Safety, 1*(3), 187–202.
10. Bohrerova, Z., Shemer, H., Lantis, R., Impellitteri, C. A., & Linden, K. G. (2008). Comparative disinfection efficiency of pulsed and continuous-wave UV irradiation technologies. *Water Research, 42*(12), 2975–2982.
11. Cacace, D., & Palmieri, L. (2014). High-intensity pulsed light technology. In *Emerging Technologies for Food Processing* (pp. 239–258). Elsevier.
12. Cheigh, C. I., Hwang, H. J., & Chung, M. S. (2013). Intense pulsed light (IPL) and UV-C treatments for inactivating *Listeria monocytogenes* on solid medium and seafoods. *Food Research International, 54*(1), 745–752.
13. Chen, B. Y., Lung, H. M., Yang, B. B., & Wang, C. Y. (2015). Pulsed light sterilization of packaging materials. *Food Packaging and Shelf Life, 5*, 1–9.
14. Chen, D., Mosher, W., Wiertzema, J., Peng, P., Min, M., Cheng, Y., An, J., Ma, Y., Fan, X., Niemira, B. A., Baumler, D. J., Chen, C., Chen, P., & Ruan Chen, R. (2021). Effects of intense pulsed light and gamma irradiation on *Bacillus cereus* spores in mesquite pod flour. *Food Chemistry, 344*, 128675.
15. Commission, I. E. (2010). Medical electrical equipment-Part 2-33: Particular requirements for the basic safety and essential performance of magnetic resonance equipment for medical diagnosis. IEC 60601-2-33 Ed. 3.0. 302 page edition 3.
16. Craighead, S., Huang, R., Chen, H., & Kniel, K. E. (2021). The use of pulsed light to inactivate *Cryptosporidium parvum* oocysts on high-risk commodities (Cilantro, mesclun lettuce, spinach, and tomatoes). *Food Control, 126*, 107965.
17. Deeth, H. C., Datta, N., Ross, A. I. V, & Dam, X. T. (2007). Chapter 13: Pulsed electric field technology: Effect on milk and fruit juices. *Advances in Thermal and Non-Thermal Food Preservation, 241,* 249–261, Blackwell Publishing.
18. Demirci, A., & Krishnamurthy, K. (2011). Pulsed ultraviolet light. *Nonthermal Processing Technologies for Food*, 249–261. Publisher: Wiley-Blackwell.
19. Dunn, J., Bushnell, A., Ott, T., & Clark, W. (1997). Pulsed white light food processing. *Cereal Foods World (USA), 42*(7), 510–515.
20. Dunn, J. E., Clark, R. W., Asmus, J. F., Pearlman, J. S., Boyer, K., Painchaud, F., & Hofmann, G. A. (1989). Methods for preservation of foodstuffs. *Google Patents*. U.S. Patent No. 4,871,559. Washington, DC.
21. Elmnasser, N., Guillou, S., Leroi, F., Orange, N., Bakhrouf, A., & Federighi, M. (2007). Pulsed-light system as a novel food decontamination technology: A review. *Canadian Journal of Microbiology, 53*(7), 813–821.

22. Esbelin, J., Mallea, S., J Ram, A. F., & Carlin, F. (2013). Role of pigmentation in protecting *Aspergillus niger* conidiospores against pulsed light radiation. *Photochemistry and Photobiology, 89*(3), 758–761.
23. Farrell, H., Garvey, M., & Rowan, N. (2009). Studies on the inactivation of medically important Candida species on agar surfaces using pulsed light. *FEMS Yeast Research, 9*(6), 956–966.
24. Farrell, H. P., Garvey, M., Cormican, M., Laffey, J. G., & Rowan, N. J. (2010). Investigation of critical inter-related factors affecting the efficacy of pulsed light for inactivating clinically relevant bacterial pathogens. *Journal of Applied Microbiology, 108*(5), 1494–1508.
25. Fernández, M., Manzano, S., de la Hoz, L., Ordonez, J. A., & Hierro, E. (2009). Pulsed light inactivation of *Listeria monocytogenes* through different plastic films. *Foodborne Pathogens and Disease, 6*(10), 1265–1267.
26. Ferrario, M., Alzamora, S. M., & Guerrero, S. (2015). Study of the inactivation of spoilage microorganisms in apple juice by pulsed light and ultrasound. *Food Microbiology, 46*, 635–642.
27. Gómez-López, V. M., & Bolton, J. R. (2016). An approach to standardize methods for fluence determination in bench-scale pulsed light experiments. *Food and Bioprocess Technology, 9*(6), 1040–1048.
28. Gomez-Lopez, V. M., Devlieghere, F., Bonduelle, V., & Debevere, J. (2005). Intense light pulses decontamination of minimally processed vegetables and their shelf-life. *International Journal of Food Microbiology, 103*(1), 79–89.
29. Gómez-López, V. M., Koutchma, T., & Linden, K. (2012). Chapter 8 – Ultraviolet and pulsed light processing of fluid Foods. In P. J. Cullen, B. K. Tiwari, & V. P. B. T. N. T., & N. T. T. for F. F. Valdramidis (Eds.), *Nonthermal Processing Technologies for Food* (pp. 185–223). Academic Press.
30. Grapperhaus, M., Schaefer, R., & Linden, K. (2007). Modeling of a new UV test cell for evaluation of lamp fluence rate effects in regard to water treatment, and comparison to collimated beam tests. *Journal of Environmental Engineering and Science, 6*(3), 271–276.
31. Hayes, J., Kirf, D., Garvey, M., & Rowan, N. (2013). Disinfection and toxicological assessments of pulsed UV and pulsed-plasma gas-discharge treated-water containing the waterborne protozoan enteroparasite *Cryptosporidium parvum*. *Journal of Microbiological Methods, 94*(3), 325–337.
32. Hillegas, S. L., & Demirci, A. (2003). Inactivation of *Clostridium sporogenes* in clover honey by pulsed UV-light treatment. *Agricultural Engineering International: The CIGR Journal, 5*, 1–7.
33. Holck, A. L., Liland, K. H., Drømtorp, S. M., Carlehög, M., & McLeod, A. (2017). Comparison of UV-C and pulsed UV light treatments for reduction of *Salmonella, Listeria monocytogenes*, and enterohemorrhagic *Escherichia coli* on eggs. *Journal of Food Protection, 81*(1), 6–16.
34. Huang, R., & Chen, H. (2018). Evaluation of inactivating *Salmonella* on iceberg lettuce shreds with washing process in combination with pulsed light, ultrasound and chlorine. *International Journal of Food Microbiology, 285*, 144–151.
35. Ibarz, A., & Barbosa-Cánovas, G. V. (2002). Unit operations in food engineering. In *Unit Operations in Food Engineering*. CRC Press. 467–485.

36. Kramer, B., Wunderlich, J., & Muranyi, P. (2017). Pulsed light decontamination of endive salad and mung bean sprouts in water. *Food Control, 73*, 367–371.
37. Krishnamurthy, K., Demirci, A., & Irudayaraj, J. M. (2007). Inactivation of *Staphylococcus aureus* in milk using flow-through pulsed UV-light treatment system. *Journal of Food Science, 72*(7), M233–M239.
38. Krishnamurthy, K., Demirci, A. L. I., & Irudayaraj, J. (2004). Inactivation of *Staphylococcus aureus* by pulsed UV-light sterilization. *Journal of Food Protection, 67*(5), 1027–1030.
39. Lasagabaster, A., & Martínez de Marañón, I. (2013). Impact of process parameters on *Listeria innocua* inactivation kinetics by pulsed light technology. *Food and Bioprocess Technology, 6*(7), 1828–1836.
40. Lee, S., Ju, C., Lee, T., Na, H., Lee, I., & Park, C. (2017). Development of evaluation methods for the approval and review of Intense pulsed light (IPL). *2017 39th Annual International Conference of the IEEE Engineering in Medicine and Biology Society (EMBC)*, 4525–4528.
41. Levy, C., Aubert, X., Lacour, B., & Carlin, F. (2012). Relevant factors affecting microbial surface decontamination by pulsed light. *International Journal of Food Microbiology, 152*(3), 168–174.
42. Liu, X., Fan, X., Wang, W., Yao, S., & Chen, H. (2021). Wetting raw almonds to enhance pulse light inactivation of Salmonella and preserve quality. *Food Control, 125*, 107946. https://doi.org/10.1016/j.foodcont.2021.107946.
43. Lukšiene, Ž. (2005). New approach to inactivation of harmful and pathogenic microorganisms by photosensitization. *Food Technology and Biotechnology, 43*(4), 411–418.
44. Luksiene, Z., Buchovec, I., & Paskeviciute, E. (2009). Inactivation of food pathogen *Bacillus cereus* by photosensitization in vitro and on the surface of packaging material. *Journal of Applied Microbiology, 107*(6), 2037–2046. https://doi.org/10.1111/j.1365-2672.2009.04383.x.
45. Luksiene, Z., Gudelis, V., Buchovec, I., & Raudeliuniene, J. (2007). Advanced high-power pulsed light device to decontaminate food from pathogens: Effects on *Salmonella typhimurium* viability in vitro. *Journal of Applied Microbiology, 103*(5), 1545–1552.
46. Macias-Rodriguez, B., Yang, W., Schneider, K., & Rock, C. (2014). Pulsed UV light as a postprocessing intervention for decontamination of hard-cooked peeled eggs. *International Journal of Food Science & Technology, 49*(11), 2472–2480. https://doi.org/10.1111/ijfs.12571.
47. Mandal, R., Mohammadi, X., Wiktor, A., Singh, A., & Pratap Singh, A. (2020). Applications of pulsed light decontamination technology in food processing: An overview. *Applied Sciences, 10*(10), 3606. https://doi.org/10.3390/app10103606.
48. Nicorescu, I., Nguyen, B., Moreau-Ferret, M., Agoulon, A., Chevalier, S., & Orange, N. (2013). Pulsed light inactivation of *Bacillus subtilis* vegetative cells in suspensions and spices. *Food Control, 31*(1), 151–157. https://doi.org/10.1016/j.foodcont.2012.09.047.
49. Ortega-Rivas, E. (2012). Pulsed light technology. In *Non-Thermal Food Engineering Operations* (pp. 263–273). Springer.
50. Otaki, M., Okuda, A., Tajima, K., Iwasaki, T., Kinoshita, S., & Ohgaki, S. (2003). Inactivation differences of microorganisms by low pressure UV and pulsed xenon lamps. *Water Science and Technology, 47*(3), 185–190.

51. Pataro, G., Muñoz, A., Palgan, I., Noci, F., Ferrari, G., & Lyng, J. G. (2011). Bacterial inactivation in fruit juices using a continuous flow Pulsed Light (PL) system. *Food Research International*, *44*(6), 1642–1648. https://doi.org/10.1016/j.foodres.2011.04.048.
52. Pollock, A. M., Pratap Singh, A., Ramaswamy, H. S., & Ngadi, M. O. (2017). Pulsed light destruction kinetics of *L. monocytogenes*. *LWT*, *84*, 114–121. https://doi.org/10.1016/j.lwt.2017.05.040.
53. Preetha, P., Pandiselvam, R., Varadharaju, N., Kennedy, Z. J., Balakrishnan, M., & Kothakota, A. (2021). Effect of pulsed light treatment on inactivation kinetics of Escherichia coli (MTCC 433) in fruit juices. *Food Control*, *121*, 107547. https://doi.org/10.1016/j.foodcont.2020.107547.
54. Rajkovic, A., Tomasevic, I., De Meulenaer, B., & Devlieghere, F. (2017). The effect of pulsed UV light on *Escherichia coli* O157: H7, *Listeria monocytogenes*, *Salmonella typhimurium*, *Staphylococcus aureus* and *staphylococcal enterotoxin* A on sliced fermented salami and its chemical quality. *Food Control*, *73*, 829–837.
55. Ramos-Villarroel, A. Y., Aron-Maftei, N., Martín-Belloso, O., & Soliva-Fortuny, R. (2012). Influence of spectral distribution on bacterial inactivation and quality changes of fresh-cut watermelon treated with intense light pulses. *Postharvest Biology and Technology*, *69*, 32–39.
56. Ramos-Villarroel, A. Y., Aron-Maftei, N., Martín-Belloso, O., & Soliva-Fortuny, R. (2012). The role of pulsed light spectral distribution in the inactivation of *Escherichia coli* and *Listeria innocua* on fresh-cut mushrooms. *Food Control*, *24*(1–2), 206–213.
57. Ramos-Villarroel, A. Y., Martín-Belloso, O., & Soliva-Fortuny, R. (2011). Bacterial inactivation and quality changes in fresh-cut avocado treated with intense light pulses. *European Food Research and Technology*, *233*(3), 395–402. https://doi.org/10.1007/s00217-011-1533-6.
58. Register, F. (1999). Pulsed light treatment of food. *Federal Register*, *66*, 338829–338830.
59. Rowan, N. J., MacGregor, S. J., Anderson, J. G., Fouracre, R. A., McIlvaney, L., & Farish, O. (1999). Pulsed-light inactivation of food-related microorganisms. *Applied and Environmental Microbiology*, *65*(3), 1312–1315. https://doi.org/10.1128/AEM.65.3.1312-1315.1999.
60. Rybak, K., Wiktor, A., Pobiega, K., Witrowa-Rajchert, D., & Nowacka, M. (2021). Impact of pulsed light treatment on the quality properties and microbiological aspects of red bell pepper fresh-cuts. *LWT*, *149*, 111906. https://doi.org/10.1016/j.lwt.2021.111906.
61. Schaefer, R., Grapperhaus, M., Schaefer, I., & Linden, K. (2007). Pulsed UV lamp performance and comparison with UV mercury lamps. *Journal of Environmental Engineering and Science*, *6*(3), 303–310.
62. Schottroff, F., Fröhling, A., Zunabovic-Pichler, M., Krottenthaler, A., Schlüter, O., & Jäger, H. (2018). Sublethal Injury and Viable but Non-culturable (VBNC) State in microorganisms during preservation of food and biological materials by non-thermal processes. *Frontiers in Microbiology*, *9*, 2773. https://doi.org/10.3389/fmicb.2018.02773.
63. Sharma, R. R., & Demirci, A. (2003). Inactivation of *Escherichia coli* O157 on inoculated alfalfa seeds with pulsed ultraviolet light and response surface modeling. *Journal of Food Science*, *68*(4), 1448–1453. https://doi.org/10.1111/j.1365-2621.2003.tb09665.x.

64. Stevens, C., Khan, V. A., Lu, J. Y., Wilson, C. L., Chalutz, E., Droby, S., Kabwe, M. K., Haung, Z., Adeyeye, O., & Pusey, L. P. (1999). Induced resistance of sweet potato to Fusarium root rot by UV-C hormesis. *Crop Protection, 18*(7), 463–470.
65. Takeshita, K., Shibato, J., Sameshima, T., Fukunaga, S., Isobe, S., Arihara, K., & Itoh, M. (2003). Damage of yeast cells induced by pulsed light irradiation. *International Journal of Food Microbiology, 85*(1–2), 151–158.
66. Tao, T., Ding, C., Han, N., Cui, Y., Liu, X., & Zhang, C. (2019). Evaluation of pulsed light for inactivation of foodborne pathogens on fresh-cut lettuce: Effects on quality attributes during storage. *Food Packaging and Shelf Life, 21*, 100358.
67. Wang, B., Wei, W., Aputexiakere, J., Li, Y., & Ma, H. (2022). Surface decontamination of whole eggs using pulsed light technology and shelf-life study of combined pulsed light and vaseline coating during room temperature storage. *Food Control, 137*, 108411. https://doi.org/10.1016/j.foodcont.2021.108411.
68. Wang, T., MacGregor, S. J., Anderson, J. G., & Woolsey, G. A. (2005). Pulsed ultra-violet inactivation spectrum of *Escherichia coli. Water Research, 39*(13), 2921–2925.
69. Wekhof, A. (2000). Disinfection with flash lamps. *PDA Journal of Pharmaceutical Science and Technology, 54*(3), 264–276.
70. Wekhof, A. (2003). Sterilization of packaged pharmaceutical solutions, packaging and surgical tools with pulsed UV light. *Proceedings of the Second International Congress UV Technologies*, 9–11.
71. Wekhof, A., Trompeter, F.-J., & Franken, O. (2001). Pulsed UV disintegration (PUVD): a new sterilization mechanism for packaging and broad medical-hospital applications. *The First International Conference on Ultraviolet Technologies*, 15.
72. Woodling, S. E., & Moraru, C. I. (2007). Effect of spectral range in surface inactivation of *Listeria innocua* using broad-spectrum pulsed light. *Journal of Food Protection, 70*(4), 909–916.
73. Wuytack, E. Y., Phuong, L. D. T., Aertsen, A., Reyns, K. M. F., Marquenie, D., De Ketelaere, B., Masschalck, B., Van Opstal, I., Diels, A. M. J., & Michiels, C. W. (2003). Comparison of sublethal injury induced in *Salmonella enterica* serovar Typhimurium by heat and by different nonthermal treatments. *Journal of Food Protection, 66*(1), 31–37. https://doi.org/10.4315/0362-028x-66.1.31.
74. Xu, W., & Wu, C. (2016). The impact of pulsed light on decontamination, quality, and bacterial attachment of fresh raspberries. *Food Microbiology, 57*, 135–143. https://doi.org/10.1016/j.fm.2016.02.009.
75. Zou, X. Y., Lin, Y. L., Xu, B., Cao, T. C., Tang, Y. L., Pan, Y., Gao, Z. C., & Gao, N. Y. (2019). Enhanced inactivation of *E. coli* by pulsed UV-LED irradiation during water disinfection. *Science of The Total Environment, 650*, 210–215. https://doi.org/10.1016/j.scitotenv.2018.08.367.

CHAPTER 9

FUNDAMENTALS OF MICROBIAL INACTIVATION BY PULSED LIGHT

N. U. SRUTHI, PAVULURI SRINIVASA RAO, and REWATI RAMAN BHATTARAI

ABSTRACT

With increasing concerns associated with microbial safety of food, novel decontamination techniques are on the verge of commercialization, among which pulsed light processing has surfaced as an excellent alternative to conventional processing techniques. The method involves using shorter duration high-power pulses of light with a broad spectral wavelength rich in ultraviolet (UV) content. The dominance of UV in the pulsed light spectrum results in photo-induced reactions and alterations in the microbial cell structure leading to reduced reproducibility and cell death. However, as pulsed light is a complex process with a broad spectrum of radiation, the mechanism of microbial inactivation needs to be explored in detail. The chapter deals with different types of interactions of the pulsed light spectra with cellular constituents and the varying susceptibility of the microbes against the radiation owing to their specific structure. An overview of inactivation kinetics and different modeling approaches is also given, with a brief discussion on various critical factors affecting decontamination efficacy.

9.1 INTRODUCTION

With developments in food microbiology, the discovery of more foodborne pathogens in the past few years, and constant outbreaks of foodborne maladies all across the world, the safety of food has become the most critical element of food processing and consumption. The development of modern

Nonthermal Light-Based Technologies in Food Processing.
C. K. Sunil, Preeti Birwal, Megh R. Goyal, and R. Mahendran (Eds.)

food processing techniques has been driven by the pivotal concept of ensuring food safety as conventional methods like refrigeration, heat, or chemical preservatives have shown significant drawbacks to the freshness and nutritional excellence of food [37]. With the strong demand for superior-quality foods with intact natural freshness and organoleptic characteristics and an emphasis on the need for stricter microbial control, several novel techniques are being developed and experimented with for achieving this objective. Pulsed light (PL) processing is one such emerging non-thermal processing technique with tremendous potential in the food industry with remarkable prospects in microbial decontamination in fresh food produce. The pulsed light technique employs electric arc-flash lamps based on noble gases like xenon, designed to generate short-duration, incoherent, high-power pulses of the broad spectrum of light covering the electromagnetic (EM) spectrum with wavelengths between 200 and 1,100 nm, which comprise the ultraviolet, visible (VL), and infrared (IR) regions. A much higher bactericidal effect is evident with a shorter pulse duration, which delivers higher energy with peak power. The high-power light pulses have been validated to be sufficiently strong for deactivating microbes by multiple mechanisms involving photochemical, photothermal, and photophysical pathways. Characteristically, each pulse presents an intensity almost 20,000 times that of the sunlight incident on the earth's surface but still preserves the sensitive organoleptic and nutritional attributes of food, therefore finding ample opportunities for minimal processing of food products [5].

Typically, Pulsed Light (PL) is claimed to be a modified and improved method of exposing the target food products to UV-C; unlike in UV-C, the treatment mode is continuous, and the germicidal effect is accomplished by means of low-pressure mercury-based lamps that deliver energy through monochromatic light at 254 nm [16]. Occasionally, medium-pressure UV-based lamps with polychromatic output have also been used due to their superior UV power per unit length. This includes wavelengths ranging from 200 to 300 nm. In contrast, PL operates with xenon lamps that generate several flashes per unit time with a broad spectrum. Large-scale applications of PL currently found on an industrial scale are for the sterilization of packaging material; however, food processing applications are not scaled up in large numbers even though they have had approval from the Food and Drug Administration (FDA) since 1996. Therefore, more thorough knowledge of the interaction of PL with spoilage-causing microorganisms in conjunction with studies on accompanying impacts on food quality parameters is necessary for commercializing this technique on a large scale.

The present chapter deals with the principles and kinetics of microbial inactivation by Pulsed Light in different foods by discussing the susceptibility of the different classes of organisms to Pulsed Light. Additionally, other critical process parameters affecting microbial contamination rates are elaborated with a summary of recent studies.

9.2 PRINCIPLE AND MECHANISM OF MICROBIAL INACTIVATION

PL is found to have multiple germicidal properties that are critical for a wide range of microorganisms leading to their inactivation through various physicochemical processes. The destruction of microbial population is mostly attributable to the UV part of the spectrum with moderate influence from the other regions, too, imparting an additive effect. The microbial destruction is achieved entirely by employing three crucial mechanisms: photochemical, photothermal, and photophysical effects (Figure 9.1). The specific principle and changes occurring due to these mechanisms have been discussed in detail in Chapter 9. These mechanisms are accountable for microbial inactivation, either acting individually or in combination with each other, with respect to the target microorganisms and treatment conditions. Although there is sufficient literature validating the impact of photothermal and photophysical effects on the inactivation of pathogenic microbes, a good number of studies suggest light's photochemical effect as the dominant factor that steers the destruction of microorganisms [35].

Additionally, the observed photochemical effect during PL processing was correlated by many authors to be the exact mechanism occurring during continuous UV processing of food with supplementary identified impacts on the microbial cells, making PL a multi-target process. Through this, structural changes to the DNA were seen as significant alterations and damaged membranes, proteins, and other macromolecules [16]. Thus, the antimicrobial outcome of the UV spectrum on bacteria is primarily credited to the radiation absorption by the conjugates of carbon-carbon double bond (C=C) in the proteins and nucleic acid molecules that bring about structural vicissitudes in the microbial DNA as a consequence. Physical alterations such as single-strand RNA breakage, virion structure damage, and viral protein breakage are also reported during Murine norovirus inactivation by PL. The instantaneous overheating of the microbial cells can also rupture the viral capsid due to the absorbance of UV light [23].

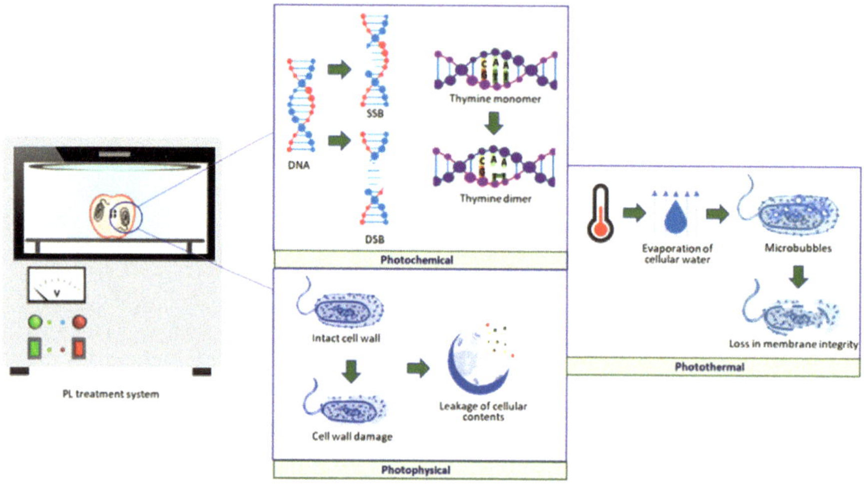

FIGURE 9.1 Mechanism of microbial inactivation by pulsed light processing showing the photochemical, photothermal, and photophysical effects. [SSB: Single-strand break; and DSB: double-strand break].

The postulation agrees with the results presented in a study where intense Pulsed Light (PL) and UV-C treatments were compared to investigate the cell injury observed in *Listeria monocytogenes* and *Escherichia coli* [7]. On electrophoresing the extracted and isolated genomic DNA and RNA from both the irradiated samples to detect double-strand breaks (DSBs), single-strand breaks (SSBs), and cyclobutane pyrimidine dimers (CPDs), it was found that higher densities of these compounds were accumulated after UV-C treatment than in PL treatment owing to the photochemical effect. However, upon analyzing the inactivation characteristics, it was found that more dramatic reductions of microbes were observed after PL treatments in a shorter duration, while UV-C treatment presented slower output. This confirms that apart from DNA/RNA damage induced by the photochemical effect, other factors are also responsible for the observed efficiency of PL processing for microbial decontamination. This was confirmed by the transmission electron microscopy examination of the treated cells, where distinct structural differences were observed following both treatments. Excepting a blurry and indistinct cell wall, the UV-C treated cell was found to be similar to the untreated control cell. In contrast, clear destruction of cellular constituents, for instance, the bacterial cell wall, the cytoplasmic membrane, and rupture of the internal organization was observed for PL treated cells of microbes which was attributed to the photothermal or photophysical effects

[27]. The thermal effects caused by light could result from the near-IR content of the PL spectrum. The instantaneous high energy of the applied pulse is eventually converted into heat, causing the emptying of the intracellular contents through explosion and evacuation. The water vapor-induced changes such as cell wall disruption, cytoplasmic membrane shrinkage, mesosome rupture, and accompanying leakage of cellular constituents fall under the photophysical effect [10]. Such drastic changes in the cellular structure eventually lead to the emptying of the cytoplasmic contents and, in due course, cell death.

The microbial destruction by PL has also been explained as an outcome of the spectral range of the light pulses apart from the thermophysical and chemical effects induced by light [5]. As mentioned before, the UV region of the PL spectrum was discerned to be the most efficient range for microbial inactivation. The UV range (100–400 nm) of the EM spectrum is further apportioned into four separate spectral areas, namely, vacuum UV (100–200 nm), UV-C (200–280 nm), UV-B (280–315 nm), and UV-A (315–400 nm), where the UV-C component was found to possess the maximum ability to inactivate microorganisms due to the strong and predominant absorption of wavelength 250–270 nm by the microbial nucleic acids with a theoretic inactivation peak at 253.7 nm [52]. The other major portions of the spectrum, i.e., the IR and VL regions, were only accountable for bacterial inactivation at higher energy levels. Though UV-C is considered the major causative factor resulting in targeted microbial destruction, the process of PL has a significantly greater advantage compared to conventional continuous UV processing. Although similar germicidal efficiency was achieved with both xenon flash lamp and low-pressure mercury lamps (continuous UV) operated at the identical wavelength, the intense light pulse, which lasts for typically a few 100 microseconds, resulted in targeted reduction in a much shorter processing duration compared to continuous UV, which required significantly longer exposure times [38].

9.3 SUSCEPTIBILITY OF MICROORGANISMS

The relationship between the light, the substrate, and the microbial cells significantly influences the extent of reflection, refraction, absorption, and scattering of the light. The occurrence of multiple internal reflections and redirections is also inevitable depending on the internal structure of the matrix. The susceptibility of microorganisms in terms of cell morphology and associated resistance to the incident radiation is a crucial factor from the point of view of microbial decontamination by pulsed light processing. This includes several inherent cellular factors, which mainly include the cell's genetic characteristics and structural properties. The former regulates the adeptness of the cell to counterweight or mend the PL-induced damages, while the latter is accountable for the absorption of the incident light spectra [11]. The structural features include the cell membrane composition and definite pigmentation, which can absorb the radiation providing a shielding effect to the internal structures.

Gram-positive bacteria are more tolerant and unaffected by PL than Gram-negative ones [13, 41, 49]. This observed difference in the sensitivity of each category of the organism can be attributed to the dissimilarities in their structure and constituents of the cell envelopes. The possible reasons also include the integral mechanisms in each type of cell to safeguard itself from several molecular and cellular damages like the leakage of the cell membrane, lipid hydroperoxidation, and the generation of intracellular reactive oxygen species (ROS) [6]. ROS, for instance, superoxide anion, hydrogen peroxide, and the hydroxyl ion, can exert intense stress on the aggravated membrane proteins making them more vulnerable to other effects. In different organisms like the Gram-positive, Gram-negative bacteria, and yeasts, the sensitivity of cellular components to ROS varies differently with the sequence of observed damage taking place in the order phospholipids > peptidoglycan in Gram-positive; lipopolysaccharides > phospholipids > peptidoglycan in Gram-negative, and glycoproteins > glucans > chitin in yeast [26].

In Gram-positive bacteria, such as *Listeria monocytogenes, Listeria innocua, Staphylococcus aureus,* and *Enterococcus faecium*, there are several thick peptidoglycan layers in the cell wall that are joined together to form a thick and rigid complex of 20–80 nm [44]. This protects the DNA located in the cytoplasm of the cells from photochemical products generated by the UV component in the PL spectrum. This could protect the DNA from structural distortion and prevent cell death [49]. Also, the presence of many polymers related to the cell wall, such as teichuronic and teichoic acids, could obstruct the penetration of UV photons into the cell wall layer, providing relatively strong resistance to Gram-positive bacteria against PL [32]. Additionally, the environment where these organisms naturally grow can also affect their resistance to PL. Gram-positive microbes are abundant in various environments, making them more likely to be exposed to UV light, thus increasing their resistance [45].

On the other hand, Gram-negative microbes are more predisposed to photodynamic therapy treatments due to the presence of a thin peptidoglycan layer of 1–2 nm and an outer membrane [44]. Furthermore, the cellular components in the organisms and some extracellular production can absorb UV, causing severe destruction of the integral structure. Additionally, these are primarily confined to survival in dark regions of animal/human alimentary tracts, making them less exposed to UV and impeding their evolution to UV radiation resistance [45].

However, in some other cases, Gram-negative bacteria were found to have better resistance to photodynamic therapy under specific treatment and matrix conditions. For instance, mucoid and pigment-forming microbes demonstrate higher resistance to photodynamic therapy at increased initial contamination levels. *Pseudomonas fluorescens*, a Gram-negative bacterium devoid of a thick peptidoglycan layer, can synthesize pyoverdine, a fluorescent blue pigment that absorbs light in the UV range, thus diminishing the effective photodynamic therapy dose [43]. On top of this potential defense mechanism, the bacterium can form biofilms that are complex matrices of extracellular polymeric substances that can offer physical protection to the enclosed bacterial population. Some studies also reported a porous outer layer of peptidoglycan and lipoteichoic acid in Gram-positive microbes, making the cell wall components interact easily with the photosensitizers, promoting their accumulation inside the cell [25].

Alternatively, Gram-negative bacteria, such as *Escherichia coli*, are stated to have an outer membrane comprising of negatively charged lipopolysaccharides localized outside the peptidoglycan layer together with some major proteins that make them more resistant to photodynamic inactivation [34]. Thus, variable intensities of bactericidal efficacy were demonstrated by different porphyrin derivatives upon illumination following the cell membrane composition of Gram-positive and Gram-negative bacteria. Resembling bacteria, the color of the spores also plays a weighty role in the resistance of fungal spores towards PL exposure. Pigments found in fungal spores, such as melanin, are presumed to be engaged in various cellular reactions, like adhesion and virulence, in addition to protecting the cells from light-induced stress and activity of ROS [9]. For instance, *Aspergillus niger* spores with specific pigments are more tolerant than *Fusarium culmorum* spores because more absorption in the UV-C region protects the spore against UV [16].

Similar to color, the difference in bacterial size could also affect the sensitivity of microbes to PL. It was perceived that a smaller bacterium is more resilient to PL than a larger one owing to the greater surface-to-volume ratio that alleviates the heat dissipation inside the cell [19]. However, it is also dependent on the surrounding matrix. Comparably, higher population densities, over 6.9 log/cm^2, can result in a shielding phenomenon as a result of overcrowding and congestion and light attenuation. The phenomenon of overcrowding usually results in protection of underlying cells to a specific extent, thus reducing the inactivation rate. The same occurrence is evident in spore-forming stalks, where a copious number of spores are encompassed in conglomerates, thus affecting the inactivation rate. Therefore, the appropriateness of PL to inactivate pathogens also depends on the different classes of microorganisms like vegetative bacteria, bacterial endospores, mold spores, viruses, parasites, or yeasts. From additional available data, it can be concluded that the sensitivities of different species and strain levels in descending order can be vegetative bacteria, yeasts, and viruses, vegetative molds, bacterial endospores, and fungal and mold spores [14, 28]. However, for making more conclusive remarks, methodical and biological variations in all different microbial groups must be compared within a single study that has not been done so far.

9.4 MICROBIAL INACTIVATION KINETICS

Pulsed light processing being a milder processing treatment, controlling and maintaining the attained microbial reduction, and avoiding any cell survival becomes utterly important. A combination of experimental data, knowledge of microbial action, and certain mathematical techniques allows the accurate prediction and control of microbial inactivation by PL processing. Also, a certain knowledge of the kinetics of microbial inactivation is a helpful tool in analyzing and describing the quality and safety kinetics through theoretical analysis of the process and interpreting the experimental results. The kinetic model responsible for any process is decided on the grounds of information that is extracted from the factors-responses relationship, which includes the primary models such as first and n^{th} kinetic order, Weibull model, Biphasic model, and their modified ones [10]. These explain the change in any particular response with respect to time for a corresponding set of processing parameters. In comparison, the secondary models characterize the parameters obtained from primary models, such as representing the inactivation rate constant as a function of the intrinsic or extrinsic processing factors or surrounding conditions. On the other hand, tertiary models combine the primary and secondary model parameters into software packages [21].

In general, the kinetics of microbial inactivation using PL acting on any target surface obeys a first-order logarithmic trend, in which the rate of inactivation is a direct function of the applied fluence and the treatment time, which is typically represented by:

$$\frac{N_f}{N_i} = e^{-kIt} \quad (1)$$

where; N_i and N_f represent the total count of viable microorganisms before and after the treatment, respectively; k is the 1st-order inactivation rate coefficient; I is the PL dosage intensity; and t shows the time of exposure, which denotes the total number of pulses [39].

However, with multiple explanations found in the literature on the mechanism of action of PL and the induced lethality, a complete characterization of the process has not been done so far, which limits the availability of quantitative calculative measures that could be employed for formulating equations. Consequently, a more inductive approach has been followed in the literature to develop models describing the inactivation curves by PL processing since a priori knowledge of the causative biological mechanisms is absent. Also, products like fruit juices and beverages are thermodynamically unstable complex systems that continuously change. Under such circumstances, the materials inside the product matrix have an affinity to shift to a high entropy-low enthalpy state from a state of low entropy-high enthalpy which enhances the spoilage by microbes and enzymes, and other physical forces [10]. After treatment with PL, there is also a high chance of occurrence of sublethal cells or viable-but-nonculturable organisms. The number of these organisms is usually underestimated by conventional counting methods like traditional agar plates [46]. Consequently, there is a need to elucidate a proper inactivation kinetic model since it is of utmost importance to public health. In such cases, the prevailing modeling approaches can be utilized to quantitatively elucidate and obtain valuable information regarding the effect of different processing parameters on the properties of the studied matrix and the inactivation of microorganisms.

9.4.1 KINETIC MODELING

Generally, different experiments are performed at varying time intervals to generate kinetic data by keeping the other variable quantities constant. In PL experiments, the kinetics could also be explained by considering the total fluence replacing the time value since total fluence represents the product of fluence rate and time or number of pulses and fluence per pulse [10]. For instance, microbial counts are recorded for varying PL exposure times at a given fluence rate. The data will be used to generate a mathematical equation relating the total count as a function of exposure period at a definite rate of fluence. This is a simple and classic example of a primary kinetic model.

The most commonly studied models to describe the inactivation curve of microbes via PL include the Log-linear model [1, 24, 50], the Weibull model [20, 42, 51], the Biphasic model [8, 13], the sigmoidal model [3, 36], etc. These models relate the change observed in the response with respect to the related factors by acquiring the characteristic trend in the model constants and coefficients such as rate constant (s^{-1}) and activation energy (J/mol).

In a log-linear model, also referred to as the D-value model, the trend obeys first-order kinetic behavior with the original treatment time as presented in Eqn. (1) replaced by the fluence (F, J/cm^2), giving the following modified equation:

$$\log_{10}\left(N_f\right) = \log_{10}\left(N_f\right) - _{k\max} \frac{F}{\text{In}\left(10\right)} \tag{2}$$

where; k_{max} is the inactivation rate (cm^2/J). The model assumes that all the vegetative cells or spores present in a population display comparable tolerance to the applied PL treatment, and the number of surviving organisms decrease in a semi-logarithmic fashion over the exposure time.

However, not always do the microbes present a similar kind of susceptibility or resistance towards the applied stress. For this, modified models were further developed to accommodate the tailing phenomenon, indicating the occurrence of sub-populations that are resistant to treatments. Moreover, the shouldering phenomenon, which represents the cells that must be subjected to a certain level of dosage before becoming lethally damaged, was also considered before refining the model. A non-log-linear model with greater biological significance was proposed to model the efficacy of PL as per the following equation:

$$\log N = \log\left(10^{\log N_i} - 10^{\log N_{res}}\right)e^{(k_{maxF})} + 10^{\log N_{res}} \tag{3}$$

where; N_{res} accounts for the residual microbial count that can possibly remain after the processing, which is responsible for the tailing phenomenon in the inactivation curve; and F is the fluence rate (W/cm^2).

It was observed that on increasing the fluence rate, k_{max} was found to decrease and $\log(N_{res})$ increased, which shows a reduced linear inactivation rate [42]. Thus, upon the increase of fluence, a tailing effect was evident in PL treatment of fruit juices.

Often, microbial population in food shows varying levels of inactivation owing to different microbial resistances of the sub-populations towards the processing method. Considering this, a biphasic model was proposed where the total population is divided into two sub-populations in which one is more sensitive to the treatment, and the other is resistant.

$$\log_{10}(N_f) = \log_{10}(N_i) + \log_{10}(f.\exp(-k_{max1.}t) + (1-f).\exp-(k_{\max 2.}t)) \quad \textbf{(4)}$$

where; $k_{\max 1}$ and $k_{\max 2}$ represent specific inactivation rates of the initial fraction of the most sensitive population represented by the fraction f and the most tolerant population represented by $(1 - f)$ [13].

According to the model equation, it was observed that with an increase in fluence rate, the number of the most sensitive microbial subpopulations was found to decrease in reaction rates. It can be understood that, even though the sensitive fraction of the microbes increased, it was difficult to inactivate the small remaining resistant population at an increased fluence rate. Furthermore, non-linear kinetics are usually described using a Weibull model where the heterogeneity of a microbial population in natural systems was considered. The model provided a probabilistic approach for studying the microbial inactivation process, also accommodating the shoulder, and tailing effects as described by the equation:

$$log(N_f) = log(N_i) - \left(\frac{F}{\delta}\right)^p \quad (5)$$

where; δ represents the fluence required for the initial decimal reduction time; and p is a dimensionless shape index expressing the upward and downward concavity of the inactivation curve.

Weibull model is the most commonly used and fitted model for *in vitro* inactivation of Gram-positive and Gram-negative bacteria in fruits, vegetables, milk, meat products, and food contact surfaces. However, the difference in the inactivation patterns and fitted kinetic model of specific microorganisms for the substrate is still not clearly explained. The observed tailing effect in irregular opaque solid food samples subjected to PL could be due to the more evident shadow effects in solids compared to stirred liquids. However, other significant factors responsible for the observed deviation of the lethality curve from linearity need to be explored.

9.4.2 MICROBIAL INACTIVATION CURVE

Usually, the PL inactivation kinetics give rise to a sigmoidal pattern with a definite shoulder, an exponential phase showing loss of cell culturability, and a final tail phase. The occurrence of a shouldering behavior can be due to many reasons. The difference in the proportions of distinct UV regions (UV-C, UV-B, and UV-A) in the light spectrum from the PL source used in different experiments could, in one way, explain the absence or presence of a shoulder phase in the inactivation curve [30]. The microorganisms show a certain amount of resistance to the PL radiation until a particular intensity threshold explains the shoulder phase. Once this intensity threshold is reached, there occurs a rapid reduction in the microbial population. Different studies showed that more resistant cells exhibit a shouldering behavior, which is mostly absent in the inactivation kinetic data of sensitive species [2]. For instance, on comparing spores and vegetative cells, the shoulder phase was evident for spores, whereas the curve does not present a shouldering nature in the sensitive vegetative cells. Similarly, Gram-positive bacteria reported an initial shoulder phase with a distinct kinetic behavior in some cases, whereas a log-linear curve was observed for Gram-negative bacteria. As discussed earlier, the resistance of any particular strain of bacteria is a result of its inherent structure and the incidence of protective substances outside or inside the cells that can delay the onset of microbial inactivation phase, resulting in a sigmoidal inactivation curve with the occurrence of a shoulder and tail. In addition, the apparent DNA damage and repair mechanism following any treatment can also result in the shouldering effect [12]. To attain the targeted inactivation, the total microbial injuries should exceed the capacity of DNA repair mechanisms to ensure a decline in survivors. For this, various photon events should penetrate the microbial cells according to the "multiple-hit target theory;" however, light attenuation reduces the number of photons reaching the microbe, allowing the DNA repair mechanisms to progress, generating the shoulder region. The existing literature shows that shoulders are only observed occasionally, besides being too short to be given relevance in practice. This may be due to the high fluences observed for the initial pulses, which shows the need for specific experiments using very few fluencies that could resolve shoulders.

After the shoulder phase, the inactivation phase occurs, resulting in loss of cell culturability and, ultimately, cell death. The inactivation can be essentially attributed to the creation of CPDs, which leads to clonogenic death, i.e., the loss of cells' capacity to replicate. In predictive microbiology, the mechanistic and the vitalistic concepts are used to explain any deviation in the linearity [47]. As per the vitalistic concept, the phenotypic variation between the cells renders the cells non-identical. This behavior persists during the application of stress which results in a deviation from the log-linear kinetics of inactivation at a population level. According to the mechanistic theory, the deviations were explained to have occurred owing to the processing parameters and the inaccessibility of some microbes towards the applied light dosage. Some explanations were also found relating this to the acquired resistance by the specific microbial strain or clumping of microorganisms and other experimental protocol issues.

Following the shoulder and inactivation phase, a plateau-like tailing stage is often seen in many inactivation curves, and various reasons have been established for this in the literature. The major cause of this tailing phenomenon is the lack of homogeneity in the population which shows the varying ability of each cell type to carry out DNA damage repair [20]. In *in vitro* experimental setups, the edge of the Petri dishes could create a shadow effect in some cases, which can also result in a tail in the inactivation curve. Also, when approaching the maximum detectable level of inactivation, tailing is evident owing to the existence of a comparatively resistant subpopulation that is impacted by the processing or owing to the treated matrix properties, which may have imparted some protective effects to the microbial population [18]. Other notable reasons for tailing found in the literature include the multi-hit phenomenon, existence of suspended solids in the treated matrix, clustering of microorganisms, and shadowing effects due to inadequate light penetration towards the depth of the treated sample [41]. On the other hand, considering the tailing phenomenon as an explanation of the saturation effect of the lethal agent while modeling the log-linear inactivation kinetics improved the model's fitness and accuracy.

9.5 CRITICAL FACTORS FOR PULSED LIGHT PROCESSING

Numerous studies have been reported in the literature over the past few years to elucidate the relevant factors associated with Pulsed Light processing influencing microbial inactivation. While designing a Pulsed Light treatment module for any specific food commodity, these critical parameters are often chosen as constraints for optimizing the processing parameters to obtain maximum efficacy against microorganisms and minimize any quality alterations in the product. With the standardization of these parametrical configurations being difficult, the output of Pulsed Light treatments varies significantly, negating interpretable, comparable, and reproducible decontamination outcomes. The following sections describe the crucial physical (pulse fluence rate, input voltage of emission, UV content in spectra), biological factors (microbial strain type, initial population size, pigments), and environmental factors (sample characteristics) associated with Pulsed Light processing that vary the decontamination efficacy of the process.

9.5.1 FLUENCE AND PEAK POWER DEPENDENCE

The incident fluence on the sample has been documented as the most significant factor determining the efficiency of the PL process against microbes. Though several other experimental factors such as treatment time, pulse duration and number, and discharge voltage of the lamp significantly describe the effects of PL, the photon fluence value aptly describes the different photochemical and physico-thermal effects occurring on the bacterial cell based on the overall average irradiance responsible for germicidal action. Theoretically, the fluence is defined as the cumulative radiant energy per unit area reaching from every direction comprising all wavelengths incident on an imaginary infinitesimal sphere [35]. However, the emitted energy from the flash lamp differs in magnitude from the energy incident or absorbed by the sample. Several factors, for example, sample properties and lamp-sample configuration, affect this observed difference in the energy values. The average rate of fluence (F_{avg}) is calculated from Beer-Lambert's Law as follows:

$$F_{avg} = \frac{F_0(1-e^{-A_e L})}{A_e L}$$

where; F_0 is the incident fluence rate; A_e represents the absorbance coefficient at wavelength 254 nm; and L is the path length.

Some studies showed varied energy efficiencies despite the same input voltages when the fluence rate was increased [51]. While a fixed fluence was chosen as the operating parameter, various degrees of inactivation could be observed owing to the different probabilities of the irradiation that makes contact with the target microorganisms. The fluence showed higher energy incidence with a decreasing pulse repetition and increasing pulse width.

For any photochemical process, the Bunsen-Roscoe reciprocity law is used to describe the validity of the theory of equi-effectivity of the outcome of rate of fluence and exposure time [16]. As per the law, the radiation effectiveness is independent of the mode of fluence delivery, i.e., the incidence of fluence on the product to be treated could be either at a higher proportion and rapid exposure or via a lower fluence rate and longer treatment time. Multiple studies have attempted to prove this law for the PL-assisted microbial inactivation process, and diversity was seen in the findings. A majority of results point towards possible peak dependence or indicate a peak power effect [22]. In certain studies, the difference in inactivation efficiency was observed when the fluence was provided by either a single light pulse with high radiant energy or multiple light pulses of lower energy when the applied voltage and spectral emission were kept constant. This is contradictory to the reciprocity law whereupon the same photo-response was expected while recurrently switching on and off a light source or supplying a high energy flux for a shorter duration. To conclude, it was discerned that viruses or spores typically go along with the Bunsen-Roscoe law as they are incapable of using the repair systems, while vegetative organisms show a violation of the reciprocity law when exposed to different treatment intensities [28]. However, increasing fluence to achieve targeted microbial inactivation should only consider the associated thermal effects and possibilities of nutrient degradation.

9.5.2 SAMPLE THICKNESS AND DISTANCE FROM THE LAMP

The intensity of the light source and the fluence rate are functions of distance [4]. According to the inverse square law, the incident fluence is inversely proportional to the square of the distance. With an increase in distance from the lamp, the penetration depth significantly decreases compared to the initial fluence supplied by the light source. This decreased penetration leads to weaker lethality of the PL spectrum and reduced inactivation of microorganisms. This can be attributed to the decrease in light scattering and absorption with an increase in substrate depth, resulting in diminished light intensity when it passes through the substrate [17]. The phenomenon of optical penetration depth could be used to describe this effect, which signifies the distance across which light declines to 37% of its initial value, showing a quantitative distribution of light dose within a substrate. This depth depends on the wavelength of the incident spectra, and it was stated that shorter wavelengths resulted in better penetration in foods compared to longer wavelengths.

The position of the sample in the treatment chamber and its orientation with respect to the light source also influence the energy received and absorbed. Usually, microbial inactivation studies employing PL are carried out by stationing the sample under a lamp. Certain studies were also conducted to analyze the effect of placing the products at other positions inside the PL chamber, and it was found that those located vertically below the sample showed a certain degree of decontamination while the rest exhibited almost no decontamination [16]. However, the intensity of inactivation decreased with an increase in vertical distance. As a result of the limited penetrability of the PL spectra, the opaque samples shield the surfaces from direct radiation exposure. In liquid samples, light is attenuated at the top surfaces.

9.5.3 SPECTRAL RANGE OF PL

Usually, 20% of the total fluence in PL spectra is made up of UV portion, which can vary depending on the applied voltage. As already mentioned, the UV-C fraction (200–280 nm) of the PL radiation has been proven to be the most significant portion of the PL spectrum responsible for microbial inactivation. For instance, it was displayed that the spectral range of 225–280 nm caused maximum inactivation of *Bacillus subtilis* spores over an agar surface, whereas above 280 nm, literally no effect was observed on viability [31]. These outcomes largely agree with the absorption spectra of microbial DNA and proteins that show peaks in the range of 250–280 nm. The emitted light spectra' characteristics are known to depend on the applied discharge voltage in Xenon flash lamps to generate light flashes. An increased germicidal effect is observed at lower wavelengths with higher proportions of UV-C while higher voltages are used in the process [15]. However, a reduced lethality of PL is observed when microorganisms are embedded within a complex matrix owing to varied absorption of light. This is usually found in beverages or liquids having suspended particles like proteins or other dispersed substances that have light-absorbing properties. Similarly, light absorption due to different pigments in the microorganisms has also been shown to affect the efficacy of the process [29].

9.5.4 TREATED MATRIX

Food products are complex substances with varied proportions of solid and liquid matter. The efficacy of PL treatment against microbes largely depends on the composition of the treated matrix, which includes the proportion of fat, carbohydrates, proteins, and water [19]. A decreased efficiency is observed in the presence of certain food constituents that can absorb the effective wavelengths. Consequently, food with high protein and fat content has less potential to be effectively decontaminated by PL, making vegetables suitable for PL treatment. Owing to its limited penetration power, solid and opaque liquid samples are only suitable for surface decontamination treatments as light experiences scattering, refraction, reflection, and absorption as it passes through the sample medium from the air. An opaque sample will only absorb on the surface, and the rest will be reflected; no radiation will penetrate the deeper layers. The optically transparent liquid will be fully treated, but translucent liquid samples will be partially treated, depending on the turbidity and presence of suspended solids. On evaluating the feasibility of PL for reducing *Escherichia coli* and *Listeria innocua* in matrices with different transparencies, it was observed that the rate of inactivation reduced with a decrease in transparency of the media [40]. A similar trend was reported in the decontamination study of *Listeria innocua* in culture media with different colorations, where higher microbial resistance was observed in the colored media [1]. Overall higher effectiveness was observed in a clearer media treated with PL [33].

The phenomenon of shielding of microorganisms, as already discussed in the previous sections, is a significant consequence of the surface topography of the sample. The occurrence of vast asymmetries and light-absorbing organic/inorganic materials between the light source and the treated sample could shield the microorganisms in the inner layers from photon incidence. Such materials could also include dust particles that may have accumulated both on the food surface as well as on the xenon lamps. Similarly, sample thickness also plays a crucial role as thin items or slices, and pieces have shown to positively influence the inactivation as a result of lower total fluence needed for penetration. Additionally, surface hydrophobicity has been found to vary inactivation effectiveness. Less hydrophobic surfaces have shown better reductions of microbial population upon PL exposure, presumably as a consequence of the broader spatial distribution of bacterial cells that resulted in greater exposure to PL [48].

9.5.5 PHOTOREACTIVATION

The almost lethal photochemical effects of PL on some forms of microorganisms are reversed by illumination with longer wavelengths, especially VL. This reverse mechanism resulting in repairing the damaged cells is called photoreactivation. The occurrence of photoreactivation has been extensively reported in continuous UV treatment studies on inactivating microbial populations in food products. The process is catalyzed by the photolyase enzyme, which utilizes light energy to split the CPDs in damaged DNA formed during initial photochemical reactions by means of a radical mechanism [16]. Photolyase is a flavoprotein containing two non-covalently bound chromophores. One of these chromophores is a completely reduced flavin-adenine dinucleotide, the catalytic cofactor that undertakes the repairing process upon excitation by either direct absorption of photons or transfer of resonance energy from the second chromophore that harvests sunlight and improves the repair efficacy. During the repair mechanism, the excited flavin cofactor passes an electron to the CPD to produce a charge-separated radical pair. By this, the anionic ring of the dimer gets split, and the excess electron reverts to the flavin radical to reinstate the catalytically competent flavin-adenine dinucleotide and close the catalytic photocycle. However, the rate of photoreactivation in flashed cells, though evident, was significantly lower than continuous UV treatment [38]. Still, the process of photoreactivation of microorganisms after PL can pose a potential limitation to the process. Thus, the wavelength of the emitted spectrum needs to be maintained in such a way as to prevent this occurrence, as wider wavelength light of the PL was pondered to have some impact on photoreactivation. Controlling the treatment time for reactivating light and the darkness prior to illumination can also be a regulatory step [10].

9.6 SUMMARY

In summary, pulsed light technology is a potential food decontamination technique with a wide range of lethality against different classes of microbial organisms. The central inactivation mechanism is associated with the photochemical effect induced by the UV part of the light spectrum. Evidence of photothermal and photophysical effects is also documented, making PL a multi-target process, which is more advantageous than continuous UV or any other non-thermal technique. With the exact mechanism of lethal action unclear, different kinetic models are used to describe the inactivation process, where consideration of shouldering and tailing effects has been found to improve the model accuracy. Given the complexity of the PL process, the efficiency of microbial destruction depends on several critical factors that require the optimization of processing conditions specific to a particular sample matrix. Additionally, most studies are *in vitro*, leading to limited information on real foods with indigenous microbial populations.

KEYWORDS

- **inactivation kinetics**
- **kinetic models**
- **microbial decontamination**
- **microbial inactivation**
- **microbial load**
- **photochemical effect**
- **pulsed light**

REFERENCES

1. Aguirre, J. S., Hierro, E., Fernández, M., & De Fernando, G. D. G. (2014). Modelling the effect of light penetration and matrix color on the inactivation of *Listeria innocua* by pulsed light. *Innovative Food Science & Emerging Technologies, 26*, 505–510.
2. Artíguez, M. L., & de Marañón, I. M. (2015). Inactivation of spores and vegetative cells of *Bacillus subtilis* and *Geobacillus stearothermophilus* by pulsed light. *Innovative Food Science & Emerging Technologies, 28*, 52–58.
3. Artíguez, M. L., & Martínez de Marañón, I. (2014). Process parameters affecting *Listeria innocua* inactivation by pulsed light. *Food and Bioprocess Technology, 7*(9), 2759–2765.
4. Bhalerao, P., Dhar, R., & Chakraborty, S. (2021). Pulsed light technology applied in food processing. In *Food Processing* (pp. 91–121). CRC Press.
5. Bhavya, M. L., & Umesh Hebbar, H. (2017). Pulsed light processing of foods for microbial safety. *Food Quality and Safety, 1*(3), 187–202.
6. Cassar, J. R., Mills, E. W., & Demirci, A. (2022). Characterization of pulsed light for microbial inactivation. *Journal of Food Engineering, 335,* 111152.
7. Cheigh, C. I., Park, M. H., Chung, M. S., Shin, J. K., & Park, Y. S. (2012). Comparison of intense pulsed light and ultraviolet (UVC)-induced cell damage in *Listeria monocytogenes* and *Escherichia coli* O157. *Food Control, 25*(2), 654–659.
8. Chen, D., Wiertzema, J. R., Peng, P., Cheng, Y., Wang, Y., Liu, J., Ma, Y., Mosher, W., Kang, M., & Min, M. (2020). Catalytic intense pulse light inactivation of *Cronobacter sakazakii* and other pathogens in non-fat dry milk and wheat flour. *Food Chemistry, 332*, 127420.
9. Cortesão, M., De Haas, A., Unterbusch, R., Fujimori, A., Schütze, T., Meyer, V., & Moeller, R. (2020). *Aspergillus niger* spores are highly resistant to space radiation. *Frontiers in Microbiology, 11*, 560.
10. Dhar, R., Basak, S., & Chakraborty, S. (2022). Pasteurization of fruit juices by pulsed light treatment: A review on the microbial safety, enzymatic stability, and kinetic

approach to process design. *Comprehensive Reviews in Food Science and Food Safety, 21*(1), 499–540.

11. Farrell, H. P., Garvey, M., Cormican, M., Laffey, J. G., & Rowan, N. J. (2010). Investigation of critical inter-related factors affecting the efficacy of pulsed light for inactivating clinically relevant bacterial pathogens. *Journal of Applied Microbiology, 108*(5), 1494–1508.
12. Fenoglio, D., Ferrario, M., García Carrillo, M., Schenk, M., & Guerrero, S. (2020). Characterization of microbial inactivation in clear and turbid juices processed by short-wave ultraviolet light. *Journal of Food Processing and Preservation, 44*(6), e14452.
13. Ferrario, M., Alzamora, S. M., & Guerrero, S. (2013). Inactivation kinetics of some microorganisms in apple, melon, orange and strawberry juices by high intensity light pulses. *Journal of Food Engineering, 118*(3), 302–311.
14. Franco-Vega, A., Reyes-Jurado, F., González-Albarrán, D., Ramírez-Corona, N., Palou, E., & López-Malo, A. (2021). Developments and advances of high intensity pulsed light and its combination with other treatments for microbial inactivation in food products. *Food Engineering Reviews, 13*(4), 741–768.
15. Gómez-López, V. M., & Bolton, J. R. (2016). An approach to standardize methods for fluence determination in bench-scale pulsed light experiments. *Food and Bioprocess Technology, 9*(6), 1040–1048.
16. Gomez-Lopez, V. M., Ragaert, P., Debevere, J., & Devlieghere, F. (2007). Pulsed light for food decontamination: A review. *Trends in Food Science & Technology, 18*(9), 464–473.
17. Gunjal, M., Tazeen, H., Gundu, P. M., Birwal, P., & Krishna, A. (2021). Pulsed light technology in food processing and preservation. In *Handbook of Research on Food Processing and Preservation Technologies: Volume 5: Emerging Techniques for Food Processing, Quality, and Safety Assurance* (p. 115). CRC Press.
18. Harguindeguy, M., & Gómez-Camacho, C. E. (2021). Pulsed light (PL) treatments on almond kernels: *Salmonella enteritidis* inactivation kinetics and infrared thermography insights. *Food and Bioprocess Technology, 14*(12), 2323–2335.
19. Heinrich, V., Zunabovic, M., Varzakas, T., Bergmair, J., & Kneifel, W. (2016). Pulsed light treatment of different food types with a special focus on meat: A critical review. *Critical Reviews in Food Science and Nutrition, 56*(4), 591–613.
20. Hwang, H. J., Seo, J. H., Jeong, C., Cheigh, C. I., & Chung, M.-S. (2019). Analysis of bacterial inactivation by intense pulsed light using a double-Weibull survival model. *Innovative Food Science & Emerging Technologies, 56*, 102185.
21. Van Impe, J., Smet, C., Tiwari, B., Greiner, R., Ojha, S., Stulić, V., Vukušić, T., & Režek Jambrak, A. (2018). State of the art of nonthermal and thermal processing for inactivation of microorganisms. *Journal of Applied Microbiology, 125*(1), 16–35.
22. Jeon, M. S., Park, K. M., Yu, H., Park, J. Y., & Chang, P. S. (2019). Effect of intense pulsed light on the deactivation of lipase: Enzyme-deactivation kinetics and tertiary structural changes by fragmentation. *Enzyme and Microbial Technology, 124*, 63–69.
23. John, D., & Ramaswamy, H. S. (2018). Pulsed light technology to enhance food safety and quality: A mini-review. *Current Opinion in Food Science, 23*, 70–79.
24. John, D., & Ramaswamy, H. S. (2020). Comparison of pulsed light inactivation kinetics and modeling of *Escherichia coli* (ATCC-29055), *Clostridium sporogenes* (ATCC-7955) and *Geobacillus stearothermophilus* (ATCC-10149). *Current Research in Food Science, 3*, 82–91.

25. Kairyte, K., Lapinskas, S., Gudelis, V., & Luksiene, Z. (2012). Effective inactivation of food pathogens *Listeria monocytogenes* and *Salmonella enterica* by combined treatment of hypericin-based photosensitization and high-power pulsed light. *Journal of Applied Microbiology, 112*(6), 1144–1151.
26. Kim, D. K., Kim, S. J., & Kang, D. H. (2017). Bactericidal effect of 266 to 279 nm wavelength UVC-LEDs for inactivation of gram-positive and gram-negative foodborne pathogenic bacteria and yeasts. *Food Research International, 97*, 280–287.
27. Kramer, B., & Muranyi, P. (2014). Effect of pulsed light on structural and physiological properties of *Listeria innocua* and *Escherichia coli*. *Journal of Applied Microbiology, 116*(3), 596–611.
28. Kramer, B., Wunderlich, J., & Muranyi, P. (2017). Impact of treatment parameters on pulsed light inactivation of microorganisms on a food simulant surface. *Innovative Food Science & Emerging Technologies, 42*, 83–90.
29. Kramer, B., Wunderlich, J., & Muranyi, P. (2017). Recent findings in pulsed light disinfection. *Journal of Applied Microbiology, 122*(4), 830–856.
30. Lasagabaster, A., & Martínez de Marañón, I. (2013). Impact of process parameters on *Listeria innocua* inactivation kinetics by pulsed light technology. *Food and Bioprocess Technology, 6*(7), 1828–1836.
31. Levy, C., Aubert, X., Lacour, B., & Carlin, F. (2012). Relevant factors affecting microbial surface decontamination by pulsed light. *International Journal of Food Microbiology, 152*(3), 168–174.
32. Liu, Z., Hu, S., Soteyome, T., Bai, C., Liu, J., Wang, Z., Kjellerup, B. V., & Xu, Z. (2021). Intense pulsed light for inactivation of foodborne gram-positive bacteria in planktonic cultures and bacterial biofilms. *LWT, 152*, 112374.
33. Mahendran, R., Ramanan, K. R., Barba, F. J., Lorenzo, J. M., López-Fernández, O., Munekata, P. E. S., Roohinejad, S., Sant'Ana, A. S., & Tiwari, B. K. (2019). Recent advances in the application of pulsed light processing for improving food safety and increasing shelf life. *Trends in Food Science & Technology, 88*, 67–79.
34. Maisch, T., Spannberger, F., Regensburger, J., Felgenträger, A., & Bäumler, W. (2012). Fast and effective: Intense pulse light photodynamic inactivation of bacteria. *Journal of Industrial Microbiology and Biotechnology, 39*(7), 1013–1021.
35. Mandal, R., Mohammadi, X., Wiktor, A., Singh, A., & Pratap Singh, A. (2020). Applications of pulsed light decontamination technology in food processing: An overview. *Applied Sciences, 10*(10), 3606.
36. Mukhopadhyay, S., Sokorai, K., Ukuku, D. O., Jin, T., Fan, X., Olanya, O. M., Leng, J., & Juneja, V. (2021). Effects of direct and in-package pulsed light treatment on inactivation of *E. coli* O157 and reduction of microbial loads in romaine lettuce. *LWT, 139*, 110710.
37. Ngadi, M. O., Latheef, M. Bin, & Kassama, L. (2012). Emerging technologies for microbial control in food processing. In *Green Technologies in Food Production and Processing* (pp. 363–411). Springer.
38. Oms-Oliu, G., Martín-Belloso, O., & Soliva-Fortuny, R. (2010). Pulsed light treatments for food preservation: A review. *Food and Bioprocess Technology, 3*(1), 13–23.
39. Ortega-Rivas, E. (2012). Pulsed light technology. In *Non-Thermal Food Engineering Operations* (pp. 263–273). Springer.
40. Palgan, I., Caminiti, I. M., Muñoz, A., Noci, F., Whyte, P., Morgan, D. J., Cronin, D. A., & Lyng, J. G. (2011). Effectiveness of high intensity light pulses (HILP) treatments

for the control of *Escherichia coli* and *Listeria innocua* in apple juice, orange juice and milk. *Food Microbiology, 28*(1), 14–20.

41. Pataro, G., Muñoz, A., Palgan, I., Noci, F., Ferrari, G., & Lyng, J. G. (2011). Bacterial inactivation in fruit juices using a continuous flow pulsed light (PL) system. *Food Research International, 44*(6), 1642–1648.
42. Preetha, P., Pandiselvam, R., Varadharaju, N., Kennedy, Z. J., Balakrishnan, M., & Kothakota, A. (2021). Effect of pulsed light treatment on inactivation kinetics of *Escherichia coli* (MTCC 433) in fruit juices. *Food Control, 121*, 107547.
43. Proulx, J., Hsu, L. C., Miller, B. M., Sullivan, G., Paradis, K., & Moraru, C. I. (2015). Pulsed-light inactivation of pathogenic and spoilage bacteria on cheese surface. *Journal of Dairy Science, 98*(9), 5890–5898.
44. Ramos-Villarroel, A. Y., Aron-Maftei, N., Martín-Belloso, O., & Soliva-Fortuny, R. (2012). The role of pulsed light spectral distribution in the inactivation of *Escherichia coli* and *Listeria innocua* on fresh-cut mushrooms. *Food Control, 24*(1–2), 206–213.
45. Ramos-Villarroel, A. Y., Aron-Maftei, N., Martín-Belloso, O., & Soliva-Fortuny, R. (2012). Influence of spectral distribution on bacterial inactivation and quality changes of fresh-cut watermelon treated with intense light pulses. *Postharvest Biology and Technology, 69*, 32–39.
46. Ren, M., Yu, X., Mujumdar, A. S., Yagoub, A. E. G. A., Chen, L., & Zhou, C. (2021). Visualizing the knowledge domain of pulsed light technology in the food field: A scient metrics review. *Innovative Food Science & Emerging Technologies, 74*, 102823.
47. Rowan, N. J., Valdramidis, V. P., & Gómez-López, V. M. (2015). A review of quantitative methods to describe efficacy of pulsed light generated inactivation data that embraces the occurrence of viable but non-culturable state microorganisms. *Trends in Food Science & Technology, 44*(1), 79–92.
48. Syamaladevi, R. M., Lu, X., Sablani, S. S., Insan, S. K., Adhikari, A., Killinger, K., Rasco, B., Dhingra, A., Bandyopadhyay, A., & Annapure, U. (2013). Inactivation of *Escherichia coli* population on fruit surfaces using ultraviolet-C light: Influence of fruit surface characteristics. *Food and Bioprocess Technology, 6*(11), 2959–2973.
49. Tao, T., Ding, C., Han, N., Cui, Y., Liu, X., & Zhang, C. (2019). Evaluation of pulsed light for inactivation of food-borne pathogens on fresh-cut lettuce: Effects on quality attributes during storage. *Food Packaging and Shelf Life, 21*, 100358.
50. Valdivia-Nájar, C. G., Martín-Belloso, O., Giner-Seguí, J., & Soliva-Fortuny, R. (2017). Modeling the inactivation of *Listeria innocua* and *Escherichia coli* in fresh-cut tomato treated with pulsed light. *Food and Bioprocess Technology, 10*(2), 266–274.
51. Yi, J. Y., Bae, Y. K., Cheigh, C. I., & Chung, M. S. (2017). Microbial inactivation and effects of interrelated factors of intense pulsed light (IPL) treatment for *Pseudomonas aeruginosa*. *LWT, 77*, 52–59.
52. Yin, R., Dai, T., Avci, P., Jorge, A. E. S., de Melo, W. C. M. A., Vecchio, D., Huang, Y. Y., Gupta, A., & Hamblin, M. R. (2013). Light based anti-infectives: Ultraviolet C irradiation, photodynamic therapy, blue light, and beyond. *Current Opinion in Pharmacology, 13*(5), 731–762.

CHAPTER 10

APPLICATION OF PULSED LIGHT IN LIQUID FOODS

NENE MELTEM KEKLIK and ALI DEMIRCI

ABSTRACT

Pulsed light offers rapid inactivation of a wide range of microorganisms in foods. This technology involves short pulses of high-energy electromagnetic radiation in the spectral range from UV to infrared delivered by inert-gas lamps. This chapter focuses on the application of pulsed light in various liquid foods. After a brief overview of pulsed light technology, the equipment specifically designed for the treatment of liquid foods and critical processing parameters are discussed. Then, research conducted on the effects of pulsed light on the microbial loads and quality parameters of liquid foods is summarized. Finally, remarks on the potential commercial application of pulsed light in liquid foods are made.

10.1 INTRODUCTION

The consumption of food and water contaminated with pathogenic microorganisms results in a large number of disease cases each year. Moreover, with the increase in human population and change in dietary habits, it becomes more difficult to ensure food safety. According to the World Health Organization (WHO), about 600 million people (i.e., almost 1 in 10 people in the world) become ill after consuming contaminated food, which results in 4,20,000 deaths every year. About 40% of the foodborne disease burden is carried by small children under the age of 5 with 1,25,000 deaths every year. These diseases not only threaten human life but also cause huge economic losses in the healthcare system, trade, and tourism.

Nonthermal Light-Based Technologies in Food Processing.
C. K. Sunil, Preeti Birwal, Megh R. Goyal, and R. Mahendran (Eds.)

Numerous methods are used in the food industry to control pathogenic microorganisms in foods. The most common techniques used to decontaminate foods are mostly thermal and/or chemical methods. However, not all these methods can be applied to all types of food production. For example, thermal techniques are not suitable for the processing of raw or minimally processed foods due to the altering of the products' quality by high temperatures (e.g., cooking). Chemical methods, on the other hand, are not favored by consumers due to health concerns. Thus, alternative methods to thermal and chemical applications have been developed and evaluated for the decontamination of various types of foods. These methods are commonly known as non-thermal food preservation methods.

Pulsed light (PL) is an innovative technology, which is acknowledged as a non-thermal method if short processing times are used. It neither involves chemical addition nor water usage. Also, pulsed light is not an ionizing radiation like gamma irradiation or X-rays. Basically, pulsed light comprises high-energy light pulses generated by an inert gas (e.g., Xenon) flashlamp. These pulses consist of electromagnetic radiation in the wavelength range of 100–1,100 nm, involving UV, visible, and near-infrared spectra [35].

The germicidal properties of UV-C in the UV range of 200–280 nm is well-known and include photochemical effects on microbial DNA, which eventually results in the inactivation of the microorganisms [3, 29, 45]. Since the UV-C content of pulsed light plays a major role in microbial inactivation [61], pulsed light is also known as pulsed UV (PUV) light. When compared with conventional continuous UV light, pulsed light offers many advantages such as better inactivation efficacy, increased penetration ability, lower costs, and greater lamp safety.

This chapter covers the basic principles and applications of pulsed light technology to decontaminate liquid foods. The specific equipment, critical processing parameters, and factors affecting the applications of pulsed light in liquid foods are discussed. A detailed literature review is provided by overviewing the findings of various research conducted on the effects of pulsed light on the microbial loads and quality parameters of liquid foods. Final remarks are made on future commercial applications of pulsed light in liquid foods.

10.2 AN OVERVIEW OF PULSED LIGHT TECHNOLOGY

UV light, which is a part of natural sunlight, constitutes the range of 100–400 nm in the electromagnetic spectrum. It is generally divided into four parts: UV-A, UV-B, UV-C, and vacuum-UV, which correspond to the wavelength ranges of 315–400 nm, 280–315 nm, 200–280 nm, and 100–200 nm, respectively. UV-A, UV-B, UV-C, and vacuum-UV have photon energies in the ranges of 3.94–3.10, 4.43–3.94, 12.40–4.43, and 12.40–124 eV, respectively [37]. Thus, photon energy increases as the wavelength decreases.

UV-C has germicidal characteristics to efficiently destroy pathogens and other harmful microorganisms. The effectiveness of UV light on microorganisms such as bacteria, viruses, protozoans, and fungi has been demonstrated by numerous studies. The highest damage to microbial DNA occurs at about 260 nm, at which the maximum absorbance by nucleic acids occurs [10, 30]. The microbial inactivation mechanism of UV light mainly includes alteration of nucleic acid structure of microorganisms, i.e., formation of photodimers between adjoining pyrimidine bases (Figure 10.1), cytosine photohydration, and, rarely, crosslinks within the same or between opposite strands, crosslinks of protein and DNA, as well as rare base adducts [34, 45]. There are also secondary effects of UV light, which include the production of highly reactive oxygen species reacting with DNA, damaged bases, strand breaks, and crosslinks between DNA and proteins [45].

In general, the lamps used in conventional UV-C light equipment designed for disinfection purposes deliver UV-C light at about 254 nm. Ultraviolet light can be produced by different kinds of lamps such as low-pressure and medium-pressure mercury-vapor lamps, low-pressure amalgam UV lamps, microwave-powered UV lamps, excimer lamps, and LEDs [34, 58]. Germicidal UVC lamps have long been utilized in many areas for surface, water, and air sterilization. Due to its simplicity, UV-C sterilization is preferred in many places with a high risk of microbial infections such as hospitals, the cosmetics sector, schools, hotels, restaurants, and airports. Especially, UV light application in hospitals carries a critical role in preventing sepsis in patients and other infections by purifying the ventilated air and disinfecting surgical equipment, rooms, hospital kitchens, and water.

UV-C lamps are used safely to remove most forms of microbiological contamination. UV disinfection is well-established for the treatment of drinking water and wastewater and has been an alternative to chlorine and other chemical disinfectants and can also be combined with them for the decontamination of drinking water [63]. In the food industry, not only is UV-C light used for the elimination of pathogens from foods and food-contact surfaces, but it is also used to prevent early spoilage of foods, especially fruits and vegetables, prolong shelf life, and preserve or increase nutritional value [5].

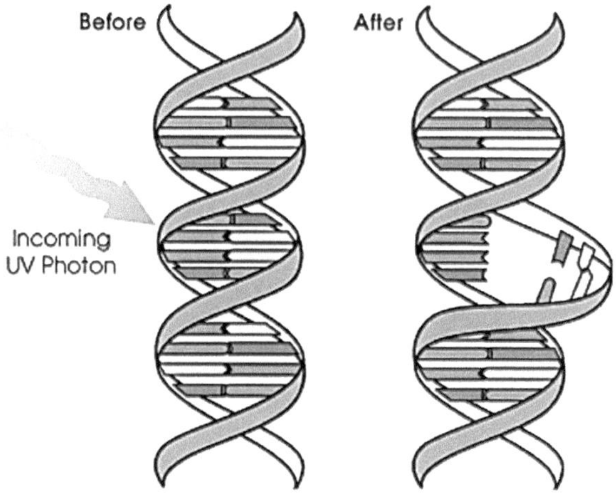

FIGURE 10.1 UV light causes the formation of pyrimidine dimers, which distort the DNA molecule and disrupt its proper functioning.

Source: Reprinted with permission. Illustration by David Herring/NASA's Earth Observatory.

Another method with potential for effective food decontamination is pulsed light, which has gained attention in recent years as a nonthermal processing technology. This technique consists of short high-energy light pulses containing a broad electromagnetic spectrum from the UV band to the infrared band, including UV-C light with germicidal effects. These pulses are produced by an inert gas lamp (generally xenon). Pulsed light is a type of nonionizing radiation, i.e., it does not remove electrons from materials like gamma and X-rays. Pulsed light holds potential as an alternative to heat and chemical treatments.

Pulsed light treatment for food disinfection was approved by the FDA in 1996, and the conditions required for the safe treatment of foods with pulsed light are summarized below [20]:

- The radiation source consists of xenon flashlamps and emits broadband radiation (200–1,100 nm). The pulse duration is ≤ 2 milliseconds.
- The treatment is used for the surface control of microorganisms.
- Foods receive the minimum required treatment to achieve the intended effect.
- The total cumulative treatment does not exceed 12.0 J/cm^2.

There are several advantages of pulsed light over conventional UV light. The efficiency of pulsed light for microbial inactivation is known to be four times more efficient than continuous UV light [7]. This is most likely caused by the higher peak of power reached by pulsed light irradiation in a short period of time than the conventional continuous UV light. Also, as inert gases are not toxic like mercury, pulsed light lamps do not pose health or environmental risks in case of a lamp leakage or breakage, and thus are safer than mercury-vapor lamps. The ability of pulsed light to penetrate into opaque materials is limited but still better than that of continuous UV light [18, 48]. The infrared part of pulsed light causes heat generation and may cause food to heat significantly over time [7]. Thus, it is recommended that the treatment times should be kept as short as possible.

Since pulsed light involves UV, visible, and infrared spectra, microbial inactivation mechanisms of pulsed light are also more complex compared to conventional continuous UV light. Apart from the effects of pulsed light on the nucleic acid of microbial cells which take place as a result of photochemical reactions, there are also photothermal and photophysical effects of pulsed light. All parts of UV light contribute to the photothermal effect.

According to Wekhof [61], overheating and subsequent rupture of bacterial cells occur when the degree of UV absorption by the bacteria is greater than that by the surrounding medium at high flux densities (>0.5 J/cm^2). On the other hand, the photophysical effect of pulsed light is about the constant physical disturbance of microbial cells by high-energy pulses, which results in damage to the cell structure. High-energy light pulses induce physical changes in microbial cells such as the destruction of the cell wall, shrinkage of cytoplasm, and leakage of cellular contents from the cytoplasm [13, 38]. Due to the physical damage done by photothermal and photophysical effects, reactivation or self-repair of microbial cells exposed to pulsed light becomes harder. On the other hand, cells exposed to conventional UV light are capable of photoreactivation [34, 45].

10.3 PULSED LIGHT EQUIPMENT DESIGNED FOR TREATMENT OF LIQUID FOODS

The basic parts of a pulsed light system include lamp/lamps and a power supply. A controller unit can be used for pulse configuration. The alternating current is converted by the power source into high voltage direct current, which is used to charge a capacitor. The capacitor releases the energy of light to the lamp in case the circuit is closed by an electrical switch [49]. The inert gas (xenon or krypton) inside the lamp is ionized by this energy, and as a result, intense and instantaneous flashlight is produced. Pulsed light flashes at a rate of 1–20 flashes per second are obtained by opening and closing the switch [49].

Pulsed light lamps, which may contain xenon, argon, krypton, or other inert gas or mixtures, emit intense flashes in a broadband spectrum [41]. Two types of lamps that produce pulsed light are the flashlamp and the surface discharge lamp [9]. In a flashlamp, an inert gas that is confined between two electrodes in a small envelope generates pulses in a broadband spectrum of 100–1,100 nm. Flashlamps generally have a UV efficiency of about 9% and intensity of about 600 W/cm^2 [34]. In a surface discharge lamp, a high-power electrical discharge produces a plasma along the surface of a dielectric substrate (fused silica tube) inside an envelope containing inert gas [56].

Surface discharge lamps have higher UV efficiency (17%) and intensity (30,000 W/cm^2) compared to flashlamps [34]. Also, the surface discharge lamps can have a longer lamp life due to the large diameter of the envelope in surface discharge lamps [56].

Although pulsed light has limited penetration depth in foods, it has a greater penetration ability in liquid foods compared to solid foods, with a steeper decline in light intensity in solids [44]. However, the treatment of liquids is still challenging as microorganisms can dwell in the whole volume of liquids. Because light energy decreases with depth, thin film or profiles are suitable for PL treatment of liquid foods. For example, a turbulent and thin-layer flow configuration was used by Sauer & Moraru [55] for the decontamination of apple juice and apple cider with pulsed light. The physicochemical properties such as viscosity and density should be considered while designing a continuous PL treatment system as these properties affect the flow characteristics of liquid foods [44].

In flow-through pulsed light systems, centrifugal and peristaltic pumps can be used to pump the liquid into the pulsed light system [17, 24]. A sight glass can be used to observe the liquid flow and to measure the UV-light intensity [17]. The flow rate can be adjusted by a control valve, and a flowmeter can be used to measure and adjust the liquid flow [17]. In a study by Demirci & Krishnamurthy [17], the flow-through pulsed system was cleaned to eliminate any cross-contamination in the system. In order to do so, pulsed light was coupled with chlorine solution before rinsing with sterile deionized (DI) water to remove any chlorine residues.

A schematic diagram of a continuous flow-through pulsed light system is given in Figure 10.2. This system by Ferrario & Guerrero [24] is summarized as follows: Juice was kept in a vessel and was circulated by using a peristaltic pump through two consecutive quartz tubes which were placed below the lamp. Only a part of the total length of each tube was exposed to pulsed light (white part of the quartz tubes in Figure 10.2), while the remaining area of the tubes was covered to block light exposure (black part of the quartz tubes in Figure 10.2). Autoclavable flexible hoses were used to circulate the juice and discharge it into an agitated vessel that was immersed in a water-ice bath. Juice samples were taken from the vessel at regular intervals. The PL-exposed area of the quartz tubes was countersunk in grooves in an aluminum unit that contained a circulating coolant. In this way, juice temperature was kept below 25C. The maximum fluence and the maximum total energy input were 0.73 J/cm^2 and 0.0175 J/mL, respectively [24]. Ferrario & Guerrero [24] calculated different pulsed light doses by multiplying the dose of a single cycle by the number of cycles corresponding to each treatment time, assuming that the whole volume of juice was treated in just a single cycle.

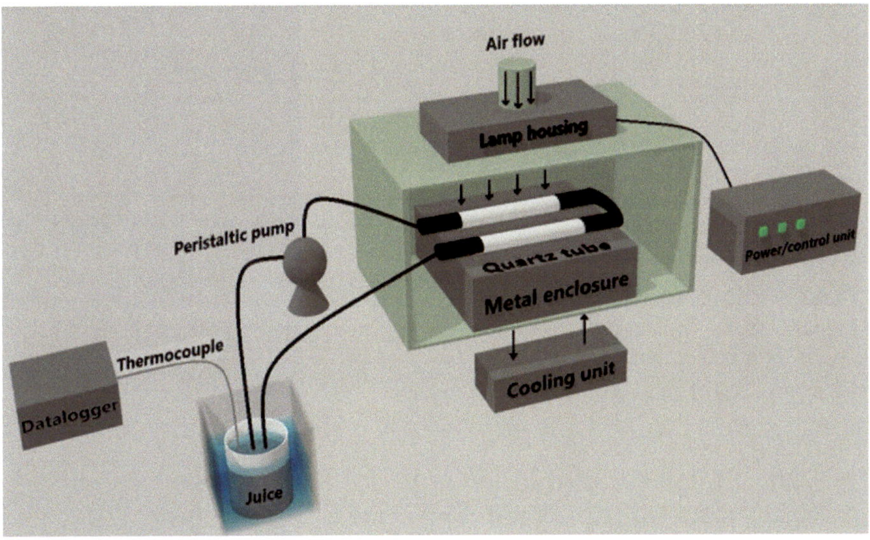

FIGURE 10.2 Schematic diagram of a continuous flow-through pulsed light system.
Source: Adapted from Ref. [24].

A laboratory-scale flow-through pulsed light system in Figure 10.3 was used by Uslu et al. [59] to disinfect synthetic real municipal wastewater effluent. The system consisted of a controller, a power supply, and a chamber containing the lamp, which produced polychromatic light in the wavelength range of 100–1,100 nm (54% in the UV-light region). Using an input voltage of 3,800 V, the system generated 3 pulses of 360 μs duration per second. The lamp was located at the center of the system chamber, which had an annular cylinder arrangement. The flow-through system was made of stainless steel. A sight glass facilitated the observation of water flow and to measure the UV-light intensity. A quartz sleeve separated the annular space between the lamp and the outer wall of the vessel to enhance the transmission of UV-light. The water from a 10 or 20 L carboy container was pumped through the water vessel using a centrifugal pump. Uslu, Demirci, & Regan [59] also used a flow meter to measure and adjust the flow rate of water.

FIGURE 10.3 A laboratory scale flow-through pulsed light system (SteriPulse®-RS 4000, Xenon Corporation, Wilmington, MA).

Source: Reprinted with permission from Ref.[66]

Currently, the commercial use of pulsed light involves sterilization of packaging materials (e.g., bottle caps) in the beverage industry [2]. Considering the complex food structure and the limitations of pulsed light, it is clear that more extensive research needs to be done in order to fully adapt the pulsed light to the industry. The future research should include pilot-scale and full-scale trials as well as focus on hurdle technology by combining pulsed light with other intervention methods.

10.4 CRITICAL PROCESSING PARAMETERS

The efficacy of pulsed light depends on light-matter interactions, e.g., the absorbance and transmittance of light by the food. Thus, the characteristics of both the incident light and the target food influence the efficacy of pulsed light treatment. A target molecule needs to absorb light in order for a photochemical reaction to proceed [8]. According to the Stark-Einstein law (the second law of photochemistry), only one molecule is activated for reaction for each photon absorbed [39]. This means that the food needs to transmit the light so that the light reaches the microbial cell and gets absorbed by the target molecule (e.g., DNA) before a photochemical reaction takes place.

Photon energy is inversely proportional to wavelength, and is expressed by the following Eqn. (1) (Planck–Einstein relation or Planck's equation) [39]:

$$E = h.\frac{c}{\lambda} \tag{1}$$

where; E is the photon energy (eV); h is Planck constant (~4.14×10^{-15} eV·s); c is the speed of light (~3.00×10^{8} m/s); and λ is the wavelength (m).

The light-matter interaction (Figure 10.4) occurs when the electromagnetic field of light interacts with the charged particles [40]. The classical model suggests that electromagnetic waves which travel through a medium force the charged particles to oscillate. According to quantum mechanics, a higher state of energy can be reached by electrons, which have absorbed quanta of light energy. In order to accomplish state transitions, the outer electrons of an atom require energy levels typically in the range of 1–10 eV, which corresponds to visible and UV light. On the other hand, the absorption of infrared light takes place in the energy transitions between different states of molecular vibrations and rotations [40].

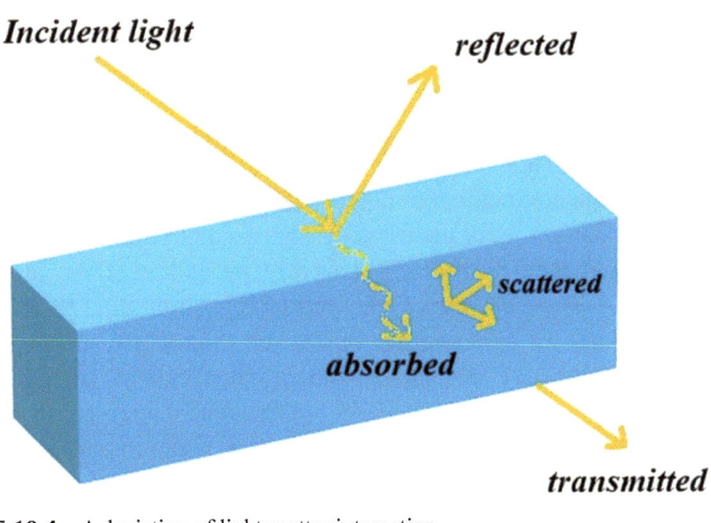

FIGURE 10.4 A depiction of light-matter interaction.

As stated earlier, pulsed light has limited penetration ability in foods. The penetration ability indicates how far light can be transmitted through a material. Thus, transparency of liquid foods becomes an important factor in pulsed light processing. It is also known that darker materials absorb more light energy. Therefore, color and transparency of liquid foods are important factors affecting the efficacy of PL treatments. High levels of destruction of selected foodborne pathogens can be obtained with pulsed light treatment; however, the efficiency of pulsed light treatment strongly depends on the energy dose that is actually absorbed by the microorganisms, which in turn depends on the absorption properties of the liquid food [50]. Unfortunately, measurement of the actual energy dose delivered to the food is difficult, as the behavior of pulsed light in a particular food is complex (i.e., what percentage of light is absorbed, reflected, or transmitted?). Generally, power and energy meters with sensors (e.g., pyroelectric or photodiode) are used for measurement, which are taken at certain distances from the lamp, before calculating the total fluence.

Gómez-López & Bolton [26] proposed that the method using a source of polychromatic light (e.g., a pulsed-light lamp or a medium-pressure UV lamp) must rely on the irradiance of the incident photons (Einstein/(m^2s)) as the reactions of photochemical and photobiological interactions are controlled by the photon flux that is absorbed but not by the energy flux. Therefore, the authors indicated that reporting the results with regard to the photon fluence (based upon the assessment of 'overall average germicidal weighted photon irradiance' per pulse) would help standardize the conditions and make it possible to compare the results of different studies and to make calculations regarding the scale-up process.

For this purpose, Gómez-López & Bolton [26] proposed a protocol for a dosimetry method in their study. The critical process factors of PL processing for liquid foods can be divided into three sections (some factors are interrelated): Parameters of pulsed light lamp and system (applied voltage, peak power, pulse width, pulse frequency, number of pulses (treatment time), fluence, wavelength range and distribution, light intensity, system geometry, position of the lamp, distance of target food from the PL source, type of flow, flow rate, number of passes, etc.), characteristics of the food (composition, transparency, color, turbidity, thickness or depth of the fluid), and microorganisms (type, form, growth phase, number of cells, Gram staining status, etc.) [4, 16, 53].

In addition to the studies carried out to determine the critical factors in the pulsed light process, the limitations defined for UV light applications, which are among the light treatments of foods, can also provide some useful information. The FDA described the limitations of UV irradiation as follows [21]:

For the treatment of potable water:

- The production of ozone may not be permitted;
- The absorption coefficient is 0.19 per cm or less;
- The flow rate is 100 gal/h (per watt of 253.7 nm radiation);
- The water depth is 1 cm or less;
- The lamp-operating temperature is between 36?C and 46?C.

For the treatment of juice products:

- Turbulent flow with a minimum Reynolds number of 2200 is obtained through the tubes.

These limitations imply that the wavelength range, fluence, flow rate, the depth of liquid, temperature, and the type of flow need to be carefully controlled.

The wavelength range and spectral distribution of pulsed light affect microbial inactivation, food quality, and process safety. Bacterial inactivation and quality parameters are highly influenced by the UV component of pulsed light, especially UV-C light [52]. The UV-C part of the spectrum is known for its germicidal and photolytic abilities, while UV-B and UV-A can cause changes in microorganisms, nucleic acids, or tissues [61]. Filtering out short wavelengths of the UV part can help prevent ozone production. Similarly, including lower frequency visible and near-infrared radiation can lead to a temperature increase [12], and narrowing the spectrum may reduce the heating effect.

The total pulsed light fluence, also known as the total energy dose, delivered to the food, depends on the energy of a pulse, pulse frequency, and treatment time (Eqn. (2)) [42]:

$$D = E_p * t * f \quad (2)$$

where; D is the total dose of the light energy (J/cm^2); E_p is the pulse energy (J/cm^2 pulse); t is the treatment time (s); and f is the pulse frequency (pulse/s).

In Eqn. (2), the total fluence varies in direct proportion with the pulse energy, pulse frequency, and the treatment time. The irradiance (I, J/cm^{2}s) can be expressed as "$E_p\ x\ f$" in Eqn. (2). The inverse square law states that light irradiance decreases by the square of the distance from the point source of light [54]. Thus, the total fluence is also affected by the proximity to the light source. The occurrence of photoreactivation should be taken into account if low PL energy doses and wavelengths of 360–380 nm are used [53].

The thermal effect of pulsed light increases with higher fluence applied, and the temperature rise may contribute to microbial inactivation as a confounding variable. Pataro et al. [50] highlighted that the temperature of both the juice and air inside the pulsed light chamber may increase significantly during processing. This uncontrolled temperature increase may lead to quality deterioration in liquid foods. Moreover, if the temperature of liquids flowing in tubes while being exposed to pulsed light significantly increases, it may accelerate the fouling on the surface of the tubes and consequently decrease the amount of PL-energy absorption by the liquid [35]. Therefore, it is critical to keep the processing time short to avoid significant heating. If an efficient cooling system can be integrated into pulsed light processing, it may prevent the rapid temperature rise in the PL chamber.

Color, viscosity, and opaqueness are among the factors that affect the inactivation of microorganisms in liquid foods [14]. The shadowing effect in liquid foods is caused by the particles/particulates in turbid or non-clear liquids [55]. The limited penetrability of pulsed light into liquid foods is challenging, but there are some ways to enhance the uniform exposure such as increasing the number of lamps, using reflectors, increasing the number of passes/cycles, using turbulent flow, etc.

10.5 APPLICATION OF PULSED LIGHT IN LIQUID FOODS AND OTHER FLUIDS

Many researchers have studied the microbial inactivation efficiency of pulsed light in liquid foods. These studies demonstrate the performance of different PL configurations and treatment methods applied to liquid foods to inactivate various types of microorganisms. In addition, many of these studies and others have investigated the effects of pulsed light on the quality attributes and shelf life of liquid foods. Studies on the effects of pulsed light on microbial inactivation efficacies and quality parameters in liquid foods and other fluids are summarized below. Some studies used different terminology for 'pulsed light' such as 'intense pulsed light,' 'high intensity light pulses,' and 'pulsed UV light,' which were mentioned here as 'pulsed light' to provide simplicity.

10.5.1 MICROBIAL INACTIVATION EFFICACIES

1. **Pure Water:** Demirci & Krishnamurthy [17] studied the effect of pulsed light on the inactivation of *Bacillus subtilis* (ATCC 6633) in water. The PL-system chamber consisted of an arrangement of an annular cylinder

with the lamp located at the center. The PL system was turned on and the inoculated water was pumped through the system at the flow rates of 2, 4, 6, 8, 10, and 14 L/min, corresponding to the residence times of 88, 44, 29, 22, 18, and 13 s in the chamber, respectively. The PL treatments using the above-mentioned flow rates resulted in complete destruction of *B. subtilis*, yielding more than a log reduction of 5.5. Demirci & Krishnamurthy [17] indicated that UV light was partly absorbed by the microbial cells in a direct manner and partly by water causing up to a 4?C increase in the bulk temperature after the treatments.

2. **Baby Foods:** Choi et al. [14] investigated the pulsed light treatment for the inactivation of *L. monocytogenes* in infant foods and the feasibility of this technique for commercial applications. Infant foods inoculated with *L. monocytogenes* were spread to a thickness of 2 mm on a petri dish and treated with pulsed light. The triggering voltage pulse, pulse frequency, pulse width, lamp operating time, and surface distance from the lamp were 15 kV, 10 Hz, 1.5 μs, 0–600 s, and 60 mm, respectively. In a light-colored thin infant beverage, 1 and 5 log reduction of *L. monocytogenes* were obtained for 630 and 3,500 μs of treatment at the same voltage, respectively. In a dark-colored viscous infant meal, PL treatments produced 1 and 3 log reduction of *L. monocytogenes* during 900 and 4,800 μs, respectively. The death rate of *L. monocytogenes* in both infant foods increased exponentially with treatment time [14].

3. **Honey:** Hillegas & Demirci [28] studied the inactivation of Clostridium sporogenes in clover honey by pulsed light treatment. Inoculated honey samples of different depths were exposed to pulsed light (5.6 J/cm^2 per pulse on the strobe surface) at varying proximities from the lamp using different numbers of pulses. A reduction of 87.6% was obtained for the spores in a 2-mm depth of honey after pulsed light treatment at 8 cm using 135 pulses. The spore inactivation increased as the treatment distance from the lamp decreased, and the number of pulses increased [28].

4. **Juices:** Sauer & Moraru [55] investigated the treatment of apple juice and apple cider with pulsed light for the inactivation of *E. coli*. The treatments were carried out in static and turbulent modes. In the case of the static mode, *E. coli* ATCC 25922 (nonpathogenic) and *E. coli* O157:H7 (pathogenic) were reduced by about 2.66 and 2.52 log in apple juice, respectively, and about 2.32 log and 3.22 log in apple

cider, respectively. In the case of the turbulent mode, the turbulence was created by placing an orbital shaker in the pulsed light unit. The velocities of 500 and 3,000 rpm yielded a low turbulence and a high turbulence, respectively. Below the energy dose of 12 J/cm^2, the treatment by using highly turbulent flow resulted in reductions of 5.76 in apple cider and 7.15 log in apple juice for *E. coli*. The reason behind higher inactivation levels achieved under turbulence was in part that the turbulence could reduce the shadowing effect of the particulates present in the nonclear liquid substrates. Thus, Sauer & Moraru [55] suggested that the controlled turbulence in a flow-through system may be a consideration in potential commercial applications of pulsed light for processing of liquid foods. Muñoz et al. [47] applied pulsed light and thermosonication alone and in combination to inactivate *E. coli* in orange juice. In the pulsed light system, the liquid was pumped using a peristaltic pump through two quartz tubes located below the xenon flashlamp. The tubes exposed to pulsed light were located in grooves cut in an aluminum unit containing a recirculating coolant (ethylene glycol) kept at –10?C to provide cooling of the juice during treatment. The product was also cooled immediately before and after PL exposure by cooling coils submerged in iced water. The fluences of 4.03 and 5.10 J/cm^2 were delivered by selecting the residence times as 2.22 and 2.81 s. The results indicated that the inactivation ranged from 1.10 to 2.42 log CFU/mL for the individual treatments, and in the range of 2.5–3.93 log CFU/mL for the treatments in combination [47]. Muñoz et al. [46] applied pulsed light and thermosonication, alone or in combination, using a continuous system to inactivate *Escherichia coli* in apple juice. The juice was exposed to pulsed light at the fluences of 4.03 and 5.1 J/cm^2, which corresponded to 51.5 and 65.4 J/mL, respectively. The highest microbial reductions were about 4.9 and 6 log CFU/mL, which were achieved when pulsed light alone (at higher fluence) and pulsed light combined with thermosonication were applied, respectively. Muñoz et al. [46] reported that combining these technologies regardless of the sequence applied shows an additive effect on the microorganisms. In a study of Ferrario et al. [22], pulsed light was applied to natural and commercial apple, melon, orange, and strawberry juices to inactivate *E. coli* ATCC 35218, *Salmonella* Enteritidis MA44, *Listeria innocua* ATCC 33090, and Saccharomyces cerevisiae. Pulsed light treatments were carried out after inoculating the refrigerated juice in a petri dish with a sample depth of 1×10^{-3} m. Treatment times were between 2 and 60 s, which corresponded to fluencies between 2.4 J/cm^2 and 71.6 J/cm^2, respectively. Accordingly, pulsed light treatment was not found to be effective in natural strawberry and orange juices. In general, log-reductions between 0.6 and 6.2 cycles were obtained after 60 s (71.6 J/cm^2) of pulsed light treatment for commercial juices and between 0.3 and 6.9 cycles for natural ones. The microorganism, type, and source of fruit juice affected the microbial inactivation. Pulsed light effectiveness decreased as the UV-C absorbance of juice increased. Ferrario et al. [22] reported that the temperature increases of samples treated for 60 s ranged from 7.4 to 16.8?C, on average, depending on the type of fruit juice. Maftei et al. [43] applied pulsed light to apple juice to inactivate *Penicillium expansum*. The effects of critical processing parameters on microbial inactivation were studied. These parameters included fluence [0.2 and 0.4 J/cm^2 per pulse], pulse number [5, 10, 15, 20, 30, and 40 pulses applied from opposite sides of the product], juice layer depth [6, 8, and 10 mm], and level of inoculation [2.3×10^4 and 3×10^5 CFU per mL]. The inactivation increased up to 3.76 log CFU/mL as the number of pulses and fluence increased. Inactivation levels obtained on thin layers of juice (6 mm) were substantially higher than those observed on deeper layers (8 or 10 mm). Also, juices with lower initial counts yielded higher inactivation of *P. expansum* compared to the samples with higher counts [43]. Ferrario et al. [23] investigated how several inoculated strains (*E. coli* (ATCC 35218), *L. innocua* (ATCC 33090), *S. Enteritidis* (MA 44), and *S. cerevisiae* (KE 162)) and natural flora responded to pulsed light treatment (Xenon lamp, 3 pulses/s, 10 cm distance from the lamp, 71.6 J/cm^2) in apple, orange, and strawberry fresh juices with different absorbance, turbidity, and particle size. Also, microbial growth dynamics of PL-treated juices were evaluated during storage at 5?C for 12 days. Pulsed light treatments led to 0.3–2.6 log reductions for inoculated microorganisms and 0.1–0.7 for native flora. The high turbidity and particles with high UV absorbance were the two major factors affecting the PL efficiency. The inactivation of *S. Enteritidis* and *L. innocua* increased (5.0–8.0 log reductions) by the cold storage of PL-treated juices, while there was no recovery of *E. coli* or retardation for yeast growth observed [23]. Ferrario & Guerrero [25] investigated the effect of single ultrasound or the combined treatment (with pulsed light) under conditions in which the heat build-up was controlled [Xenon lamp (3 pulses per s, 71.6 J per cm^2), temperature levels (2–20?C and 44–56?C)] on the destruction of *S. cerevisiae* KE 162 in commercial and freshly pressed apple juices. The treatment of ultrasound followed by pulsed light at the highest heat build-up resulted in yeast reduction of up to 6.4 log in commercial apple juice and 5.8 log-cycles in freshly pressed apple juice. Images of transmission electron microscopy (TEM) for the treated cells demonstrated major damage, loss of cell wall, and coagulation of inner content. Also, flow cytometry showed that yeast cells that have esterase activity and intact membrane shifted to those with permeabilized membrane [25]. Karaoglan et al. [31] investigated the effect of pulsed light on wild yeast *Candida inconspicua* in Turnip juice (Shalgam), a traditional Turkish beverage produced by the lactic acid fermentation of black carrot. Pulsed light was applied to turnip juice samples at three different distances (5, 8, and 13 cm) from the quartz window (located at 5.8 cm below the lamp) and for five different times (5, 15, 30, 45, and 60 s) using a bench-type pulsed light system. The applied fluences ranged from 0.77 to 19.71 J/cm^2. The 60-s treatment applied at 5 cm (19.71 J/cm^2) yielded a maximum of 2.80 log reduction (CFU/mL). The authors reported that the temperature increase of samples ranged from 6 to 44?C, depending on time and distance [31]. Xu et al. [64] studied a continuous spiral flow-through pulsed light system to inactivate *E. coli* in red grape juice. Different pulse numbers, pulsed light intensities, and flow rates were used in the treatment of grape juice. The correlation among the factors was analyzed by the response surface methodology. Higher pulse numbers and PL intensities were associated with higher inactivation. The treatment (100 pulses, flow rate of 30 mL per min) with intensities of 0.13, 0.40, and 0.66 J/cm^2 per pulse resulted in 2.1, 2.6, and 3.7 log reductions, respectively. Verifying tests from the optimization study demonstrated that the number of *E. coli* decreased by 4.89 log CFU/mL at the optimal levels of 80 pulses, intensity of 0.66 J/cm^2/pulse, and flow rate of 40 mL/min [64]. Zhu et al. [65] investigated the antimicrobial mechanism of pulsed light against *Escherichia coli* O157:H7 and its inactivation in carrot juice. *E. coli* O157:H7 after PL treatment was examined in terms of the changes in morphology, endoenzyme, respiratory metabolism, and entocyte (protein, DNA, and ATP). SEM and TEM images revealed that pulsed light destructed the cell integrity severely and decreased the entocyte (proteins, ATP, and DNA) considerably. Furthermore, pulsed light treatment inhibited not only the activity of ATPase, alkaline phosphatase, β-galactosidase, and topoisomerase but also the metabolism of *E. coli* O157:H7. The production of hydroxyl radical by pulsed light was validated by the electron spin resonance spectrum, hence leading to the elevation of intracellular reactive oxygen and the acceleration of cell death. Pulsed light (0.5 Hz, 30 pulse numbers, 500 J of single pulse energy) treatment of carrot juice reduced the bacterial load to an undetectable level after a 3-day incubation, without a significant effect on its quality [65].

5. **Milk:** Krishnamurthy et al. [35] used a flow-through pulsed light system to investigate its effect on the inactivation of *Staphylococcus aureus* in milk. The inoculated milk was pumped into the system through a quartz tube. A V-groove reflector setup was utilized not only to adjust the distance between the quartz tube and the light source but also to reflect the light back into the quartz window to increase the absorption of PL energy by the microbial cells in milk. Milk was exposed to pulsed light at distances of 5, 8, or 11 cm from the strobe and flow rates of 20, 30, or 40 mL/min. Also, the milk was recirculated up to three times to evaluate how the number of passes affected the efficiency of inactivation. Microbial reduction was in the range of 0.55–7.26 log CFU/mL. Complete destruction of *S. aureus* was observed after two treatments: (i) Single pass, flow rate of 20 mL per min, a distance of 8 cm from the quartz window of the lamp (at 0.98 W/cm^2); and (ii) Double pass, flow rate of 20 mL per min, a distance of 11 cm (0.80 W/cm^2). Temperature increases up to 38?C were observed in milk due to PL treatment. Krishnamurthy, Demirci, & Irudayaraj [35] noted that although the temperature increase was not high enough to cause inactivation of *S. aureus*, it might have played a role in microbial inactivation by PL treatment. Krishnamurthy et al. [36] applied pulsed light to inactivate *Staphylococcus aureus* (ATCC 25923) in milk and milk foam. Varying amounts of static cell suspensions in milk or milk foam were treated with pulsed light for 30, 105, or 180 s at various distances (8, 10.5, and 13 cm for milk, and 5, 8, and 11 cm for milk foam) from the quartz window of the lamp. Microbial reductions in milk were in the range of 0.16–8.55 log CFU/mL. *S. aureus* was completely inactivated in milk after treatment at the combination of 8-cm distance, 30-mL sample

volume, and 180-s time, as well as after treatment at the combination of 10.5-cm distance, 12-mL sample volume, and 180 s time. Reductions of 1.05 to 6.61 log CFU/g in *S. aureus* populations in milk foam were obtained after pulsed light treatments. The maximum reduction (6.61 log CFU/g) was achieved after PL treatment of 3 g of milk foam for 180 s at an 8-cm distance from the quartz window. The authors noted that pulsed light treatments increased the temperature of milk (up to 90?C in some treatments), which might have had a synergistic effect on microbial inactivation [36]. Kasahara et al. [33] investigated the efficacy of pulsed light for inactivating *Escherichia coli* in goat milk. After the inoculation with *E. coli*, the samples of goat milk were treated with varying doses of pulsed light produced by an exciplex laser unit. The pulsed-light treatment at 10,000 mJ/cm^2 yielded a 6-log reduction. The penetration of UV light into cells containing milk samples was less than 5 mm [33].

6. **Other Fluids:** Vimont et al. [60] investigated the efficacy of pulsed light for inactivating murine norovirus 1 (MNV-1) as a human norovirus surrogate in phosphate-buffered saline, hard water, mineral water, turbid water, and sewage treatment effluent. More than 3 log reduction in viral infectivity was obtained in less than 3 s (3.45 J/cm^2) in clear suspensions. Viral inactivation was affected by the presence of protein or bentonite. Pulsed light seemed to disrupt MNV-1 structure and cause degradation of viral protein and RNA [60]. Uslu et al. [59] characterized the efficacy of a flow-through PL system for the destruction of *E. coli* and *B. subtilis* spores in synthetic (SMWE) and real municipal wastewater effluent (RMWE). The authors observed total bacterial destruction with a 10 L/min flow rate for *E. coli* and 6 L/min flow rate for *B. subtilis* using SMWE and one-pass PL treatment. Total bacterial destruction was obtained in SMWE after two-pass PL treatment using a 16 L/min flow rate for *E. coli* and 10 L/min flow rate for *B. subtilis*, respectively. In RMWE for one pass, total destruction of *E. coli* was obtained by using a 10 L/min flow rate, while a 4.15 log decrease was achieved with 6 L/min for *B. subtilis.* Complete inactivation was also obtained after PL treatment of the raw wastewater at a 10 L/min flow rate [59]. Even though this study was conducted on municipal wastewater, it is also an important study for the food industry not only for recommending pulsed light for the treatment of its wastewater, but also suggesting that pulsed light can be utilized for recycling the process water during food manufacturing.

10.5.2 EFFECTS ON QUALITY PARAMETERS OF FOODS

Aside from microbial inactivation efficacies, many studies have focused on the effects of pulsed light treatment on the quality parameters such as color, antioxidant activity, sensory attributes, and some physicochemical properties of various liquid foods and other fluids, some of which are summarized in this section.

1. **Juices:** Caminiti et al. [11] investigated the effect of pulsed light (3.3 J/cm^2) in combination with manothermosonication on selected physicochemical and sensorial properties of a blend of orange and carrot juice. The Hunter Lab color values were not affected by pulsed light treatment, but the combined treatment increased the color parameters significantly. The application of pulsed light did not induce any significant change in the phenolic content or antioxidant activity compared to a fresh control. However, a significant decrease in the level of total phenols was observed in samples after combined treatment. The treatments did not affect the non-enzymatic browning of the orange and carrot juice significantly. Results for pectin methyl-esterase (PME) residual activity in orange and carrot juice processed by pulsed light showed significant inactivation of PME under the selected conditions, which reached about 78% after the combined treatment. Sensory analysis demonstrated that the color of the juice processed by the combined treatment was preferred to that of a pasteurized sample (at 72?C for 26 s), while the flavor was adversely affected by the combined treatment. No differences in the odor, sweetness, or acidity of the product were found by the panelists [11]. Muñoz et al. [46] studied the effects of pulsed light and its combination with thermosonication on the selected quality attributes (pH, ?Brix, color, non-enzymatic browning, and antioxidant activity) of apple juice before or after processing. Pulsed light was applied to apple juice at 4.03 and 5.1 J/cm^2, corresponding to 51.5 and 65.4 J/mL, respectively. The color of apple juice was found to be significantly affected by all treatments, and the sequence in which the technologies were applied also significantly influenced the color. The treatment of pulsed light followed by thermosonication induced the highest decrease in the color parameters 'L' and 'b' and the lowest decrease in the parameter 'a' compared to the treatment of thermosonication followed by pulsed light. The energy level applied did not have an effect on any of the measured quality attributes [46].

Maftei et al. [43] evaluated the effects of pulsed light treatments on the color, pH, and soluble solids of apple juice. A total color difference (ΔE) of no greater than 0.5 was observed after treatments of 0.2 J/cm^2/pulse, whereas treatments of 0.4 J/cm^2/pulse did not result in significant color changes, with ΔE values ranging from 0.5 to 1.5. However, slight darkening of apple juice was observed after applying an overall fluence of 32 J/cm^2. No major changes in pH and soluble solids were noticed after pulsed light treatments [43]. Karaoglan et al. [32] studied how pulsed light treatment of turnip juice affected the degradation of anthocyanin and the physicochemical characteristics. Turnip juice was exposed to pulsed light at three distances from the quartz window of the xenon lamp (5, 8, and 13 cm) for five different times (5, 15, 30, 45, and 60 s). The PL fluence of the treatments ranged from 0.77 to 19.71 J/cm^2. Pulsed light treatments at each level significantly changed the total acidity (as % lactic acid), pH, monomeric anthocyanin content, color density, brightness, hue, and percent color components (yellow, red, blue) of turnip juice. Karaoglan et al. [32] reported that the maximum percentage of anthocyanin degradation after PL treatments was approximately 63%. The anthocyanin degradation, yellow and blue color (%), and brightness showed an increase, while a decrease was observed in red color (%) with longer treatment time and shorter distance from the quartz window [32]. Zhu et al. [65] studied the effects of different pulsed light treatments (0.5 Hz, 500 J of single pulse energy, and the pulse numbers of 20, 25, and 30) on the sensory characteristics of carrot juice. The results indicated that color, taste, flavor, and comprehensive acceptability of carrot juice were not significantly affected by the pulsed light treatments, implying that pulsed light has no adverse effect on the quality of carrot juice [65].

2. **Milk:** Kasahara et al. [33] evaluated the effect of pulsed light on milk's organoleptic properties. The product was treated with pulsed light using two doses: a medium (5,000 mJ/cm^2) and a high (10,000 mJ/cm^2) level. The sample that was pasteurized (70?C for 20 min) but not PL-treated was used as the control. Irradiation at both medium and high doses resulted in aromatic changes compared to the untreated control sample. However, the parameters related to the physical or compositional characteristics did not significantly differ between the PL-treated samples and the control. Although the PL-treated samples showed a small increase in viscosity, this increase was not significant compared to that of the untreated fresh

sample. On the other hand, the PL-treated and pasteurized samples had lower pH values compared to the fresh sample [33].

3. **Oils:** Abuagela et al. [1] used a pilot-scale pulsed light system to reduce aflatoxin levels in peanut oil. Peanut oil samples (5 mL) with a 10-mm depth were treated for durations in the range of 1–10 minutes in the presence or absence of temperature reduction. To provide temperature reduction, the sample containers were placed in an ice water bath. The color, physical, and chemical quality parameters (acidity value, free fatty acid, peroxide value, and oxidative stability index (OSI)) were determined for peanut oil. Abuagela et al. [1] achieved significant aflatoxin destruction after using pulsed light. With temperature reduction, peanut oil samples treated for 400, 600, and 800 seconds resulted in an AFT reduction of 48.4%, 55.6%, and 78%, respectively. The quality of peanut oil samples that were treated with temperature reduction exhibited lower degradation compared to the samples of peanut oil treated without temperature reduction when equivalent treatment times were used [1].

4. **Whey Protein Isolate:** Siddique et al. [57] investigated the whey protein isolate (WPI) solutions in terms of aggregation after pulsed light treatments. WPI (1% w/v) solutions in sodium phosphate buffer (pH = 7.5) were treated at PL fluences of 4–16 J/cm^2. Free SH-groups and UV-absorption spectra were determined to evaluate whey protein structural modification and aggregation. In addition, turbidity, aggregation index, particle size distribution, and SDS-PAGE were measured to identify the covalent and non-covalently linked protein-protein interactions. Structural changes including the instant rise in the content of free SH-group (unfolding) and the following formation of a small aggregation fraction of unfolded proteins were observed in WPI after PL treatment, caused by both hydrophobic interactions and the formation of disulfide bonds. Siddique et al. [57] observed an increase in turbidity, mean particle size, and aggregation index of samples treated with pulsed light (4–16 J/cm^2). WPI dimer dissociation and formation of larger particles likely occurred based on the findings of the particle size distribution analysis of samples treated at higher PL fluence [57]. Additionally, the gel electrophoresis analysis indicated a relationship between molecules of intermediate and larger protein in addition to the soluble aggregate formations between α-lactalbumin and β-lactoglobulin. In summary, this study

[57] demonstrated that the PL treatments can induce protein denaturation with a minimal formation of soluble protein aggregates.

5. **Other Suspension Solutions:** Pellicer et al. [51] investigated the effect of pulsed light on the inactivation of pectin-methyl esterase (PME), which is a degradative enzyme of the cloudiness of fruit juices. A suspension of the enzyme was subjected to pulsed light (up to 128 J/cm^2). The structural changes were assessed by far-UV circular dichroism, intrinsic and extrinsic fluorescence, and spectrophotometry. In addition, the presence/absence of unfolding intermediates was determined by phase-diagram analysis. Pulsed light treatment at a fluence of 128 J/cm^2 reduced 90% of PME. The inactive protein appeared to have a molten-globule structure due to the conserved secondary structure and modified tertiary structure. A significant increase in absorbance at 420 nm demonstrated that pulsed light promoted enzyme aggregation [51]. Uslu et al. [59] evaluated how a flow-through PL system affected the decrease of suspended solids, total organic carbon (TOC), chemical oxygen demand (COD), and turbidity of synthetic municipal wastewater effluent (SMWE) and raw municipal wastewater effluent (RMWE). The following findings were obtained for SMWE after two passes: Decrease in COD was found to be 31.21% at 10 L/min and 12.5% at 20 L/min for *E. coli*, while 20.77% at 10 L/min and 12.15% at 20 L/min for *B. subtilis*. TOC reductions were 35.45% at 10 L/min and 11.19% at 20 L/min for *E. coli*, while 33.74% at 10 L/min and 6.54% at 20 L/min for *B. subtilis*. The suspended solids were reduced by 27.35% at 10 L/min and 12.19% at 20 L/min for *E. coli*, while 17.25% at 10 L/min and 8.39% at 20 L/min for *B. subtilis*, respectively. Turbidity was reduced by 24.58% at 10 L/min and 2.83% at 20 L/min for *E. coli*, while 14.03% at 10 L/min and 7.38% at 20 L/min for *B. subtilis*. On the other hand, the following findings were observed for RMWE at 10 L/min: COD reductions were obtained as 49.46% and 41.38% for *E. coli* and *B. subtilis,* respectively. Reductions of TOC in RMWE were found to be 16.25% and 6.98% for *E. coli* and *B. subtilis*, respectively. The decreases in suspended solids were 57.83% for *E. coli* and 39.30% for *B. subtilis*. The turbidity was reduced by 20.03% for *E. coli* and 12.45% for *B. subtilis*. Overall, Uslu, Demirci, & Regan [59] indicated pulsed light decontaminated the wastewater effluent, and also decreased the organic load of municipal wastewater effluent.

10.6 LIMITATIONS AND CHALLENGES OF PL

For the photochemical reactions to take place, photon energy delivered by pulsed light needs to be absorbed by microbial cells. The more the microbial cells absorb the pulsed light, the greater its efficiency will be. Thus, the complex composition of foods presents another challenge, as various compounds present in foods, which absorb light, may interfere with microbial inactivation. For example, the proteins and oils in food absorb UV light and reduce the effective radiation dose available for microbial inactivation [19].

As mentioned before, the penetration ability of pulsed light is low and therefore the transparency of liquid foods is a critical process parameter. This makes it difficult to control the pulsed light process because the interaction of light with molecules in the liquid and the amount of shadowing of microbial cells by these molecules will vary according to the liquid composition and concentration. This can be expected to make standardization difficult in the pulsed light process.

Another problem is that pulsed light causes food to heat up during long operations. The reason for this heating is the infrared light present in the pulsed light, and the amount of heating will change depending on the proximity to the lamp, treatment time, and the composition of the food. Temperature rise appears to be a confounding factor in pulsed light processing. Therefore, processing times should be kept short in order to prevent quality deteriorations in foods associated with heat. However, this problem can be overcome by using intermittent processing and efficient cooling methods.

Pulsed light is also difficult to apply to pre-packaged liquid foods because of its limited penetrability. Packaging increases the thickness of the liquid, decreasing the efficiency of the process even if the package (e.g., bottle) is highly PL-permeable. For this reason, liquid foods can be packaged after pulsed light treatment in a flow-through system. This process should be done carefully with aseptic packaging to prevent post-processing contamination.

Regarding liquid foods, the FDA permits the use of UV radiation for decontaminating drinking water used in food processing and reducing microbial load in juice products under certain conditions [21]. The FDA has approved the application of pulsed light for foods but has not made any statements regarding its use on liquid foods [20]. Therefore, the conditions for applying pulsed light to liquid foods need to be specified.

Another issue with pulsed light is the production of ozone during PL processing due to the presence of UV light. Wavelengths below 240 nm readily generate ozone [15]. Ozone is a highly reactive chemical that can adversely affect the respiratory, cardiovascular, and central nervous systems. The safe treatment of food, food products, and potable water with UV irradiation defines the lack of ozone production as a requirement [21]. In pulsed light applications, the accumulation of high levels of ozone should be prevented by using an efficient ventilation system.

10.7 SUMMARY AND FUTURE TRENDS

Pulsed light technology has the potential to decontaminate liquid foods. However, the adverse effects of PL treatment on food quality tend to increase with higher fluence associated with longer treatments. It should also be noted that pulsed light also has the potential to produce photochemical reactions that yield beneficial compounds such as vitamins and antioxidants.

Pulsed light has a limited penetration ability and is only considered for the surface decontamination of solid foods. However, with liquid foods, most of which contain a respectable amount of turbidity, the number of passes through the system can be increased to expose as many microbial cells to the pulsed light as possible. Thus, pulsed light treatment of liquid foods seems to be more versatile in terms of the process design compared to solid foods.

Although the efficacy of pulsed light in inactivating microorganisms in liquid foods has been demonstrated by many studies, there is still a lack of system parameters that should be optimized with respect to the physical design requirements and the characteristics of the target food product. In the optimization process, the aim should be to prevent adverse changes in quality while ensuring the safety of liquid foods.

In this context, the processing of liquid foods with pulsed light can be efficiently optimized using mathematical models, thereby accelerating the adaptation of this technology in the food industry. The use of mathematical models to describe the PL inactivation of microorganisms helps us better understand the behavior of microbial cells under the effects of pulsed light. Similarly, mathematical modeling of quality changes in liquid foods would assist in controlling the process more efficiently. It is needless to say that the mathematical models should be validated by comparing model predictions to physical measurements.

Since the penetration ability of pulsed light is limited, PL treatment of bottled or other in-package liquid foods is less feasible compared to bulk processing. In an industrial-scale production, continuous-flow pulsed light processing should be integrated with aseptic packaging so that PL-treated liquid food is pumped into a section (a holding tube or vessel) where it is aseptically filled into containers and sealed. Overall, pulsed light technology continues to be an innovative and promising technology for future processing of foods.

KEYWORDS

- **chemical oxygen demand**
- **colony-forming unit**
- **critical processing parameters**
- **deoxyribonucleic acid**
- **microbial inactivation**
- **oxidative stability index**
- **pulsed light**
- **ribonucleic acid**

REFERENCES

1. Abuagela, M. O., Iqdiam, B. M., Baker, G. L., & MacIntosh, A. J. (2018). Temperature-controlled pulsed light treatment: Impact on aflatoxin level and quality parameters of peanut oil. *Food and Bioprocess Technology, 11*(7), 1350–1358.
2. Arnold, N., Yang, L., & Boyer, R. R. (2020). How is pulsed-light used to process foods? *Virginia Cooperative Extension Bulletin FST-357*, 1–2.
3. Bank, H. L., John, J., Schmehl, M. K., & Dratch, R. J. (1990). Bactericidal effectiveness of modulated UV-light. *Applied Environmental Microbiology, 56*, 3888–3889.
4. Barbosa-Canovas, G. V., Schaffner, D. W., Pierson, M. D., & Zhang, Q. H. (2000). Pulsed light technology. *Journal of Food Science, 65*, 82–85.
5. Begum, M., Hocking, A. D., & Miskelly, D. (2009). Inactivation of food spoilage fungi by ultra violet (UVC) irradiation. *International Journal of Food Microbiology, 129*, 74–77.
6. Berk, Z. (2018). *Food Process Engineering and Technology* (3rd ed., pp. 421–438). London, UK: Academic Press, Elsevier.
7. Bialka, K. L. (2006). Decontamination of berries with ozone and pulsed UV-light. PhD thesis, The Pennsylvania State University, University Park, PA. pp. 65–68.
8. Blatchley III, E. R., & Peel, M. M. (2001). Disinfection by Ultraviolet irradiation. In: Block, S. S. (Ed.), *Disinfection, Sterilization, and Preservation* (5th ed., pp. 823–851). Philadelphia, PA: Lippincott Williams & Wilkins.

9. Bohrerova, Z., Shemer, H., Lantis, R., Impellitteri, C. A., & Linden, K. G. (2008). Comparative disinfection efficiency of pulsed and continuous-wave UV irradiation technologies. *Water Research, 42*(12), 2975–2982.
10. Bowker, C., Sain, A., Shatalov, M., & Ducoste, J. (2011). Microbial UV fluence-response assessment using a novel UV-LED collimated beam system. *Water Research, 45*(5), 2011–2019.
11. Caminiti, I. M., Noci, F., Morgan, D. J., Cronin, D. A., & Lyng, J. G. (2012). The effect of pulsed electric fields, ultraviolet light or high intensity light pulses in combination with manothermosonication on selected physicochemical and sensory attributes of an orange and carrot juice blend. *Food and Bioproducts Processing, 90*(3), 442–448.
12. Cassar, J. R., Mills, E. W., & Demirci, A. (2022). Characterization of pulsed light for microbial inactivation. *Journal of Food Engineering, 334*, Article ID 111152.
13. Cheigh, C. I., Hwang, H. J., & Chung, M. S. (2013). Intense pulsed light (IPL) and UV-C treatments for inactivating *Listeria monocytogenes* on solid medium and seafoods. *Food Research International, 54*, 745–752.
14. Choi, M. S., Cheigh, C. I., Jeong, E. A., Shin, J. K., & Chung, M. S. (2010). Nonthermal sterilization of *Listeria monocytogenes* in infant foods by intense pulsed-light treatment. *Journal of Food Engineering, 97*(4), 504–509.
15. Claus, H. (2021). Ozone generation by ultraviolet lamps. *Photochemistry and Photobiology, 97*(3), 471–476.
16. Cudemos, E., Izquier, A., Medina-Martinez, M. S., & Gomez-Lopez, V. M. (2013). Effects of shading and growth phase on the microbial inactivation by pulsed light. *Czech Journal of Food Sciences, 31*(2), 189–193.
17. Demirci, A., & Krishnamurthy, K. (2007). Disinfection of water by flow-through pulsed UV light sterilization system. *Ultrapure Water, 2007*(January/February), 24-29.
18. Dunn, J. (1996). Pulsed light and pulsed electric field for foods and eggs. *Poultry Science, 75*, 1133–1136.
19. Elmnasser, N., Guillou, S., Leroi, F., Orange, N., Bakhrouf, A., & Federighi, M. (2007). Pulsed-light system as a novel food decontamination technology: A review. *Canadian Journal of Microbiology, 53*(7), 813–821.
20. FDA (Food and Drug Administration). (2022). Pulsed light for the treatment of food. Code of Federal Regulations, 21CFR179.41. Title 21, Volume 3, https://www.accessdata.fda.gov/scripts/cdrh/cfdocs/cfcfr/CFRSearch.cfm?fr=179.41 (accessed on 4 June 2024).
21. FDA (Food and Drug Administration). (2022). Ultraviolet radiation for the processing and treatment of food. Code of Federal Regulations, 21CFR179.39. https://www.accessdata.fda.gov/scripts/cdrh/cfdocs/cfcfr/cfrsearch.cfm?fr=179.39 (accessed on 4 June 2024).
22. Ferrario, M., Alzamora, S. M., & Guerrero, S. (2013). Inactivation kinetics of some microorganisms in apple, melon, orange and strawberry juices by high intensity light pulses. *Journal of Food Engineering, 118*(3), 302–311.
23. Ferrario, M., Alzamora, S. M., & Guerrero, S. (2015). Study of pulsed light inactivation and growth dynamics during storage of *Escherichia coli* ATCC 35218, *Listeria innocua* ATCC 33090, *Salmonella enteritidis* MA 44 and *Saccharomyces cerevisiae* KE 162 and native flora in apple, orange and strawberry juices. *International Journal of Food Science & Technology, 50*(11), 2498–2507.

24. Ferrario, M., & Guerrero, S. (2016). Effect of a continuous flow-through pulsed light system combined with ultrasound on microbial survivability, color and sensory shelf life of apple juice. *Innovative Food Science & Emerging Technologies, 34*, 214–224.
25. Ferrario, M., & Guerrero, S. (2017). Impact of a combined processing technology involving ultrasound and pulsed light on structural and physiological changes of *Saccharomyces cerevisiae* KE 162 in apple juice. *Food Microbiology, 65*, 83-94.
26. Gómez-López, V. M., & Bolton, J. R. (2016). An approach to standardize methods for fluence determination in bench-scale pulsed light experiments. *Food and Bioprocess Technology, 9*(6), 1040–1048.
27. Gómez-López, V. M., Koutchma, T., & Linden, K. (2012). Ultraviolet and pulsed light processing of fluid foods. In P. J. Cullen, B. K. Tiwari, & V. P. Valdramidis (Eds.), *Novel Thermal and Non-thermal Technologies for Fluid Foods* (pp. 185–223). London, UK: Academic Press, Elsevier.
28. Hillegas, S., & Demirci, A. (2003). Inactivation of *Clostridium sporogenes* in clover honey by pulsed UV-light treatment. *Agricultural Engineering International: The CIGR Journal of Scientific Research and Development (Manuscript FP-03-009), V*(Dec), 87–99.
29. Jay, J. M. (1997). Radiation preservation of foods and nature of microbial radiation resistance. In J. M. Jay (Ed.), *Modern Food Microbiology* (pp. 304–323). New York, NY: Chapman and Hall.
30. Kalisvaart, B. F. (2004). Re-use of wastewater: Preventing the recovery of pathogens by using medium-pressure UV lamp technology. *Water Science and Technology, 50*(6), 337–344.
31. Karaoglan, H. A., Keklik, N. M., & Develi Işikli, N. (2017). Modeling inactivation of *Candida inconspicua* isolated from turnip juice using pulsed UV light. *Journal of Food Process Engineering, 40*(2), e12418.
32. Karaoglan, H. A., Keklik, N. M., & Develi-Isıklı, N. (2019). Degradation kinetics of anthocyanin and physicochemical changes in fermented turnip juice exposed to pulsed UV light. *Journal of Food Science and Technology-Mysore, 56*(1), 30–39.
33. Kasahara, I., Carrasco, V., & Aguilar, L. (2015). Inactivation of *Escherichia coli* in goat milk using pulsed ultraviolet light. *Journal of Food Engineering, 152*, 43–49.
34. Koutchma, T. N., Forney, L. J., & Moraru, C. I. (2009). *Ultraviolet Light in Food Technology – Principles and Applications* (p. 265). Boca Raton, FL: CRC Press (Taylor & Francis Group).
35. Krishnamurthy, K., Demirci, A., & Irudayaraj, J. M. (2007). Inactivation of *Staphylococcus aureus* in milk using flow-through pulsed UV-light treatment system. *Journal of Food Science, 72*(7), M233–M239.
36. Krishnamurthy, K., Demirci, A., & Irudayaraj, J. (2008). Inactivation of *Staphylococcus aureus* in milk and milk foam by pulsed UV-light treatment and surface response modeling. *Transactions of the ASABE, 51*(6), 2083–2090.
37. Krishnamurthy, K., Irudayaraj, J., Demirci, A., & Yang, W. (2008). UV pasteurization of food materials. In S. Jun & J. M. Irudayaraj (Eds.), *Food Processing Operations Modeling: Design and Analysis* (2nd ed., pp. 281–299). Boca Raton, FL: CRC Press (Taylor and Francis Group).
38. Krishnamurthy, K., Tewari, J. C., Irudayaraj, J., & Demirci, A. (2010). Microscopic and spectroscopic evaluation of inactivation of *Staphylococcus aureus* by pulsed UV light and infrared heating. *Food and Bioprocess Technology, 3*(1), 93–104.

39. Kusuma, P., Pattison, P. M., & Bugbee, B. (2020). From physics to fixtures to food: Current and potential LED efficacy. *Horticulture Research, 7*, 56–62.
40. Lee, H. C. (2005). Light–matter interaction. In *Introduction to Color Imaging Science* (pp. 66–67). Cambridge, UK: Cambridge University Press.
41. Linden, K. (2004). Introduction. In *Disinfection Efficiency and Dose Measurement of Polychromatic UV Light* (pp. 1–5). London, UK: IWA Publishing.
42. Luksiene, Z., Gudelis, V., Buchovec, I., & Raudeliuniene, J. (2007). Advanced high-power pulsed light device to decontaminate food from pathogens: Effects on *Salmonella typhimurium* viability in vitro. *Journal of Applied Microbiology, 103*(5), 1545–1552.
43. Maftei, N. A., Ramos-Villarroel, A. Y., Nicolau, A. I., Martín-Belloso, O., & Soliva-Fortuny, R. (2014). Influence of processing parameters on the pulsed-light inactivation of *Penicillium expansum* in apple juice. *Food Control, 41*, 27–31.
44. Mandal, R., Mohammadi, X., Wiktor, A., Singh, A., & Singh, P. A. (2020). Applications of pulsed light decontamination technology in food processing: An overview. *Applied Sciences, 10*(10), 3606–3611.
45. Miller, R. V., Jeffrey, W., Mitchell, J. D., & Elasri, M. (1999). Bacterial responses to ultraviolet light. *ASM News, 65*, 535–541.
46. Muñoz, A., Caminiti, I. M., Palgan, I., Pataro, G., Noci, F., Morgan, D. J., Cronin, D. A., Whyte, P., Ferrari, G., & Lyng, J. G. (2012). Effects on *Escherichia coli* inactivation and quality attributes in apple juice treated by combinations of pulsed light and thermosonication. *Food Research International, 45*(1), 299–305.
47. Muñoz, A., Palgan, I., Noci, F., Morgan, D. J., Cronin, D. A., Whyte, P., & Lyng, J. G. (2011). Combinations of high intensity light pulses and thermosonication for the inactivation of *Escherichia coli* in orange juice. *Food Microbiology, 28*(6), 1200–1204.
48. Oms-Oliu, G., Martín-Belloso, O., & Soliva-Fortuny, R. (2010). Pulsed light treatments for food preservation: A review. *Food and Bioprocess Technology, 3*(1), 13–23.
49. Palmieri, L., Cacace, D., & Dall'Aglio, G. (1999). Non-thermal methods of food preservation based on electromagnetic energy. *Food Technology and Biotechnology, 37*(2), 145–149.
50. Pataro, G., Muñoz, A., Palgan, I., Noci, F., Ferrari, G., & Lyng, J. G. (2011). Bacterial inactivation in fruit juices using a continuous flow pulsed light (PL) system. *Food Research International, 44*(6), 1642–1648.
51. Pellicer, J. A., Navarro, P., & Gómez-López, V. M. (2020). Pectin methylesterase inactivation by pulsed light. *Innovative Food Science & Emerging Technologies, 62*, 102366.
52. Ramos-Villarroel, A. Y., Aron-Maftei, N., Martín-Belloso, O., & Soliva-Fortuny, R. (2012). The role of pulsed light spectral distribution in the inactivation of *Escherichia coli* and *Listeria innocua* on fresh-cut mushrooms. *Food Control, 24*(1–2), 206–213.
53. Rowan, N. J. (2019). Pulsed light as an emerging technology to cause disruption for food and adjacent industries–Quo vadis? *Trends in Food Science & Technology, 88*, 316–332.
54. Ruston, C., Holtkamp, D., Li, P., Koziel, J. A., Stephan, A., Loesekann, T., Torremorell, M., Murray, D., Wedel, K., Johnson, C., Zaabel, P., & Sundberg, P. (2022). Effectively using ultraviolet-C light for supply decontamination on swine farms. *Journal of Swine Health and Production, 30*(2), 101–106.

55. Sauer, A., & Moraru, C. I. (2009). Inactivation of *Escherichia coli* ATCC 25922 and *Escherichia coli* O157 in apple juice and apple cider using pulsed light treatment. *Journal of Food Protection, 72*(5), 937–944.
56. Schaefer, R., Grapperhaus, M., Schaefer, I., & Linden, K. (2007). Pulsed UV lamp performance and comparison with UV mercury lamps. *Journal of Environmental Engineering and Science, 6*(3), 303–310.
57. Siddique, M. A. B., Maresca, P., Pataro, G., & Ferrari, G. (2017). Influence of pulsed light treatment on the aggregation of whey protein isolate. *Food Research International, 99*, 419–425.
58. Song, K., Mohseni, M., & Taghipour, F. (2016). Application of ultraviolet Light-Emitting Diodes (UV-LEDs) for water disinfection: A review. *Water Research, 94*, 341–349.
59. Uslu, G., Demirci, A., & Regan, J. M. (2016). Disinfection of synthetic and real municipal wastewater effluent by flow-through pulsed UV-light treatment system. *Journal of Water Process Engineering, 10*, 89–97.
60. Vimont, A., Fliss, I., & Jean, J. (2015). Efficacy and mechanisms of murine norovirus inhibition by pulsed-light technology. *Applied and Environmental Microbiology, 81*(8), 2950–2957.
61. Wekhof, A. (2000). Disinfection with flash lamps. *PDA Journal of Pharmaceutical Science and Technology, 54*(3), 264–276.
62. World Health Organization (WHO). (2022). Food safety. https://www.who.int/news-room/fact-sheets/detail/food-safety (accessed on 4 June 2024).
63. Wright, H. B., & Cairns, W. L. (1998). Ultraviolet light. In *Regional Symposium on Water Quality: Effective Disinfection* (pp. 1–26). CEPIS OPS, Lima, Peru.
64. Xu, F., Wang, B., Hong, C., Telebielaigen, S., Nsor-Atindana, J., Duan, Y., & Zhong, F. (2019). Optimization of spiral continuous flow-through pulse light sterilization for *Escherichia coli* in red grape juice by response surface methodology. *Food Control, 105*, 8–12.
65. Zhu, Y., Li, C., Cui, H., & Lin, L. (2019). Antimicrobial mechanism of pulsed light for the control of *Escherichia coli* O157 and its application in carrot juice. *Food Control, 106*, e106751.
66. Xenon. (2014). XENON RC-800-modular UV curing system. 1 -8. Xenon Corporation, Wilmington, MA. https://xenoncorp.com/wp-content/uploads/2019/11/RC-800_Brochure.pdf.

CHAPTER 11

APPLICATION OF PULSED LIGHT IN SOLID AND SEMI-SOLID FOODS

NENE MELTEM KEKLIK and ALI DEMIRCI

ABSTRACT

As the fight against foodborne diseases continues, pulsed light (PL) continues to be a promising technology for the preservation and safe consumption of foods. Pulsed light technology comprises short high-power pulses of a broad spectrum ranging from UV to infrared radiation generated via inert-gas flash lamps. In this chapter, the applications of pulsed light in solid and semi-solid foods are discussed. Basic principles of pulsed light technology, equipment designed and used for the treatment of solid and semi-solid foods, critical parameters of pulsed light processing, effects of pulsed light on solid and semi-solid foods in terms of microbial inactivation and quality changes, and potential commercial applications of pulsed light in solid and semi-solid foods are mainly focused on.

11.1 INTRODUCTION

Food safety issues continue to be a threat to human health and economies globally. An estimated 600 million cases of foodborne illnesses occur every year, resulting in 110 billion US dollars lost annually in productivity and medical expenses due to consuming contaminated foods globally [106]. Thus, ensuring food safety is a public health priority and a crucial step to achieving food security. Effective and efficient food safety and quality control systems are required not only to protect human health and well-being, but also to support economic development and improve livelihoods

Nonthermal Light-Based Technologies in Food Processing.
C. K. Sunil, Preeti Birwal, Megh R. Goyal, and R. Mahendran (Eds.)

by promoting access to domestic, regional, and international markets [33]. Pathogenic bacteria, viruses, parasites, or chemical substances present in unsafe foods cause more than 200 diseases, particularly affecting infants, young children, the elderly, and the sick [106]. Ensuring food safety is a matter of effective communication and collaboration between governments, producers, and consumers.

The main process used by the food industry to eliminate harmful microorganisms from foods to ensure food safety is conventional thermal methods. It is undeniable that these methods must be effective and reliable. Heat treatments are the most commonly used intervention methods in the food industry. However, the major disadvantage of heat treatments is their effect on food quality. The high heat required to kill microorganisms also degrades nutrients such as heat-labile vitamins and adversely affects the physical, chemical, and sensory quality of foods. Although various types of chemicals are also used in food processing as antimicrobials, there has been an ongoing debate on the health effects of these chemical treatments. Considering the increasing demands of consumers for fresh and raw/minimally processed products, it is clear that non-thermal techniques are needed in the food industry to ensure food safety.

Pulsed light (PL) is an emerging food processing technology, which is considered to be non-thermal if treatment times are kept short. Pulsed light consists of high-intensity light flashes which consist of electromagnetic radiation ranging from UV to infrared light. These light flashes are generated by inert gas lamps, e.g., xenon flashlamps. During the application of pulsed light in foods, microbial inactivation occurs when the food absorbs the energy of these light flashes. The microbial DNA absorbs UV light mainly at 200–280 nm in the UV-C range, which causes photochemical alterations in the DNA structure, leading to the inactivation of microbial cells [11]. Apart from photochemical effects, pulsed light was also demonstrated to have photophysical and photothermal effects on microorganisms [62, 103].

This chapter focuses on the main principles, equipment, and critical processing parameters of pulsed light technology with an emphasis on solid and semi-solid foods.

11.2 BASIC PRINCIPLES OF PULSED LIGHT TECHNOLOGY

UV light constitutes a part of the electromagnetic spectrum (Figure 11.1) from 100 to 400 nm and is basically subdivided into 4 categories: 315–400 nm as UV-A, 280–315 nm as UV-B, 200–280 nm as UV-C, and 100–200 nm as vacuum-UV. In some literature, vacuum-UV is not mentioned, and the wavelength range of 100–280 nm is named as UV-C [83, 89, 93]. In addition to eliminating microorganisms, there are other uses of UV light based on the wavelength such as illumination, UV curing, and phototherapy.

The main part of UV light used in the inactivation of microorganisms is UV-C. Although UV-A and UV-B also have a lethal effect on microorganisms, this effect is much lower than UV-C [104]. In fact, in order to increase the germicidal effect of UV-A, photosensitive compounds (e.g., furocoumarins) that diffuse into the microbial cell before irradiation are required [11]. The main reason UV-C has higher germicidal properties compared to UV-A and UV-B is that absorption of UV light by microbial DNA is maximal and fairly constant across UV-C, reaching a peak at about 260 nm [59]. UV-C induces lethal dimers on the DNA of microbial cells, which disrupt DNA replication and eventually lead to cell inactivation [22].

Lamps designed to produce UV-C radiation are proven to have good sterilization and disinfection properties, and UV technology is now used for water, air, or surface sterilization all around the world. Low-pressure (monochromatic at about 254 nm) mercury lamps, high-pressure (polychromatic) mercury lamps, and excimer lamps (207 nm, 222 nm) are designed for disinfection with UV-C light [12, 14]. UV-C is able to kill bacteria, viruses, protozoa, mycelial fungi, yeasts, and algae [11].

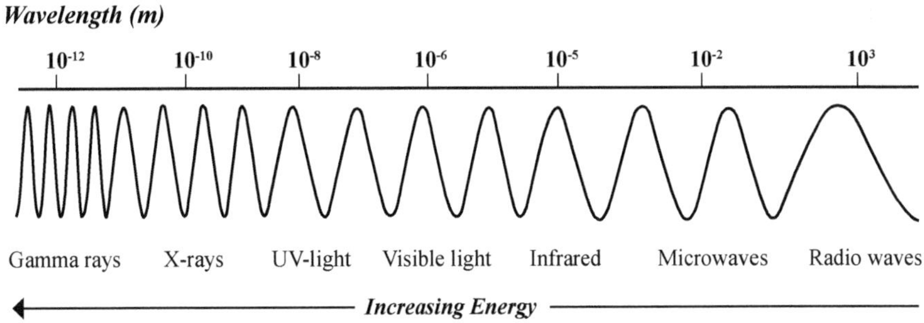

FIGURE 11.1 Electromagnetic spectrum.

On the other hand, pulsed light, also known as pulsed UV (PUV) light, is one of the emerging food processing technologies that promises efficient disinfection and decontamination of foods, water, air, packaging materials, and equipment. Pulsed light was first used for disinfection in the late 1970s in Japan [103]. Pulsed light (PL) is produced by inert gas lamps, particularly xenon gas lamps, which emit short pulses of broad-spectrum light including

UV, visible, and infrared. Light pulses are formed when the electrical energy stored in a capacitor is released in pulses of electrical energy [25]. These pulses then excite an inert gas (e.g., xenon), which in turn releases short and high-intensity light pulses. Typically, a flashlamp produces intense light pulses at a rate of about 1 to 20 pulses per second [25]. Each pulse contains electromagnetic radiation in the wavelength range of 100–1,100 nm (UV, visible, and infrared) [61]. The duration of each pulse is 200–300 μs [25]. One of the main characteristics of pulsed light is the high peak power, which differentiates pulsed light from continuous UV light [96]. The FDA approved pulsed light treatment of foods in 1996 [29].

One of the strongest aspects of pulsed light is that it is effective on a wide range of microorganisms. The resistance of microbial species and strains to PL irradiation decreases roughly in the following order [87]: fungal spores, bacterial endospores > parasites > viruses > and > vegetative bacteria. The germicidal effects of pulsed light depend on the absorption of the energy by the microorganism and its surroundings. Once absorbed by living cells, UV light induces certain photochemical reactions in the nucleic acids of microbial cells. The direct effects of UV light include photodimerization between adjacent pyrimidine bases and photohydration of cytosine [73]. Inter- and intra-strand crosslinks, protein-DNA crosslinks, DNA strand breaks, and rare base adducts are also induced by UV-B at much lower frequencies [91]. UV light also indirectly plays a role in the generation of reactive oxygen species, which react with DNA and proteins [73]. Cyclobutyl pyrimidine dimers including thymine dimers are the most common UV photoproducts [37]. The replication of the microbial DNA is prevented by these lethal photodimerizations. As a result, cellular functions are inhibited, and eventually the cell becomes inactivated.

Physical destruction also occurs in cells exposed to pulsed UV light. Two mechanisms have been proposed to explain this type of destruction: photothermal and photophysical mechanisms. Photothermal effects of pulsed light were explained by Wekhof [104] as follows: "*Rupture and disintegration of microbial cells occur when the incident UV photons are absorbed by the microorganisms at a faster rate than the surrounding medium.*" Fine & Gervais [32] supported this by proposing that PL disinfection at higher flux densities (0.5 J/cm^2) is achieved through bacterial rupture caused by overheating related to UV absorption.

Photophysical effect of pulsed light was demonstrated by Krishnamurthy et al. [62], who observed structural damages in *Staphylococcus aureus* cells in a transmission electron microscope (TEM) after a 5-s PL treatment. Since the treatment time was too short for a thermal effect, the authors attributed these changes (cell wall damage, cytoplasmic membrane shrinkage, internal cellular structure collapse, and leakage of cellular contents from the cytoplasm) to a constant physical disturbance caused by high-energy pulses. Similar structural changes of yeast (*Saccharomyces cerevisiae*) cells, such as expanded vacuoles, cell membrane distortion, and change in cell shape, were observed after pulsed light irradiation in TEM [96].

A major part of pulsed light that is responsible for the microbial inactivation is the UV region, which is why lamps generating more UV light with a higher peak power are preferred [103]. As mentioned before, the maximum absorption of UV light by the microbial DNA occurs at around 260 nm, thus it is clear that UV-C light (200–280 nm) causes most of the DNA damage. Wekhof [103] reported that half of the flash disinfection is caused by only UV-C, while UV-B and UV-A are also capable of inducing some changes in microorganisms. Another factor affecting the germicidal properties of UV irradiation is the absorption of UV light by water which hinders the lethality below 230 nm [58].

Miller et al. [73] reported that the DNA absorbance decreases more than three orders of magnitude in the UV-B compared to UV-C and is almost negligible in the UV-A region. Therefore, UV-C seems to play the main role in pulsed light disinfection, and other parts of UV light can also show germicidal action to an extent. In addition, visible and infrared parts of pulsed light may also contribute to the microbial inactivation [12, 27] due to the photothermal and photophysical inactivation mechanisms of pulsed light. The wavelengths known to cause undesirable changes in foods can be filtered through glass or liquid filters [5].

11.3 PULSED LIGHT EQUIPMENT DESIGNED FOR TREATMENT OF SOLID AND SEMI-SOLID FOODS

Pulsed light can be used for decontamination of surfaces or very thin layers of solid and semi-solid foods. This is because pulsed light has limited transmittance even though it still has better transmittance than conventional continuous UV light due to its high peak energy. The equipment specifications may vary depending on the requirements of particular manufacturing operations [24]. The basic parts of a pulsed light system are lamp, a power supply, and a controller (i.e., a pulse configuration device) [27].

Pulsed lamps are categorized as flashlamps and surface discharge lamps [12]. The flashlamp has a discharge which is confined in a small envelope. On the other hand, the surface discharge lamps involve a discharge of plasma out of a dielectric material located in a wide tube. Some of the advantages of surface discharge lamps over flashlamps are their higher UV efficiency and better tolerance of stress levels [90]. Power supply can be used to adjust input voltage, while the controller unit enables to configure pulse number and pulse rate. All of these factors are critical in terms of achieving a target microbial inactivation. Some producers incorporated the power supply and the controller into a cabinet with a front-panel control, which makes the setup and operation easier and more practical [24]. Pulsed light is instantaneous, i.e., it does not require warm-up time. A blower kit is generally used to provide air into the lamp housing to prevent overheating of the lamp during operation. A schematic diagram of a laboratory-scale pulsed light system is given in Figure 11.2.

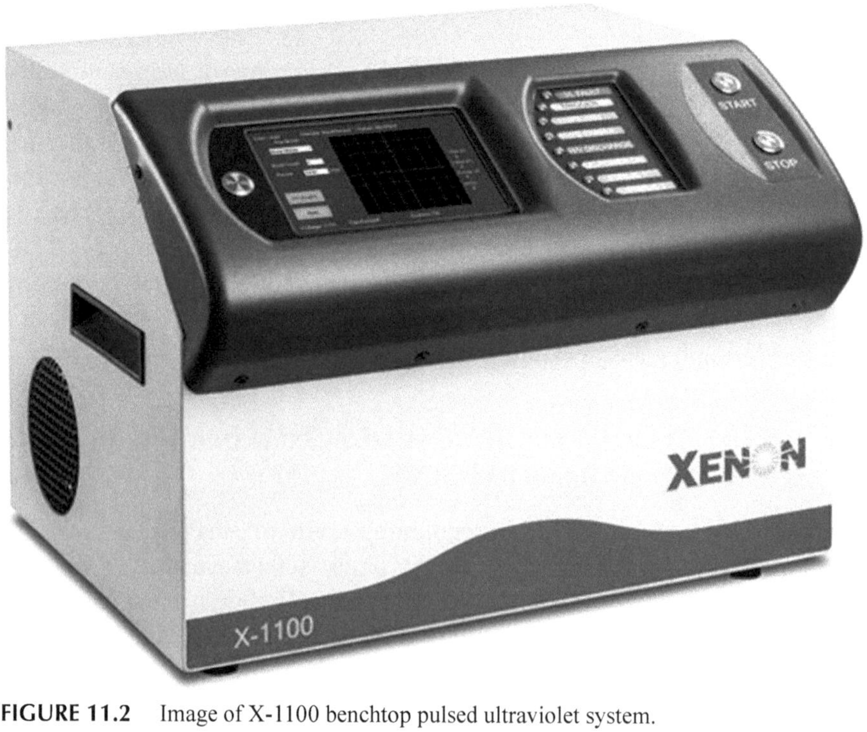

FIGURE 11.2 Image of X-1100 benchtop pulsed ultraviolet system.

Source: Reprinted with permission from Ref. [107]

Pulsed light offers some advantages over conventional continuous UV light such as the peak power, the lamp safety, the wavelength range, the inactivation mechanisms, the processing time, and the penetration ability [7]. The lamp is considered to be safe in pulsed light systems because of the utilization of inert gas. However, in conventional continuous UV light, mercury is often used in the lamps, which has potential hazards in case of a leakage [36]. The treatment or processing times required for the same level of microbial inactivation are much shorter in pulsed light processing compared to conventional continuous UV light. This is mainly due to the high peak power of pulsed light. While the power emission from continuous UV light ranges from 100 to 1,000 W [23], a pulsed light system can generate a peak power as high as 35 MW [68, 71, 77]. Pulsed light also has better penetration ability than continuous UV-C light [20].

In the lamp housing, a metal reflector is placed around the lamp except the bottom part to redirect the light emitted by the lamp toward the target food. It is also known as a light reflector cone, which provides guidance for the direction of light towards the food [55]. Another important factor is the angle of light incident on the food. As the surface roughness and the incidence angle of light increase, the shadow areas on an irregular surface also increase [39]. Thus, due to the irregular surface properties of solid and semi-solid foods, it may also be necessary to randomly apply pulsed light from different directions to the food for a sufficient period of time. For this purpose, dispersing reflectors may be used to achieve better light exposure [64].

The control unit may also have an interweave system, in which a single power supply and controller operate two lamps that flash alternately to improve the exposure of the target substance to pulsed light. In this configuration, which is also known as the 'dual-lamp system,' each lamp is sequenced by the single controller's high-speed logic [107]. This helps reduce overall hardware requirements, and thus may be useful in continuous processing. In Figure 11.3, an interweave pulsed light system is demonstrated. This pilot system (RC-802 Interweave Pulsed Ultraviolet System, Xenon Corporation, Wilmington, MA) includes two 16-inch, linear 'C' type xenon flashlamps, which are positioned on top of a conveyor. The lamp housings can be adjusted by 90? to 180? with respect to the long axis of the conveyor. The conveyor speed ranges from 4.0 to 72.0 ft/min and the distance between the conveyor and lamp housing ranges from 3 to 15 cm [17].

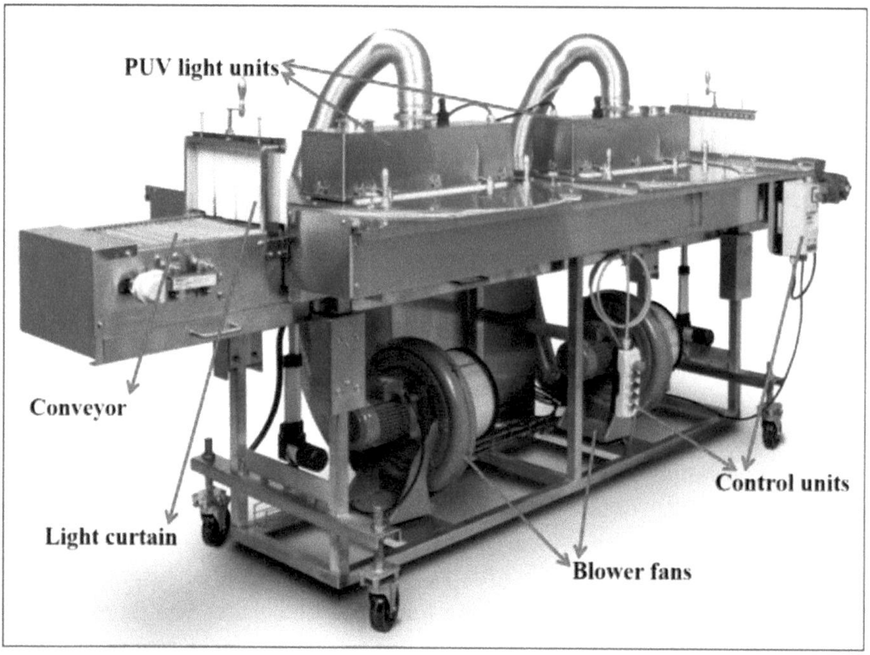

FIGURE 11.3 Z-5000 conveyor pulsed ultraviolet system.

Source: Reprinted with permission from Ref.[107]

Proper maintenance practices are also important for a safe and efficient process. Periodic cleaning is required, especially for the equipment that may come into contact with food and the surfaces through which light passes (e.g., quartz windows). The fact that pulsed light is a dry method, i.e., no water is involved, actually provides some advantages in terms of hygiene. The number of lamps, lamp position, and characteristics of the moving components (e.g., conveyor configuration and speed) of the pulsed light system should be adjusted precisely to maximize the pulsed light efficiency. The performance of the system should be monitored with regular energy measurements.

11.4 CRITICAL PROCESSING PARAMETERS

Pulsed light inactivates microorganisms if only its energy reaches the microorganism. For this to happen, first the food being processed needs to absorb the light. At this point, light absorption becomes a critical factor in pulsed light irradiation of foods. According to physics, three things can happen to a light wave which is incident upon an object: it can be absorbed by the object, it can be transmitted through the object, or it can be reflected from the object. All these events depend on the characteristics of light such as its frequency and intensity, and the optical properties of the object, i.e., food, such as its color and transparency. Light that is absorbed by the food may transform into heat, localized vibrations, and/or initiate photochemical reactions.

Three most important factors affecting the efficiency of the pulsed light process are the characteristics of light pulses, the properties of food, and the type of the target microorganism. The spectral distribution and fluence of light pulses are important and should be carefully selected for the target food and microorganism [79]. The input voltage, the total number of pulses, and the distance to the light source are among the factors that affect the total fluence (energy per area). The total fluence is directly proportional to the input voltage and the total number of pulses, but inversely proportional to the distance. The higher the total flux, the higher the inactivation is expected. Another way to increase the total fluence is to use more than one lamp (e.g., a dual-lamp system).

Color, transmittance, and depth of the food are among the most important food properties that affect the pulsed light process. The color of the food is actually an optical parameter that depends on the wavelengths in the visible spectrum leaving the surface of the food. The reflectivity of the surface is related to the angle of incidence, the polarization of the light, the properties of the food at the surface onto which the radiation falls, and the wavelength of the incident light [74]. The food's ability to absorb light may be modified by adding a colored component or dye.

The transmittance changes depending on the interaction between light and food. Reflectivity of light by opaque materials is generally high. Metals may reflect over 90% of the incident light [74]. On the other hand, transparent materials (e.g., glass, many crystals, and water) are able to transmit the bulk of the incident radiation, and thus reflectivity is low. The thickness of the food is one factor that affects the light transmittance of the food.

Bialka et al. [10] used two different gelled materials, agar, and denatured whey protein isolate (DWPI) based on their opaqueness to characterize the penetration depth of pulsed UV-light. Two mm-thick gel samples were analyzed for the amount of absorption of UV radiation. This study reported that as the material depth increased from 2 to 10 mm, the energy level decreased by 30% and 45% in agar and DWPI, respectively [10].

The type of microorganism and its resistance to pulsed light is also another critical factor in pulsed light treatment of foods. The microbial resistance to PL irradiation decreases roughly in the order of fungal spores, bacterial endospores, parasites, viruses, and vegetative bacteria [87]. There can even be differences between strains of the same species. Lagunas-Solar et al. [64] studied the pulsed-light inactivation of fungal pathogens on fresh fruits and found that *Aspergillus niger* was particularly resistant to pulsed light. The authors linked this to the presence of UV-absorbent melanin-type pigments on *A. niger* cells, which acted as a surface shield. Thus, microorganisms with such surface pigmentation appear to be potentially more resistant to pulsed light. A higher energy dose would be required for these microorganisms to destroy the outer pigment layer in order for the photons to reach the cytoplasm and nucleus of the cell [64].

11.5 APPLICATIONS OF PULSED LIGHT IN SOLID AND SEMI-SOLID FOODS

Pulsed light (PL) is also known as pulsed ultraviolet (PUV) light, intense light pulses (ILP), high-intensity pulsed UV light (HIPL), high-intensity broad-spectrum UV light (BSPL), and pulsed white light (PWL) [87]. In some of the following examples of research from literature, different names were used for pulsed light. However, for the sake of simplicity, we will mention the method as pulsed light (PL) in these studies.

11.5.1 EFFICIENCY OF MICROBIAL INACTIVATION

There are many researchers who have investigated the effects of pulsed light on the microbial populations in/on solid and semi-solid foods. In general, the studies show diversity in terms of the type of food, the target microorganism, and pulsed light configuration. As the total fluence depends on the distance from the lamp and the total number of pulses (pulse frequency x time), these are generally utilized by researchers as process parameters. Researchers compare the microbial inactivation levels based on different microorganisms, foods, and process parameters. Below are some examples of studies on the decontamination of solid and semi-solid foods using pulsed light technology.

11.5.1.1 MEAT AND POULTRY PRODUCTS

Söbeli et al. [94] applied pulsed light to beef loin steak and observed that a fluence of 4.2 J/cm^2 yielded about a 3.49 log (CFU per g) reduction in the counts of total aerobic mesophilic bacteria compared to the natural microbiota in the control group. Bryant et al. [13] investigated the effectiveness of pulsed light on the inactivation of *Escherichia coli* K12 on the surfaces of goat meat and beef. Maximum log reductions of 1.66 and 1.74 CFU/mL rinse solution were obtained on goat meat and beef, respectively, after a 60-second treatment at a 4.47 cm distance. Pulsed light was more effective on beef compared to goat meat. It was stated that beef had a relatively smooth, even surface than goat meat, which had an irregular surface structure that may have had shadowing effects. The authors suggested that the difference in composition, particularly fat content, between goat meat and beef might have an effect on the efficacy of pulsed light treatment [13].

Koch et al. [56] applied pulsed light at fluences of 0.52 to 19.11 J/cm^2 on pork skin and loin to inactivate *Salmonella typhimurium* and *Yersinia enterocolitica*. Reductions were obtained in the range of 1.73–3.16 log for *Salmonella* and 1.48–4.37 log for *Yersinia* on pork skin. Microbial reduction was found to be significantly lower on pork loin, ranging from 0.4 to 1.7 log for both pathogens. Koch et al. [56] stated that the differences in contour, porosity, and surface roughness between the skin and muscle may have caused this observation.

The 60-second treatment at an 8 cm distance (from the quartz window) in a pulsed UV-light system resulted in about 1.6 and 1.5 log10 CFU/cm^2 reductions of *Listeria monocytogenes* on unpackaged and vacuum-packaged (in polypropylene) chicken frankfurters, which corresponded to the energy doses of 55.9 and 48.4 J/cm^2, respectively [51]. Uesugi & Moraru [99] applied pulsed light and nisin to ready-to-eat sausages to inactivate *Listeria innocua*. Individual treatments, pulsed light (9.4 J/cm^2) and nisin (5,000 IU/ml), resulted in log10 reductions of 1.37 and 2.35, respectively. However, the combined treatment yielded a reduction of 4.03 $\log_{10}$ following storage at 4?C for 48 h, which indicates the combination of the treatments created an additive effect [99].

After just 15 and 30-second treatments at a 5 cm distance from the quartz window of the lamp, about 2 $\log_{10}$ reductions (CFU/cm^2) of *Salmonella* Typhimurium were obtained on both unpackaged and vacuum-packaged (in polypropylene) chicken breast, respectively [52]. Keklik et al. [49] worked later on the design of a pulsed UV light system that was designed at a pilot scale (Figure 11.4), in which whole chicken carcasses were transferred by a linear rail conveyor between two PUV lamps on each side. Treatment times of 30 and 180 seconds corresponded to a conveyor speed of 78 and 13 cm per minute, respectively, which resulted in 0.87 to 1.43 $\log_{10}$ (CFU/ml rinse solution) reductions of *E. coli* K12 [49].

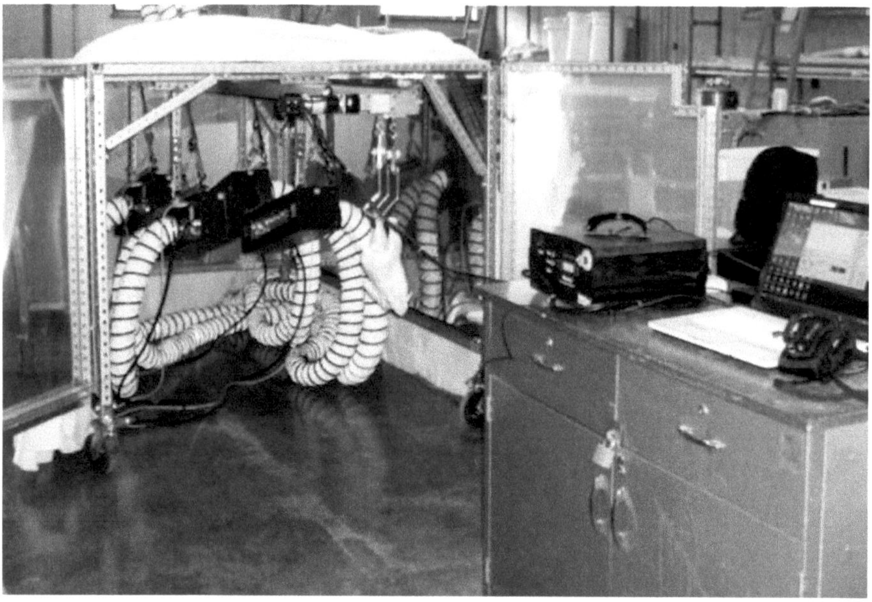

FIGURE 11.4 A pilot-scale pulsed light system designed and built for the treatment of whole chicken carcasses used in the study by Keklik et al. [49].

Source: Author's own work.

Liu et al. [66] investigated the influences of pulsed light-UV treatment on the microbial loads, quality, and storage period of Chinese traditional dry-cured meat products. The pulsed ultraviolet light treatment device included both a xenon lamp and a UV lamp. After the irradiation (pulse energy: 8 J, pulse number: 300, power: 6 W, effective exposure distance: 10 cm from the pulsed light source and 11 cm from the UV source), the total bacterial count was reduced from 6.89 to 4.53 lg (CFU per g), *Micrococcus* and *Staphylococcus* counts from 6.49 to 4.10 lg (CFU per g), and the counts of molds and yeasts from 5.45 to 4.28 lg (CFU per g) in dry-cured meat [66].

Kramer et al. [60] studied the pulsed light disinfection of sliced boiled ham, chicken cold cuts, and frankfurter sausages, which were packaged in PA/PE vacuum bags. After the treatments (0.35 J/cm^2–3.6 J/cm^2), the authors obtained about 1 log reductions of *L. innocua* on the surface of sliced boiled ham and chicken cold cuts, while 3–4 log reductions could be achieved on frankfurter sausages. Moreover, the initial count reduction was maintained over 12 days of storage at 5?C. The authors attributed the higher reductions on frankfurter sausages to their smoother surface microstructure compared to sliced ham or chicken cold cuts [60].

Cassar et al. [18] studied the effectiveness of pulsed light for inactivation of *Salmonella*, *E. coli*, and *Campylobacter* on the surface of chicken thigh meat by using a lab-scale system. Pulsed light treatments of lean surface thighs for 5 and 45 s yielded log10 (CFU/cm^2) reductions of 1.22 and 2.02 for *E. coli*, 1.45 and 2.09 for *Campylobacter*, and 1.55 and 2.42 for *Salmonella*, respectively. The total energy levels ranged from 3.38 to 62.24 J/cm^2. On the other hand, pulsed light treatments of skin surface thighs for 5 and 45 s yielded log10 (CFU/cm^2) reductions of 1.19 and 1.96 for *E. coli*, 1.08 and 1.85 for *Campylobacter*, and 0.90 and 1.82 for *Salmonella*, respectively. The authors suggested that the higher reductions observed on the lean surface compared to the skin surface may be due to the fact that skin has a more complex surface than lean muscle, and thus may provide some protection for the microorganisms [18].

Cassar et al. [19] also evaluated the effectiveness of pulsed light applied to chicken on a moving conveyor for inactivation of *Escherichia coli* on the surface of raw boneless/skinless chicken breasts, boneless/skinless chicken thighs, and bone-in/skin-on chicken thighs. The authors used the pulsed light system (Figure 11.3). Total energy fluences of 5, 10, 20, and 30 J/cm^2 were delivered to the surface of the products by adjusting the conveyor height (distance from the flashlamp) and speed. Exposure to pulsed light for 5 and 30 J/cm^2 yielded Log10 reductions of 0.29 and 1.04 for boneless/skinless breasts, 0.34 and 0.94 for boneless/skinless thighs, and 0.10 and 0.62 for bone-in/skin-on thighs, respectively [19].

In the study of Rajkovic et al. [86], pulsed light applied at a fluence of 3 J/cm^2 was effective in reducing Listeria monocytogenes, *Escherichia coli* 0157:H7, *Salmonella Typhimurium,* and *Staphylococcus aureus* by 2.24, 2.29, 2.25, and 2.12 log CFU/g, respectively, on the surface of dry fermented salami. The authors indicated that further increase in fluence did not increase microbial inactivation levels. Also, it was stated that the time interval between contamination and PL treatment should be kept as short as possible. The authors also reported that preformed *S. aureus* enterotoxin A was not affected by pulsed light treatment; thus, staphylococcal intoxication by fermented sausages must be prevented by inhibiting *S. aureus* growth and enterotoxin formation [86].

McLeod et al. [72] studied the efficacy of pulsed light on chicken fillet surfaces inoculated with microorganisms including *Staphylococcus aureus, Listeria monocytogenes, Pseudomonas spp., Salmonella Enteritidis, enterohemorrhagic Escherichia coli, Carnobacterium divergens, Brochothrix thermospacta,* and *E. coli* that produce extended-spectrum beta-lactamase.

Fluences applied in the range of 1.25–18.0 J/cm^2 resulted in reductions varying from 0.9 to 3.0 log CFU/cm^2 [72].

Ganan et al. [35] evaluated the efficacy of pulsed light to reduce *Listeria monocytogenes* and *Salmonella* Typhimurium on the surface of salchichón and loin, two ready-to-eat (RTE) dry-cured meat products. PL treatments at a fluence of 11.9 J/cm^2 yielded maximum log reductions in the range of 1.5–1.8 CFU/cm^2 for both microorganisms [35]. In the study of Hierro et al. [40], pulsed light at 0.7, 2.1, 4.2, and 8.4 J/cm^2 was applied to vacuum-packaged ham and bologna slices to inactivate *Listeria monocytogenes*. PL treatment at 8.4 J/cm^2 decreased the counts of *L. monocytogenes* by 1.78 and 1.11 CFU/cm^2 in cooked ham and bologna, respectively [40].

Keklik [48] studied the efficacy of pulsed light treatments in a wide range of fluence on pastirma (a dry-cured meat product) to inactivate *Staphylococcus aureus*. Even though the 45-s treatment at 5 cm distance (72.3 J/cm^2) resulted in a maximum decrease of 2.99 log (CFU per cm^2) in *S. aureus* counts, some quality parameters of pastirma were altered significantly after this treatment. The quality deterioration was avoided with the 8 cm/15 s treatment (18 J/cm^2), which resulted in 1.3 log CFU/cm^2 reduction in *S. aureus*.

As high as 8 $\log_{10}$ reduction of *Salmonella enteritidis* could be obtained on the surface of shell-eggs after applying 8 pulses (4 J/cm^2) [25].

Keklik et al. [50] reported a complete destruction of *S. Enteritidis* cells on the surface of shell eggs treated with pulsed light at 23.6 J/cm^2. A maximum 3.6 $\log_{10}$ (CFU/egg) reduction of *S. Enteritidis* was obtained on eggs (unwashed) treated with pulsed light at fluences up to 12 J/cm^2, whereas pulsed light treatment of eggs (washed) at a fluence of 12 J/cm^2 resulted in a maximum 1.8 $\log_{10}$ reduction of the same microorganism [42]. Hierro et al. [42] suggested that PL-treatment effectiveness for washed eggs was reduced likely because of the cuticle removal during washing, which facilitated the migration of the microbial cells into deeper layers of the eggs where UV light could not reach.

11.5.1.2 SEAFOOD

Ozer & Demirci [78] studied the effect of pulsed light treatment on the inactivation of *E. coli* O157:H7 and *Listeria monocytogenes* on salmon fillets. For *E. coli* O157:H7, the maximum $\log_{10}$ reductions were 1.09 $\log_{10}$ CFU/g on the muscle side and 0.86 $\log_{10}$ CFU/g on the skin side, which were obtained after the 60-second treatment at 8 cm and 30-second treatment at

5 cm, respectively. For *L. monocytogenes* Scott A, the maximum reductions were 0.74 $\log_{10}$ CFU/g on the muscle side and 1.02 $\log_{10}$ CFU/g on the skin side, both obtained after the 60-second treatment at 8 cm [78].

Cheigh et al. [20] studied the inactivation effects of pulsed light on *Listeria monocytogenes* surface-inoculated on seafood, including flatfish, salmon, and shrimp fillets. Pulsed light treatment produced about 2.2, 1.9, and 1.7 log reductions of *L. monocytogenes* cells on shrimp, salmon, and flatfish fillets, respectively, at a total fluence of 6.3 J/cm^2 (720 s), and about 2.4, 2.1, and 1.9 log reductions, respectively, at a total fluence of 12.1 J/cm^2 (1,380 s) [20]. Figueroa-Garcia et al. [31] investigated the effects of pulsed light on the Psychrotrophic counts on raw channel catfish (*Ictalurus punctatus*) fillets. Psychrotrophic plate counts on catfish fillets subjected to pulsed light at a total of 0.5, 1, and 2.0 J/cm^2 (2–4 flashes) were reduced by 1.5 log (CFU/g).

Pedrós-Garrido et al. [82] applied pulsed light (for 9 s at 3.5 cm) to raw salmon fillets to inactivate *Listeria* spp. (LIS), lactic acid bacteria (LAB), *Brochothrix thermosphacta* (BRT), *Photobacterium phosphoreum* (PHP), psychrophilic and mesophilic viable counts (PVC and MVC), Enterobacteriaceae (ENT), and *Pseudomonas* spp. (PSE). Inactivation levels ranged from about 0.6 log CFU/g for PSE to 1.6 log CFU/g in the LIS group. Hence, PSE was the most resistant bacterial group to PL treatments on salmon fillets [82].

Nicorescu et al. [76] applied pulsed light to raw salmon and obtained a maximum log reduction of 0.7 log (CFU/g) in naturally occurring aerobic flora after treatment at 30 J/cm^2. During storage at 4?C, both raw salmon controls and PL-treated samples showed constantly increasing microbial loads, which reached the level of 10^6 CFU/g on the 4th day. Also, exposure to 3 J/cm^2 yielded about 1 log CFU/g reduction in *P. fluorescens*. The authors attributed the low germicidal effectiveness of PL treatments on raw salmon to intrinsic factors such as the high content of lipids (protective against the pulsed light), high content of proteins (strong UV absorption characteristics which decrease the availability of photons for microbial decontamination), high water activity (a_w) (consequent rapid bacterial growth), and surface roughness (shield bacteria from the light) [76].

Hierro et al. [41] investigated the efficacy of pulsed light to improve the safety of carpaccio. Tuna slices inoculated with about 3 log (CFU/cm^2) of *L. monocytogenes*, *E. coli*, *S. Typhimurium*, and *Vibrio parahemolyticus* were subjected to fluences ranging from 0.7 to 11.9 J per cm^2. Treatments with 8.4 and 11.9 J per cm^2 reduced the counts of the above-mentioned microorganisms by about 1 log (CFU/cm^2) [41].

11.5.1.3 FRUITS AND VEGETABLES

Lagunas-Solar et al. [64] reported that the treatment of fresh fruits (apples, kiwi, lemon, nectarines, oranges, peaches, pears, raspberries, and grapes) with PUV (248-nm beams from excimer lasers) yielded more than 5 $\log_{10}$ reductions in plant (fungal) pathogens under 10 s. The threshold PUV energy dose in order to control fungal inactivation ranged from 100 to 1,900 mJ/cm^2 for fruit surfaces depending on the microorganism, e.g., *Aspergillus niger* was the most resistant and required the highest energy dose [64]. Ignat et al. [43] studied the effects of pulsed light applied at fluences of 17.5, 52.5, 105.0 and 157.5 kJ/m^2 on the microbial loads of fresh-cut apple slices. Independent of the applied fluence, inoculated bacteria and viable counts were decreased by 3 and 1 log, respectively.

A PL treatment for 60 s at doses in the range of 25.7–72 J/cm^2 reduced *Salmonella* spp. and *E. coli* O157:H7 by 3.4 and 3.9 $\log_{10}$ CFU/g on raspberries, while by 2.8 and 2.1 log10 CFU/g on strawberries, respectively [9]. In another study by Bialka & Demirci [8], maximum reductions of 4.3 and 2.9 $\log_{10}$ CFU/g were obtained for *Salmonella* and *E. coli* O157:H7 on blueberries, respectively, after treatment for 60 s at 8 cm from the light source. Cao et al. [16] applied dry pulsed light (6 J/cm^2) and water-assisted pulsed light (with agitation of samples in water during PL treatment; 9 J/cm^2) to blueberries. While the dry treatment yielded 0.9 log reduction of *Salmonella* for spot inoculation and 0.6 log for dip inoculation, the water-assisted treatment resulted in 4.4 log reduction in *Salmonella* for spot inoculation and 0.8 log for dip inoculation [16].

In the study of Escott et al. [28], red grapes (*Vitis vinifera* variety Tempranillo) were treated with pulsed light (PL) (5 or 10 flashes applied at medium (300 J/pulse) or maximum energy (600 J/pulse)) before fermentation. PL treatments significantly reduced wild microbial populations in grapes. After treatment, the grapes were inoculated and fermented with *Saccharomyces cerevisiae* at 22?C for 2 weeks with skins and seeds. The authors stated that wild organisms, especially yeasts, can be controlled by PL treatments, which in turn facilitate the growth and development of the yeast starter, and eventually affect wine quality [28].

Marquenie et al. [70] investigated the effect of pulsed light to destruct conidia produced by the fungi *Botrytis cinerea* and *Monilia fructigena*, which cause economic loss throughout postharvest storage and transportation of sweet cherries and strawberries. The authors indicated the conidia inactivation increased with higher intensity of treatment, although complete destruction was not obtained. However, complete destruction of the conidia of *M. fructigena* was achieved after combination of pulsed light with mild heat treatment (e.g., 40 s (PL treatment) + 15 min (41?C), or 80 s (PL treatment) + 10 min (41?C)) [70].

Aguiló-Aguayo et al. [1] reported that the treatments with pulsed light at 2.68 and 5.36 J per cm^2 decreased microbial counts of whole tomatoes during 15-day storage at 20?C. A fluence of 4 J/cm^2 yielded 1 log_{10} decrease in the microflora of tomato present in both the skin and the peduncle scar parts. Treatment at 2.2 J per cm^2 resulted in a 2.3 log_{10} reduction of *Saccharomyces cerevisiae* inoculated onto the tomato surface [1].

Leng et al. [65] studied the effectiveness of pulsed light (PL) treatment (1–63 J/cm^2), a novel antimicrobial (LAPEN) wash (2 min), and their combination in inactivating *Salmonella enterica* on stem scars of cherry tomato. The treatment effects on background microbiota were also evaluated during storage at 10?C for 21 days. A 30-s treatment (31.5 J/cm^2) resulted in a 2.3 log reduction of *Salmonella*, while a 2-min wash in LAPEN sanitizer yielded about a 2.1 log CFU/g reduction on stem scars. Two sequences of PL and LAPEN combinations were tested: Inoculated tomatoes were initially exposed to the optimum PL dose (31.5 J/cm^2) and then washed for 2 min in LAPEN sanitizer (PL-LAPEN) or vice versa (LAPEN-PL). No *Salmonella* was detected after treatment with PL-LAPEN, indicating > 5 log reduction, whereas LAPEN-PL provided a 4.5 log reduction of the pathogen. The PL-LAPEN treatment also reduced the native microbiota of the tomato and hindered their growth during storage for three weeks [65].

Luksiene et al. [67] studied the efficacy of pulsed light treatment (1,400 V, 1,000 pulses, total UV light dose 5.4 J/cm^2) in controlling microorganisms on the surface of various fruits and vegetables with different irregular surfaces (plums, tomatoes, cauliflowers, sweet peppers, and strawberries). Mesophilic bacteria that were naturally distributed decreased by 1.0–1.3 log based on the surface properties of the fruits and vegetables. *Bacillus cereus* was found to be less resistant to treatment with pulsed light, with an inactivation level of 1.3–2.0 log based on the surface [67].

Kaack & Lyager [46] inoculated carrot slices with diluted dispersions of yeasts (*Saccharomyces cerevisiae*) and applied pulsed light at a fluence of 0.7 J/cm^2 per pulse. The authors reported that the majority of the yeast cells were inactivated using just two pulses, and pulsed light may decrease the counts of yeast cells up to 6 log [46].

Tao et al. [97] evaluated the bactericidal efficacy of pulsed light (4.0–16.8 J/cm^2) against Staphylococcus aureus, *E. coli*, *Salmonella* Enteritidis, and *Listeria monocytogenes* inoculated on lettuce surfaces. PL treatments were effective in reducing the load of pathogens on the lettuce leaves in the range of 2.03–6.56 $\log_{10}$ (CFU per g) depending on the fluence applied. *L. monocytogenes* and *S. aureus* were the most resistant and the most sensitive strains to PL, respectively, while *E. coli* and *S. Enteritidis* exhibited moderate susceptibility. Right after refrigerated storage for 8 days, all lettuce treated with pulsed light-maintained counts of both total bacteria and total yeast and mold significantly lower than the control [97].

Agüero et al. [2] evaluated the bactericidal effect of pulsed light (up to 120 kJ/m^2) against *Listeria innocua* and *Escherichia coli* inoculated on spinach leaves. Reductions of 1.85 and 1.72 log CFU/g were obtained for *L. innocua* and *E. coli*, respectively, after the application of 2 light pulses (8 kJ/m^2). For both microorganisms, fluencies much lower than 10 kJ/m^2 were enough to achieve higher than 1 log reduction in microbial counts. However, increasing the fluence did not result in a proportional increase in microbial reduction. The microbial load naturally present on the raw material was reduced by 0.4–2.2 Log (CFU/g) after PL treatments depending on the fluence applied. The growth rates of microbial populations also decreased after pulsed light treatments during refrigerated storage [2].

The efficacies of pulsed light (PL), a new formula of sanitizer (HEN) consisting of hydrogen peroxide, EDTA, and nisin, as well as the combined treatment of PL and HEN sanitizer (PL-HEN) wash in inactivating *E. coli* O157:H7 on spinach leaves were investigated in the study of Mukhopadhyay et al. [75]. The treatment effect on microbial loads during 13 days of storage at 4C was also determined. The optimal treatment dose was identified as 15.75 J/cm^2, equivalent to 15-s intense PL treatment, above which quality was adversely affected. The optimal PL dose yielded a 2.7 log CFU/g reduction of *E. coli* O157:H7 while a 2-min wash in sanitizer formulation HEN resulted in a comparatively low 1.8 log CFU/g reduction of the pathogen. The authors also evaluated two different sequences of PL and HEN treatment combinations, in which inoculated leaves were first treated at the optimal PL dose (15.75 J/cm^2) followed by 2 min immersion in HEN (PL-HEN) or vice versa (HEN-PL). HEN-PL treatment resulted in a 4.6 log reduction, while *E. coli* cells were found to be undetectable after PL-HEN treatment indicating a strong synergistic inactivation (>5 log reduction). The PL-HEN treatment of spinach also slowed the growth of spoilage microorganisms during storage [75].

11.5.1.4 DAIRY PRODUCTS

Dunn et al. [26] investigated the effect of pulsed light (16 J/cm^2 per pulse) on the inactivation of *Pseudomonas* spp. in cottage cheese. The cell population was reduced by 78.9% and 96.7% after 1 and 2 pulses, respectively. Fernández et al. [30] investigated the efficacy of pulsed light on *Listeria innocua* on two cheese varieties with different surface features: Gouda (smooth and shiny) and Manchego (porous and matte). The slices of cheese were treated with pulsed light at fluence levels ranging from 0.9 to 8.4 J per cm^2. The PL treatment was found to be less successful in Manchego compared to Gouda, for which a reduction of 3 log (CFU per cm^2) was achieved at 0.9 J per cm^2. On the other hand, *L. innocua* counts were only reduced by < 1 log CFU/cm^2 on Manchego slices, regardless of the applied fluence [30].

In the study of Can et al. [15], the effect of pulsed light on *Penicillium roqueforti* and *Listeria monocytogenes* inoculated onto packaged and unpackaged white American Cheese was investigated. Maximum reductions of 1.32 and 1.24 log CFU/cm^2 were obtained for *P. roqueforti* after 40-s treatment at 5 cm (3 pulses/s, 0.445 J/cm^2/pulse) on unpackaged and packaged cheese, respectively. Reductions of about 2.9 and 2.8 log CFU/cm^2 were achieved for *L. monocytogenes* under the same treatment conditions on packaged and unpackaged cheeses, respectively [15].

Proulx et al. [84] investigated the effect of pulsed light treatment on the destruction of the spoilage bacteria *Pseudomonas fluorescens*, the non-enterohemorrhagic *E. coli* (ATCC 25922 – nonpathogenic surrogate of *E. coli* O157:H7), and *L. innocua* (nonpathogenic surrogate of *L. monocytogenes*) on white cheddar and processed cheeses. Maximum inactivation levels of 3.37 log, 3.74 log, and 5.41 log were obtained for *L. innocua, P. fluorescens*, and *E. coli*, respectively. The PL treatments without packaging and through UV-transparent packaging consistently yielded similar inactivation levels [84].

As the effectiveness of PL treatment is limited by light-cheese interactions, Proulx et al. [85] investigated the effectiveness of pulsed light by combining it with the antimicrobials nisin and natamycin to inactivate *P. fluorescens*, the non-enterohemorrhagic *E. coli,* and *L. innocua* on white cheddar and processed cheeses (white American singles). Cheese slices received antimicrobial treatment either before or after pulsed light (1.02 to 12.29 J/cm^2). Treatment with antimicrobials before PL decreased the effectiveness of PL for all the bacteria tested. The authors attributed this to the absorption of UV light by nisin and natamycin, which reduced the UV fluence received by the bacteria. On the other hand, the inactivation levels increased in the case of the application of antimicrobials after pulsed light. A maximum decrease of 3.73 log was achieved on processed cheese for *L. innocua* after treatment with PL at 9.22 J/cm^2 followed by nisin, in comparison to 3.01 log by PL alone [85].

In the study of Lacivita et al. [63], the effects of pulsed light (0.39–28.0 J/cm^2) on *Pseudomonas fluorescens* and *Enterobacteriaceae* inoculated onto mozzarella cheese were assessed. A reduction of about 1 log cycle was obtained for both microorganisms on the inoculated cheese. In the second phase of the study, the effectiveness of PL treatments (7.02 and 14.04 J/cm^2) on the native flora of mozzarella cheese was evaluated. After PL treatments, both the control (untreated) and PL-applied mozzarella cheese samples were packed and kept for 2 weeks at 10?C. While there was an increase of *Pseudomonas* spp. population in the control samples after a few days of storage (4 d) and the microbiological limit was exceeded after 5.5 d, the microbial load of PL-treated samples remained below the detection limit until the 6th day of storage [63]. Keklik et al. [53] evaluated the effectiveness of pulsed light for the inactivation of *S. aureus* and *E. coli* O157:H7 on the surface of fresh kashar cheese. The authors reported that depending on the level of inactivation, treatment time, and appearance 'the most favorable treatment' was found to be the 45-s treatment at 13 cm (about 44 J per cm^2), which resulted in a decrease (CFU per cm^2) by about 1.62 log for *S. aureus* and 3.02 log for *E. coli* O157:H7.

In another study by Keklik et al. [54], the efficacy of pulsed light in inactivating selected pathogens on white cheese was investigated. The 60 s-8 cm treatment (91.22 J/cm^2), which resulted in the highest inactivation level without causing a visual alteration in the appearance of white cheese, produced about 1.31 and 2.20 log10 reductions (CFU/cm^2) for *S. aureus* and *E. coli* O157:H7, respectively. The lower inactivation levels obtained for white cheese compared to kasar cheese may be due to the differences in the surface characteristics of these two types of cheeses. While the surface of kasar cheese is smoother, white cheese has a more irregular and porous surface. Additionally, it is evident that the pulsed light treatments resulted in higher reductions for *E. coli* O157:H7 compared to *S. aureus* in both studies [53, 54]. The reason behind this observation may be that Gram-negative cells are less resistant to pulsed light compared to Gram-positive cells, as reported in several studies [88, 100].

Chen et al. [21] investigated the effect of using a vibratory-assisted PL system to inactivate *Cronobacter sakazakii* and *Salmonella* surrogate *Enterococcus faecium* in different powdered foods. The authors reported that *C. sakazakii* and *E. faecium* on non-fat dry milk (NFDM) were reduced by 5.27 and 3.67 $\log_{10}$ CFU/g, respectively, after 3 or 4 passes of PL treatments [21].

11.5.1.5 OTHER SOLID AND SEMI-SOLID FOODS

A pulsed light treatment of cornmeal for 100 s at 3 cm from the lamp was able to decrease the counts of *Aspergillus niger* spores by 4.93 $\log_{10}$ (CFU/g) at most [45]. In another study, alfalfa seeds inoculated with *E. coli* O157:H7 were exposed to pulsed UV light, and a $\log_{10}$ reduction of 4.80 (CFU/g) was achieved for 1.02-mm thick seed layers after a 30-s treatment [92].

Harguindeguy & Gómez-Camacho [38] studied the inactivation of *Salmonella enteritidis* on the surface of almond kernels using pulsed light at different operative conditions, namely power (1,000, 1,250, and 1,500 W), frequency (1.8, 3.0, and 100.0 Hz), and different treatment times (from 5 to 250 s), which corresponded to fluences in the range of 0–100 J/cm^2. About 1.8, 3.0, and 100.0 Hz systems yielded the maximum log reductions of 6.02, 4.69, and 3.66, respectively [38].

Koçer & Keklik [57] investigated the effect of pulsed light on *Listeria monocytogenes* inoculated onto the surface of sliced button mushrooms (*Agaricus bisporus*). The mushroom samples were treated with pulsed light at different distances from the quartz window (5, 8, 13 cm) for different times (5, 15, 30, 45, 60 s) corresponding to the energy doses in the range of 4.00–101.26 J/cm^2. The log reductions ranging from 1.96 log CFU/g (5 cm–60 s) to 0.48 log CFU/g (13 cm–5 s) were obtained [57].

In the study of Chen et al. [21], efficacy of a vibratory-assisted PL system in inactivating *Cronobacter sakazakii* and *Salmonella surrogate Enterococcus faecium* in different powdered foods was investigated. *C. sakazakii* on wheat flour and egg white powder were significantly inactivated by 4.92 and 5.30 $\log_{10}$ CFU/g, respectively, after 3 or 4 passes of PL treatments. On the other hand, 3–4 passes of PL treatments reduced the *E. faecium* counts on wheat flour and egg white by 2.79 and 2.74 $\log_{10}$ CFU/g, respectively. The authors highlighted that this PL system produces higher microbe inactivation compared to immobilized pasteurization and does not cause obvious undesirable powder agglomeration [21].

Fine & Gervais [32] evaluated the efficiency of pulsed light on the destruction of microbial viability in wheat flour and black pepper. Initial cell counts of mainly sporulated forms of *Bacillus* spp. were 1×10^7 and 4×10^6 CFU/g for wheat flour and black pepper, respectively. After treatment at a fluence of 31.12 J per cm^2 (20 mm from the lamp, 64 pulses at 220 J), residual viability was maintained around 89.9% and 55.5% for wheat flour and black pepper, respectively [32].

11.5.2 EFFECTS OF PULSED LIGHT ON QUALITY PARAMETERS

Pulsed light can only be applied to foods if it does not cause undesirable changes in the quality of the final product. The treatment of foods with pulsed light does not typically involve chemical addition. However, the irradiation involving the UV, visible, and infrared portions of the electromagnetic spectrum may lead to photodegradation and other chemical changes in foods. For example, high-intensity light may cause the formation of chemicals such as hydroxyl radicals, ozone, or hydrogen peroxide [69]. The extent of quality alteration depends not only on specific properties of food such as its composition and optical characteristics, but also on the process parameters such as the distance from the lamp and treatment time. Furthermore, heat generation during long treatments due to the infrared portion of pulsed light may bring additional heat-related quality changes in foods.

11.5.2.1 MEAT AND POULTRY PRODUCTS

In the study of Söbeli et al. [94], the application of pulsed light at fluences of 0.525, 1.05, 2.1, and 4.2 J/cm^2 did not affect total moisture, total protein, total lipid, and ash content of beef loin steaks. The treatments, however, reduced the a* and b* values of the samples significantly. Pulsed light treatments caused an increase in some volatile compounds found in raw beef loin steaks such as hexanal, 2-heptenal, 2-octenal, 2,4-nonadienal, 2,4-decadienal, 3-octen-2-one, and limonene. On the other hand, the amounts of some compounds coming from microbial metabolism such as 1-heptanol, 1-octen-3-ol, and 2-octen-1-ol showed a decrease after pulsed light treatments. The authors attributed most of these changes in the volatile profile of raw beef loin steaks to the lipid oxidation due to the photochemical effects of UV light or the microbial death with higher UV-C fluence levels [94].

Koch et al. [56] reported that pulsed light treatments $\geq$7.36 J/cm^2 changed the color parameters of pork skin, and fluences $\geq$9.66 J/cm^2 decreased the redness of pork loin samples. Fluences of 0.52 to 12.81 J/cm^2 resulted in odor changes, except for the treatment of pork skin at 0.52 J/cm^2. The authors also reported that the malondialdehyde concentration was significantly higher in pork skin at fluences of 0.84, 9.66, and 14.56 J/cm^2 compared to the control and other fluences applied, but still remained beneath the threshold concentration of 0.5 μg/g. In pork loin, however, no statistically significant difference was observed between the control and treated samples [56].

Keklik [48] evaluated the effect of pulsed light on the quality parameters of pastirma exposed to a wide range of pulsed light fluence. With the use of higher fluences, the lightness of pastirma tended to increase, while the redness and yellowness decreased. The 45-s PL treatment at 8 cm (≥54.0 J/cm^2) decreased the moisture content of pastirma significantly. While the 5 and 25-s treatments at any distance (5, 8, or 13 cm) did not result in a significant rise in the TBARS levels, the treatments for 45 s increased the TBARS values significantly compared to the control. Overall, the quality of pastirma tended to change above the fluence of 20 J/cm^2 [48].

Wambura et al. [102] monitored the moisture loss, texture, color, and oxidative stability of sliced ham during storage (at 4C for up to 14 days) following pulsed light treatments (for 60, 90, and 120 s at 4.5, 8.3, and 14.6 cm from the UV lamp). The authors observed that during storage, the moisture loss of sliced ham samples treated with pulsed light at shorter distances from the lamp was higher compared to longer distances. It was suggested that the high PL energy close to the lamp might have resulted in damage to the tissues of ham, which then caused a greater amount of moisture loss during storage [102]. Texture evaluation demonstrated that the hardness of most PL-treated samples decreased during storage. Although the color parameters of sliced ham showed some fluctuations during storage, the lightness and yellowness decreased while the redness increased toward the end of the 14-day storage. The authors reported that although all samples became rancid during storage, there was a more rapid reduction of oxidative stability in sliced ham samples treated with pulsed light. Wambura et al. [102] suggested that the heat produced by PL treatment might have accelerated the oxidation.

Hierro et al. [40] observed significant increases in the TBARS levels for cooked ham that were PL-treated (above 2.1 J per cm^2). Nevertheless, the extent of lipid oxidation was regarded as very low by the authors even after the highest fluence applied (8.4 J/cm^2). While pulsed light treatment at 8.4 J/cm^2 did not change the sensory quality of cooked ham, treatments above 2.1 J/cm^2 negatively affected the sensory properties of bologna [40].

Ganan et al. [35] observed that the lightness of a ready-to-eat (RTE) dry-cured meat product named *Salchichón* increased significantly after being treated with pulsed light at 11.9 J per cm^2, however these alterations were absent during storage. Conversely, the treatments did not change the redness and yellowness of *Salchichón* significantly [35]. Also, the sensory analysis of salchichón indicated no changes either after treatment or during 30-day storage. However, some changes in the sensory quality of loin, another RTE dry-cured meat product, were perceived by the panelists immediately after treatment, which disappeared along the storage [35].

Rajkovic et al. [86] reported that among the treatments only the highest fluence (15 J/cm^2) applied caused unacceptable lipid oxidation values for fermented salami slices packed in modified atmosphere and kept in cold storage for 9 weeks. The significant increase in protein oxidation of fermented salami slices treated with pulsed light, which was discerned as a 28% increase in carbonyl content in comparison to the samples untreated, was only found following the cold storage of samples packed in both vacuum and modified atmosphere for 9 weeks [86].

Keklik et al. [51] observed that lipid peroxidation of chicken frankfurters did not change significantly after 'mild' (2.9 J/cm^2) (5-second treatment at 13 cm for the quartz window of the lamp) and 'moderate' (27.9 J/cm^2) (treatment at 8 cm for 30 seconds) pulsed light treatments, while color parameters differed significantly following the treatments involving both unpackaged and vacuum-packaged frankfurters. In another study conducted by the same authors [52], neither malonaldehyde contents nor color parameters of unpackaged and vacuum-packaged raw chicken breast changed significantly after 'mild' and 'moderate' treatments.

Kramer et al. [60] did not observe a significant difference in the general appearance or color parameters of pulsed-light treated boiled ham, chicken cold cuts, and frankfurters in comparison to the untreated reference samples throughout storage (5?C, up to 12 days) for all investigated intensities (0.35 J/cm^2–3.6 J/cm^2). However, the authors reported that intense pulsed light treatments induced off-flavor formation directly after the treatments, and the highest applied fluence (3.6 J/cm^2) caused significant differences in smell [60].

Cassar et al. [19] investigated the application of pulsed light to the surface of raw boneless/skinless chicken breasts, boneless/skinless chicken thighs, and bone-in/skin-on chicken thighs on a moving conveyor (Figure 11.3). Lipid oxidation and changes in color of chicken samples were evaluated after exposure to 30 J/cm^2. Measurements of lipid oxidation were taken at 0, 24, 48, and 120 hours after the treatment. Neither lipid oxidation nor the color parameters were significantly affected by PL treatments regardless of the product type [19].

Keklik et al. [51] also analyzed the effect of pulsed light on the packaging material (polypropylene) used for chicken frankfurters. The mechanical properties, i.e., elastic modulus, maximum tensile strength, yield strength, percent elongation at yield point, and percent elongation at break did not alter significantly after the 'mild' treatment (2.9 J/cm^2) (5-second treatment at 13 cm).

In Kramer et al. [60]'s study, the physical properties including oxygen permeability, water vapor permeability, tensile strength, tear strength, and sealing strength of the PA/PE bags were not affected by pulsed light treatments under the applied conditions (up to 3.6 J/cm^2).

In Keklik et al. [52]'s study, the extreme pulsed light treatment (67.0 J/cm^2) (60-second treatment at 5 cm) significantly changed the elastic modulus of polypropylene (used for packaging raw boneless chicken breast) in both the perpendicular-to-machine and along-machine directions, and yield strength only in the perpendicular-to-machine direction.

After examining the eggshell using scanning electron microscopy, Hierro et al. [42] reported that the damaged cuticles can create shadow zones and pores which can facilitate the protection of bacteria from pulsed light. Thus, the authors indicated that the egg cuticle needs to be intact in order to decontaminate shell eggs effectively by pulsed light [42]. Keklik et al. [50] reported that the treatment of shell eggs for 20 s at a total fluence of 23.6 J/cm^2, which resulted in complete inactivation of *Salmonella enteritidis*, did not significantly change ($p>0.05$) the albumen height, eggshell strength, or cuticle presence [50].

11.5.2.2 SEAFOOD

Cheigh et al. [20] did not observe significant color changes on flatfish, salmon, and shrimp fillets exposed to pulsed light at 12.1 J/cm^2. Nicorescu et al. [76] observed that pulsed light treatment at 30 J/cm^2 increased the MDA content of raw salmon by 39.3%, probably partly due to the presence of a high content of polyunsaturated fatty acids. However, the pulsed light treatments did not affect the color parameters significantly [76].

Tomašević [98] evaluated the effects of pulsed light (intense light pulses) on the sensory quality of seafood including tuna, salmon, flounder, and crab, which were treated with 1 and 5 light pulses (pulse duration: 300 μs, pulse intensity: 3.4 J/cm^2, pulse rate: 1 pulse/2 s). All the seafood samples were found to be very acceptable. The sensory attribute that was the most sensitive to the pulsed light treatment was odor. The lower dose of pulsed light (3.4 J/cm^2) did not affect the color parameters of tuna, while the higher treatment (17 J/cm^2) decreased the lightness. Even the lower dose of pulsed light reduced the lightness of salmon samples significantly, while a* and b* values remained unchanged by the pulsed light treatment. The yellowness of flounder and crab meat samples was more sensitive to pulsed light treatments compared to lightness and redness. The authors indicated that pulsed light could affect the color of seafood but not to the extent that it significantly alters total sensory scores [98].

Figueroa-Garcia et al. [31] treated raw catfish fillets with pulsed light at 0.25 and 0.5 J/cm^2 per flash using 2 and 4 flashes at each energy level. The fillets were stored at 2?C and analyzed to determine quality changes after 2, 4, 6, 9, and 13 days of treatment. The authors observed increased Hunter 'L' and 'b' values after 4 and 2 days of storage, respectively, in PL-treated channel catfish fillets. Thiobarbituric acid oxidation values increased for all PL-treated fish after 2 days of storage. Gapping, which indicates tearing of the connective tissue between muscle layers [44], was not affected by PL treatment and storage time [31]. However, water holding capacity and moisture content decreased in all catfish fillets during storage [31].

In the study by Hierro et al. [41], the fluence of $\geq$8.4 J/cm^2 resulted in significant differences in lightness and redness of tuna carpaccio. Yellowness was more sensitive to pulsed light because fluences above 0.7 J/cm^2 significantly decreased the b* value. The sensory analysis demonstrated that the fluences of 8.4 and 11.9 J/cm^2 affected color or odor scores adversely. The authors reported that although the shelf-life was not extended by pulsed light treatments, a fluence of 2.1 J/cm^2 was sufficient to Improve the microbial quality and retain the sensory properties of the product for one week [41].

11.5.2.3 FRUITS AND VEGETABLES

Treatment with pulsed light at an 8-cm distance from the quartz window of the lamp for 30 and 60 s did not significantly change the sensory quality and color of blueberries, respectively [8]. In the study of Cao et al. [16], dry PL (6 J/cm^2) and water-assisted PL (9 J/cm^2) were applied to blueberries, which were subsequently stored at room temperature (3 days) or 5?C (7 days). None of the treatments improved the shelf life of blueberries. Both treatments reduced the surface lightness, while they did not decrease the firmness of blueberries. An increased weight loss was observed for only the dry PL-treated samples. Both PL treatments delayed anthocyanin accumulation and reduced total antioxidant activity at the end of storage at room temperature. The water-assisted PL treatment resulted in a slight enhancement in total phenolics content [16].

Avalos-Llano et al. [4] applied pulsed light treatments (4, 8, 12, and 16 J/cm^2) combined with a stabilizing dip to fresh-cut strawberries. Quality changes and antioxidant properties were monitored for 14 days at 5?C. The authors reported that the surface color of treated fresh-cut strawberries was preserved, while the softening was reduced by the doses of 4 and 8 J/cm^2 over storage. The total phenolic contents of untreated and treated fresh-cut strawberries did not show significant differences during storage. While Vitamin C and total anthocyanin contents of the samples treated at low energy doses were unaffected, those of samples treated at the highest energy dose were reduced by between 20% and 30%, respectively. Regardless of the applied dose, the initial antioxidant capacity was better maintained in all of the samples during storage. Avalos-Llano et al. [4] reported that the doses of 4 and 8 J/cm^2 were the most effective for maintaining the quality and antioxidant properties of fresh-cut strawberries.

Escott et al. [28] monitored the effect of pulsed light treatment (5 or 10 flashes applied at medium (300 J/pulse) or maximum energy (600 J/pulse)) before fermentation on grape external structure, phenol extraction, and wine composition. There were no changes to the external appearance of grapes after PL treatments of 5 or 10 pulses. The pH values also did not differ significantly among treatments, which means PL energy did not affect the extraction of cations that can modify acidity/pH [28]. The level of methanol, which indicates the enhancement of extraction, was found to increase slightly after PL treatments. The authors suggested that this may be due to the photothermic effect of PL radiation on skin structure, as methanol is a constituent of the pectins, which mainly form the cell wall of grape skins. The levels of anthocyanins at the end of fermentation were similar to controls or slightly lower in PL treatments of 5 flashes at medium energy and 10 flashes at maximum energy. The effect of PL treatment was not found to be significant in the sensory profile of red grapes. Escott et al. [28] indicated that PL treatments can assist in prolonging the shelf-life of grapes in post-harvest technology, in addition to enhancing the fermentative purity and the quality of wine.

Ignat et al. [43] determined the effects of pulsed light treatment on the quality of fresh-cut apple slices during storage. The PL fluence did not change the firmness of the apple slices but significantly affected weight loss, color, and sensory attributes during storage at 6?C. While the low-intensity pulsed light (17.5 kJ/m^2) did not affect the fresh appearance of apple slices, pulsed light treatment above 17.5 kJ/m^2 caused dehydration, browning, and flavor depletion [43].

Avalos Llano et al. [3] studied the effects of pulsed light (PL) treatments combined with a quality-stabilizing dip on the quality and antioxidant properties of fresh-cut 'Golden Delicious' apples. After the apple wedges were dipped into a solution of N-acetylcysteine and $CaCl_2$, they were flashed with broad-spectrum light at 4, 8, 12, and 16 J/cm^2. Quality attributes were evaluated over 15 days at 5?C. The authors indicated that the quality-stabilizing pre-treatment was effective in preventing browning phenomena on the cut-tissue surface. Moreover, pulsed light treatment did not promote browning and oxidation in samples. The initial contents of phenolic compounds and vitamin C were found to be better maintained than in untreated samples. Avalos Llano et al. [3] indicated the best PL treatments were at 8 and 16 J per cm^2 for preserving the antioxidant properties and quality of fresh-cut apples.

In the study of Leng et al. [65], exposure of cherry tomatoes to pulsed light (31.5 J/cm^2) followed by a novel antimicrobial (LAPEN) wash (2 min) (PL-LAPEN) did not significantly affect the firmness and visual appearance quality of tomatoes during storage for three weeks, while the texture of treated tomatoes was significantly affected. In the study of Aguiló-Aguayo et al. [1], softening, wrinkles, and increased weight loss were observed on the PL-treated tomato surface three days later. The levels of ascorbic acid did not change during storage. The authors also observed that the contents of α-carotene, β-carotene, total lycopene, and lycopene isomerization percent were increased in tomato extracts obtained from fresh tomato fruit PL-treated at 30 J per cm^2. The lycopene bio-accessibility increased in hot-break purees made from fresh tomatoes that were PL-treated at 5.36 J per cm^2 and kept in storage for 15 days [1].

Pataro et al. [81] applied pulsed light or UV-C irradiation (1–8 J/cm^2/side) to immature green tomatoes, and then stored them at 20 ± 2?C for up to 21 days. The authors reported that pH and ?Brix of all samples did not significantly change after treatments or during the storage period. However, the content of lycopene, total carotenoids, phenolic compounds, and antioxidant activity of light-treated samples increased during storage up to 6.2, 2.5, 1.3, and 1.5-fold, respectively, in comparison to the untreated samples [81].

Valdivia-Nájar et al. [101] studied how pulsed light treatment affected the bioactive compounds and antioxidant capacity in fresh-cut tomatoes (*Lycopersicon esculentum* Mill. cv. Daniela). Tomato slices were exposed to pulsed light at fluences of 4, 6, and 8 J/cm^2 and stored for 18 days. Pulsed light treatment caused an increase in the contents of lycopene (up to 26.1% higher) and phenolic compounds (up to 4.3% higher) compared to the untreated tomato slices, while a significant depletion of vitamin C was observed. Pulsed light treatment and subsequent storage significantly reduced the antioxidant capacity of tomato slices. The changes in the antioxidant capacity and bioactive compound contents of tomato slices over storage increased with higher fluence applied [101]. Luksiene et al. [67] reported that there were no significant alterations observed in total phenolics, vitamin C, antioxidant capacity, color, and firmness in plums, tomatoes, cauliflowers, sweet peppers, or strawberries that were treated with pulsed light (1,400 V, 1,000 pulses, total UV light dose 5.4 J/cm^2).

Tao et al. [97] evaluated the impact of pulsed light treatment (4.0–16.8 J/cm^2) on the quality characteristics of lettuce during 8-day storage at 4?C. The product quality could be maintained after PL treatments during storage for 8 days by keeping weight loss at a minimum, maintaining color and a good level of ascorbic acid, chlorophyll, and total soluble solids. By considering the quality and safety of the product in addition to the feasibility of the treatment, pulsed-light fluence levels of 8.2–12.5 J/cm^2 were suggested for surface disinfection and extending the shelf-life of fresh-cut lettuce [97].

Agüero et al. [2] evaluated the impact of pulsed light treatments (20 and 40 kJ/m^2) on headspace gas composition, microbial quality, antioxidant properties, and color of spinach leaves after treatment and during refrigerated storage. PL treatments significantly affected the changes in the package headspace composition. While O_2 concentrations decreased, in-package production of CO_2 increased at a higher rate in comparison to the untreated spinach leaves. The authors reported that the total polyphenolic content and antioxidant capacity of spinach increased significantly in the range of 5–10% and 32–34% after treatments at 20 or 40 kJ/m^2, respectively. However, these quality indicators of the PL-treated spinach leaves showed an accelerated decrease during refrigerated storage. At the end of storage, slightly lower phytochemical quality yet significantly better microbial quality was observed for the PL-treated samples compared to the control samples [2].

11.5.2.4 DAIRY PRODUCTS

Dunn et al. [26] reported that there was no visible discoloration of the curds of cottage cheese exposed to 1 or 2 pulses of flashlamp light. Also, a panel of testers rated the taste of cottage cheese, and the taste was found to be substantially unaffected by the treatment with 2 pulses.

Fernández et al. [30] investigated how pulsed light treatment affected the sensory quality and volatile profile of two cheese varieties (Gouda and

Manchego). No changes in sensory quality were observed with 0.9 J/cm^2. In both cheeses, the treatments at fluences $\geq$4.2 J/cm^2 resulted in differences in odor and flavor probably due to the increase of sulfur compounds which were shown to be increased by volatile analysis. However, the concentration of sulfur volatiles and the corresponding sensory notes disappeared during cold storage at 4?C for 30 days [30].

Can et al. [15] evaluated the changes in color and lipid oxidation of white American cheese, as well as the mechanical properties of the packaging material, after treatments for 5 s at 13 cm (mild), for 30 s at 8 cm (moderate), and for 40 s at 5 cm (extreme) PL treatments. The mild treatment of cheeses did not affect color or chemical quality significantly. The analysis of polypropylene (the plastic packaging material) exhibited a decreasing trend for elastic modulus, which, however, was not significantly different between untreated samples and those samples after mild and moderate treatments [15].

In Keklik et al.'s [53] study, no significant difference in moisture content or lipid oxidation levels was observed between the control and PL-treated kashar cheese after pulsed light treatments (5–60 s, 5–13 cm from quartz window, 4.88–123.25 J/cm^2). The pH of kashar samples treated at 123.25 J/cm^2 (60 s–5 cm) decreased significantly compared to that of the control samples. The lightness decreased and the redness increased with fluence, while the yellowness showed a fluctuation. Overall, the total color difference (ΔE) increased significantly as the severity of the treatment increased [53]. Similarly, Keklik et al. [54] evaluated the effect of pulsed light treatments (5–60 s, 5–13 cm from quartz window, 4.89–123.69 J/cm^2) on pH, color, moisture content, and lipid oxidation levels of white cheese. The treatments did not produce significant differences in lipid oxidation, pH, or moisture content of white cheese. However, fluctuations in the color parameters were observed, and the total color difference (ΔE) was found to be significantly different between the treated and untreated samples [54].

11.5.2.5 OTHER SOLID AND SEMI SOLID FOODS

Fine & Gervais [32] evaluated the colorimetry (ΔE) of colored food powders (wheat flour and black pepper) exposed to pulsed light. A visual change was observed as ΔE=4.3 and ΔE=4.5 for wheat flour and black pepper, respectively. Wheat flour and black pepper increasingly burned, while the content of piperine fell from 5.12 to 3.40%, which was observed at the fluences of

15.56 and 3.89 J/cm^2 for wheat flour and black pepper, respectively. Thus, visual and flavor attributes significantly changed even before an efficient decontamination level could be reached. The authors concluded that the thermal effect of pulsed light dominated the UV effect for these colored food powders [32].

The treatment of 1.02 mm and 1.92 mm-thick alfalfa seeds with pulsed UV-light (5.6 J/cm^2/pulse at the lamp surface) for >30 and 75 s, respectively, reduced the germination capability of the seeds due to the excessive heating [92]. It was also reported in this study that the treatment of alfalfa seeds of constant layer thickness (6.25 mm) at the distances of 3 and 5 cm from the lamp reduced their germination ability after 30 and 60 s, respectively. Accordingly, Sharma & Demirci [92] reported that the heating effect of pulsed light on the viability of the alfalfa seeds increased with lower seed layer thickness, longer treatment time, and shorter distance from the lamp.

According to Fine & Gervais [32], the thermal effect of pulsed light dominated the UV effect on colored food powders including black pepper and wheat flour. The visual appearance and flavor of these food powders changed significantly before completing microbial inactivation was reached at 58 J/cm^2. Black pepper was more sensitive to pulsed light than wheat flour, which was attributed to the greater absorption of light energy by darker materials [32].

UV light is known to have 'hormetic' effects on plants, i.e., it stimulates a beneficial response termed as 'hormesis' [95]. The beneficial response includes increased accumulation of certain compounds such as bioactive compounds and antioxidants [80]. A similar effect has been seen in mushrooms treated with pulsed light. Vitamin D contents of portabella and sliced white whole mushrooms were greater than 100% Daily Value after pulsed-light treatment for less than 1 s [105].

In another study conducted by Kalaras et al. [47], pulsed light rapidly increased vitamin D2 content in fresh button mushrooms. The authors reported an increase of D2 to over 100% RDA/serving following 3 pulses (1 s). D2 started to approach a maximum concentration of 27 μg/g dry weight after 12 pulses. The authors reported that the D2 level with 3 pulses decreased from 11.9 to 9.05 μg/g dry weight after 3 days of storage but remained nearly constant afterwards throughout the 11-day shelf-life study. It was also indicated that more D2 production was observed for sliced mushrooms than whole mushrooms, and also for white buttons than brown buttons [47].

Koçer & Keklik [57] studied the effects of pulsed light treatments (4.00–101.26 J/cm^2) on the quality attributes of sliced button mushrooms (*Agaricus bisporus*). As the treatment time and proximity to the quartz window in the pulsed light system increased, the browning on the mushroom surface increased while the moisture content and the water activity decreased. A significant increase in antioxidant activity of the mushroom was obtained after the 5 cm–60 s treatment [57].

11.6 LIMITATIONS AND CHALLENGES OF PULSED LIGHT

Pulsed light is a nonthermal technology for short-time treatments, which can be considered an alternative to thermal and chemical methods for preserving foods. Obviously, pulsed light has advantages over thermal and chemical methods, as it minimally involves heat (for a short time) and does not require chemicals. In addition, as pulsed light is a dry method, it does not leave wet surfaces that can facilitate microbial growth. This makes the process more sustainable and practical. Among nonthermal food processing technologies, pulsed light stands out for its effectiveness on a broad range of microorganism types and its very rapid inactivation ability. Pulsed light is nonionizing, unlike gamma radiation or X-rays. With high peak power, pulsed light also offers many advantages over conventional continuous UV light, such as better penetration ability, lamp safety, and inactivation mechanisms.

However, pulsed light still has limited penetration depth in opaque media and thus can mostly target the surface microorganisms in solid and semisolid foods [6].

Bialka et al. [10] reported that pulsed light is capable of penetrating 10 mm deep in opaque materials with declining energy levels. After treating kashar cheese with a thickness of 0.5 cm at 5 cm (the shortest treatment distance from the quartz window) in a PL system, only about 9.16% of the energy could be transmitted [53].

In another study, only 1.75% of pulsed light could reach a depth of 0.5 cm in white cheese at a treatment distance of 5 cm [54]. There was no pulsed light transmission from white cheese at treatment distances of 8 and 13 cm [54]. Thus, the optical properties of the food become much more critical in terms of light absorption to achieve a desired level of inactivation.

Complex properties of food surfaces bring the challenge of uniform exposure of foods to pulsed light. The pores and crevices of a food surface can have a shadowing effect and may protect the microorganisms present in these areas from light [64]. This suggests that pulsed light is more suitable for foods with smooth surfaces such as some of the fresh fruits and vegetables, hard cheeses, and certain meat products [77].

Another concern with pulsed light technology is the generation of heat. As pulsed light involves a broad spectrum including visible and infrared light in addition to UV light, heat generation is expected for longer treatment times when the food absorbs this energy. Especially, infrared light is well known for producing a heating effect. Pulsed light involves a part of the near-infrared spectra at the wavelength range of 700–1,100 nm. As a result, heat may build up in the system uncontrollably, which can confound the microbial inactivation and lead to deteriorations in food quality. So how is pulsed light considered a nonthermal processing technology? The pulsed light technology is nonthermal only if very short treatment times are used. A few seconds are not expected to raise the temperature of the food significantly. The rate at which the temperature increases during pulsed light treatment depends on the wavelength range, distance from the lamp, treatment time, and thermal properties of food. Therefore, reducing the infrared portion of light or cooling the treatment chamber can be effective to minimize this problem.

Heat capacity of fats is almost half of that of water [108], which causes fats to heat more quickly than water. Furthermore, fats have very high boiling points and can reach much higher temperatures than water does [34]. Also, higher fat content can cause higher lipid oxidation. Therefore, the fat content of food may be a critical factor in controlling temperature as well as the quality during the pulsed light processing.

11.7 SUMMARY AND FUTURE TRENDS

Pulsed light is a nonthermal processing technology for short-time treatments, which has potential for effective microbial inactivation. It is an FDA-approved technology for the decontamination of foods. The pulsed light is generated by inert gas lamps and is delivered as flashes of high peak power that involve electromagnetic radiation in the UV, visible, and infrared spectra. Pulsed light has effectiveness on a broad range of organisms such as bacteria, viruses, fungi, spores of bacteria and fungi, algae, and protozoa.

Pulsed light offers certain advantages over thermal and chemical methods, as well as other nonthermal processing technologies including conventional continuous UV light. Pulsed light does not cause heating of food if the duration of treatment is kept short. Thus, it has the potential to decontaminate minimally processed and/or raw foods. No water or chemicals are added to the process unless combinations of different techniques are to be used.

The operation and maintenance of the process are straightforward. Pulsed light can also be applied to packaged foods if photodegradation-resistant and light-permeable packaging materials are used. This makes pulsed light suitable as a postprocessing intervention for decontamination of solid and semi-solid foods.

Compared to continuous UV light, pulsed light provides a much faster and thus more efficient inactivation of microorganisms. There are also health and environmental concerns regarding mercury-based lamps which are used to generate continuous UV light. Lamps used to produce pulsed light are based on inert gases, which do not pose such risks. The inactivation mechanism of pulsed light is complex and involves photochemical, photothermal, and photophysical activities. Photochemical activities mainly involve the formation of pyrimidine dimers on the DNA of the microbial cells exposed to UV-C light. Photothermal and photophysical activities damage the structure of microbial cells by localized heating and constant physical disturbance, respectively. Thus, microbial injuries are mostly expected to be irreversible unlike continuous UV light.

Many researchers have evaluated the effects of pulsed light on various microorganisms in or on solid and semi-solid foods. Mostly, these studies have focused on microbial inactivation and quality changes brought by pulsed light treatment. Also, some of these studies evaluated mathematical models to describe or estimate microbial survival and/or quality alterations. There has been scarce research into the development of a commercial setup of a pulsed light system or the limited evaluation of pilot or full-scale pulsed light systems for the processing of solid and semi-solid foods.

The research demonstrates that there are two main concerns with pulsed light treatment: low penetrability and heating of food during long periods of treatment. The low penetration ability of pulsed light indicates that the process is only limited to the surface decontamination of solid and semi-solid foods. This is also a challenge because most foods have rough surfaces, i.e., contain many surface irregularities, which may prevent the light from reaching the microorganism, and thus hinder the inactivation. The other concern, heating, is addressed by using short periods of time for treatment.

In conclusion, pulsed light is a promising technology for the non-thermal decontamination of solid and semi-solid foods. Although many studies have evaluated the effectiveness of pulsed light on the microorganisms in/on various solid and semi-solid foods, more effort is required to develop commercial applications of this technology. Since the principle of pulsed light is based on the interaction of light and matter, the commercial success of pulsed light depends on the development of technology-driven methods to better understand these interactions and overcome the above-mentioned shortcomings and/or limitations.

KEYWORDS

- **microbial inactivation**
- **microorganism**
- **non-thermal decontamination**
- **pulsed light**
- **quality effects**
- **semi-solid foods**
- **short-time treatments**
- **solid food**

REFERENCES

1. Aguiló-Aguayo, I., Charles, F., Renard, C. M., Page, D., & Carlin, F. (2013). Pulsed light effects on surface decontamination, physical qualities and nutritional composition of tomato fruit. *Postharvest Biology and Technology, 86,* 29–36.
2. Agüero, M. V., Jagus, R. J., Martín-Belloso, O., & Soliva-Fortuny, R. (2016). Surface decontamination of spinach by intense pulsed light treatments: Impact on quality attributes. *Postharvest Biology and Technology, 121,* 118–125.
3. Avalos Llano, K. R., Marsellés-Fontanet, A. R., Martín-Belloso, O., & Soliva-Fortuny, R. (2016). Impact of pulsed light treatments on antioxidant characteristics and quality attributes of fresh-cut apples. *Innovative Food Science & Emerging Technologies, 33,* 206–215.
4. Avalos Llano, K. R., Martín-Belloso, O., & Soliva-Fortuny, R. (2018). Effect of pulsed light treatments on quality and antioxidant properties of fresh-cut strawberries. *Food Chemistry, 264,* 393–400.
5. Barbosa-Canovas, G. V., Schaffner, D. W., Pierson, M. D., & Zhang, Q. H. (2000). Pulsed light technology. *Journal of Food Science, 65,* 82–85.
6. Bhavya, M. L., & Umesh Hebbar, H. (2017). Pulsed light processing of foods for microbial safety. *Food Quality and Safety, 1*(3), 187–202.
7. Bialka, K. L. (2006). Decontamination of berries with ozone and pulsed UV-light (p. 239). *PhD Thesis, Pennsylvania State University, College Park–PA.*

8. Bialka, K. L., & Demirci, A. (2007). Decontamination of *Escherichia coli* O157 and *Salmonella enterica* on blueberries using ozone and pulsed UV-light. *Journal of Food Science, 72*(9), M391–M396.
9. Bialka, K. L., & Demirci, A. (2008). Efficacy of pulsed UV-light for the decontamination of *Escherichia coli* O157 and *Salmonella* spp. on raspberries and strawberries. *Journal of Food Science, 73*(5), M201–M207.
10. Bialka, K. L., Walker, P. N., Puri, V. M., & Demirci, A. (2008). Pulsed UV-light penetration of characterization and the inactivation of *Escherichia coli* K_{12} in solid model systems. *Transactions of the ASABE, 51*(1), 195–204.
11. Bintsis, T., Litopoulou-Tzanetaki, E., & Robinson, R. K. (2000). Existing and potential application of ultraviolet light in the food industry: A critical review. *Journal of the Science of Food and Agriculture, 80*(6), 637–645.
12. Bohrerova, Z., Shemer, H., Lantis, R., Impellitteri, C. A., & Linden, K. G. (2008). Comparative disinfection efficiency of pulsed and continuous-wave UV irradiation technologies. *Water Research, 42*(12), 2975–2982.
13. Bryant, M. T., Degala, H. L., Mahapatra, A. K., Gosukonda, R. M., & Kannan, G. (2021). Inactivation of *Escherichia coli* K_{12} by pulsed UV light on goat meat and beef: microbial responses and modelling. *International Journal of Food Science & Technology, 56*(2), 563–572.
14. Buonanno, M., Welch, D., Shuryak, I., & Brenner, D. J. (2020). Far-UVC light (222 nm) efficiently and safely inactivates airborne human coronaviruses. *Scientific Reports, 10*(1), E-article ID 10285.
15. Can, F. O., Demirci, A., Puri, V. M., & Gourama, H. (2014). Decontamination of hard cheeses by pulsed UV light. *Journal of Food Protection, 77*(10), 1723–1731.
16. Cao, X., Huang, R., & Chen, H. (2017). Evaluation of pulsed light treatments on inactivation of *Salmonella* on blueberries and its impact on shelf-life and quality attributes. *International Journal of Food Microbiology, 260,* 17–26.
17. Cassar, J. R. (2018). *Utilization of Pulsed Ultraviolet Light as a Microbial Reduction Intervention on Raw Chicken*. MSc Thesis, The Graduate School, College of Agricultural Sciences, The Pennsylvania State University, State College, PA, 16801, USA.
18. Cassar, J. R., Mills, E., Campbell, J., Demirci, A., Mills, E. W., & Campbell, J. A. (2019). Decontamination of chicken thigh meat by pulsed ultraviolet light. *Meat and Muscle Biology, 3*(1), 479–487.
19. Cassar, J. R., Mills, E. W., Campbell, J. A., & Demirci, A. (2021). Pulsed ultraviolet light treatment of chicken parts. *Meat and Muscle Biology, 5*(1), 28, 1–8.
20. Cheigh, C. I., Hwang, H. J., & Chung, M. S. (2013). Intense pulsed light (IPL) and UV-C treatments for inactivating *Listeria monocytogenes* on solid medium and seafoods. *Food Research International, 54,* 745–752.
21. Chen, D., Cheng, Y., Peng, P., Liu, J., Wang, Y., Ma, Y., Anderson, E., Chen, C., Chen, P., & Ruan, R. (2019). Effects of intense pulsed light on Cronobacter sakazakii and *Salmonella* surrogate *Enterococcus faecium* inoculated in different powdered foods. *Food Chemistry, 296,* 23–28.
22. Dai, T., Vrahas, M. S., Murray, C. K., & Hamblin, M. R. (2012). Ultraviolet C irradiation: An alternative antimicrobial approach to localized infections? *Expert Review of Anti-Infective Therapy, 10*(2), 185–195.
23. Demirci, A. (2002). Novel processing technologies for food safety. *Association of Food Drug Officials, 64*(4), 1–8.

24. Demirci, A., & Keklik, N. M. (2012). Process design involving pulsed UV light. In A. Jasim & S. R. Mohammad (Eds.), *Handbook of Food Process Design* (1st ed., pp. 1166–1187). West Sussex, UK: Wiley-Blackwell.
25. Dunn, J. (1996). Pulsed light and pulsed electric field for foods and eggs. *Poultry Science, 75,* 1133–1136.
26. Dunn, J. E., Clark, R. W., Asmus, J. F., Pearlman, J. S., Boyer, K., Painchaud, F., & Hofmann G. A. (1989). Methods for preservation of foodstuffs. *US Patent* 4871559A, 1.
27. Elmnasser, N., Guillou, S., Leroi, F., Orange, N., Bakhrouf, A., & Federighi, M. (2007). Pulsed-light system as a novel food decontamination technology: A review. *Canadian Journal of Microbiology, 53*(7), 813–821.
28. Escott, C., Vaquero, C., del Fresno, J. M., Bañuelos, M. A., Loira, I., Han, S. Y., Bi, Y., Morata, A., & Suárez-Lepe, J. A. (2017). Pulsed light effect in red grape quality and fermentation. *Food and Bioprocess Technology, 10*(8), 1540–1547.
29. FDA (Food and Drug Administration). (1996). Pulsed light for the treatment of food. Code of Federal Regulations, Title 21, Vol. 3, 21-CFR179.41. https://www.accessdata.fda.gov/scripts/cdrh/cfdocs/cfcfr/CFRSearch.cfm?fr=179.41.
30. Fernández, M., Arias, K., & Hierro, E. (2016). Application of pulsed light to sliced cheese: Effect on Listeria inactivation, sensory quality and volatile profile. *Food and Bioprocess Technology, 9*(8), 1335–1344.
31. Figueroa-García, J. E., Silva, J. L., Kim, T., Boeger, J., & Cover, R. (2002). Use of pulsed-light to treat raw channel catfish fillets. *Journal of the Mississippi Academy of Sciences, 47,* 114–120.
32. Fine, F., & Gervais, P. (2004). Efficiency of pulsed UV light for microbial decontamination of food powders. *Journal of Food Protection, 67*(4), 787–792.
33. Food and Agriculture Organization of the United Nations (FAO). (2022). Food safety and quality. https://www.fao.org/food-safety/background/en/.
34. Fu, Y. C. (2004). Fundamentals and industrial applications of microwave and radio frequency in food processing. In J. S. Smith & Y. H. Hui (Eds.), *Food Processing: Principles and Applications* (p. 89). Ames, IA: Blackwell Publishing.
35. Ganan, M., Hierro, E., Hospital, X. F., Barroso, E., & Fernández, M. (2013). Use of pulsed light to increase the safety of ready-to-eat cured meat products. *Food Control, 32*(2), 512–517.
36. Gomez-Lopez, V. M., Ragaert, P., Debevere, J., & Devlieghere, F. (2007). Pulsed light for food decontamination: A review. *Trends in Food Science and Technology, 18,* 464–473.
37. Guerrero-Beltran, J. A., & Barbosa-Canovas, G. V. (2004). Advantages and limitations on processing foods by UV light. *Food Science and Technology International, 10*(3), 137–147.
38. Harguindeguy, M., & Gómez-Camacho, C. E. (2021). Pulsed Light (PL) treatments on almond kernels: *Salmonella enteritidis* inactivation kinetics and infrared thermography insights. *Food and Bioprocess Technology, 14*(12), 2323–2335.
39. Hébert, M., Hersch, R. D., & Emmel, P. (2015). Fundamentals of optics and radiometry for color reproduction. In M. Kriss (Ed.), *Handbook of Digital Imaging* (pp. 1021–1077). West Sussex, UK: Wiley.
40. Hierro, E., Barroso, E., De la Hoz, L., Ordóñez, J. A., Manzano, S., & Fernández, M. (2011). Efficacy of pulsed light for shelf-life extension and inactivation of *Listeria*

monocytogenes on ready-to-eat cooked meat products. *Innovative Food Science & Emerging Technologies, 12*(3), 275–281.
41. Hierro, E., Ganan, M., Barroso, E., & Fernández, M. (2012). Pulsed light treatment for the inactivation of selected pathogens and the shelf-life extension of beef and tuna carpaccio. *International Journal of Food Microbiology, 158*(1), 42–48.
42. Hierro, E., Manzano, S., Ordóñez, J. A., de la Hoz, L., & Fernández, M. (2009). Inactivation of *Salmonella enterica* serovar Enteritidis on shell eggs by pulsed light technology. *International Journal of Food Microbiology, 135*(2), 125–130.
43. Ignat, A., Manzocco, L., Maifreni, M., Bartolomeoli, I., & Nicoli, M. C. (2014). Surface decontamination of fresh-cut apple by pulsed light: Effects on structure, color, and sensory properties. *Postharvest Biology and Technology, 91*, 122–127.
44. Jacobsen, A., Joensen, H., & Eysturskarð, J. (2017). Gaping and loss of fillet firmness in farmed salmon (*Salmo salar* L.) closely correlated with post-slaughter cleaning of the abdominal cavity. *Aquaculture Research, 48*(1), 321–331.
45. Jun, S., Irudayaraj, J. M., Demirci, A., & Geiser, D. (2003). Pulsed UV-light treatment of corn meal for inactivation of *Aspergillus niger* spores. *International Journal of Food Science & Technology, 38*(8), 883–888.
46. Kaack, K., & Lyager, B. (2007). Treatment of slices from carrot (*Daucus carota*) using high intensity white pulsed light. *European Food Research and Technology, 224*(5), 561–566.
47. Kalaras, M. D., Beelman, R. B., & Elias, R. J. (2012). Effects of postharvest pulsed UV light treatment of white button mushrooms on vitamin D2 content and quality attributes. *Journal of Agricultural and Food Chemistry, 60*(1), 220–225.
48. Keklik, N. M. (2020). Treatment of pastirma with pulsed UV light: Modeling of *Staphylococcus aureus* inactivation and assessment of quality changes. *Food Science and Technology International, 26*(2), 185–198.
49. Keklik, N. M., Demirci, A., & Bock, R. G. (2011). Surface decontamination of whole chicken carcasses using a pilot-scale pulsed UV light system. *Transactions of the ASABE, 54*(3), 993–1000.
50. Keklik, N. M., Demirci, A., Patterson, P. H., & Puri, V. M. (2010). Pulsed UV light inactivation of *Salmonella enteritidis* on eggshells and its effects on egg quality. *Journal of Food Protection, 73*(8), 1408–1415.
51. Keklik, N. M., Demirci, A., & Puri, V. M. (2009). Inactivation of *Listeria monocytogenes* on unpackaged and vacuum-packaged chicken frankfurters using pulsed UV-light. *Journal of Food Science, 74*(8), 431–439.
52. Keklik, N. M., Demirci, A., & Puri, V. M. (2010). Decontamination of unpackaged and vacuum-packaged boneless chicken breast with pulsed ultraviolet light. *Poultry Science, 89*(3), 570–581.
53. Keklik, N. M., Elik, A., Salgın, U., Demirci, A., & Koçer, G. (2019). Inactivation of *Staphylococcus aureus* and *Escherichia coli* O157 on fresh Kashar cheese with pulsed ultraviolet light. *Food Science and Technology International, 25*(8), 680–691.
54. Keklik, N. M., Elik, A., Salgın, U., Demirci, A., & Koçer, G. (2020). Surface decontamination of white cheese by pulsed UV light treatment. *Journal of Food Safety and Food Quality (Archiv Fur Lebensmittelhygiene), 71*, 86–92.
55. Keklik, N. M., Krishnamurthy, K., & Demirci, A. (2012). Microbial decontamination of food by ultraviolet (UV) and pulsed UV light. In A. Demirci & M. O. Ngadi (Eds.),

Microbial Decontamination in the Food Industry – Novel Methods and Applications (1st ed., pp. 344–369). Cambridge, UK: Woodhead Publishing Limited.
56. Koch, F., Wiacek, C., & Braun, P. G. (2019). Pulsed light treatment for the reduction of *Salmonella typhimurium* and *Yersinia enterocolitica* on pork skin and pork loin. *International Journal of Food Microbiology, 292*, 64–71.
57. Koçer, G., & Keklik, N. M. (2018). Inactivation of *Listeria monocytogenes* on the surface of cultivated mushroom by pulsed UV light. *Gıda – The Journal of Food, 43*(6), 896–905.
58. Koutchma, T. N., Forney, L. J., & Moraru, C. I. (2009). UV processing effects on quality of foods. In D. W. Sun (Ed.), *Ultraviolet Light in Food Technology – Principles and Applications* (pp. 104–105). Boca Raton, FL: CRC Press.
59. Kowalski, W. (2009). UVGI disinfection theory. In *Ultraviolet Germicidal Irradiation Handbook* (pp. 17–50). Berlin–Heidelberg: Springer.
60. Kramer, B., Wunderlich, J., & Muranyi, P. (2019). Inactivation of *Listeria innocua* on packaged meat products by pulsed light. *Food Packaging and Shelf Life, 21*, Article 100353.
61. Krishnamurthy, K., Demirci, A., & Irudayaraj, J. M. (2007). Inactivation of *Staphylococcus aureus* in milk using flow-through pulsed UV-light treatment system. *Journal of Food Science, 72*(7), M233–M239. https://doi.org/10.1111/j.1750-3841.2007.00484.x.
62. Krishnamurthy, K., Tewari, J. C., Irudayaraj, J., & Demirci, A. (2010). Microscopic and spectroscopic evaluation of inactivation of *Staphylococcus aureus* by pulsed UV light and infrared heating. *Food and Bioprocess Technology, 3*(1), 93–104. https://doi.org/10.1007/s11947-008-0103-0.
63. Lacivita, V., Conte, A., Lyng, J. G., Arroyo, C., Zambrini, V. A., & Del Nobile, M. A. (2018). High intensity light pulses to reduce microbial load in fresh cheese. *Journal of Dairy Research, 85*(2), 232–237. https://doi.org/10.1017/S0022029918000214.
64. Lagunas-Solar, M. C., Pina, C., MacDonald, J. D., & Bolkan, L. (2006). Development of pulsed UV light processes for surface fungal disinfection of fresh fruits. *Journal of Food Protection, 69*(2), 376–384. https://doi.org/10.4315/0362-028X-69.2.376.
65. Leng, J., Mukhopadhyay, S., Sokorai, K., Ukuku, D. O., Fan, X., Olanya, M., & Juneja, V. (2020). Inactivation of *Salmonella* in cherry tomato stem scars and quality preservation by pulsed light treatment and antimicrobial wash. *Food Control, 110*, 107005. https://doi.org/10.1016/j.foodcont.2019.107005.
66. Liu, N., Zhu, Q., Zeng, X., Yang, B., Liang, M., Hu, P., He, L., Deng, L., Liang, C., Zhang, R., & Zhou, J. (2019). Influences of pulsed light-UV treatment on the storage period of dry-cured meat and shelf-life prediction by ASLT method. *Journal of Food Science and Technology, 56*(4), 1744–1756. https://doi.org/10.1007/s13197-019-03669-5.
67. Luksiene, Z., Buchovec, I., Kairyte, K., Paskeviciute, E., & Viskelis, P. (2012). High-power pulsed light for microbial decontamination of some fruits and vegetables with different surfaces. *Journal of Food, Agriculture and Environment, 10*, 162–167. https://doi.org/10.1234/4.2012.2596.
68. MacGregor, S. J., Rowan, N. J., Mcllvaney, L., Anderson, J. G., Fouracre, R. A., & Farish, O. (1997). Light inactivation of food-related pathogenic bacteria using a pulsed power source. *Letters in Applied Microbiology, 27*, 67–70. https://doi.org/10.1046/j.1472-765X.1997.00118.x.

69. Malley, J. P. (2002). Ultraviolet disinfection. In S. Lingireddy (Ed.), *Control of Microorganisms in Drinking Water* (pp. 213–235). Reston, VA: American Society of Civil Engineers. https://doi.org/10.1061/9780784406181.ch10.
70. Marquenie, D., Geeraerd, A. H., Lammertyn, J., Soontjens, C., Van Impe, J. F., Michiels, C. W., & Nicolai, B. M. (2003). Combinations of pulsed white light and UV-C or mild heat treatment to inactivate conidia of *Botrytis cinerea* and *Monilia fructigena*. *International Journal of Food Microbiology, 85*(1–2), 185–196. https://doi.org/10.1016/S0168-1605(02)00547-8.
71. McDonald, K. F., Curry, R. D., Clevenger, T. E., Unklesbay, K., Eisenstark, A., & Golden, J. (2000). A comparison of pulsed and continuous ultraviolet light sources for the decontamination of surfaces. *IEEE Transactions on Plasma Science, 28*, 1581–1587. https://doi.org/10.1109/27.902229.
72. McLeod, A., Hovde Liland, K., Haugen, J. E., Sørheim, O., Myhrer, K. S., & Holck, A. L. (2018). Chicken fillets subjected to UV-C and pulsed UV light: Reduction of pathogenic and spoilage bacteria, and changes in sensory quality. *Journal of Food Safety, 38*(1), 12421. https://doi.org/10.1111/jfs.12421.
73. Miller, R. V., Jeffrey, W., Mitchell, J. D., & Elasri, M. (1999). Bacterial responses to ultraviolet light. *ASM News, 65*, 535–541.
74. Microscopy Society of America. (2005). Absorption, transmission and reflection of light by the specimen. *Microscopy Today, 5*(5), 14–15.
75. Mukhopadhyay, S., Sokorai, K., Ukuku, D. O., Fan, X., Olanya, M., & Juneja, V. (2019). Effects of pulsed light and sanitizer wash combination on inactivation of *Escherichia coli* O157, microbial loads and apparent quality of spinach leaves. *Food Microbiology, 82*, 127–134. https://doi.org/10.1016/j.fm.2019.01.006.
76. Nicorescu, I., Nguyen, B., Chevalier, S., & Orange, N. (2014). Effects of pulsed light on the organoleptic properties and shelf-life extension of pork and salmon. *Food Control, 44*, 138–145. https://doi.org/10.1016/j.foodcont.2014.03.032.
77. Oms-Oliu, G., Martín-Belloso, O., & Soliva-Fortuny, R. (2010). Pulsed light treatments for food preservation. A review. *Food and Bioprocess Technology, 3*(1), 13–23. https://doi.org/10.1007/s11947-008-0177-8.
78. Ozer, N., & Demirci, A. (2006). Inactivation of *Escherichia coli* O157 and *Listeria monocytogenes* inoculated on raw salmon fillets by pulsed UV-light treatment. *International Journal of Food Science & Technology, 41*, 354-360. https://doi.org/10.1111/j.1365-2621.2005.01061.x.
79. Palmieri, L., & Cacace, D. (2005). High intensity pulsed light technology. In D. W. Sun (Ed.), *Emerging Technologies for Food Processing* (pp. 279–304). London, UK: Elsevier Academic Press. https://doi.org/10.1016/B978-012676757-5/50012-6.
80. Papoutsis, K., Vuong, Q. V., Pristijono, P., Golding, J. B., Bowyer, M. C., Scarlett, C. J., & Stathopoulos, C. E. (2016). Enhancing the total phenolic content and antioxidants of lemon pomace aqueous extracts by applying UV-C irradiation to the dried powder. *Foods, 5*(3), 55–63. https://doi.org/10.3390/foods5030055.
81. Pataro, G., Sinik, M., Capitoli, M. M., Donsì, G., & Ferrari, G. (2015). The influence of post-harvest UV-C and pulsed light treatments on quality and antioxidant properties of tomato fruits during storage. *Innovative Food Science & Emerging Technologies, 30*, 103–111.
82. Pedrós-Garrido, S., Condón-Abanto, S., Clemente, I., Beltrán, J. A., Lyng, J. G., Bolton, D., Brunton, N., & Whyte, P. (2018). Efficacy of ultraviolet light (UV-C) and pulsed

light (PL) for the microbiological decontamination of raw salmon (*Salmo salar*) and food contact surface materials. *Innovative Food Science & Emerging Technologies, 50*, 124–131.
83. Ploydaeng, M., Rajatanavin, N., & Rattanakaemakorn, P. (2021). UV-C light: A powerful technique for inactivating microorganisms and the related side effects to the skin. *Photodermatology, Photoimmunology & Photomedicine, 37*(1), 12–19.
84. Proulx, J., Hsu, L. C., Miller, B. M., Sullivan, G., Paradis, K., & Moraru, C. I. (2015). Pulsed-light inactivation of pathogenic and spoilage bacteria on cheese surface. *Journal of Dairy Science, 98*(9), 5890–5898.
85. Proulx, J., Sullivan, G., Marostegan, L. F., VanWees, S., Hsu, L. C., & Moraru, C. I. (2017). Pulsed light and antimicrobial combination treatments for surface decontamination of cheese: Favorable and antagonistic effects. *Journal of Dairy Science, 100*(3), 1664–1673.
86. Rajkovic, A., Tomasevic, I., De Meulenaer, B., & Devlieghere, F. (2017). The effect of pulsed UV light on *Escherichia coli* O157: H7, *Listeria monocytogenes*, *Salmonella typhimurium*, *Staphylococcus aureus* and *Staphylococcal enterotoxin* A on sliced fermented salami and its chemical quality. *Food Control, 73*, 829–837.
87. Rowan, N. J. (2019). Pulsed light as an emerging technology to cause disruption for food and adjacent industries–Quo vadis? *Trends in Food Science & Technology, 88*, 316–332.
88. Rowan, N. J., MacGregor, S. J., Anderson, G., Fouracre, R. A., Mcllvaney, L., & Farish, O. (1999). Pulsed-light inactivation of food-related microorganisms. *Applied and Environmental Microbiology, 65*, 1312–1315.
89. Sadeghianfar, P., Nazari, M., & Backes, G. (2019). Exposure to ultraviolet (UV-C) radiation increases germination rate of maize (*Zea maize* L.) and sugar beet (*Beta vulgaris*) seeds. *Plants, 8*(2), 49–54.
90. Schaefer, R., Grapperhaus, M., Schaefer, I., & Linden, K. (2007). Pulsed UV lamp performance and comparison with UV mercury lamps. *Journal of Environmental Engineering and Science, 6*, 303–310.
91. Seckmeyer, G., Zittermann, A., McKenzie, R., & Greinert, R. (2013). Solar radiation and human health. In E. A. Laws (Ed.), *Environmental Toxicology: Selected Entries from the Encyclopedia of Sustainability Science and Technology* (p. 547). New York, NY: Springer Science & Business Media.
92. Sharma, R. R., & Demirci, A. (2003). Inactivation of Escherichia coli O157: H7 on inoculated alfalfa seeds with pulsed ultraviolet light and response surface modeling. *Journal of Food Science, 68*(4), 1448-1453.
93. Shimoda, H., Matsuda, J., Iwasaki, T., & Hayasaka, D. (2021). Efficacy of 265-nm ultraviolet light in inactivating infectious SARS-CoV-2. *Journal of Photochemistry and Photobiology, 7*, 100050.
94. Söbeli, C., Uyarcan, M., & Kayaardı, S. (2021). Pulsed UV-C radiation of beef loin steaks: Effects on microbial inactivation, quality attributes and volatile compounds. *Innovative Food Science & Emerging Technologies, 67*, 102558.
95. Stevens, C., Khan, V. A., Lu, J. Y., Wilson, C. L., Chalutz, E., Droby, S., Kabwe, M. K., Huang, Z., Adeyeye, O., Pusey, L. P., & Tang, A. Y. A. (1999). Induced resistance of sweet potato to Fusarium root rot by UV-C hormesis. *Crop Protection, 18*, 463–470.

96. Takeshita, K., Shibato, J., Sameshima, T., Fukunaga, S., Isobe, S., Arihara, K., & Itoh, M. (2003). Damage of yeast cells induced by pulsed light irradiation. *International Journal of Food Microbiology, 85*(1–2), 151–158.
97. Tao, T., Ding, C., Han, N., Cui, Y., Liu, X., & Zhang, C. (2019). Evaluation of pulsed light for inactivation of foodborne pathogens on fresh-cut lettuce: Effects on quality attributes during storage. *Food Packaging and Shelf Life, 21*, 100358.
98. Tomašević, I. (2015). The sensory quality of seafood as affected by intense light pulses. *MESO: Prvi Hrvatski Časopis O Mesu, 17*(3), 232–237.
99. Uesugi, A. R., & Moraru, C. I. (2009). Reduction of *Listeria* on ready-to-eat sausages after exposure to a combination of pulsed light and nisin. *Journal of Food Protection, 72*(2), 347–353.
100. Valdivia-Nájar, C. G., Martín-Belloso, O., Giner-Segui, J., & Soliva-Fortuny, R. (2017). Modeling the inactivation of *Listeria innocua* and *Escherichia coli* in fresh-cut tomato treated with pulsed light. *Food and Bioprocess Technology, 10*, 266–274.
101. Valdivia-Nájar, C. G., Martín-Belloso, O., & Soliva-Fortuny, R. (2018). Kinetics of the changes in the antioxidant potential of fresh-cut tomatoes as affected by pulsed light treatments and storage time. *Journal of Food Engineering, 237*, 146–153.
102. Wambura, P., & Verghese, M. (2011). Effect of pulsed ultraviolet light on quality of sliced ham. *LWT-Food Science and Technology, 44*(10), 2173–2179.
103. Wekhof, A. (2000). Disinfection with flash lamps. *PDA Journal of Pharmaceutical Science and Technology, 54*(3), 264–276.
104. Wekhof, A. (2003). Ultra-fast sterilization by disintegration of microorganisms with intense pulsed UV light. *Business Briefing: Global Surgery. Technology & Services, 2003*, 1–5.
105. Williams, R. (2008). Vitamin D formation from post-harvest pulsed light treatment of mushrooms. 1–5. Xenon Corporation, Wilmington, MA. https://xenoncorp.com/wp-content/uploads/2019/10/PulsedLight-for-Mushroom-VitaminD-Enhancement-WP.pdf.
106. World Health Organization (WHO). (2022). Food safety-fact sheet. https://www.who.int/news-room/fact-sheets/detail/food-safety.
107. Xenon. (2014). XENON RC-800-modular UV curing system. 1–8. Xenon Corporation, Wilmington, MA. https://xenoncorp.com/wp-content/uploads/2019/11/RC-800_Brochure.pdf.
108. Zulkurnain, M., Maleky, F., & Balasubramaniam, V. M. (2016). High pressure processing effects on lipids thermophysical properties and crystallization kinetics. *Food Engineering Reviews, 8*(4), 393–413.

CHAPTER 12

APPLICATION OF PULSED LIGHT IN MEAT, FISH, AND POULTRY PRODUCTS

MARWA MAYAN, SMRUTHI BHAT, JAYAPRAKASH APOORVA, N. A. NANJE GOWDA, HUMEERA TAZEEN, and JEYAMKONDAN SUBBIAH

ABSTRACT

In this chapter, the application of pulsed light technology for decontamination of meat, fish, and poultry products has been discussed. The principles and mechanisms of pulsed light, factors (fluence, exposure time, distance, pulse duration, etc.) affecting the effectiveness of surface decontamination, and its limitations are also elucidated. The effects of pulsed light on the quality, safety, shelf-life, microbial decontamination (log reduction), and changes in physico-chemical and organoleptic properties of treated foods are briefly discussed. The effectiveness of pulsed light alone and in combination with other techniques, as well as the limitations of pulsed light technique in the decontamination process, are also deliberated.

12.1 INTRODUCTION

The demand for safe and contaminant-free meat, seafood, and poultry food products is persistently growing across the globe. Consumption of meat has increased by 58% globally in the last 20 years and has reached about 360 million tons in 2018 [46]. Meat is one of the most widely consumed foods due to its protein-rich source in the human diet. Meat is also considered a staple in many communities, holding a significant value and often being the

Nonthermal Light-Based Technologies in Food Processing.
C. K. Sunil, Preeti Birwal, Megh R. Goyal, and R. Mahendran (Eds.)

main dish around which meals are planned. Various types of meat are central to many social and festive gatherings around the world, and from both a general and scientific perspective, it is considered a high-nutrient-value food [3]. In this regard, the value of meat is high as it is a relatively abundant provider of high-quality protein (NPU 0.75–0.8), which is highly digestible. The net protein unit (NPU) of meat is 0.95 compared to plant-based foods where NPU is 0.8–0.9, and it also provides a relative surplus of one vital amino acid, such as lysine, which is in limited supply in most grains [3, 15].

Despite meat and meat products being widely consumed, they entail many limitations such as being highly perishable, having a short shelf life, microbial contamination during storage, concerns over food safety, and health hazards [48]. The biotic agents responsible for these challenges are a variety of microorganisms that lead to spoilage and wastage of proteinaceous foods in meat, poultry, seafood, etc. [36]. The need to combat food safety issues while also delivering quality products offers a wide scope for research on innovative meat processing and packaging technologies to decontaminate or arrest the growth of pathogens in meat foods. Conventionally, the meat processing industry has employed chemical and thermal processing methods to overcome these problems. Despite the fact that these traditional methods rely on the idea of applying heat or chemical preservatives to reduce microbial growth and arrest the development of foodborne pathogens [43]. The conventional methods, such as curing using salts of nitrides and nitrates, smoking, aging, and canning, preserve these foods for consumption. However, these methods could also lead to some undesirable changes in organoleptic properties, loss of product, loss of heat-sensitive nutrients, and even have a detrimental effect on consumers' health [16]. On the contrary, various innovative processing methods are being researched to tackle these challenges, such as "hydrostatic pressure processing (HPP)," ultrasound, "pulsed electric field (PEF)," cold plasma, irradiation technology, and pulsed light technology (PLT), which do not primarily rely on heat treatment to achieve sterilization of food products. Pulsed light technology (PLT) processing is one such novel method that meets consumer expectations and market trends by being minimally processed, easy to handle, free from chemical preservatives, and offering a fresh appearance to foods after treatment and packaging [17]. Pulsed light technology in food processing is a novel surface decontamination technique that kills microorganisms by delivering a series of high-intensity, rapid bursts of light pulses at different frequencies. Depending on the individual process parameters used, pulsed light can be studied under types such as "Pulsed UV light," "High-intensity broad spectrum pulsed light (BSPL)," "Intense light pulse (ILP)," "High-intensity pulsed UV light (HIPL)," and "Pulsed white light (PWL)" [17].

Several studies have reported on the application of pulsed light to decontaminate fresh and processed fruits and vegetables such as grated carrot, shredded white cabbage, and shredded iceberg lettuce [14], fresh-cut avocado [38], fresh-cut watermelons [37], cantaloupe melon [24], strawberry and blueberries [20], fresh tomatoes [44], and spinach [32]. Limited work on the preservation of meat with pulsed light technology has been reported so far, and among the meat products, studies on the decontamination of beef [5], pork [33], and chicken [29] using pulsed light have been reported. Subsequently, the effect of pulsed light on sensory attributes of fish, meat, seafood, and other processed foods is also studied [28]. In addition, a vast amount of research is underway to decontaminate dense and opaque liquids, ready-to-eat foods, powdered seeds, and many other heat-sensitive food products [28]. Therefore, pulsed light, which is a simple white light similar to sunlight, aside from having extremely high intensity, is applied in short pulses to destroy microbial growth on food surfaces.

The objective of this chapter is to highlight the recent advancements in pulsed light technology for food surface decontamination, specifically meat and related food products. Additionally, to review the effect of pulsed light on microbial inactivation in different meat products, quality, shelf-life, and changes in physicochemical and organoleptic properties of treated foods.

12.2 APPLICATIONS OF PULSED LIGHT IN DECONTAMINATION OF MEAT, SEAFOOD, AND POULTRY PRODUCTS

Pulsed light technology (Figure 12.1) is a non-thermal processing technique to facilitate the inactivation and destruction of a great diversity of micro-organisms when used in various intensities of pulses. The FDA's approval of "Pulsed UV light in production, processing, and handling" of food has sparked various scientific investigations in the field of food processing, specifically sterilization of food and food contact surfaces [22]. The concept of pulsed light technology works on the principle of electromagnetic energy [17]. A capacitor stores accumulated energy every half second. The power is amplified because the energy is unconstrained in the form of light within a short period of time, and it also provides minor amounts of extra energy consumption. For food processing applications, a pulsed light generally between 1 and 20 flashes per second with an energy density (about 0.01 to 50 J/cm^2) at the surface is preferred [34].

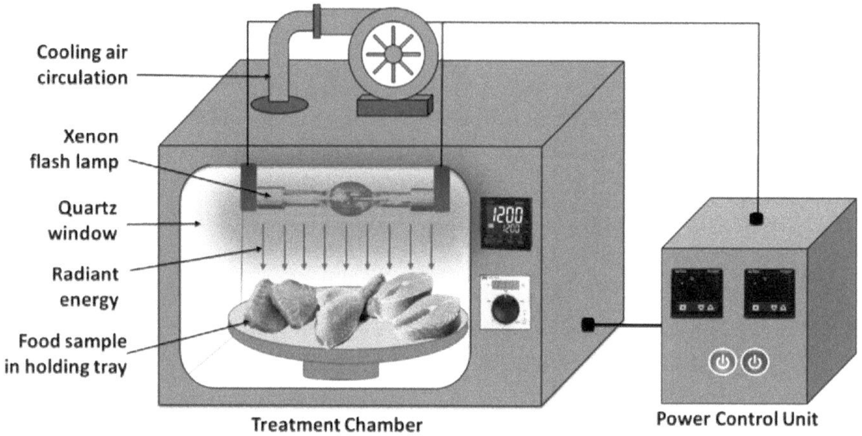

FIGURE 12.1 Pulse light generating and application system.

The pulsed light is found to be efficient for decontaminating surfaces of fresh meat, seafood, and poultry products according to research studies conducted recently. The pulsed light application at various intensities and its effect on different meat products have been discussed further. The effect of pulsed light on microbial decontamination, physico-chemical properties, and quality of meat, poultry, and seafood has been summarized (Figure 12.2).

12.2.1 MEAT

12.2.1.1 BEEF

Escherichia coli, found in the human and endothermic animal intestines, is a common microbiological contaminant in raw meat, chicken, and seafood. Some strains like *E. coli* (STEC) produce certain harmful toxins that result in food poisoning and foodborne illness. Hence, it is important to eliminate these harmful pathogenic organisms. The inactivation of *E. coli* bacteria by UV light on beef using the "Z-1000 Modular Sterilization System (XENON Corporation, Wilmington, MA)" with a frequency of 3 pulses/s producing energy of 505 J/pulse was reported [5]. The linear lamp delivered 1.27 J/cm^2 of radiant energy at 19.33 mm from the window. The beef samples were placed at distances of 4.470, 8.280, and 12.090 centimeters from the light source and treated for 5, 15, 30, 45, and 60 s. For a treatment duration of 60 s, it was observed that the log reduction declined from 1.740 to 1.140 CFU/ml as the distance from the light source increased from 4.470 to 12.090

centimeters. These results were in agreement with the findings of Ouyang [35], indicating that a higher log reduction could be achieved for a shorter distance between the food sample and the light source due to greater energy absorbance by the meat surface.

The log reduction increases with an increase in treatment time irrespective of meat type and the distance between the light and the object. The short treatment time of 5s for a sample placed at 12.09 cm distance offered the least log reduction (0.45 CFU ml^{-1}), while the sample placed at 4.47 cm and treated for 60s offered the highest log reduction (1.74 CFU ml^{-1}). This proved that PUVL was significantly effective in decontaminating the beef samples [5]. However, an important recurring change in the textural properties (loss of water from the meat) and sensory properties (light-yellow discoloration) were also observed. The decontamination of *L. monocytogenes, E. coli* strain named 0157:H7, and *S. enterica serovar Typhimurium* in cow meat was studied using pulsed light [18]. The device named "Steribeam SBS-XeMatic-2L-A" (Kehl am Rhein, Germany), with the specification of a lamp having a 0.1750 J cm^{-2} of fluence and 250 μs of pulse duration was used. The

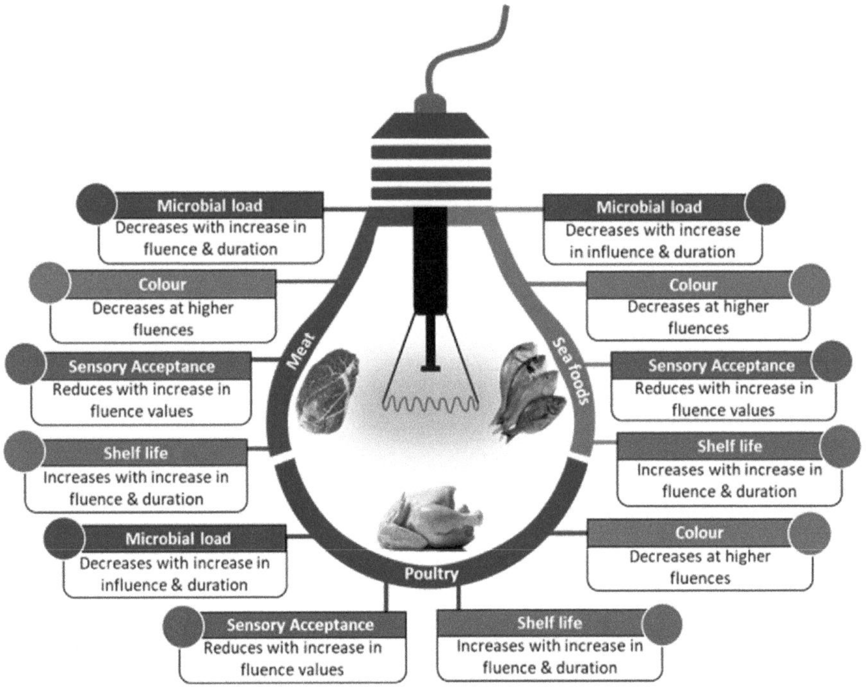

FIGURE 12.2 Summary of effect pulse light on decontamination and quality of meat, poultry, and sea foods.

decontamination was experimented with fluence values of 2.10, 4.20, 0.70, 11.90, 8.40 J cm^{-2} on the beef samples. It was observed that higher fluences resulted in greater lethality of the treatment. At 11.9 J cm^{-2}, a significant decontamination was observed, with the highest log reduction of 1.0 CFU cm^{-2} for *S. Typhimurium*, 1.2 CFU cm^{-2} for *E. coli* and 0.8 CFU cm^{-2} for *L. monocytogenes*. The pulse light treatment efficacy was relatively low due to surface irregularities. Similar observations were reported in studies [18, 47]. The PL treatments of only higher fluences significantly improved shelf-life of beef by a week. However, the samples were rejected by panelists for off color and odor. The color analysis showed that redness decreased with increase in fluence, yellowness and a higher threshold and decreased at higher fluences of 8.40 J cm^{-2}, with none of the significant difference was seen for lightness. The sensory analysis done by panelists revealed that low fluence treatments did not alter much of the sensory aspects of the meat whereas higher fluence treatments were considered unacceptable or had affected the meat freshness [18].

12.2.1.2 GOAT

Pathogenic bacteria like Campylobacter, *Salmonella*, *Listeria,* and *E. coli* can enter healthy goats without showing any pre-mortem and/or post-mortem clinical symptoms. This leads to zoonotic foodborne illness. To rectify this, decontamination of microorganisms is a compulsory pre-treatment. The inactivation of *E. coli* K12 strain in goat meat by using the "Z-1000 modular sterilization system" (XENON, MA, Wilmington) was reported. The treatment parameters used were a pulse energy of 1.27 J cm^{-1} by placing samples at distances of 4.470, 8.280, and 12.090 cm from the source of light. The goat meat treated for 60 seconds recorded a log reduction of 1.66 CFU ml^{-1} and 1.28 CFU ml^{-1} for *E. coli* when samples were placed at 4.47 cm and 12.090 cm from the source of light, respectively. It was inferred that a shorter distance between the sample and the light source, irrespective of treatment duration, was greater. However, since goat meat is less tender, juicy, and has lower fat content in muscles compared to other meats, it results in more surface irregularities. This creates a shadow effect that reduces the efficacy of PUVL treatment. Therefore, the most substantial result recorded was a 1.66 CFU ml^{-1} log reduction at 4.47 cm distance, and the least substantial result was a 0.32 CFU ml^{-1} log reduction at 12.09 cm distance. This was compared with another study [10] where the goat meat treated with ultraviolet radiation for 12 minutes resulted in a 1.18 CFU ml^{-1} log reduction, which proved that pulsed UV light (PUVL) is quite effective compared to Ultraviolet-C type when the lethality of microbial inactivation was considered. However, a treatment time of a longer duration and much less interspace distance resulted in lower sensory acceptance rates.

12.2.1.3 PORK

Pork is one of the most susceptible meats to biological contamination. It is the most common host for several parasites and pathogens like *Salmonella, Campylobacter, Trichinella spiralis, Taenia solium, Toxoplasma gondii, Listeria monocytogenes* that led to various illnesses and toxicity. The effect on the flora of aerobic and *Pseudomonas fluorescens* by the PL present in pork was studied [33]. The raw pork roast (RPR) and roast pork (RP) samples were treated using a clarnor (France) pilot-scale experimental setup [33]. The pork samples were placed 3 cm away from the light source and exposed to 3, 10, 30 J cm^{-1} fluences in each trial. For RPR samples, the lowest log reduction for aerobic flora was 0.51 CFU g^{-1} at 3 J cm^{-1} and the highest were 0.96 CFU g^{-1} at 10 J cm^{-1}. As the storage days increased, the highest level of decontamination was observed at 30 J cm^{-1}. Similar results were described for RP samples, with the highest log reduction of 0.99 CFU g^{-1} at 30 J cm^{-1}. As the storage period increased, the log reduction increased to 1.6 CFU g^{-1} on the fourth day and eventually reached the highest value of 3.4 CFU g^{-1} on the seventh day. In the case of *P. fluorescens*, only a limited level of decontamination was observed with a 1.5 log CFU value per gram at a dosage of 10 and 30 joules per centimeter. This is due to pork meats having a high-water activity (aw) of 0.99, which is an optimum condition for microbial growth. This may lead to easier penetration of bacteria into deeper layers of meat, where pulsed light cannot penetrate efficiently. The color analysis revealed a decrease in redness and yellowness, especially for treatment at 30 J cm^{-1}. Higher fluences and treatment time contribute to lipid peroxidation that accelerates the rate of myoglobin oxidation in meats. The color change can also be brought about by overheating and resulting oxidation (photoreaction). These observations are in concurrence with studies conducted by Hierro et al. [18] and Wambura & Verghese [45]. The shelf life was significantly improved for samples treated with higher fluencies of 10 J cm^{-1} and above but resulted in low sensory scores.

12.2.2 SEA FOODS

12.2.2.1 SALMON

Most of the bacterial pathogens present in salmon are gram-negative in nature. These include *Aeromonas, Pseudomonas*, and *Vibrio,* while the foodborne gram-positive pathogens include *Clostridium, Renibacterium*, and *Streptococcus.* Additionally, a pilot-scaled experimental design was conducted by Clarnor (France) [33] regarding the decontamination of aerobic flora and *Pseudomonas fluorescens*. The treatment was applied to raw salmon samples 3 cm below the lamps at fluencies of 3, 10, and 30 J cm^{-1}. It was observed that even at a higher fluence of 30 J cm^{-1}, a significant log reduction in microbial count was not achieved (0.7 CFU g^{-1}), indicating a lower germicidal effect of the treatment. Salmon meats are high in lipids and protein, which have natural UV absorption effects. They also have a higher water activity (a_w), ideal for bacterial proliferation and penetration, while irregular and uneven surfaces create a shadow effect when pulsed light is applied. This impacted the shelf-life, leading to an increase in the microbial count to an unacceptable level of 10^6 CFU per gram on the fourth day of storage. Color analysis conducted using the calorimeter CR-200 from Tokyo, Japan, revealed no significant change in the color parameters of salmon due to its naturally occurring pinkish-orange pigment, astaxanthin. However, the pigment was degraded due to oxidation during the treatment process.

12.2.2.2 TUNA

Due to its high moisture and water activity, tuna is one of the most susceptible foods to spoilage by bacteria and pathogens (mesophiles) like *Klebsiella, Moraxella, Pseudomonas, Staphylococcus, Vibrio, Weissella, Escherichia, Listeria,* and *Salmonella.* This prompted [18] to conduct a study on tuna carpaccio decontamination using pulsed light for the microorganisms *S. typhimurium, E. coli, V. parahemolyticus,* and *L. monocytogenes*. This was demonstrated using pulsed light with the help of a German-made machine named Steribeam SBS-Xematic-2L-A (Kehl). The fluences considered for the experiment were 0.7, 2.1, 4.2, 8.4, and 11.9 J cm^{-2}. The highest log reduction was recorded as 1.0 CFU cm^{-2} for *V. parahemolyticus* and 0.7 CFU cm^{-2} for *L. monocytogenes* at fluence 11.9 J cm^{-2}. Color analysis showed that yellowness was more susceptible to the PL treatment even at lower fluences. The higher fluences showed decreased redness and increased lightness that caused a loss of muscle transparency immediately after the treatment.

During the storage period, the higher fluence-treated samples developed a sulfur odor and turned pale, resulting in unacceptable sensory scores. On the other hand, the lower and moderate fluences (up to 4.2 J cm^{-2}) had an insignificant effect on the sensory properties of meat samples. Although the shelf-life of treatment with lower fluences was shorter, it was concluded that the lower fluences of 2.1 J cm^{-2} gave notable microbial decontamination while preserving much of the sensory attributes. Ananthanarayanan [1] experimented with chill-stored yellowfin tuna that were packaged in polypropylene (PP) bags. XENON Steripulse (XL 3000 RC847) was the device used to treat chill-stored yellowfin tuna packed in polypropylene (PP) bags [1]. The treatment time was set for 6s with pulse generation of 3 pulses/second and fluence of 1.27 J cm^{-2}. The total plate count was 4.3 log CFU g^{-1} for treated samples compared to 4.7 log CFU g^{-1} for untreated samples. Total volatile base nitrogen (TVBN) was used as a parameter to test efficacy as it represents "the total number of nitrogenous compounds released by the decomposition of stored fish due to microbial action," with a common cause being muscle degradation due to putrefaction [9, 21]. About 5–20 mgN $100g^{-1}$ is set as the TVBN value for fresh fish, and 25–35 mgN $100g^{-1}$ is set as the elimination limit for commercial fresh fishes. On the 20th day of storage, the treated samples had an acceptable TVBN value (16.89 mgN $100g^{-1}$), while the untreated sample's TVBN value (25 mgN $100g^{-1}$) was beyond the rejection limit.

Free fatty acids (FFA) production during storage due to hydrolytic rancidity in fish was also reduced in PL-treated samples when compared to the samples that were not treated. The peroxide values (PV) were observed to be higher in treated samples right after the treatment, which was suspected because of photo-oxidative effects of PL [39]. The color analysis showed that the lightness increased, and yellowness slightly decreased with an increase in the period of storage. But there wasn't any important change in redness throughout the storage period. Oxidative browning due to myoglobin conversion was also not seen in PL-treated samples. The sensory evaluation was done based on the degradation of lipids. The control/untreated samples were deemed unacceptable after 15 days (i.e., on the 16th day) whereas the PL-treated samples were in the acceptable range throughout the storage period of 26 days. Microbiological analysis after 25 days (i.e., on the 26th day) of storage of PL-treated samples was also in the acceptable range. These observations collectively proved that the pulsed light treatment was effective against microbial decontamination in packaged yellowfin tuna by preserving the sensory and accepted microbiological characteristics.

12.2.3 POULTRY PRODUCTS

12.2.3.1 CHICKEN

The most common spoilage pathogens that like to grow in poultry are *Salmonella* spp., *E. coli, Campylobacter* spp., *L. monocytogenes*, and *S. aureus* spp. Among these, *Salmonella, Campylobacter,* and *E. coli* are the most associated foodborne pathogens observed with poultry [7, 29], which necessitates inactivation of these pathogens. Cassar [7] demonstrated the depuration of chicken thigh meat with the help of a UV light pulsed in nature (PUVL). The setup used was a machine from Wilmington, MA named SteriPulse-XL 2000 pulsed light system belonging to the Xenon corporation having a xenon bulb that delivered 1.27 J/cm^2 per pulse of energy output at a distance of 1.8 cm away from the light source. The xenon bulb produced 3 pulses per second in an estimated duration of 0.36s per pulse with a wavelength of 100–1,100 nm. A log reduction of 1.55, 1.76, 1.89, and 2.42 for *Salmonella,* 1.45, 1.69, 1.82, and 2.09 for *Campylobacter*, and 1.22, 1.50, 1.74, and 2.02 for *E. coli* was observed for 5, 15, 30, and 45s exposure, respectively, for an interspace distance of 8 cm away from the quartz window. It was observed that the microbial reduction after 45s of PUVL exposure was significantly greater compared to a shorter duration of exposure. Whereas the increase in proximity or energy fluence parameters in this experiment did not equate to greater microbial reduction. This finding was quite similar to an earlier report [23]. Another study on skinless chicken breast was exposed to pulsed ultraviolet light with energy fluences between 1.25 and 18 J/cm^2 [29]. This resulted in an average log reduction of 0.9 to 3.0 log CFU/cm^2 for similar food spoilage pathogens discussed earlier. The raw chicken subjected to PUVL at low intensity with an energy fluence of 1.25 J/cm^2 produced a pale color with low myoglobin muscle pigment. In earlier studies, chicken breast when exposed to high energy fluence of UV light appeared darker and showed more redness and slight yellow coloration. The outcomes of this study are contrary to other results reported by Chun et al. [8] and McLeod et al. [29]. This is because the dosages used for treating chicken are different in the studies; the color of the chicken filet was not affected at lower values of PL treatment dosage in their respective studies. These findings also indicate that changes in the sensory region have a negligible effect on the PL treatment process of chicken filets.

12.2.3.2 EGGS

In the United States, about 9% of the total illness caused by food is due to table eggs, as reported by the Centers for Disease Control and Prevention

[11]. Furthermore, some pathogenic strains can significantly harm poultry health, leading to economic losses, including during the disposal process [30, 42]. The major pathogens associated are *Salmonella enterica* serovars *Typhimurium, Enteritidis*, *E. coli*, and *Enterococcus faecium*. Taking this into consideration, Cassar [6] conducted research on inactivating *Escherichia coli* and *Enterococcus faecium* on table and hatching eggs. The graded eggs were conveyed on a candling conveying system and exposed to PUVL generated by a xenon flashlamp coupled with an RC-802 pulsed light Interweave system of a model belonging to Z-5000 of Xenon Corporation, Wilmington. The modified conveying system ensured exposure of the egg surface almost 360? for 3 times per second at 360 μs per pulse during a single pass time of 26.7 seconds. For table eggs, the fluences used were 4.9, 3.1, and 2.1 J cm^{-2} at distances of 9.5, 14.5, and 19.5 cm from the flashlamp in parallel with the conveyor. A fluence of 1.0 J cm^{-2} was delivered at 14.5 cm from the flashlamp perpendicular to the conveyor. A total microbial reduction of ≥3.43 log CFU cm^{-2} at 1.0 J cm^{-2}, ≥4.00 log CFU cm^{-2} at 2.4 J cm^{-2}, ≥3.76 log CFU cm^{-2} at 3.1 J cm^{-2}, and ≥4.54 log CFU cm^{-2} at 4.9 J cm^{-2} were observed for *E. coli* K12. The total log reduction of 2.03 log CFU cm^{-2} at 1.0 J cm^{-2}, 2.81 log CFU cm^{-2} at 2.4 J cm^{-2}, 2.98 log CFU cm^{-2} at 3.1 J cm^{-2}, 3.52 log CFU cm^{-2} at 4.9 J cm^{-2} was observed for *E. faecium*. At 4.9 J cm^{-2} an effective surface decontamination was observed, and the lowest decontamination was observed at 1.0 and 3.1 J cm^{-2} for both *E. coli* K12 and *E. faecium*. Another interesting inference was noted regarding the nature of bacteria and them efficacy with reference to pulsed light. Even though *E. coli* is gram-negative, it had a greater log reduction (4.54 log CFU cm^{-2}) than *E. faecium*, which is gram-positive and had a 3.52 log CFU cm^{-2} reduction. This difference may be due to a thicker peptidoglycan layer inside the gram-positive bacteria, acting as a protective layer to pulsed light, making gram-negative bacteria more susceptible [2, 6, 25, 40]. Hatch eggs were also subjected to similar PUVL treatments with fluences of 4.9, 24.4, and 48.8 J cm^{-2}. They were then incubated at a temperature of 37.6?C (Chick Master, Englewood, NJ) for 18 days and transferred to a hatcher at 36.8?C (Chick Master) until they hatched. Parameters such as fertility, percent hatch, hatchability, and livability were not significantly altered by the PUVL treatment, even at high fluences. This indicates that the pulsed light treatment was effective in surface decontamination and did not alter the functional properties of the eggs.

12.3 COMBINATION OF PULSE LIGHT TREATMENT WITH OTHER TECHNOLOGIES

Pulsed light has proven potential for effective surface decontamination of meat and meat products. The combination of PL with other conventional and/or non-thermal processing techniques to produce minimally processed foods has been researched in the recent past. The non-conventional treatment using HPP (high hydrostatic pressure processing) and MAP (modified atmospheric packaging) in combination with pulsed light treatment has proven compatibility [17]. Other chemical treatment methods in combination with pulsed light include malic acid dips [31], antimicrobial edible coating [19] with an advanced level of oxidation with a process also containing hydrogen peroxide. Salinas [41] has also provided promising results in the sterilization of pure vegetables and fruits. Ultrasound along with other disturbances like UV light (pulsed), antimicrobial additives (natural), and medium temperature processing has shown to increase the inactivation of microbes [13]. The primary target sites inside bacteria for inactivation through pulsed light radiation are DNA molecules, resulting in a photochemical pathway that eventually leads to the destruction of microbial cells. However, there is also a potential hurdle along with ultrasound under consideration that there is a multitargeted concept of preservation for the inactivation of microbes [22].

To increase the efficiency of treatment done by photosensitization is also an option. This technology aims to further enhance the sensitivity of microorganisms in the treatment done by light. Photosensitizing agents such as Hematoporphyrin, sodium chlorophyllin, psoralen, and riboflavin are used. These substances have the potential to release free radicals upon exposure to light, which leads to microbial cell disintegration in the food product [17, 26]. Although several combinations of non-thermal preservation techniques involving pulsed light have been proposed and tested for a variety of foods in recent years, most studies seem to focus on microbiological safety and food safety rather than quality aspects [13]. The literature on the use of pulsed light in conjunction with other novel processing techniques for meat and meat products was found to be scarce, and it is a less explored area that requires much deeper research in the future.

12.4 LIMITATIONS

Despite being considerably investigated in recent years, the pulsed light technique has certain limitations. The effectiveness of pulsed light treatment depends on exposure time and distance between the light source and food samples. Generally, most foods are opaque and consist of surface irregularities, cracks, and crevices. These irregular and uneven surfaces limit the microbial exposure to pulsed light, leaving free spaces for microbial growth, which is called the shadow effect [28]. To overcome this issue, an additional machine, like a light system (pulsed), is used to create pulsed light containing multiple directions with the potential of decontaminating all surfaces uniformly [12]. Due to the fact that pulsed light is only a surface decontamination phenomenon, harmful pathogenic organisms such as trichinae and other helminths are not impacted during the process [17]. Another drawback of PL treatment is its effectiveness depends on food composition. Proteinaceous and oil-rich foods are not ideal for pulsed light decontamination because they absorb parts of the radiation, reducing the effective radiation dosage required for microbial inactivation. This poses an obstacle during the decontamination of meat and other meat products which are naturally high in protein and fats [12]. The pulsed light treatment device has a high initial investment cost and a limited lifespan of its components, such as bulbs/lamps required to create the light, which makes the design faulty. In certain cases, it has been observed that pulsed light treatment has undesirable effects on food compounds, leading to the creation of different flavors and a change in the sensory profile of the food product. It can also induce the degradation of naturally occurring pigments, browning, and a foul smell [27]. To be considered a non-thermal process, the Pulsed Light treatment process must be completed in a short duration while complying with various requirements. These include a short pulse duration, low average power and high pulse peak power, and a sufficient cooling period between pulses to reduce treatment time, which inevitably reduces the risks of rapid temperature elevations [12]. These limitations are being researched and eventually overcome with the concept of a combination of pulsed light and other novel techniques.

12.5 SUMMARY

Pulsed light is a kind of technology that is non-thermal and non-conventional in nature. It is a food technique that has shown promising *in-vitro* results in the decontamination of various meats and their products, such as beef, goat, pork, salmon, tuna, chicken, and eggs. It distinguishes itself from other conventional methods as it preserves the original sensory and quality parameters to a considerable extent. The efficacy and changes, or lack thereof, in sensory qualities in meat often depend on the animal species, type of meat, duration of treatment, interspace between the considered sample, and the source of light. However, there are various challenges such as shadow effects, limitations on the nature of the organism, higher investment costs, and the gap between *in-vitro* and practical application of the treatment itself. This makes PL a processing method that needs deeper research in the future with a huge scope. Many researchers are experimenting with the combination of PL with other novel technologies to achieve significantly effective results. A continuous effort is underway with the aim of bridging the gap in the application of pulsed light for the surface decontamination of meat and meat products.

KEYWORDS

- **decontamination**
- **investment cost**
- **microbial inactivation**
- **novel techniques**
- **pulsed light**
- **sensory qualities**

REFERENCES

1. Ananthanarayanan, T. R., Nithin, C. T., Joseph, T. C., Bindu, J., & Srivivasa Gopal, T. K. (2019). Effect of pulsed light on shelf life of chill stored yellow fin tuna (*Thunnus albacares*) steaks. *Indian Journal of Fisheries, 66*(4). DOI:10.21077/ijf.2019.66.4.82781–16.
2. Anderson, J. G., Rowan, N. J., MacGregor, S. J., Fouracre, R. A., & Parish, O. (2000). Inactivation of food-borne enteropathogenic bacteria and spoilage fungi using pulsed-light. *IEEE Transactions on Plasma Science, 28*(1). **DOI:** 10.1109/27.842870.
3. Bender, A. (1992). Meat and meat products in human nutrition in developing countries. *FAO Food Nutr Pap., 53*, 1–91. PMID: 1300286.
4. Bhavya, M. L., & Umesh Hebbar, H. (2017). Pulsed light processing of foods for microbial safety. *Food Quality and Safety, 1*(3), 187–202.
5. Bryant, M. T., Degala, H. L., Mahapatra, A. K., Gosukonda, R. M., & Kannan, G. (2021). Inactivation of *Escherichia coli* K_{12} by pulsed UV light on goat meat and beef: microbial

responses and modelling. *International Journal of Food Science and Technology, 56*(2), 563–572.
6. Cassar, J. R., Bright, L. M., Patterson, P. H., Mills, E. W., & Demirci, A. (2021). The efficacy of pulsed ultraviolet light processing for table and hatching eggs. *Poultry Science, 100*(3), 100923.
7. Cassar, J. R., Mills, E., Campbell, J., & Demirci, A. (2019). Decontamination of chicken thigh meat by pulsed ultraviolet light. *Meat and Muscle Biology, 3*(1), 479–487.
8. Chun, H. H., Kim, J. Y., Lee, B. D., Yu, D. J., & Song, K. B. (2010). Effect of UV-C irradiation on the inactivation of inoculated pathogens and quality of chicken breasts during storage. *Food Control, 21*(3), 276–280.
9. Dalgaard, P. (2000). Fresh and Lightly Preserved Seafood In C. M. D. Man, & A. A. Jones (Eds.), *Shelf Life Evaluation of Foods* (pp. 110–139). Aspen Publishers.
10. Degala, H. L., Mahapatra, A. K., Demirci, A., & Kannan, G. (2018). Evaluation of non-thermal hurdle technology for ultraviolet-light to inactivate *Escherichia coli* K12 on goat meat surfaces. *Food Control, 90*, 113–120.
11. Dewey-Mattia, D., Manikonda, K., Hall, A. J., Wise, M. E., & Crowe, S. J. (2018). Surveillance for foodborne disease outbreaks – United States, 2009–2015. *MMWR Surveillance Summaries, 67*(10).
12. Elmnasser, N., Guillou, S., Leroi, F., Orange, N., Bakhrouf, A., & Federighi, M. (2007). Pulsed-light system as a novel food decontamination technology: a review. *Canadian Journal of Microbiology, 53*(7), 813–821.
13. Ferrario, M., & Guerrero, S. (2016). Effect of a continuous flow-through pulsed light system combined with ultrasound on microbial survivability, color and sensory shelf life of apple juice. *Innovative Food Science & Emerging Technologies, 34*, 214–224.
14. Gómez-López, V. M., Devlieghere, F., Bonduelle, V., & Debevere, J. (2005). Intense light pulses decontamination of minimally processed vegetables and their shelf-life. *International Journal of Food Microbiology, 103*(1), 79–89.
15. Grujić, R. (2010). Meat in human nutrition. *Quality of Life*, 1(1), 16–25.
16. Grzegorzewski, F. (2011). *Influence of non-thermal plasma species on the structure and functionality of isolated and plant-based 1, 4-benzopyrone derivatives and phenolic acids.* Doctoral Thesis. Dept. of Food Chemistry and Food Analysis, Institute of Food Technology and Food Chemistry, Technical university Berlin.
17. Heinrich, V., Zunabovic, M., Varzakas, T., Bergmair, J., & Kneifel, W. (2016). Pulsed light treatment of different food types with a special focus on meat: A critical review. *Critical Reviews in Food Science and Nutrition, 56*(4), 591–613.
18. Hierro, E., Ganan, M., Barroso, E., & Fernández, M. (2012). Pulsed light treatment for the inactivation of selected pathogens and the shelf-life extension of beef and tuna carpaccio. *International Journal of Food Microbiology, 158*(1), 42–48.
19. Huang, Y., Sido, R., Huang, R., & Chen, H. (2015). Application of water-assisted pulsed light treatment to decontaminate raspberries and blueberries from *Salmonella. International Journal of Food Microbiology, 208*, 43–50.
20. Huang, Y., Ye, M., Cao, X., & Chen, H. (2017). Pulsed light inactivation of murine norovirus, *Tulane virus*, *Escherichia coli* O157 and *Salmonella* in suspension and on berry surfaces. *Food Microbiology, 61*, 1–4.
21. Huss, H. H. (1995). Quality and quality changes in fresh fish. In FAO. *Fisheries Technical Paper.*

22. John, D., & Ramaswamy, H. S. (2018). Pulsed light technology to enhance food safety and quality: A mini-review. *Current Opinion in Food Science, 23*, 70–79.
23. Keklik, N. M., Demirci, A., & Puri, V. M. (2010). Decontamination of unpackaged and vacuum-packaged boneless chicken breast with pulsed ultraviolet light. *Poultry Science, 89*(3), 570–581.
24. Koh, P. C., Noranizan, M. A., Karim, R., & NurHanani, Z. A. (2016). Microbiological stability and quality of pulsed light treated cantaloupe (*Cucumis melo* L. reticulatus cv. Glamour) based on cut type and light fluence. *Journal of Food Science and Technology, 53*(4), 1798–1810.
25. Koutchma, T. (2009). Advances in ultraviolet light technology for non-thermal processing of liquid foods. *Food and Bioprocess Technology, 2*(2), 138–155.
26. Kreitner, M., Wagner, K.-H., Alth, G., Ebermann, R., Foißy, H., & Elmadfa, I. (2001). Haematoporphyrin- and sodium chlorophyllin-induced phototoxicity towards bacteria and yeasts – a new approach for safe foods. *Food Control, 12*(8), 529–533.
27. Mahendran, R., Ramanan, K. R., Barba, F. J., Lorenzo, J. M., López-Fernández, O., Munekata, P. E. S., Roohinejad, S., Sant'Ana, A. S., & Tiwari, B. K. (2019). Recent advances in the application of pulsed light processing for improving food safety and increasing shelf life. *Trends in Food Science & Technology, 88*, 67–79.
28. Mandal, R., Mohammadi, X., Wiktor, A., Singh, A., & Pratap Singh, A. (2020). Applications of pulsed light decontamination technology in food processing: An overview. *Applied Sciences, 10*(10), 3606.
29. McLeod, A., Hovde Liland, K., Haugen, J., Sørheim, O., Myhrer, K. S., & Holck, A. L. (2018). Chicken fillets subjected to UV-C and pulsed UV light: Reduction of pathogenic and spoilage bacteria, and changes in sensory quality. *Journal of Food Safety, 38*(1), e12421.
30. Mellata, M. (2013). Human and avian extra intestinal pathogenic *Escherichia coli*: Infections, zoonotic risks, and antibiotic resistance trends. *Foodborne Pathogens and Disease, 10*(11), 916–932.
31. Moreira, M. R., Álvarez, M. V, Martín-Belloso, O., & Soliva-Fortuny, R. (2017). Effects of pulsed light treatments and pectin edible coatings on the quality of fresh-cut apples: A hurdle technology approach. *Journal of the Science of Food and Agriculture, 97*(1), 261–268.
32. Mukhopadhyay, S., Sokorai, K., Ukuku, D. O., Fan, X., Olanya, M., & Juneja, V. (2019). Effects of pulsed light and sanitizer wash combination on inactivation of *Escherichia coli* O157, microbial loads and apparent quality of spinach leaves. *Food Microbiology, 82*, 127–134.
33. Nicorescu, I., Nguyen, B., Chevalier, S., & Orange, N. (2014). Effects of pulsed light on the organoleptic properties and shelf-life extension of pork and salmon. *Food Control, 44*, 138–145.
34. Oms-Oliu, G., Martín-Belloso, O., & Soliva-Fortuny, R. (2010). Pulsed light treatments for food preservation: A review. *Food and Bioprocess Technology, 3*(1), 13–23.
35. Ouyang, B., Demirci, A., & Patterson, P. H. (2020). Inactivation of *Escherichia coli* K-12 in liquid egg white by a flow-through pulsed UV light treatment system. *Journal of Food Protection, 83*(3), 418–425.
36. Petruzzi, L., Corbo, M. R., Sinigaglia, M., & Bevilacqua, A. (2017). Microbial spoilage of foods: Fundamentals. In A. Bevilacqua, & M. Sinigaglia (Eds.), *The Microbiological Quality of Food: Foodborne Spoilers, 1–21*. Cambridge: Woodhead Publishing.

37. Ramos-Villarroel, A. Y., Aron-Maftei, N., Martín-Belloso, O., & Soliva-Fortuny, R. (2012). Influence of spectral distribution on bacterial inactivation and quality changes of fresh-cut watermelon treated with intense light pulses. *Postharvest Biology and Technology, 69*, 32–39.
38. Ramos-Villarroel, A., Aron-Maftei, N., Martín-Belloso, O., & Soliva-Fortuny, R. (2014). Bacterial inactivation and quality changes of fresh-cut avocados as affected by intense light pulses of specific spectra. *International Journal of Food Science and Technology, 49*(1), 128–136.
39. Rodriguez-Turienzo, L., Cobos, A., & Diaz, O. (2013). Effects of microbial transglutaminase added edible coatings based on heated or ultrasound-treated whey proteins in physical and chemical parameters of frozen Atlantic salmon (*Salmo salar*). *Journal of Food Engineering, 119*(3), 433–438.
40. Rowan, N. J., MacGregor, S. J., Anderson, J. G., Fouracre, R. A., McIlvaney, L., & Farish, O. (1999). Pulsed-light inactivation of food-related microorganisms. *Applied and Environmental Microbiology, 65*(3), 1312–1315.
41. Salinas-Roca, B., Soliva-Fortuny, R., Welti-Chanes, J., & Martín-Belloso, O. (2016). Combined effect of pulsed light, edible coating and malic acid dipping to improve fresh-cut mango safety and quality. *Food Control, 66,* 190–197.
42. Shane, S. M., & Faust, A. (1996). Evaluation of sanitizers for hatching eggs. *Journal of Applied Poultry Research, 5*(2), 134–138.
43. Singla, M., & Sit, N. (2021). Application of ultrasound in combination with other technologies in food processing: A review. *Ultrasonics Sonochemistry, 73*, 105506.
44. Valdivia-Nájar, C. G., Martín-Belloso, O., & Soliva-Fortuny, R. (2018). Impact of pulsed light treatments and storage time on the texture quality of fresh-cut tomatoes. *Innovative Food Science and Emerging Technologies, 45*, 29–35.
45. Wambura, P., & Verghese, M. (2011). Effect of pulsed ultraviolet light on quality of sliced ham. *LWT – Food Science and Technology, 44*(10), 2173–2179.
46. Whitnall, T., & Pitts, N. (2019). Meat consumption: analysis of global meat consumption patterns. Dept. of Agriculture, Fisheries and Forestry, Australian Government. http://www.agriculture.gov.au/abares/research-topics/agricultural commodities/mar-2019/meat-consumption.
47. Woodling, S. E., & Moraru, C. I. (2005). Influence of surface topography on the effectiveness of pulsed light treatment for the inactivation of *Listeria innocua* on stainless-steel surfaces. *Journal of Food Science, 70*(7), m345–m351.
48. Zagorec, M., & Champomier-Vergès, M. C. (2017). Meat Microbiology and Spoilage. *Lawrie's Meat Science*: Eighth Edition, Woodhead Publishing, 187–203. DOI:10.1016/B978-0-08-100694-8.00006-6.

CHAPTER 13

MODELING OF PULSED LIGHT IN FOOD PROCESSING

ATHAKKANAKATH JAMNA SHERIN, ELEONORA SHYLLA, PONRAJAN HIMASHREE, R. MAHENDRAN, and C. K. SUNIL

ABSTRACT

Pulsed light technology decontaminates and achieves sterilization by applying high-intensity pulses of light energy at short durations on food product surfaces. Much research is being done showing the utilization of pulsed light to inactivate microorganisms and sterilize food products. However, in-depth studies on the principles behind microbial inactivation, factors influencing it, and inactivation kinetics have to be conducted before the adoption of pulsed light technology in commercial food industries. Modeling can help better understand how various factors affect the inactivation process and help optimize it. Various models have been designed by scientists to understand the kinetics of pulsed light processing and have been used in several studies on microbial inactivation in food products. This chapter discusses the importance of pulsed light technology modeling and describes the different modeling approaches used to evaluate microbial inactivation by pulsed light processing. The interpretation of the models is also outlined, showing various applications in food research.

13.1 INTRODUCTION

Non-thermal technologies are used in food processing as a sustainable alternative to thermal processing [35]. The food temperature is kept lower during non-thermal processing than in traditional thermal processing. Pulsed

Nonthermal Light-Based Technologies in Food Processing.
C. K. Sunil, Preeti Birwal, Megh R. Goyal, and R. Mahendran (Eds.)

light (PL), a non-thermal technology, is used for the rapid inactivation of microorganisms and the deactivation of spores and vegetative cells [11, 25]. PL is composed of a broad spectrum of "white light," which has short and intense pulses. The PL treatment uses electromagnetic energy ranging from 100 to 1,100 nm, covering ultraviolet, visible, and infrared wavelengths. Microbial cell inactivation is likely due to DNA damage induced by PL's photochemical, photothermal, and photophysical (structure damage) effects [49]. PL has been defined as a low-cost, non-thermal method that produces no undesirable residues on treated food surfaces [37]. PL has also been referred to as high-intensity pulsed UV light (HIPL), pulsed UV light (PUV), high-intensity broad-spectrum UV light (BSPL), intense light pulsed (ILP), and pulsed white light (PWL) [71].

Food-borne pathogens must be killed using appropriate fluence to achieve safe food. Loss of culturability is often used as the only criterion for assessing cell death, with no further inquiry into the molecular or physiological processes that contribute to the PEF-mediated killing of microbial cells. The treatment must meet a performance objective, such as the FDA's requirement that cider and juice processors achieve a minimum 5-log decrease in a relevant microorganism [1]. To achieve a safe food product and increase its shelf life, it is essential to inactivate spoiling microorganisms and enzymes and attain kinetic stability. The demand and requirements keep increasing from the food quality regulatory bodies to the consumers. It is very challenging to design a product, process, and achieve the quality that satisfies all their demands and requirements and needs flexibility [69]. The trial-and-error method may not always be feasible for design and optimization to attain a satisfactory quality. A quantitative approach is thus needed to be incorporated for determining suitable quality indicators. The study of kinetic modeling is the best systematic method to satisfy the requirements of the regulatory bodies and consumers. A food processing method has to be designed using mathematical models and kinetic relations for consumer satisfaction, acceptance, safety regulations, and enzymatic stability. Searching for more suitable methods for each process, product, and quality indicator is always encouraged to get better performances and processes [77].

This chapter describes the strategies for designing a PL technology for various food products and the sequential steps for PL process kinetic modeling. Various modeling approaches used for different circumstances and products are discussed here. The interpretation of the applied models and their applications are also highlighted.

13.2 EXPERIMENTAL DESIGN OF PULSED LIGHT TECHNOLOGY

Formulating a mathematical model necessitates a sufficient number of experiments in the complex parts of the PL technology experimental field, which can be accomplished by an appropriate experimental design. Developing primary models and secondary models for PL technology from the data obtained through experimental runs is explained below:

Kinetic models in terms of processing parameters can be developed for critical responses involved in pulsed light technology for the application of food products. This is done on two levels. The first one is pre-planning, and the second level is after experimentation and calculation. The first step in the preplanning is to choose the food matrix and processing. The pH and transmittance are different based on the type of the food, whether it is a solid or a liquid product. The next step is choosing the required responses, such as the target microorganism or an enzyme, the microbial count, enzymatic activity, bioactive content, and the physicochemical and sensorial properties. Then the deciding factors are considered, including the number of pulses applied or the exposure time, the volume of the sample or layer thickness, flow rate, initial microbial count or the enzymatic activity or the bioactive content, and distance between the PL source and sample, voltage, fluence, fluence rate, and the geometric arrangement. The design of the experiment is selected that may be full factorial or central composite or Box-Behnken, etc., or the equal level or different or mixed levels. After experimentation, the primary modeling is done, which includes various modeling types discussed later in this chapter. They are 1st order kinetics, 1st order with tail, nth order kinetics, Biphasic, Weibull, Modified Weibull, Polynomial, Beer-Lambert law-based, etc. Secondary empirical models, such as polynomial, or explicit, such as combined Beer-Lamberts and Ohm's law can be used. Practical or statistical values like p-value, R^2, Adj. R^2, RMSE, SSE, etc., are used for validation. Modeling can be improved in several ways:

- A large number of runs;
- Choose a suitable experiment design;
- Follow adequately sensitive analysis for response;
- Minimum number of essential factors;
- Model reparameterization;
- Select and compare suitable models.

The primary model describes the change in reaction for a single parameter (like temperature and pressure that remain constant) over a time period. The response of primary models, such as the rate constant (k) which is the function of intrinsic element or extrinsic element or environmental elements, is studied in secondary models [56]. An example of a secondary kinetic model is the Arrhenius equation. The sampling procedure is an essential aspect of data collection. Primary samples should be collected at random to accurately indicate the variety and number of organisms present in the product and associated process. At the same time, the variations between samples are minimized.

The mathematical modeling of microbial inactivation is influenced by several elements, including the microbe species, treatment settings, and matrix features. The effect of PL on solids is relatively superficial. Therefore, the surface features are expected to condition significantly microbial inactivation; for instance, a high degree of roughness may protect microorganisms from light [46]. The microorganisms present on the food and the nature of the food surface impact the success of high intensity pulsed light treatment [3]. Therefore, several microbial loads and distance of penetration data should be collected before going for any mathematical modeling intensity.

All the microbial counts CFU per g were log-transformed ($\log_{10}$ CFU/ml). The log-transformed data are fitted for the modeling of microbial inactivation [66]. At least 10 data points must be taken for a particular process parameter combination. Most points must be placed at the inflection area at which the rate of change of the microbial kinetics is maximum [78].

13.3 IMPORTANCE AND TYPES OF PULSED LIGHT TECHNOLOGY MODELLING

Pulsed light (PL) technology is an emerging non-thermal technology for decontaminating surfaces by killing microbes with powerful broad-spectrum light pulses, with UV-C being the highly lethal part [33]. Inactivating the cells is by the principle of photochemical damage of DNA, and other photothermal and photophysical effects of the light pulses [49]. One of the most significant components of characterizing novel processes is the dose-response studies. Here, the models offer numerical descriptions that are crucial for making precise process simulations. The Weibull model has been used by numerous researchers to characterize non-linear inactivation curves.

The three primary inactivation curves are – (a) shoulder; (b) log-linear inactivation component; and (c) tail. Only the first two or last two parts of the curve are visible sometimes. On agar plates, *Salmonella typhimurium* inactivation was described using the shoulder, log-linear decay, and tail models [29, 52]. Because the shoulder is present, no inactivation occurs at low fluences. Since treating food once tailing starts would increase food deterioration and be a time and energy loss, tailing is critical for food stability.

Pure bacterial cultures implanted into liquids or fruit juices were studied for their inactivation model by PL. The survivor ratios for *Listeria innocua* inoculated in a clear buffer solution and in broth after PL treatment were correctly predicted by the Weibull model, but the effectiveness of PL for stainless steel substrates was greatly overestimated [76]. The Weibull model described the *Escherichia coli* O157:H7 and non-pathogenic *E. coli* inactivation in buffer, tryptic soy broth, apple juice, and apple cider [70]. Liquid substrates are agitated to achieve this fundamental requirement, allowing for a better exposition while limiting the liquid's attenuation effect. PL's ability to inactivate microorganisms in solid foods, on the other hand, is a more complex issue. Microorganisms must be exposed to light in order to be rendered inactive. Microbes might be hidden from light by cuticle degradation, internalization, surface structures, and biofilms on solid matters like cut vegetables, making PL inactivation more difficult. Inactivating inoculated *E. coli* O157:H7 and *Salmonella enterica* by PL on raspberries and strawberries was also described using the Weibull model [7]. While tests with injected pathogens are helpful for safe food investigations, inoculation allows bacteria under PL exposure on the food surface to be more easily inactivated. On the other hand, naturally occurring microbes are found in the phyllosphere of plant products. They can attach for long periods, creating microcolonies and biofilms, and even becoming absorbed, reducing PL efficacy.

13.4 MODELS DESCRIBING THE MICROBIAL INACTIVATION BY PULSED LIGHT TECHNOLOGY IN FOOD PRODUCTS

13.4.1 MODELS EVALUATING THE INACTIVATION PROCESS

13.4.1.1 LOG-LINEAR MODEL

Log-linear model is also called the D-value model, showing first-order kinetic behavior. This model assumes that all microorganisms (or any target response) provide similar resistance to PL technology, and survival count shows semi-log reduction.

$$Log\ (\frac{N}{N0}) = \frac{t}{D};\ (t \geq 0)$$

$$\text{Slope} = \frac{1}{D} = \text{k} \quad (1)$$

where; N_0 (CFU mL^{-1}) refers to initial viable microbial count; N (CFU mL^{-1}) refers to the survivors' number after PL exposure for a specific time t (s). The PL treatment time (s) needed to inactivate the first 90% of microbial population is represented as D.

Log-linear kinetics is applied to PL studies to decontaminate vegetable surfaces [8, 41]. A combination of both log-linear and the tailing phenomena has been used for modeling the *E. coli* inactivation by PL in different fruit juices [66].

$$log(N) = log((10^{(logN_0}) - 10^{log(N_res)})).e^{k_max}.EPL + 10^{logN_{res}} \quad (2)$$

Here; k_{max} (s^{-1}) refers to inactivation constant of the curve's linear portion; N_0 (CFU mL^{-1}) represents the initial inoculum size; and N_{res} (CFU mL^{-1}) refers to residual microbial count remaining at the treatment end making a tailing effect of the inactivation curve.

13.4.1.2 BIPHASIC MODEL

Due to the susceptibility of the two subpopulations to the processing method, microorganisms are frequently observed to have changed inactivation levels. In this model, the population is divided into two different sub-populations, each having separate log-linear inactivation rates.

$$1 \log 10N_f = log_{10}(N_o) + log_{10}(\text{f}.\text{e}^{-k_{max1}.t)+}(1-f).e^{-k_{max2}.t}) \quad (3)$$

where; f represents the initial population fraction corresponding to the subpopulation that is more treatment-sensitive; (1 – f) represents the initial population fraction corresponding to the sub-population that is more treatment-resistant; and k_{max1} and k_{max2} refers to specific inactivation rates of the two populations, respectively [10, 25].

13.4.1.3 SIGMOIDAL MODEL

The responses of microorganisms can be convoluted and follow the sigmoidal pattern, with three separate stages: a shoulder phase, a log-linear inactivation phase, and a tail phase. The mathematical model established [29] to characterize this behavior is as follows:

$$log_{10}(N_f) = log_{10}\left(10^{log_{10}(N_o)} - 10^{log_{10}(N_{res})}\right) + \log_{10}(e^{(-k_m ax.t)}. \frac{e(k_{max}.s_l)}{(1+(e^{((}k_m ax.S_l)-1).e^{((}k_{max}.t)} 10^{log_{10}(N_r es)}) \quad (4)$$

13.4.1.4 WEIBULL MODEL

The Weibull model is the most extensively utilized non-linear model for quantifying microbe and enzyme inactivation behavior. Each cell of the microbial population has resistance for inactivation (heterogeneity), and each cell's resistance manifests itself only while the cell is still alive (shoulder or tailing effect) [9]. The equation given below presents the Weibull model [62]:

$$log \frac{N}{N_o} = -\alpha.t^{\beta} \quad (5)$$

Weibull distribution takes into account the shape parameter (β) and scale parameter (α,$s^{-\beta}$). The shape parameter describes the concavity of the survival curve ($\beta > 1$ implies a downward concavity, while $\beta < 1$ indicates an upward concavity). Since total fluence can be calculated as the product of fluence rate (W/cm^2) and treatment time (s), the above-mentioned Weibull model equation has also been utilized with fluence (F, J/cm^2) in place of time (t) (s).

$$log \frac{N}{N_o} = -\alpha.F^{\beta} \quad (6)$$

The scale parameter $\propto$, which denotes the inactivation per unit fluence, is expressed in the unit $(cm^2\ J^{-1})^{\beta}$. The shape parameter (δ, min) of the re-parametrized Weibull equation) is presented similarly to the D-value of the log-linear model [9].

$$log \frac{N}{N_0} = -\left[\left(\frac{t}{\delta}\right)^{\beta}\right] \quad (7)$$

The time needed to get the first decimal reduction is shown here as δ. Another modified form of the above equation, by substituting the time parameter (t) with pulsed light intensity or fluence rate parameter(E_{PL},W. cm^{-2}) and given as below [66]:

$$log \frac{N}{N_0} = -\left[\left(\frac{E_{PL}}{\delta}\right)^{\beta}\right] \tag{8}$$

The δ has a unit of (E_{PL},W.cm^{-2})

13.4.1.5 WEIBULL WITH TAIL

Weibull modeling with tailing effect where F is an independent variable is given by the model:

$$log_{10}N_f = log_{10}N_0\left[10^{log_{10}(N_0)} - 10^{log_{10}(N_{res})}.10 - \left(\left(\frac{F}{F_1}\right)^{\beta}\right) + 10^{log_{10}(N_res)}\right] \tag{9}$$

Here; N_f is the CFU number after treatment at a fluence 'F;' N_0 is the initial number of the tested microbes (in CFU/mL or CFU/g depending upon experimental data); N_{res} is the surviving cells number; F is the applied fluence (J/cm^2); F_1 is the fluence allowing first log_{10} reduction; and β is a determines the curve concavity direction [2].

13.4.1.6 MIXED WEIBULL MODEL

Two mixed Weibullian distributions can define the kinetics of subpopulations that have different resistance, in which F is replaced by t (seconds), refers to the shape parameter, refers to the log_{10} proportion between the sensitive fraction (f) and the resistant one (1–f), δ_1 and δ_2 are the time required for the first decimal reduction of the subpopulation 1 and 2, respectively [12, 25].

$$log_{10}N_f = log_{10}N_0 + log_{10}\left(\frac{1}{1+10^{\alpha}}\right) + log_{10}\left[10^{\left(-\left(\frac{t}{\delta_1}\right)^{\beta+\alpha}\right)} + 10^{\left(-\left(\frac{t}{\delta_2}\right)^{\beta}\right)}\right] \tag{10}$$

13.4.1.7 NTH ORDER KINETIC MODEL

In nth order kinetic model, there is no fixed order of kinetic model, and it is kept as a variable for the detention of the existing overall order of inactivation.

$$-\frac{dA}{dt} = k.A^n \tag{11}$$

The above equation is the nth order kinetic model. The equation we get on integrating this is:

$$\frac{A}{A_0} = [1 + A_0^{(n-1)}.(n-1).k.t]^{\left(\frac{1}{1-n}\right)} \quad (12)$$

where; n is the inactivation order; A refers to the enzymatic activity after the exposure for 't' minutes; and A_0 is the initial enzymatic activity or the control. In special circumstances, the 'n' reduces to 1 and the model converts to the first-order kinetic model [15].

13.4.1.8 POLYNOMIAL MODEL

Polynomial models are helpful when it is required to study the effect of various factors (X_1, X_2, ...) on a particular response Y simultaneously. They are, however, not precisely kinetic models, but when time becomes a factor for the residual microbial count response, then a part of the polynomial equation will contain a kinetic relation in it. The polynomial equation is expressed below:

$$Y = \beta_0 + \Sigma_{i=1}^{2} \beta_i X_i + \Sigma_{i=1}^{2} \beta_{ii} X_i^2 + \beta_{ij} X_i X_j \quad (13)$$

where; $_{\beta 0}$ refers to the intercept Y≠0 while X = 0; $_{\beta i}$ represents the ith linear term coefficient; $_{\beta ii}$ represents the ii^th^ quadratic term coefficient; and $_{\beta ij}$ represents the ij^th^ interaction term coefficient. The corresponding regression coefficients (β) act as the inactivation rate 'k.' A quadratic polynomial model was used for the inactivation time modeling of *A. acidoterrestris* in PL-treated apple juice. This method of modeling was used to determine the PL treatment time ($Y = t_{PL}$) in relation with the survival counts of the microbe ($X_1 = \log (N/N_0)$) and initial inoculum size IS ($X_2 = \log (IS)$) [27].

13.4.1.9 BEER-LAMBERT'S LAW-BASED MODEL

The light intensity weakens when it is propagated through some medium. The loss of intensity is proportional to the original light intensity or the fluence rate (I_0 or E_0, W cm^{-2}), medium properties like the absorption or attenuation coefficient α (cm^{-1}), and the path length of the medium ($x = r - r_0$). The mathematical expression of this phenomenon is:

$$Ir = I_0.\frac{r_0}{r}.\exp\left(-a.(r-r_0)\right)$$

$$E_x = E_0.\ \mathrm{Exp}\ (-a.\ x) \tag{14}$$

where; I_r or E_x represents the light intensity or fluence rate at a distance 'x' from the light source; r_0 is the start of the medium path while 'r' is the end of the medium path. The attenuation coefficient (α) is dependent on the light absorbance parameter (A_b) as shown below:

$$A_b = \varepsilon.\ C.\ x$$

$$\alpha = \mathrm{In}10 \times \left(\frac{A_b}{x}\right) = 2.303 \times \left(\frac{A_b}{x}\right) \tag{15}$$

where; ε represents the molar absorptivity or the molar extinction coefficient, expressed in L $mol^{-1}cm^{-1}$; and 'C' refers to the light-absorbing species concentration in the medium expressed in mol L^{-1} [45].

13.4.1.10 GEERAERD MODEL

The model is intended to define a non-log-linear survivor curve of microorganisms and comprises an initial shoulder, then a log-linear decline of cell count, and finally results in the form of a tailing region. The model predicts survivor curves in terms of time, but here the time was re-parameterized to total fluence F (Jcm^{-2}). Moreover, it is as follows:

$$log(N) = log\left[(10^{N_0} - 10^{log(N_{res})}) \times e^{-k_{max}F} \times \left(\frac{e^{k_{max}s1}}{1+(e^{k_{max}s1}-1)\times e^{-k_{max}F}} + 10^{log(N_{res})}\right)\right] \tag{16}$$

where; log ($_{N0}$) is initial population density; log ($_{Nres}$) is the residual population density; $_{\kappa max}$ is the inactivation rate; and S1 is shoulder length that is the degrees of freedom for parameter estimation [47].

13.4.1.11 RESPONSE SURFACE MODEL (RSM)

Empirical models were also designed to forecast microorganism log reduction in solid material surfaces. Solid layer thickness (H), distance from UV strobe (D), and the treatment time (t) are the variables used (Table 13.1). The general equation for the model was:

$$y = \alpha_0 + \alpha_1 x_1 + \alpha_2 x_2 + \alpha_3 x_1^2 + \alpha_4 x_2^2 + \alpha_5 x_1 x_2 \quad (17)$$

where; 'y' is the log reduction; log α_0 is the constant term; $\alpha_{1....5}$ are the coefficient; x_1 is the pulsed light treatment time; and x_2 is the distance from UV strobe.

1. **Thickness Model of RSM:** The thickness model contained solid thickness (H) and treatment time (t), the quadratic variables and their interactions. A response surface figure demonstrates that the microbial population decreases as thickness and time increase.
2. **Distance Model of RSM:** The distance method makes use of the distance from UV strobe and treatment time, quadratic effect of these variables, and the interaction between them. The following is the distance model equation:

$$\text{Log}_{10}\ \text{reduction} = -0.5485 + 0.0472t + 0.0826D + 0.0006\ t^2 + 0.0012\ D^2 - 0.0065Dt \quad (18)$$

13.4.2 INTERPRETATION OF MODELS

It is well-known that the food matrix affects the efficiency of PL due to bacteria's competition for light absorption. However, other extrinsic factors like pH may have synergistic or contradictory impacts on every multi-target lethal inactivation process [34]. The below-listed possible explanations are for the incidence of the features in the PL inactivation curve.

13.4.2.1 THE SHOULDER PHASE OF INACTIVATION KINETICS GENERATED BY PULSED LIGHT

While several models with shoulders exist, including the log-linear model with shoulder and the biphasic with shoulder [30], only sigmoidal model is employed to characterize PL inactivation kinetics. Complete microbial inactivation is very exceptional, and the existence of tailing is very common, and the shoulders and tails provide a way to a sigmoidal pattern.

TABLE 13.1 Comparison of Various Models Used to Evaluate the Microbial Inactivation by PL Treatment

Treated Sample	Microorganism Inactivated	Mathematical Model	Parameters Fitted into the Model	R2 Value	RMSE	References
Almond kernels	*Salmonella enteritidis*	Log-linear model	k_{max} ($cm^{-2}\cdot J$) to describe inactivation rate	0.79–0.95	0.69–0.33	[36]
		Log-linear model with tailing	k_{max} ($cm^{-2}\cdot J$) to describe log-linear region and N_{res} represents the tailing section.	0.87–0.99	0.63–0.11	
		One-parameter Weibull model	Fluence required δ ($J\cdot cm^{-2}$) and shape parameter p.	0.87–0.99	0.46–0.08	
		Two-parameter Weibull model	Shape parameter p and assumes two subpopulations having fluence constants δ_1 and δ_2.	>0.98	<0.33	
Orange juice	*Escherichia coli*	Log linear plus tail mode.	Residual population density N_{res} (log, CFU/mL) and inactivation rate K_{max} (cm^2/J).	0.94–0.97	0.36–0.19	[66]
		Weibull model	Number of survivors N (CFU/g); initial inoculum size No (CFU/g); dosage (W/cm^2); time required for first decimal reduction of sub-population δ; shape parameter p.	>0.99	<0.08	
		Biphasic model	Fraction of the initial population corresponding to the subpopulation.	>0.98	0.10–0.03	
			More sensitive to the treatment f.			
			Fraction of the initial population corresponding to the subpopulation.			
			More resistant to the treatment 1-f.			
			Specific inactivation rates of two populations K_{max1} and K_{max2}.			

TABLE 13.1 *(Continued)*

Treated Sample	Microorganism Inactivated	Mathematical Model	Parameters Fitted into the Model	R2 Value	RMSE	References
Cell culture	*Clostridium sporogenes*	Log-linear model	Initial viable count N_0 (CFU/mL); number of survivors after PL treatment Nt (CFU/mL); and treatment time required to destroy 90% of the microbial population D.	0.94–0.97	0.56–0.39	[42]
		Two parameter Weibull model	Time required for first decimal reduction α; shape parameter β.	0.93–0.96	0.52–0.45	
Cow milk and concentrated milk.	*Escherichia coli* ATCC 25922	First order kinetics	Inactivation curves showed both linear and concave shapes.	Model not appropriate		
		Weibull model	Number of survivors before and after treatment; scale parameter α; shape parameter β.	$R^2 = 0.65–1.00$		[57]
Food processing contact surface – stainless steel coupons.	*Listeria monocytogenes*	Weibull model	Scale parameter α= 3.01 and Shape parameter β = 0.33.	–		[76]

Although the Weibull model may implicitly fit shoulders [30], the kinetic curves evaluated in this edition and explained by the Weibull model demonstrate a sharp decline in the survivor population subsequent to the first pulse.

Some researchers gave very high fluences in the first pulse itself, leading to the missing of shoulders in the inactivation curve; therefore, tests with less fluences can resolve shoulders.

Small shoulders can be ignored for process design during practical implementation. Examining its importance in PL microbial cultivation is more important than its biological significance. Using fluence-based microbial inactivation curves, a shoulder length of just 0.08 mJ.cm^{-2} was observed [52], while 0.045–0.073 J.cm^{-2} was reported in another study [51], which appears minor when compared to the value of 12 J.cm^{-2}, the maximum authorized by the FDA [79]. It is more important to examine its significance in the PL inactivation of microorganisms than its biological significance. Furthermore, when 13 bacteria were experimented under the same conditions, only eight Gram-positive bacteria showed a shoulder [24]. However, as reported for several Candida species, the five Gram-negative bacteria did not, except *Pseudomonas aeruginosa*, which showed a shoulder but only at the least lamp discharge [22].

13.4.2.2 THE INACTIVATION PHASE OF INACTIVATION KINETICS GENERATED BY PULSED LIGHT

Since all reported inactivation curves were acquired from culture methods, it is attributed solely to creating cyclobutene pyrimidine dimers, which results in chlorogenic death, i.e., the loss of replicability of the cells. Concepts of mechanistic and ritualistic explain the deviations from linearity of microbial inactivation [10].

As stated by the vitalistic concept, each cell is not identical [50], which is attributed to a molecular-level mechanism [9] that may vary between individuals. As a result of the non-identical behavior caused by stress, there are variations from the log-linear inactivation model at the population level. Some authors have characterized this variance in the statistical features of various underlying distributions of resistances or sensitivities (e.g., Weibull) [9, 53, 62]. Assessing the microbial resistance to more harsh treatments and comparing it to or assessing the resistance of decreasingly minor fractions of the microbial population to find out whether the continuously reducing death rate curves convert progressively into exponential as cell counts reduce are the two possible approaches to authenticate the vitalistic theory. According to the mechanistic theory [10, 30], deviations shown by some microorganisms are related to the inaccessibility of the main processing parameter, the acquired microbial resistance, like microorganisms clumping, the occurrence of genetically variable microorganisms, or any experimental protocol issues. These results must be compared with caution, specifically when using the Weibull model to analyze data with several reparameterizations.

13.4.2.3 THE TAIL PHASE OF INACTIVATION KINETICS GENERATED BY PULSED LIGHT

Sometimes, a population will survive at a constant or closely constant value independent of the continued treatment time; this is called tailing. In practice, this means that extending the treatment time does not result in additional microbial inactivation once the tail has been reached. However, it will damage the food in which the microorganism is present.

Keeping this in mind, complete or almost complete inactivation of microorganisms can be found not just in inactivation models that contain the tail obviously (sigmoidal, Weibull + tail), nonetheless it is also found in inactivation curves when the second inactivation phase has a very low inactivation rate. Tailing may appear in inactivation curves that have not been identified when greater fluences are applied. There are various theories about why tailing occurs, some universal and others specific to the PL process. When one of the sub-populations is highly resistant to treatment, the vitalistic approach suggests that the occurrence of multiple sub-populations causes tailing [54]. According to the mechanistic theory, penetration of UV light is limited when there is an opaque body in between the source of light and the microbes that can protect it from inactivation, which is called the shadow effect. This shadow effect generates a tail since a part of the microorganisms will never be affected by the light. Surface characteristics like the acne in strawberries, or droplets of raspberries [7], or surface imperfections of food contact surfaces might protect bacteria in solids [67]. Although proper mixing might increase the light exposure on all bacteria in liquid mass, turbidity, and the solids suspended are the main impediments to microbial inactivation in liquids [32]. It has also been shown that when the microbes overlap, the ones at the top become inactivated while microbes at the bottom are protected [13, 24]; the same occurs in liquids when cells agglomerate [76].

For example, when interpreting the Weibull model, an investigation found that the IPL lamp voltage affected the scale and form characteristics but not the α value. The readings for *S. Enteritidis, C. perfringens,* and *S. sonnei*, in particular, exhibited a propensity to increase with voltage. While *S. aureus, B. cereus*, and *P. aeruginosa* displayed a propensity to decrease as the voltage increased. Except for *S. enteritidis*, and *S. sonnei*, no associations between α value and the voltage were found in these four bacteria. No external variables influenced the outcome, and the total microbe value did not alter with voltage in this investigation [40].

13.4.2.4 INACTIVATION KINETICS OF ENZYMES

The inactivation of three enzymes polygalacturonase (PG), Peroxidase (POD), and Polyphenol oxidase (PPO) was compared using the same experimental setup. The inactivation of POD by PL followed a first-order kinetic model (k = 0.0255 cm^2/J, [enzyme] = 0.25–1.25 μg/mL) [63], while PPO inactivation data was best fitted to the Weibull model [64]. Yet, their fitting in the first-order kinetic model (k = 0.0175 cm^2/J, [enzyme] = 31.5 μg/mL) was also excellent (R^2 = 0.9925). When the data for these enzymes are compared, it is clear that PG is the most PL labile. Because PG was the most concentrated enzyme in these assays, the single confounding factor for this comparison, the higher opacity of the treated samples, can be ruled out [65].

13.5 CULTURE-DEPENDENT VS. CULTURE-INDEPENDENT APPROACHES FOR EVALUATING THE EFFICACY OF PULSED LIGHT

Agar-based conventional culture methods have been mostly used to enumerate survivors after PEF treatments [68]. The aim of future investigations is to show that PEF application for microbial inactivation is effective at a reasonable technology readiness level (TRL) for an update in market and implementation. Inactivation differs based on the operational factors (voltage applied, number of pulses, and distance from the source of light, i.e., dose or fluence), biological factors (nature, type, and number of microorganisms present, suspension medium's nature, occurrence of antibiotics or dyes, and shading effects), and occurrence of enrichment/resuscitation phase after treatment. As a result, measuring the microbial lethality due to PEF treatments has been far from simple [37, 68].

Proofs show that indication of a harsh environment may induce a shift to an adaptive survival state during Pulsed Light treatments [45]. Photoreactivation was reported in a few studies on Pulsed Light treatments [34, 60]. Recovery of the culturability of microbes after Pulsed Light treatment is photoreactivation. It is restricted to the use of low fluencies and small recoveries (<1 log). One of the initial methods to demonstrate that not all Pulsed Light-treated bacteria transferred to agar plates retain essential molecular and cellular activities but are unable to form colonies was through photoreactivation investigations. The fact that current data clearly demonstrates that Pulsed Light treatment destructs yeast by a multi-hit or mechanistic method which alters the cell membrane permeability as well as the stability and functionality of DNA and macromolecules liable to the applied UV dose further complicates the prediction process. A study reported various cellular response mechanisms in *Candida albicans* clinical strains to Pulsed Light treatments [23]. Propidium iodide absorption and protein leakage were used to show that UV pulses applied significantly increased the cell membrane permeability.

The latter discovery was strongly linked with an elevated amount of lipid hydroperoxidation in yeast cell membranes treated with PL. Reactive oxygen species (ROS) generation during PL treatment of yeast cells showed a distinct pattern, with initial ROS bursts being consistent with the presence of apoptotic cells. Late apoptosis and necrosis were also induced by enhanced PL treatment, which was likewise associated with a change from nuclear to cytoplasmic ROS accumulation and cell membrane leakage. The Comet assay revealed that cells treated with PL had increased nuclear damage. All yeast exposed to PL treatments sub-lethally displayed cellular repair. According to the treatment regimen, bacteria could survive PL, according to these intricate structural and physiological analyses. Understanding and appreciating all operating conditions, including the target organism(s) under inquiry, is crucial for achieving total lethality. Effective microbiological modeling of survivors following PL treatments and the interpretation of related mortality rate kinetic data will be affected by this. It was investigated how to inactivate *S. cerevisiae* utilizing flow cytometry in conjunction with several fluorescent stains and was compared to traditional plate count enumeration [26]. They discovered that the reduction in cells' ability to be cultured was considerably greater than the corresponding rise in permeabilized cells. The impact of PL treatment on the morphological and physiological characteristics of *E. coli* and *L. innocua* was investigated using a similar methodology [45].

In contrast to direct rupture of cell membrane or intracellular enzyme inactivation, oxidative stress having a consequent DNA damage was seen to be liable for the microbial culturability loss. However, it might have seemed that microbial lethality results from the accumulation of various insults inflicted on treated cells, with the onset rate being affected in part by the fluence delivered. Pathogens and microbial food spoilage present a variety of mortality rate kinetic data that indicate the complicated cellular response to treatment. When combined with various fluorescent probes, flow cytometric investigations offer insightful information about physiological states and are an appropriate method for understanding the effects of microbial disinfection process [44, 59]. Despite the data of colony count enumeration decreasing below a detection limit, flow cytometric investigations were performed to report statistically distinct levels of the metabolic activity of *E. coli* and *L. innocua* levels observable following PL treatment [5]. But applying high levels of PL energy began a slow cessation of cellular functions. In fact, high percentages of bacterial populations were able to keep their polarized cell membranes immediately after administering 0.76 J/cm^2 fluence, despite colony counts in each case were decreased by more than 99.99%. These experiments showed that, although being unable to grow on laboratory media, PL-treated bacteria that enter viable but non-culturable state may nevertheless exhibit a number of essential functions. By examining the *Streptococcus typhi* phage susceptibilities, the existence of viable but non-culturable bacteria following PL treatments was revealed [58]. Despite losing culturability, the host bacteria's infectiousness could still be detected, suggesting that it is still alive. It seemed plausible that the inactivation mechanisms like photothermal or photophysical processes are likely to be related to their cellular structure, size, and UV absorption properties due to widely variable results from several reported research addressing the probable rupture of microorganisms treated under PL [45]. In addition to evident DNA damage, microbial inactivation due to PL may also be related to changes in proteins and lipids. For example, treated yeast cell membranes have been found to include carbonylated proteins, lipid peroxides, and lipid hydroperoxidation [23]. Because significant amounts of carboxyfluorescein-stained bacteria were found when cells were non-culturable and de-energized already, the assessment of the activity of intracellular esterase revealed to be a poor parameter for examining cell viability after PL treatment [47]. Therefore, analyzing enzyme activity does not always indicate the viability of cells. Additionally, they demonstrated that eliminating the commonly used indicator of live bacteria, propidium iodide, could not be considered a reliable indicator of viability because significant numbers of cells with intact membranes were found following the administration of lethal energy doses. Additionally, it was noted that the identification of high ROS levels at 0.50 J/cm^2 corresponded to a fluence at which PL treatments caused a growing loss of culturability. This supported past research that showed elevated ROS levels were present in nonculturable cells [23]. According to the latter authors, the initiation of apoptosis in yeast treated by PL begins after lethal amounts of PL are administered, making it a potential candidate marker for microbial death.

13.6 APPLICATIONS OF PULSED LIGHT TECHNOLOGY MODELLING

13.6.1 SOLID FOODS

The log reductions obtained after the PL treatment of minimally processed vegetables like spinach, green paprika, celery, soybean sprouts, carrot, radicchio, white cabbage, and iceberg lettuce with a 7 J PL intensity ranged from 0.21 to 1.67 for 45 s/side and 0.56 to 2.04 for 180 s/side [31]. The shredded white cabbage that was kept in MAP for 9 days at 7?C had limited shelf life due to the off-odors; however, the overall appearance was good up to 3 days only. This investigation also demonstrates that PL treatment raised lettuce respiration rate by 80% while not affecting cabbage respiration rate. Using only two 0.15 J cm^{-2} pulses for each flash, aerobic reductions ranging from 1.6 log CFU mL^{-1} to greater than 2.6 log CFU mL^{-1} for carrot and paprika respectively were found [39]. According to these researchers, two log reductions effectively helped in shelf-life extension of sliced vegetables kept at 7?C by around four days. Response surface methodology was utilized to determine the optimal conditions for inactivating *A. niger* spores in maize meals [43].

Consequently, the model indicated that a treatment time of 100 seconds, a distance of 3 cm from the strobe, and maximum energy of 5.6 J cm^{-2} per pulse might result in a maximum 4.93 log reduction. A considerable reduction of *E. coli* O157:H7 up to 4.89 log CFU mL^{-1} was achieved in alfalfa seeds treated by 5.6 J cm^{-2} light for 90 s, i.e., 270 pulses from 8 cm distance, while only 1.42 log decreased from 13 cm distant treatment [72]. In colored food powders, the PL's thermal effect outweighed the UV effect (black pepper and wheat flour). Food powders' visual and flavor properties were drastically impacted, even before the attainment of 58 J cm^{-2}, which is the decontamination threshold. The color changes in wheat flour and black pepper were due to overheating and oxidation, which happened significantly faster in black pepper than in wheat flour because dark materials absorb more light energy than lighter ones. However, after a prolonged storage period, 2–5 PL flashes at 3 J cm^{-2} fluence maintained the potato color, whereas the untreated samples quickly turned brown [19].

PL treatment at 22.6 J cm^{-2} for 60 s used to decontaminate blueberries showed 4.3 log and 2.9 log CFU mL^{-1} decrease for *Salmonella*, and *E. coli*, respectively [6]. On raspberries and strawberries, *E. coli* O157:H7 was reduced by 3.9 logs at 72 J cm^{-2} and 2.1 logs at 25.7 J cm^{-2}, while *Salmonella* was reduced by 3.4 and 2.8 logs at 59.2 and 34.2 J cm^{-2}, respectively. Furthermore, the Weibull model accurately predicted the *E. coli* O157:H7 and *Salmonella* inactivation by PL rather than first-order kinetics [7]. No reduction of fungal development was found on *B. cinerea* on treated strawberries for up to 250 seconds with 7 J pulse intensity [54]. Furthermore, even when the lag period was extended by one or two days, combining PL with short heat treatments such as 40–45?C treatment for about 3–15 min or continued UV-C irradiation didn't significantly reduce fungal growth. It is also established that exposing the fruit to low amounts of continued UV radiation improves disease resistance.

13.6.2 MILK AND LIQUID FOODS

Cold Pasteurization of bovine milk was effectively done by PL treatment in 56 s with 12.6 J/cm^2 dose [74]. *Staphylococcus aureus* was completely inactivated on treating milk with PL [48]. It was concluded that the PL system designed for continuous flow type can be used to pasteurize milk in less time, thereby preventing excessive high temperatures that can cause quality changes in milk. PL can efficiently prevent oxidative reactions when compared with UV treatment, due to its short pulses ranging from 300 ns to 1 ms [28]. PL technology does not show much impact on the protein components of milk or cause any lipid oxidation after treating milk [21].

PL technology is more efficient for application on surfaces than liquid foods. PL treatment for 1 second with 5.6 J/cm^2 per pulse to inactivate *S. aureus* in a liquid medium showed a 5.0 log reduction in agar-seeded cells and a 1.35 log reduction in suspended cells [48]. Decreasing the depth of the sample or decreasing the distance between light and sample can increase the efficiency of PL technology. Studies showed that PL can result in equal microbial reductions in liquids regardless of the amount of microbial load [76]. In clear liquids, the Weibull model was sufficient to determine the microbial inactivation. However, this model does not suit when different substrate properties have a significant impact on microbial inactivation.

13.6.3 MEAT, POULTRY, AND FISH PRODUCTS

PL treatments caused considerable microbial inactivation on simple surfaces but only 1 log to 3 log reductions on complex surfaces like meats [18]. Some of the radiation might be absorbed by the proteins and lipids, resulting in a lower effective radiation dose on microorganisms [34]. In the UV-B range, proteins absorb UV at 280 nm and higher wavelengths, while lipids with isolated double bonds or conjugated double bonds absorb UV [38]. Microbial counts were reduced by two logs in beef steaks after PL treatment using 5 J cm^{-2} on each side and refrigerated for three days at 4–5?C [17]. After a PL treatment, *Listeria innocua* was reduced by two log cycles on hot dogs. Shrimp treated with PL and kept in refrigeration for 7 days remained edible, whereas untreated shrimp exhibited severe microbiological degradation. A 1-log reduction of *L. monocytogenes* or *E. coli* O157:H7 could be achieved after 60 seconds of PL treatment (at 5.6 J cm^2 per pulse) from an 8 cm distance with no effect on the quality of salmon fillets. Samples were warmed when the treatment time increased to 30 or 45 seconds at short distances (3–5 cm), resulting in visible color and quality changes [61].

When PL of 30 J cm^{-2} was applied to frankfurters, no differences in protein, nitrosamine, riboflavin, vitamin C, or benzopyrene were reported when compared to control samples. It was discovered that there was no impact on riboflavin content after PL treatments in beef, chicken, or fish, even though heat, oxygen, and light cause considerable riboflavin loss in foods. And also there was no significant change in color values or shear force values in pre-packed catfish fillets subjected to 2–4 2.5–5 J cm^{-2} pulses [73].

13.6.4 FOOD CONTACT MATERIALS AND FOOD PACKING MATERIALS

PL technology can also be used for sterilization or reducing the microbial population on food contact surfaces or the surfaces of food packaging material in the processing industry [16]. At the surface of numerous packaging materials seeded with 10–1,000 CFU cm^2, *S. aureus* was inactivated, whereas *Aspergillus* spp. and *B. cereus* spores were inactivated at intensities more than 2 J cm^{-2} [20]. With 4.103 J cm^{-2} PL and 8.103 J cm^{-2} of continued UV light treatments, Bacillus subtilis inactivation values were almost comparable [55]. Molds such as *A. niger*, *A. repens, A. cinnamomeus*, and *C. herbarum*, which are typically found on the surface of food packaging papers, were significantly reduced after a PL treatment on paper-polyethylene packaging material [75]. When subjected to 0.977 J cm^{-2} PL for 30 ms, *C. herbarum* inactivation was around a 2.7 log decrease, which is sufficient for typical contaminated packaging material.

The effect of PL altered the microbicidal characteristics of the food contact surfaces. About 1.27 J cm^{-2} pulses were used to infect four types of stainless-steel surfaces with *L. innocua.* Their findings show that surface structure had a complex effect on the performance of PL treatments for microbial reduction. PL effect on microbial inactivation reduced in the even finish surfaces because of its highly reflective and hydrophobic nature, which induces cell clumping. Whereas the roughest surface provided a more even weight distribution [46].

13.7 FUTURE PROSPECTS AND CHALLENGES

Inactivation by pulsed light can be utilized on both liquid and solid surfaces. However, a literature review shows that solid surface modeling requires more research. Many scientists have recognized the requirement for databases providing parameters of kinetic models for diverse pathogenic and spoilage bacteria as a priority [4]. Due to the absence of data on enzyme inactivation in foods, it is impossible to point out enzyme susceptibility to pulsed light treatment [14]. The lack of universal consensus and data and procedure standardization will stymie the advancement of PL technology.

Models can aid in understanding how the absorption coefficient of a sample or air medium (between the sample and light source) influences the overall inactivation, as well as how the location of the lamp and its corresponding fluence distribution affect microbial inactivation. Future research should focus on determining the mechanisms contributing to microbe transmission variances. A suitable biological marker for microbial cell death must first be found, and further research must be done to determine whether there is any correlation between the detection of this "cell death marker" and the culture-dependent plate count data now used in the food sector. However, more studies are required to identify a suitable cellular marker to tag microbial cell death and investigate the relationship between the discovery of this 'cell death marker' and current food industry culture-dependent plate count statistics.

13.8 SUMMARY

With the help of statistical heuristics, mathematical/statistical modeling promotes a true understanding of the nonlinear challenges associated with food engineering operations and assists in developing validated nonlinear regression models. This approach helps in optimizing processes through carefully planned experimental designs with fewer experiments. The mathematical models created facilitate easy extrapolation to scale-up studies. Ultimately, there is a necessity for industry and academia to collaborate to enhance modeling approaches by addressing common issues encountered in implementing modeling techniques in food engineering.

To design high-intensity pulsed light processes that achieve the required inactivation levels, it is essential to have sufficient information about process parameters (such as number of pulses, pulse width, distance from the lamp, total fluence, temperature, and layer depth) to comprehend and compare the work conducted by various laboratories and researchers.

KEYWORDS

- **food engineering**
- **inactivation kinetics**
- **log-linear**
- **microbial inactivation**
- **nonlinear regression models**
- **pulsed light technology**
- **Weibull**

REFERENCES

1. Administration, Food & Drug (FDA). (2001). Hazard analysis and critical control point (HACCP): Procedures for the safe and sanitary processing and importing of juice. *Fed. Regist., 63*(90), 20450–20486.
2. Albert, I., & Mafart, P. (2005). A modified Weibull model for bacterial inactivation. *International Journal of Food Microbiology, 100*(1–3), 197–211.
3. Anderson, J. G., Rowan, N. J., MacGregor, S. J., Fouracre, R. A., & Farish, O. (2000). Inactivation of food-borne enteropathogenic bacteria and spoilage fungi using pulsed-light. *IEEE Transactions on Plasma Science, 28*(1), 83–88.
4. Balasubramaniam, V. M., & Farkas, D. (2008). High-pressure food processing. *Food Science and Technology International, 14*(5), 413–418.

5. Berney, M., Weilenmann, H.-U., & Egli, T. (2006). Flow-cytometric study of vital cellular functions in *Escherichia coli* during solar disinfection (SODIS). *Microbiology, 152*(6), 1719–1729.
6. Bialka, K. L., & Demirci, A. (2007). Decontamination of *Escherichia coli* O157: H7 and *Salmonella enterica* on blueberries using ozone and pulsed UV-light. *Journal of Food Science, 72*(9), M391–M396.
7. Bialka, K. L., Demirci, A., & Puri, V. M. (2008). Modeling the inactivation of *Escherichia coli* O157: H7 and *Salmonella enterica* on raspberries and strawberries resulting from exposure to ozone or pulsed UV-light. *Journal of Food Engineering, 85*(3), 444–449.
8. Bigelow, W. D. (1921). The logarithmic nature of thermal death time curves. *The Journal of Infectious Diseases, 29*(5), 528–536.
9. Cerf, O., & Métro, F. (1977). Tailing of survival curves of *Bacillus licheniformis* spores treated with hydrogen peroxide. *Journal of Applied Bacteriology, 42*(3), 405–415.
10. Choi, M. S., Cheigh, C. I., Jeong, E. A., Shin, J. K., & Chung, M. S. (2010). Nonthermal sterilization of *Listeria monocytogenes* in infant foods by intense pulsed-light treatment. *Journal of Food Engineering, 97*(4), 504–509.
11. Coroller, L., Leguérinel, I., Mettler, E., Savy, N., & Mafart, P. (2006). General model, based on two mixed Weibull distributions of bacterial resistance, for describing various shapes of inactivation curves. *Applied and Environmental Microbiology, 72*(10), 6493–6502.
12. Cudemos, E., Izquier, A., Medina-Martinez, M. S., & Gomez-Lopez, V. M. (2013). Effects of shading and growth phase on the microbial inactivation by pulsed light. *Czech Journal of Food Sciences, 31*(2), 189–193.
13. Dhar, R., Basak, & S., Chakraborty, S. (2022). Pasteurization of fruit juices by pulsed light treatment: A review on the microbial safety, enzymatic stability, and kinetic approach to process design. *Comprehensive Reviews in Food Science and Food Safety, 21*(1), 499–540.
14. Dhar, R., & Chakraborty, S. (2020). Influence of voltage and distance on quality attributes of mixed fruit beverage during pulsed light treatment and kinetic modeling. *Journal of Food Process Engineering, 43*(11), e13517.
15. Dunn, J. (1996). Pulsed light and pulsed electric field for foods and eggs. *Poultry Science, 75*(9), 1133–1136.
16. Dunn, J., Bushnell, A., Ott, T., & Clark, W. (1997). Pulsed white light food processing. *Cereal Foods World, 42*(7), 510–515.
17. Dunn, J., Ott, T., & Clark, W. (1995). Pulsed-light treatment of food and packaging. *Food Technology (Chicago), 49*(9), 95–98.
18. Dunn, J. E., Clark, R. W., Asmus, J. F., Pearlman, J. S., Boyer, K., & Hoffamn, G. A. (1989). Methods for preservation of foodstuffs. US Patent 4,871,559. US4871559A – Methods for preservation of foodstuffs – Google Patents.
19. Dunn, J. E., Clark, R. W., Asmus, J. F., Pearlman, J. S., Boyer, K., & Hoffamn GA. (1991). Methods for preservation of foodstuffs. U.S. Patent No. 5,034,235. Washington, DC: U.S. Patent and Trademark Office. US4871559A – Methods for preservation of foodstuffs – Google Patents.
20. Elmnasser, N., Dalgalarrondo, M., Orange, N., Bakhrouf, A., Haertlé, T., Federighi, M., & Chobert, J. M. (2008). Effect of pulsed-light treatment on milk proteins and lipids. *Journal of Agricultural and Food Chemistry, 56*(6), 1984–1991.

21. Farrell, H., Garvey, M., & Rowan, N. (2009). Studies on the inactivation of medically important *Candida* species on agar surfaces using pulsed light. *FEMS Yeast Research, 9*(6), 956–966.
22. Farrell, H., Hayes, J., Laffey, J., & Rowan, N. (2011). Studies on the relationship between pulsed UV light irradiation and the simultaneous occurrence of molecular and cellular damage in clinically-relevant *Candida albicans*. *Journal of Microbiological Methods, 84*(2), 317–326.
23. Farrell, H. P., Garvey, M., Cormican, M., Laffey, J. G., & Rowan, N. J. (2010). Investigation of critical inter-related factors affecting the efficacy of pulsed light for inactivating clinically relevant bacterial pathogens. *Journal of Applied Microbiology, 108*(5), 1494–1508.
24. Ferrario, M., Alzamora, S. M., & Guerrero, S. (2013). Inactivation kinetics of some microorganisms in apple, melon, orange and strawberry juices by high intensity light pulses. *Journal of Food Engineering, 118*(3), 302–311.
25. Ferrario, M., Guerrero, S., & Alzamora, S. M. (2014). Study of pulsed light-induced damage on *Saccharomyces cerevisiae* in apple juice by flow cytometry and transmission electron microscopy. *Food and Bioprocess Technology, 7*(4), 1001–1011.
26. Ferrario, M. I., & Guerrero, S. N. (2018). Inactivation of *Alicyclobacillus acidoterrestris* ATCC 49025 spores in apple juice by pulsed light. Influence of initial contamination and required reduction levels. *Revista Argentina de Microbiología, 50*(1), 3–11.
27. Fine, F., & Gervais, P. (2004). Efficiency of pulsed UV light for microbial decontamination of food powders. *Journal of Food Protection, 67*(4), 787–792.
28. Geeraerd, A. H., Herremans, C. H., & Van Impe, J. F. (2000). Structural model requirements to describe microbial inactivation during a mild heat treatment. *International Journal of Food Microbiology, 59*(3), 185–209.
29. Geeraerd, A. H., Valdramidis, V. P., & Van Impe, J. F. (2005). GInaFiT, a freeware tool to assess non-log-linear microbial survivor curves. *International Journal of Food Microbiology, 102*(1), 95–105.
30. Gomez-Lopez, V. M., Devlieghere, F., Bonduelle, V., & Debevere, J. (2005). Intense light pulses decontamination of minimally processed vegetables and their shelf-life. *International Journal of Food Microbiology, 103*(1), 79–89.
31. Gomez-Lopez, V. M., Ragaert, P., Debevere, J., & Devlieghere, F. (2007). Pulsed light for food decontamination: A review. *Trends in Food Science & Technology, 18*(9), 464–473.
32. Gómez-López, V. M., Koutchma, T., & Linden, K. (2012). Ultraviolet and pulsed light processing of fluid foods. In Cullen, P. J., Tiwari, B. K., & Valdramidis, V. (Eds.), *Novel Thermal and Non-Thermal Technologies for Fluid Foods* (pp. 433–471). Academic Press.
33. Gómez-López, V. M., Devlieghere, F., Bonduelle, V., & Debevere, J. (2005). Factors affecting the inactivation of micro-organisms by intense light pulses. *Journal of Applied Microbiology, 99*(3), 460–470.
34. Guerrero-Beltrán, J. A., & Barbosa-Cánovas, G. V. (2004). Advantages and limitations on processing foods by UV light. *Food Science and Technology International, 10*(3), 137–147.
35. Harguindeguy, M., & Gómez-Camacho, C. E. (2021). Pulsed Light (PL) treatments on almond kernels: *Salmonella enteritidis* inactivation kinetics and infrared thermography insights. *Food and Bioprocess Technology, 14*(12), 2323–2335.

36. Hayes, J., Kirf, D., Garvey, M., & Rowan, N. (2013). Disinfection and toxicological assessments of pulsed UV and pulsed-plasma gas-discharge treated-water containing the waterborne protozoan enteroparasite *Cryptosporidium parvum*. *Journal of Microbiological Methods, 94*(3), 325–337.
37. Hollósy, F. (2002). Effects of ultraviolet radiation on plant cells. *Micron, 33*(2), 179–197.
38. Hoornstra, E., De Jong, G., & Notermans, S. (2002). Preservation of vegetables by light. In Conference *Frontiers in Microbial Fermentation and Preservation* (pp. 75–77). The Netherlands: Wageningen.
39. Hwang, H. J., Seo, J. H., Jeong, C., Cheigh, C. I., & Chung, M. S. (2019). Analysis of bacterial inactivation by intense pulsed light using a double-Weibull survival model. *Innovative Food Science & Emerging Technologies, 56*, 102185.
40. Izquier, A., & Gómez-López, V. M. (2011). Modeling the pulsed light inactivation of microorganisms naturally occurring on vegetable substrates. *Food Microbiology, 28*(6), 1170–1174.
41. John, D., & Ramaswamy, H. S. (2020). Comparison of pulsed light inactivation kinetics and modeling of *Escherichia coli* (ATCC-29055), *Clostridium sporogenes* (ATCC-7955) and *Geobacillus stearothermophilus* (ATCC-10149). *Current Research in Food Science, 3*, 82–91.
42. Jun, S., Irudayaraj, J., Demirci, A., & Geiser, D. (2003). Pulsed UV-light treatment of corn meal for inactivation of *Aspergillus niger* spores. *International Journal of Food Science & Technology, 38*(8), 883–888.
43. Kennedy, D., Cronin, U. P., & Wilkinson, M. G. (2011). Responses of *Escherichia coli, Listeria monocytogenes*, and *Staphylococcus aureus* to simulated food processing treatments, determined using fluorescence-activated cell sorting and plate counting. *Applied and Environmental Microbiology, 77*(13), 4657–4668.
44. Koutchma, T. (2009). Advances in ultraviolet light technology for non-thermal processing of liquid foods. *Food and Bioprocess Technology, 2*(2), 138–155.
45. Kramer, B., & Muranyi, P. (2014). Effect of pulsed light on structural and physiological properties of *Listeria innocua* and *Escherichia coli*. *Journal of Applied Microbiology, 116*(3), 596–611.
46. Woodling, S. E., & Moraru, C. I. (2005). Influence of surface topography on the effectiveness of pulsed light treatment for the inactivation of *Listeria innocua* on stainless-steel surfaces. *Journal of Food Science, 70*(7), m345–m351.
47. Krishnamurthy, K., Demirci, A., & Irudayaraj, J. M. (2007). Inactivation of *Staphylococcus aureus* in milk using flow-through pulsed UV-light treatment system. *Journal of Food Science, 72*(7), M233–M239.
48. Krishnamurthy, K., Tewari, J. C., Irudayaraj, J., & Demirci, A. (2010). Microscopic and spectroscopic evaluation of inactivation of *Staphylococcus aureus* by pulsed UV light and infrared heating. *Food and Bioprocess Technology, 3*(1), 93–104.
49. Humpheson, L., Adams, M. R., Anderson, W. A., & Cole, M. B. (1998). Biphasic thermal inactivation kinetics in *Salmonella enteritidis* PT4. *Applied and Environmental Microbiology, 64*(2), 459–464.
50. Lasagabaster, A., & de Marañón, I. M. (2014). Survival and growth of *Listeria innocua* treated by pulsed light technology: Impact of post-treatment temperature and illumination conditions. *Food Microbiology, 41*, 76–81.

51. Luksiene, Z., Gudelis, V., Buchovec, I., & Raudeliuniene, J. (2007). Advanced high-power pulsed light device to decontaminate food from pathogens: Effects on *Salmonella typhimurium* viability in vitro. *Journal of Applied Microbiology, 103*(5), 1545–1552.
52. Mafart, P., Couvert, O., Gaillard, S., & Leguerinel, I. (2002). On calculating sterility in thermal preservation methods: Application of the Weibull frequency distribution model. *International Journal of Food Microbiology, 72*(1), 107–13.
53. Marquenie, D., Michiels, C. W., Van Impe, J. F., Schrevens, E., & Nicolaï, B. N. (2003). Pulsed white light in combination with UV-C and heat to reduce storage rot of strawberry. *Postharvest Biology and Technology, 28*(3), 455–461.
54. McDonald, K. F., Curry, R. D., Clevenger, T. E., Unklesbay, K., Eisenstark, A., Golden, J., & Morgan, R. D. (2000). A comparison of pulsed and continuous ultraviolet light sources for the decontamination of surfaces. *IEEE Transactions on Plasma Science, 28*(5), 1581–1587.
55. McKellar, R. C., & Lu, X. (Eds.). (2003). Modeling Microbial Responses in Food (1st ed.). CRC Press. https://doi.org/10.1201/9780203503942.
56. Miller, B. M., Sauer, A., & Moraru, C. I. (2012). Inactivation of *Escherichia coli* in milk and concentrated milk using pulsed-light treatment. *Journal of Dairy Science, 95*(10), 5597–5603.
57. Myriam, B. S., Otaki, M., Shinobu, K., & Abdennaceur, H. (2010). Detection of active *Escherichia coli* after irradiation by pulsed UV light using a Qβ phage. *African Journal of Microbiology Research, 4*(11), 1128–1134.
58. Nocker, A., Caspers, M., Esveld-Amanatidou, A., van der Vossen, J., Schuren, F., Montijn, R., & Kort, R. (2011). Multiparameter viability assay for stress profiling applied to the food pathogen *Listeria monocytogenes* F2365. *Applied and Environmental Microbiology, 77*(18), 6433–6440.
59. Otaki, M., Okuda, A., Tajima, K., Iwasaki, T., Kinoshita, S., & Ohgaki, S. (2003). Inactivation differences of microorganisms by low pressure UV and pulsed xenon lamps. *Water Science and Technology, 47*(3), 185–190.
60. Ozer, N. P., & Demirci, A. (2006). Inactivation of *Escherichia coli* O157 and *Listeria monocytogenes* inoculated on raw salmon fillets by pulsed UV-light treatment. *International Journal of Food Science and Technology, 41*(4), 354–360.
61. Peleg, M., & Cole, M. B. (1998). Reinterpretation of microbial survival curves. *Critical Reviews in Food Science, 38*(5), 353–380.
62. Pellicer, J. A., & Gómez-López, V. M. (2017). Pulsed light inactivation of horseradish peroxidase and associated structural changes. *Food Chemistry, 237*, 632–637.
63. Pellicer, J. A., Navarro, P., & Gómez-López, V. M. (2018). Pulsed light inactivation of mushroom polyphenol oxidase: A fluorometric and spectrophotometric study. *Food and Bioprocess Technology, 11*(3), 603–609.
64. Pellicer, J. A., Navarro, P., & Gómez-López, V. M. (2019). Pulsed light inactivation of polygalacturonase. *Food Chemistry, 271*, 109–113.
65. Preetha, P., Pandiselvam, R., Varadharaju, N., Kennedy, Z. J., Balakrishnan, M., & Kothakota, A. (2021). Effect of pulsed light treatment on inactivation kinetics of *Escherichia coli* (MTCC 433) in fruit juices. *Food Control, 121*, 107547.
66. Ringus, D. L., & Moraru, C. I. (2013). Pulsed light inactivation of *Listeria innocua* on food packaging materials of different surface roughness and reflectivity. *Journal of Food Engineering, 114*(3), 331–337.

67. Rowan, N. J., MacGregor, S. J., Anderson, J. G., Fouracre, R. A., McIlvaney, L., & Farish, O. (1999). Pulsed-light inactivation of food-related microorganisms. *Applied and Environmental Microbiology, 65*(3), 1312–1315.
68. Saguy, I. S. (2016). Challenges and opportunities in food engineering: Modeling, virtualization, open innovation and social responsibility. *Journal of Food Engineering, 176*, 2–8.
69. Sauer, A., & Moraru, C. I. (2009). Inactivation of *Escherichia coli* ATCC 25922 and *Escherichia coli* O157: H7 in apple juice and apple cider, using pulsed light treatment. *Journal of Food Protection, 72*(5), 937–944.
70. Schottroff, F., Fröhling, A., Zunabovic-Pichler, M., Krottenthaler, A., Schlüter, O., & Jäger, H. (2018). Sublethal injury and Viable But Non-Culturable (VBNC) state in microorganisms during preservation of food and biological materials by non-thermal processes. *Frontiers in Microbiology, 9*, 1–19.
71. Sharma, R. R., & Demirci, A. (2003). Inactivation of *Escherichia coli* O157 on inoculated alfalfa seeds with pulsed ultraviolet light and response surface modeling. *Journal of Food Science, 68*(4), 1448–1453.
72. Shuwaish, A., Figueroa, J. E., & Silva, J. L. (2000). Pulsed light treated prepackaged catfish fillets. In *IFT Annual Meeting* (pp. 10–14). IFT, USA.
73. Smith, W. L., Lagunas-Solar, M. C., & Cullor, J. S. (2002). Use of pulsed ultraviolet laser light for the cold pasteurization of bovine milk. *Journal of Food Protection, 65*(9), 1480–1482.
74. Turtoi, M., & Nicolau, A. (2007). Intense light pulse treatment as alternative method for mold spores destruction on paper–polyethylene packaging material. *Journal of Food Engineering, 83*(1), 47–53.
75. Uesugi, A. R., Woodling, S. E., & Moraru, C. I. (2007). Inactivation kinetics and factors of variability in the pulsed light treatment of *Listeria innocua* cells. *Journal of Food Protection, 70*(11), 2518–2525.
76. Valdramidis, V. P., Taoukis, P. S., Stoforos, N. G., & Van Impe, J. F. M. (2012). Modeling the kinetics of microbial and quality attributes of fluid food during novel thermal and non-thermal processes. In P. J. Cullen, B. K. Tiwari, & V. Valdramidis (Eds.), *Novel Thermal and Non-Thermal Technologies for Fluid Foods* (pp. 433–471). Academic Press.
77. van Boekel, M. A. J. S. (2002). On the use of the Weibull model to describe thermal inactivation of microbial vegetative cells. *International Journal of Food Microbiology, 74*(1–2), 139–159.
78. Walker, S. J., & Jones, J. E. (1993). Protocols for data generation for predictive modeling. *Journal of Industrial Microbiology, 12*(3), 273–276.
79. Food and Drug Administration (2015). *Code of Federal Regulations Title 21, Sec. 179.41*. Pulsed light for treatment of food: 21 CFR170.41.

PART IV

LIGHT-EMITTING DIODES: ADVANCES IN FOOD PROCESSING

CHAPTER 14

INTRODUCTION TO LIGHT-EMITTING DIODES: PRINCIPLES, FUNDAMENTALS, ADVANTAGES, LIMITATIONS, AND APPLICATIONS

RADHIKA THEAGARAJAN, LAVANYA MALUR NARAYANASWAMY, and PINTU CHOUDHARY

ABSTRACT

Light-emitting Diode (LED) is a solid-state lighting device that emits light with a wavelength of 200 to 780 nm when electricity flows into it. In recent years, LED has emerged as a novel non-thermal processing technique due to its unique properties such as lesser radiant heat emission, high emission of monochromatic light, luminous, electrical, and photon efficiency, durability, mechanical toughness, and flexibility. It has applications in several fields; it is being used in agriculture and horticulture for plant growth, water disinfection, cold storage, and in the preservation of food. Recently, the application of LED sources during post-harvest stages has shown significant changes in nutritional quality, ripening of fruits, and also reduced fungal infection. LED in combination with UV is more effective in the inactivation of pathogenic microorganisms in food such as *Escherichia coli, Pseudomonas fluorescens,* and *Salmonella spp.* The commercial application of LED is not only limited to food; it also has wide applications including air purification, curing, and medical applications. LED has low penetration depth and high-intensity usage is the main challenge in optimal treatment that can be overcome by optimizing studies by selecting a different range of light wavelengths. In this chapter, we will discuss the fundamentals of LED, advantages, limitations,

Nonthermal Light-Based Technologies in Food Processing.
C. K. Sunil, Preeti Birwal, Megh R. Goyal, and R. Mahendran (Eds.)

and applications of LED in different areas. Nevertheless, LED technology is a worthy alternative to other thermal and chemical preservation techniques.

14.1 INTRODUCTION

A light-emitting diode, commonly known as an LED, is formed when the p-n junction is electrically biased in the forward direction, as in the conventional LED circuit, emitting incoherent narrow-spectrum light. Therefore, they are a type of luminous gadget that works on the electroluminescence principle (converting electrical current into light). Additionally, an LED is a very small area light source, and the optics on the chip are frequently used to change the emission pattern and promote reflection. They are extensively deployed as modest indicator lights in electronic devices, and they are increasingly finding use in highly powerful implementations including flashlights and area lighting. Depending on the composition and quality of the semiconducting material used, the light produced may be infrared, visible, or ultraviolet. H.J. Round discovered electroluminescence in 1907 while working on a post-epoxy lens/case. For decades, however, no practical use of the discovery was made. Nick Holonyak, Jr., while working at General Electric Company, invented the first functional LED in 1962. The original LEDs were red and became commercially accessible in the 1960s. They were extensively used as incandescent indication replacements and in seven-segment displays [25, 39].

Radio waves have relatively long wavelengths to gamma rays with relatively short wavelengths. Furthermore, the visible spectrum represents a minor percentage of the entire electromagnetic spectrum [13].

Moreover, considering its reduced power consumption, high light productivity, and minimum heat output, LED lights have since been incorporated into agricultural production and other applications, making them more eco-friendly. Likewise, LEDs provide monochromatic sources of light, with a wavelength in the middle of the spectral region that induces morphogenesis among plants. LED light produces less heat, illuminating plants without burning them, significantly increasing space utilization efficiency. For these reasons, LED lights have been widely used in plant-based facility cultivation, factories, and tissue culture [22].

Usually, for a variety of food applications, novel technologies such as increased pressure or high hydrostatic processing, ultrasound, cold plasma processing, and pulsed electric field have been researched as substitutions for heat pasteurization [14]. However, advanced functional materials are driving the rapid progress of several modern technologies. For instance, luminous semiconductors, transition-metal doped phosphors, or Lanthanides have been primarily employed in the area of lighting and display, including cathode ray screens, fluorescent lamps, current LED light sources, and backlights [46].

Recent years have witnessed a rise in the use of LED illumination for microbial inactivation among preservation techniques in the agricultural and medical sectors. In recent years, there has been an increase in interest in blue LEDs and their potential application in food preservation as a potential substitute for ultraviolet light. Additionally, several researchers have investigated the effectiveness of mixing blue wavelengths from LEDs with antimicrobials to inactivate foodborne germs. Previous studies have shown that blue LEDs have antimicrobial properties; however, more information on the antibacterial processes is still required [16]. Additionally, Kusuma et al. [19] explain that LEDs have considerably enhanced the conversion of electric energy to photons, but that this process is quickly reaching a physical limit. When all supplied energy is transformed into photons of energy, the maximum theoretical efficiency emerges. Phosphor-converted blue LEDs can be up to 93% efficient.

When employing portable LED-based photometric devices, due to certain intrinsic and physical limits of the light sources used in these devices, it can be difficult to obtain absorbance across a wide range of wavelengths in some situations. Additionally, there is a growing interest in developing portable and easy-to-use absorption devices using LEDs as a light source because of the remarkable *in situ* determination and other advantages of portable devices. Therefore, these devices offer excellent stability, low cost, reduced energy consumption, monochromaticity, high brightness, durability, and a compact size, which allows them to be used in portable analytical systems [42].

In another interesting study conducted by Aurum & Nguyen [2], LED light alone showed less significant antibacterial effects in the study; however, photoactivated curcumin demonstrated substantial bactericidal properties against *Escherichia coli*. Photodynamic microbial inactivation using visible LED light has recently emerged as a potential option for decontaminating fruits and vegetables. This is because the activation of endogenous or exogenous photosensitizers under visible spectrum LED lighting produces reactive oxygen species that could have lethal effects on bacterial cells. Furthermore, LEDs, unlike traditional light sources, are energy-efficient, cost-effective, long-lasting, and produce very little heat [17].

14.2 WORKING PRINCIPLE OF LED

LEDs are two-terminal semiconductor devices that produce light when a voltage is applied across them. The active (junction) layer is typically where positively charged carriers (holes) from the p-type layer, which is primarily composed of "holes," are driven. There, they interact with electrons from the n-type layer that are being driven in the opposite direction by the same voltage (which mainly contains "electrons"). When an electron and a hole collide, the energy created by the recombination of the electron and hole is transported away by the emission of a photon. The band gap of the material determines both the energy of the released photon and the wavelength of the light emitted. The structure of a standard LED is shown in Figure 14.1.

14.2.1 COMPONENTS OF LED

Additionally, distinct semiconductor materials are those which potentially emit at various wavelengths and colors; the band gap of a semiconductor determines the emission wavelengths. Materials for compound semiconductors, also referred to as III-V materials (They are composed of elements from groups III and V of the periodic table), have a straight band gap. Modern LED technology is built on these III-V materials. In practical systems, a very thin layer of a material with a narrow band gap is often present at the semiconductor junction. This tiny quantum well accelerates and enhances the electron and hole recombination process [14].

The study of semiconductors is explored in the branch of physics known as semiconductor theory. Materials can be categorized using the energy difference between their valence and conduction bands. In conductors, there is no energy gap because the conduction and valence bands overlap. Electrons demand a minimal amount of energy to transition from the valence band to the conduction band and, as a result, become charge carriers. This energy might come from the thermal energy of the surroundings. Semiconductors are materials having conductivities that fall between those of a conductor and an insulator. They are effective media for controlling electrical current because they can conduct electricity owing to the presence of impurities or temperature effects; Furthermore, these materials' conductance fluctuates in accordance with the current, voltage, and light-irradiation intensity. There are two types of semiconductor materials: intrinsic and extrinsic, which exist in pure form and have intentional impurities introduced, respectively [44].

14.2.2 N- AND P-TYPE SEMICONDUCTORS

The concentrations of electron and hole carriers in extrinsic semiconductors that are estimated by the types of additional impurities are classified as n- and p-types. In n-type semiconductors, the donor impurity atoms have more valence electrons than the semiconductor atoms they replace. These excess valence electrons can be transferred to the conduction band and cause the electron to exceed the hole in these materials. In contrast, holes make up the majority of charge carriers in p-type semiconductors while electrons compensate the minority carriers.

Consider the element silicon (Si). With four valence electrons, pure Si is an inherent semiconductor. Phosphorus (P), antimony (Sb), bismuth (Bi), and arsenic (As) are examples of Group V elements with five valence electrons that can dope pure Si to create n-type Si. The system is that one additional electron in the Group V element is left over after the production of four covalent bonds due to the sharing of four electrons between Group V and Group IV elements. By gathering thermal energy from the environment, a charge carrier can be generated by stimulating an electron to the conduction band. The total electron content of the material is the sum of the electrons generated thermally and the electrons delivered by donor atoms. The number of holes in the same material, on the other hand, remains constant at the same level as the number of thermally produced electrons. This indicates that the majority of charge carriers are electrons, while the tiny fractions are holes. As a result, n-type semiconductors are given their name [44].

14.2.3 P–N JUNCTION

Figure 14.1 represents the carrier transfer in a p-n junction. Free electrons in the n-type semiconductor diffuse towards the heterojunction to the p-type semiconductor when a p-type semiconductor contacts an n-type semiconductor because the concentration of electrons in the n-type region is significantly larger than in the p-type region. Electrons spread across the junction, leaving the n-type semiconductor with a band of fixed positive charge. Similarly, near the junction, a free hole in the p-type semiconductor moves into the n-type semiconductor, leaving a negative charge layer in the p-type semiconductor. This charge in the gap creates an electrostatic field that prevents further diffusion over the connection. Therefore, when the diffusion of minority carriers back across the junction, due to the integrated electrostatic field, balances the diffusion of majority carriers across the junction, equilibrium is achieved [44].

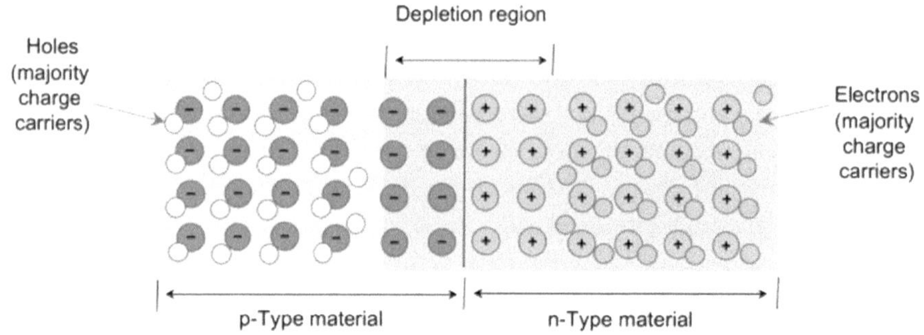

FIGURE 14.1 Schematic representation of carrier transfer in p–n junction.
Source: Reprinted with permission from Ref. [44]. Copyright 2019. Elsevier.

14.2.4 OPERATING SYSTEM

To make LEDs electroluminescent, battery connections must be made properly so that the positive terminal is connected to the p-type region and the negative terminal is connected to the n-type region. The diode will enter reverse-bias mode and turn off if the terminals are connected improperly. In the forward-bias mode, the free holes from the p-region and the free electrons from the n-region begin to flow toward the p–n junction. The junction between the p-type and n-type areas is nonconductive, despite the fact that both are conducting. The depletion region is located around this intersection. Between the p-type and n-type areas, in the depletion region (also called the active region), positively charged holes and negatively charged electrons attract and annihilate one another through recombination.

In contrast to the negative charge, which repels electrons from the n-type region to the junction, the positive charge drives electrons from the n-type area into the p-type region and repels holes from the p-type region to the junction, while the negative charge pushes electrons away from the n-type region and toward the junction. Additional electrons might transfer from the n-type to the p-type region as a result of the reduction in the potential barrier across the junction and the reduction of the depletion area. Positively charged holes in the n-type region are left behind as electrons migrate from the n-type to the p-type region. These holes mix with the free electrons that are present in the local region to emit radiation. The electrons that routed through the close barrier and proceeded to the p-type area also recombined with the holes in a radiative method. The majority of recombination occurs in the depletion zone. The LED prevents the emission of photons when it is put into a reverse-bias state. This is due to the negative terminal connected to the p-type region resisting electrons away from the junction towards the n-type region and attracting holes from the junction towards the p-type region, while the positive terminal connected to the n-type region starts to attract all electrons away from the junction. As a result, the depletion region expands and the barrier potential across the junction increases [25]. White LEDs were widely produced in huge quantities and have since become the most effective artificial light source, revolutionizing every aspect of illumination [17].

In general theory, a big lens with a short focal length positioned extremely close to the LED might gather the generated luminescence efficiently. However, such lenses, particularly those composed of fused silica, are typically heavy and expensive. Ball lenses, on the other hand, are suitable in this situation due to their extremely small effective focal length and low cost. Ball lenses are also available in a variety of sizes. To focus light onto the capillary, we discovered that a system of two ball lenses works well: the first, larger ball lens collects light from the LED and focuses it onto the smaller, focusing ball lens which is positioned close to the capillary [40].

Although an LED's basic structure is similar to that of a p–n junction diode, the LED's manufacturing is substantially different from that of a conventional diode. Three layers of semiconductor die are placed on the substrate to make up the semiconductor die. The active region is in the middle, the n-type zone is at the bottom, and the p-type region is at the top. The anvil that links the die to the anode is rested on the semiconductor die. A reflector cup is fitted beneath the die to direct all available light upward. A hemispherical-shaped clear case constructed of durable plastic epoxy resin surrounds the entire system, acting as a shield to defend the LED from vibrations or shocks, and any other influences from the environment. The dome-shaped lens on the top part of the casing focuses the light that is reflected out from the reflector cup, resulting in maximum brightness. Some LEDs have a flat-surfaced top and a cylindrical or rectangular enclosure. Two electrodes may be seen emerging from the bottom of the shell. The LED's cathode and anode terminals are represented by these electrodes. LEDs will only work if they are forward biased; therefore, it is critical to properly detect the electrodes when placing LEDs in a circuit. As a result, the anodes are usually observed to be slightly extended than the cathode terminal in order to distinguish them. On the cathode terminal, a notch or a flat area is sometimes included. A thermal absorber is also provided for heat dissipation in high-power LEDs [25].

14.3 FUNDAMENTALS OF LED

LEDs are a unique type of solid-state semiconductor diode that produces light in response to an applied current. The LED chip is made of two processed materials, p-type semiconductors, and n-type semiconductors, which come into direct contact to construct the p-n junction. The current usually flows from the p-side (anode) to the n-side (cathode). Electrons and holes enter the junction from electrodes of various voltages. The energy gap of a semiconductor determines the color (wavelength) of light, and the components employed in LEDs have energies that are comparable to near-ultraviolet, visible, or near-infrared light. Unlike incandescent and fluorescent bulbs, LEDs have a tendency to polarize. LED chips possess a negative and positive pole and could only emit light when a forward current is passed through them. LED chips are a viable substitute for a range of lighting applications due to their low voltage damage [8].

LED is a solid-state semiconductor device with a single junction. The electrical characteristics of a semiconductor can be altered by purposely introducing impurities (a process known as "doping") into its chemical structure. By doping the semiconductor with various materials at different points, a junction can be created between two sections of material that have been doped separately. Electrical current flows freely in one direction across the junction, similar to a typical diode, but not in the other direction. The LED is considered forward biased when current is flowing. LEDs emit light when they are forward biased. The type of semiconductor material and impurities used to create the junction determine the color of the emitted light. Generally, peak emission wavelengths for LEDs range from 250 nm to 1,000 nm and beyond, where the former wavelength is UVC, and the latter is infrared.

Although there are laser diodes, they are a whole different type of technology than an LED, which is not the same thing as a laser. Moreover, LEDs are not harmful in general and have a non-coherent narrow spectrum. However, caution should be exercised when using LEDs in the UVB and UVC ranges, as these powerful wavelengths can be damaging to the skin and eyes. Even though the LED was developed in the 1920s, it was not until the early 1960s that a workable visible-light version (red) was developed. These early devices had a relatively low power output and could only be used as indicator lamps. During the 1970s, shorter wavelengths (yellow, orange, and green) were introduced to the commercial sector. In 1993, the first practical blue LED was produced, followed by the first white LED created in 1996 by phosphor coating a blue LED. The advent of the high-power device (1 W or higher) in 1999 was the most recent important advancement in LED technology. This has brought a revolution in the lighting industry, including agricultural illumination [5].

14.3.1 DEVICE CONFIGURATIONS

The fundamental component of an LED is an emitting layer positioned between metal electrodes and transparent conducting metal oxide. This indicates that there is only one functional layer present because it is a conventional single-layered LED arrangement. Despite the compactness of the devices, it is hard to balance the electron and hole in single-layered LEDs in order to achieve high efficiency. In some instances, multilayered LEDs employ layers that obstruct holes or electrons because performing so can mitigate the efficacy of the device by restricting carrier leakage of the opposite charge-carrier type. Multilayered LEDs offer a suitable base for device optimization with a greater range of possibilities, irrespective of the complexity of device fabrication. The majority of the LEDs reported are bottom-emitting devices, which emit light from the substrate. Conversely, for application purposes, the light should be emanated from the final layer, which is known as a top-emitting LED, generally used in the manufacture of back-plane electronics like Si active-matrix. However, rather than the conventional opaque metal cathode, it features a clear top electrode [44].

In an LED, light is generated by the recombination of electrons and holes in the corresponding active region. When the electrons are introduced into the diode junction from the n-side and dispersed into it, the electron distribution in the active region is not uniform. Moreover, ignoring this impact and accepting that light is created consistently across the active zone and that the extent of light produced per unit volume in front- and edge-emitting LEDs is the same. The two constructions are not compared on the basis of identical electricity usage. The length of the edge emitter, which is frequently longer than the front emitter, increases the amount of electrical power consumed because the active portions of both forms of diodes are similar in breadth and thickness but not in length. As a result, they may be less efficient in terms of overall energy conversion [45].

14.3.2 LED STRUCTURES AND THEIR THERMAL EQUIVALENT MODELS

The LED device is positioned atop a heat sink; consequently, the heat produced by the LED wafer can be transmitted to the ambient environment via the heat sink. While functioning, a collection of LED systems can be installed on the same heat sink. As a result, an electro-thermal model for LED devices with one or more LED devices must be developed in order to achieve optimal thermal design. The basic equation of heat transport in steady-state conditions is:

$$P_{heat} = h_{\theta} A_{\theta} \Delta T \quad (1)$$

where; P_{heat} is the power dissipated as heat; *h* is the heat transfer coefficient; *A* is the cross-sectional area of the heat transfer; and *T* is the temperature difference between two points or regions.

14.4 ADVANTAGES AND LIMITATIONS OF LED

14.4.1 ADVANTAGES

In recent years, the application of LED is considered to be a novel innovation; in fact, their noticeable and remarkable qualities have helped them to outperform other light sources. They possess a number of inherent advantages and are intriguing candidates for future low-cost illumination. In battery-powered or energy-saving devices:

- Additional reflectors are occasionally required to gather light from incandescent and fluorescent sources and direct it in a usable fashion, whereas LEDs have a solid package that may be tailored to focus their light. In contrast to incandescent bulbs, which turn yellow as the current running through them is reduced, LEDs do not change their color tint as the current passing through them is reduced. Unlike fluorescent lamps, which burn out more quickly when cycled often or high-intensity discharge (HID) lamps, which take a long time to restart, LEDs are excellent for use in applications that demand frequent on–off cycling.
- LEDs are extremely durable electronics. There is no gentle glass wrap to disrupt, and unlike incandescent, fluorescent, and arc-discharge lamps, their repetitive on/off cycles do not have an impact on durability. Mercury is present in fluorescent bulbs, which might cause difficulties if they are accidentally broken.

- LEDs have various advantages over traditional light sources: they are energy-efficient, cost-effective, long-lasting, and generate less heat. Furthermore, when LEDs photoactivate curcumin, they generate reactive oxygen species (ROS) such as hydroxyl radicals, superoxide anions, and others, which cause bacterial cell death [24].
- LEDs' small size provides unparalleled design versatility in lighting fixtures. LEDs can be contained in low-profile fixtures that use less space than huge, bulky fixtures that store bulbs, reflectors, and ballasts [5].
- Light produced by the LEDs is more efficient than incandescent bulbs, which is advantageous.
- Moreover, LEDs do not "burn out" like conventional bulbs; estimating their endurance is comparatively subjective. Instead, they gradually fade over time. The most frequently mentioned "lifetime" characteristic for LEDs is the time it requires for the LED to decline to 70% of its initial brightness when operated at its nominal operating current and temperature. The average lifetime for high-power LEDs is 50,000 hours. This is a 50-fold improvement over standard incandescent lighting and preferable over fluorescent or HID bulbs by two to three times.
- One of the most significant advantages of LED bulbs is their extended lifespan, which can reach 50,000 hours on average if they have efficient heat dissipation and are utilized under appropriate power input and temperature settings.
- Other advantages of a long lifespan include less environmental impact and lower light replacement and maintenance expenses. In 2011, LED bulbs had 5 and 50 times longer lifespan than compact fluorescent and incandescent lamps [27].
- Traditional lighting techniques need the use of color filters to create light of a specific color, which makes LEDs more efficient and reduces initial expenses.
- Using an LED light source, several investigations have achieved photodynamic microbial inactivation of several harmful bacteria. LEDs can photoactivate curcumin, which is a natural photosensitizer and could be effective in the cleaning of food surfaces, at a wavelength of 425 nm.

14.4.2 LIMITATIONS

- Another disadvantage is the need to replace LEDs after they fail.

- If the working temperature of an LED exceeds 150?C, device malfunctions are almost certain to arise because when the ambient temperature is high, pushing the LED firmly can result in the LED module overheating.
- LEDs are usually combined into a module to achieve a high lumen output, and each unit comprises many LEDs in a unique arrangement. Because changing a single LED in a module is not possible, therefore, replacement of the complete module is necessary if even one LED in a module expires.
- To ensure the LED's long life, it requires a sufficient heat sink facility. LEDs do not altogether fail, but they do exhibit a substantial decrease in productivity with time.
- Various investigations have been undertaken to better comprehend how LEDs degrade optically. The degradation is linked to thermal or heating processes as well as optical exposure or surface damage. Material deterioration occurs during phosphor manufacture and continues throughout operation. Different degradation mechanisms lead to different rates of light output decline, hence different LED degradation values were found in research [23].

14.5 APPLICATIONS OF LED

14.5.1 AGRICULTURE

The fundamental source of energy that keeps life on Earth going is solar radiation. Solar radiation has a broad spectral distribution with wavelengths spanning from 300 to 1,000 nm. Plants, on the other hand, only get 50% of the radiant energy through photosynthetic active radiation, which spans the wavelength range of 400 to 700 nm. Plants have also evolved complex mechanisms for converting various solar light wavebands into specialized chemical signals that regulate a variety of complex growth and development processes.

Moreover, light has a significant impact on plant growth and shape. Natural light, on the other hand, is frequently insufficient for optimal plant growth in nature due to the influence of climatic conditions. In controlled facility agriculture, an artificial light source is provided to achieve high yields and quality products. Fluorescent and incandescent lamps were employed as light sources until recently, but they have some disadvantages, such as low luminous efficiency, high power consumption, and an incomplete spectrum distribution.

14.5.1.1 BENEFITS OF LEDS AS GROW LIGHTS

In addition, photomorphogenesis, photoperiodism, and phototropism are all regulated by the ambient light conditions, in addition to the photosynthetic process, which requires significant energy. The definition of photomorphogenesis is light-mediated plant development that includes cell, tissue, and organ differentiation and is dependent on a far-red wavelength between 730 and 735 nm. However, photoperiodism is defined as a plant's ability to detect changes in the photoperiod, or the lengths of day and night, and to respond accordingly. Phototropism is the migration of plants' development in the path of their light. Phototropic reactions are activated by light with wavelengths in the range of 400 and 500 nm [12].

14.5.1.2 DEVICE FOR MONITORING CROP PERFORMANCE

In general, extents of CO_2 exchange and chlorophyll fluorescence can be used to evaluate the photosynthetic interactions towards environmental circumstances. Chlorophyll fluorescence, on the other hand, provides a number of advantages over CO_2 exchange: It's less expensive, the equipment doesn't need to be calibrated on a regular basis, and it reacts instantly to changes in light conditions [12].

14.5.1.3 PHYSICAL MANAGEMENT OF INSECT PESTS BY USING LIGHT SOURCES

Based on the most recent research being carried out, improvements in the use of light for pest control are also anticipated by the National Agricultural Research Organization. It is ensured that further development of LED equipment with varied wavelengths can now be developed, and new agricultural technology based on LED is beginning to gain traction. As a result of these technological advancements in LED, improvements in the use of light for pest control are envisaged [32].

14.5.2 HORTICULTURE

Utilization of LED light systems allow for direct intervention in plant growth and metabolism; yet it necessitates extensive research because the impacts of the quality of light are complex. Furthermore, the heterogeneity in research outcomes could be attributable to a variety of experimental circumstances and plant kinds, making it impossible to compare those [28]. Likewise, plant reactions to monochromatic light are influenced by the background spectrum; hence, when combined with other distinctive light wavelengths, the effect of certain light wavelengths fluctuates. As a result, lighting solutions designed for closed environment chambers do not necessarily achieve the same outcomes in greenhouses. In fact, low levels of natural light from a greenhouse, particularly variations in spectrum, photoperiod, and amount, have a significant physiological impact on plants. As a result, greenhouse illumination and closed-environment cultivation system lighting should be tuned individually. The majority of LED lighting in greenhouses uses wavelength combinations with high photon efficiency, such as red and blue. Especially, in secured horticulture, combinations of red/white or red/blue are also ideal for use in plant factories. Green or white light, which contains a substantial quantity of green light, has significant physiological effects on plants and is also helpful for enhancing their visual impression in enclosed spaces. Protected horticulture combinations of red/blue or red/white are also ideal for use in plant factories. Green or white light, which contains a significant quantity of green wavelengths, has a beneficial physiological effect on plants and is also beneficial for improving the visual appearance of plants in enclosed spaces. Furthermore, cool white LEDs include an increased amount of blue light, which meets the actual requirement for regular plant growth and development. Plant growth is also aided by dynamic illumination that can switch from cool to warm white LEDs at various phases of development. Short and powerful hypocotyls are produced by a high concentration of blue light in the spectrum of cool white LEDs during the early stages of growth. Warm white LEDs could eventually replace cold white LEDs in subsequent stages of development. This spectrum of light would promote leaf growth; hence cool white LEDs should be deployed further in the last stages of growth to prevent excessive stem elongation. The dynamic lighting settings imitate natural lighting patterns that are attributed to biomimicry's unique technology.

Horticulture with LED lighting biomimicry is a rapidly expanding area that tries to integrate biological systems' intrinsic structures and functions into a range of applications. It has been demonstrated that dynamic management of supplemental LED illumination, by increasing the intensity, can preserve crop quality and yield while using 20% less electricity [33, 38]. However, while adopting the principles of natural lighting, modeling artificial dynamic lighting factors necessitates a broad understanding.

A detailed study of the plant's illumination requirements during different developmental stages (as part of a community in its environment) may allow the development of a regulated spectrum LED system that would be significantly beneficial for plant growth, compared to white LED treatment. Once the horticulture LED market has attained development, the multi-wavelength hybrid package LEDs, as active and customizable light sources providing a broad light spectrum, may be effective for the creation of dynamic LED lighting systems. Plant responses to the light in which they are grown have a complex impact on their growth and development. Certain photoreceptors, including cryptochromes, phytochromes, and phototropins, activate a signaling cascade that changes the expression of numerous genes in response to light quality and amount. While specific plant responses to a spectrum can be predicted in some cases based on previous research, the total plant interaction is difficult to anticipate due to the complexity of the responses obtained [15]. To examine particular plant responses to light quality, narrow-band LED spectra are utilized. In most illumination spectra, the most fundamental element is red light, and it is alone sufficient for regular plant development and photosynthesis.

Goins et al. [10] conducted a study on the effect of various red-light wavelengths (660, 670, 680, and 690 nm) on the growth and photosynthesis of the plant lettuce. Also, when the red LED light wavelength notably increased from 660 to 690 nm, the biomass production was determined to increase. Furthermore, the authors claim that red LEDs with wavelengths of 640 nm or 660 nm can be used to grow not only lettuce but other green vegetables as well. Whether used alone, or in association with fluorescent lights, or in the presence of natural light, red LED light had no discernible impact on growth and development, although it did activate the antioxidant system.

In comparison to blue or green LED wavelengths, 660 nm LED light was found to enhance anthocyanin accumulation in red leaf cabbages when used as the sole light source in a controlled environment. Supplementing cool white, fluorescent lamps with 658 nm red light resulted in a 6% increase in phenolics in baby lettuce leaves [20]. Thus, red-leaf cabbages exhibited better increases in lutein and glucosinolate sinigrin accumulation when they were subjected to short-term pre-harvest treatment with red LEDs (640 nm) under a controlled environment. The results of a series of pre-harvest experiments in greenhouses using 638 nm LED light (supplemental for High-pressure sodium illumination and natural light) show increased antioxidant capacity, increased phenolic compound and alpha-tocopherol content, and reduced nitrate content in various lettuces and other vegetables [30].

14.5.3 PRESERVATION OF FOODS

Broad research is being conducted on LED-based photodynamic inactivation in the food processing industry [47]. LED technology is safe, energy-efficient, ecologically beneficial, and long-lasting. The generation of reactive oxygen species (ROS) by intracellular photosensitizer excitation is related to the inactivation of germs by LED illumination. ROS may damage cellular components including DNA, lipids, and proteins, contributing to the death of bacteria [3].

LED treatments for liquid or solid food systems, such as red-pepper powder, fresh-cut fruits, fruit juice, dairy products, and chicken items, offer a lot of potential as alternatives to existing preservation procedures. However, just a few research studies have looked into how LEDs affect the inactivation of bacteria in food. According to Gunther et al. [11], chicken skin surfaces subjected to 405 nm LEDs had lower concentrations of *C. jejuni* and *Campylobacter coli* populations at 10?C for 7.5 hours. Additionally, Ghate et al. [9] discovered that 460 nm LEDs reduced the growth of *Salmonella* spp. in orange juice by 3.3–4.8 $\log_{10}$ CFU/mL following 14 hours of storage at 4?C, 12?C, and 20?C. Furthermore, 460 nm LEDs were discovered to be successful at sustaining color fidelity at lower temperatures. Similarly, another study found that LEDs with 405, 433, and 460 nm ranges of wavelengths for 90 minutes are efficient at reducing *E. coli* populations in skimmed milk at temperatures of 5?C, 10?C, and 15?C [43]. Physicochemical parameters of skim milk, such as carbohydrate, protein, moisture, fat, acidity, color, pH, and viscosity, were also unaffected by 405 nm LEDs ($p > 0.05$). Therefore, a number of research studies have observed the use of LEDs to inactivate harmful microorganisms on solid foods.

14.5.3.1 DELAY OF SENESCENCE IN FOODS THROUGH LEDS

Senescence is a genetically controlled process that prolongs the life of plants by migrating nutrients and macromolecules from dying plant to developing or new plant tissue, resulting in the poor quality of harvested plants. Light has been shown to delay senescence in separated flowers, leaves, and stems [34]. Yet, putting an effectual treatment in place is difficult because the appropriate supply of light is dependent on spectral composition, intensity, and duration or photoperiod factors [29]. Prolonged exposure to low-temperature light may cause photo-oxidative damage and decrease postharvest quality [7].

14.5.3.2 EXPEDITE OR DELAYING FRUIT RIPENING THROUGH LEDS

It is necessary to slow the rate of ripening for food that is transported long distances to avoid the fruits becoming overripe when they reach their destination. Different types of fruits react differently to light. Despite being stored in the dark, tomatoes can be pre-treated with blue light to extend their ripening period. When compared to storing mature green tomatoes in the dark or subjecting them to red light (650 to 660 nm) for a comparable period of time, mature green tomatoes treated with blue light (440 to 450 nm) for 7 days showed a slower rate of color change from green to red. Furthermore, blue-LED-treated tomatoes were firmer and darker than red-LED-treated tomatoes. Likewise, blue-light irradiation slowed the buildup of lycopene. As a result, blue light pre-treatment may be a viable option for reducing tomato ripening time and increasing their post-harvest commercial value. LEDs are used to prevent fungus deterioration. Fruit fungus infection is the most common cause of postharvest loss.

The impact of LED light in inhibiting fungal infections in citrus fruits has been studied in a number of ways. In general, moderately intense blue LED light was considered to be efficient in the prevention of fungal infection. Depending on the fungal species, blue light treatments at 40 mol m^2 s^1 for 5 to 7 days were able to significantly reduce the mycelial growth, soft rot area, and sporulation of several fungi (*Penicillium digitatum, Penicillium italicum*, and *Phomopsis citri*) on the exterior of fruits compared to darkness and white LED light [1]. Blue LEDs improved the expression of phospholipase A2 in citrus fruits, according to real-time qRT-PCR research. PLA2 hydrolyzes phosphatidylcholine to produce lysophosphatidylcholine, which is implicated in regulating citrus fruit tolerance to fungal infection and development. Red LEDs, on the other hand, promoted fungal infection by suppressing the phospholipase D expression, a protein that also acts as an antifungal defense. In addition, blue LEDs increased the amount of octanal in the flavedo of "Fallglo" sweet oranges and tangerines. At dosages of 0.5 mM, octanal inhibited fungal growth, sporulation, and soft rot *in vitro*. In general, the whole mechanism of fungal suppression by reduced fungal polygalacturonase activity and the application of blue LED light could both be contributing factors and overexpression of PLA2 and synthesis of octanal in the flavedo [7].

14.5.3.3 EVALUATION OF LEDS IN POSTHARVEST PRESERVATION

Only in recent years has the role of light in postharvest applications received more attention. Due to recent advancements, significant research has not been conducted on LEDs for postharvest applications in LED technology up to this point. Certain plant parts, such as fruits and edible flowers like broccoli, strawberries, and citrus fruits, have been successfully preserved and enhanced using LEDs. Several studies have even found a connection between LED light exposure and gene expression in response to bio-molecular processes. For example, blue LEDs increase flavonoid and anthocyanin content by up-regulating gene expression associated with the phenylpropanoid pathway. In Satsuma mandarin, red light increased gene expression related to carotenoid metabolism. Other products could be examined in a similar manner to confirm the connection between monochromatic light and biological pathways [11].

The impact of different LED light irradiation on strawberry post-harvest quality was investigated. The chemical analysis, which included total soluble solids (TSS), pH, and titratable acidity, proceeded smoothly. The percentage of weight loss, firmness, and color changes on the surface of strawberries are all part of the physical analysis, which can help determine the quality of the strawberries. These factors are essential in assessing consumer acceptance of fresh produce consumption. The most effective treatment for extending the shelf life and maintaining the postharvest quality of strawberries is to expose them to blue LED light during storage. Microorganisms such as *E. coli*, *Salmonella*, and mold can be inhibited and delayed by using blue LED light. Strawberry fruits could be preserved by using different LEDs in combination with chilling temperatures to reduce the risk of foodborne diseases [30].

14.5.4 LED IN FOOD SAFETY

As a result, efficient and safer approaches for controlling harmful microorganisms are required. Recently, LEDs have drawn a rational interest due to their capacity to keep food secured and sustainable; this has been proved as a revolutionary method for inactivation of bacteria and other microbes as well. When compared to traditional preservation technology, LEDs have various advantages (e.g., chemical or thermal treatment). Due to the lower infrared light emissions that transfer radiant heat, LEDs reduce unwanted effects such as sensory degeneration in food [16].

In agriculture and the food business, solutions for artificial lighting that use light energy of various wavelengths have been applied to sterilize food and water, as well as to increase plant health and growth. Reports state that several microorganisms have been inactivated using conventional methods such as pulsed light produced by xenon lamps or UV light emitted by mercury vapor lamps. The likelihood of mercury residual contamination and the equipment's limited lifespan are both disadvantages of these procedures [36].

Applications of LEDs in Food Safety: UV LEDs have been used to inactivate fungal spoilage organisms and food pathogens in both liquid media (ice) and on interact surfaces (stainless steel, cutting boards). Sliced cheese, raw chicken meat, and cabbage tissue were all treated with UV LEDs. To date, UV LEDs have been used to inactivate the following bacterial food pathogens: *Staphylococcus, Escherichia coli, Listeria, Campylobacter*, and *Salmonella*. In addition, two fungal spoilage species, *Saccharomyces* and *Pichia*, have been reported to be inactivated.

Recent studies used LEDs in the UVC range (259–280 nm), followed by other research applying LEDs in the UVB range (289 nm), yet another study administered LEDs in the UVA range (365, 370, and 395 nm). Additionally, bacterial foodborne pathogens were more UV sensitive than spoilage organisms, and they were successfully inactivated by 3.8e5 logs in clear solution or on smooth surfaces using LEDs in the UVC range. UV wavelength, microbial strain, and treatment media all influenced the ligase chain reaction (LCR) of bacterial and fungal food pathogens. UV LEDs used in studies with lower UV doses were required for moderately translucent solutions (saline) or fine surfaces (solid growth media) between 0.2 and 7.0 mJ/cm^2. When compared to UVB and UVA-LED treatments, UVC-LED treatments demonstrated a higher germicidal effectiveness (more inactivation per UV dosage). Despite this, Haughton et al. [48] found a significant decrease (>4 log) of *C. jejuni* employing UV at 395 nm on a slicing board surface exposure of 1,800 mJ/cm^2. The only food surface that was exposed to UVC light was sliced cheese. Interestingly, at 271 nm with a UV dosage of only 3.0 mJ/cm^2, both *S. typhimurium* and *L. monocytogenes* were estimated to be reduced up to 3.94 and 4.73 logs, respectively. Both *C. jejuni* and *E. coli* DH5a were detected on raw skinless chicken fillets and raw chicken tissue and were treated with UVA-LEDs in orange juice (365 nm and 395 nm).

The juice samples, on the other hand, were not agitated during the process, which may have hampered inactivation. Germicidal properties in the UVA region are worth investigating further since this LED therapy option may be more advantageous than UVC light in terms of cost and optical power output. Finally, because fungal development is the principal reason for fresh food deterioration, UV LEDs can be used to inactivate fungal food species such as *Penicillium, Fusarium, Aspergillus,* and yeasts, particularly on food surfaces [18].

14.5.5 WATER DISINFECTION

Disinfection of water. Drinking water safety is a major concern around the world, particularly in developing countries [41]. Moreover, UV-assisted water disinfection systems are reliable, economical, and energy-efficient and must be further developed to combat hazardous bacteria found in drinking water to accommodate the expanding world population and related increase in pollution. In the present UV water disinfection systems, luminescence from low- and medium-pressure mercury lamps is both monochromatic and polychromatic. UV-LEDs have been shown to have disinfectant properties in water samples in numerous investigations. However, as a result of inconsistent research, comparing outcomes between studies is challenging [14].

Safe drinking water is critical around the world, especially in areas with limited resources. Approximately 1.2 billion people lack access to clean drinking water. Every year, millions of people die due to waterborne infections. Mainly, intestinal illnesses such as diarrhea, cholera, dysentery, salmonellosis, amebiasis, and hepatitis A are caused by waterborne microbes. Traditional wastewater treatment methods involve the use of chemicals and a significant amount of energy, making them expensive and inaccessible to many communities. Advanced water treatments, including thermal treatments, chemical disinfections (chlorination, ozone, chloramination, chlorine dioxide), and metal ions (Cu and Ag) to reduce microbial content, are also costly in developed countries. Conventional water disinfection treatments are generally ineffective and unsustainable, as well as expensive. Therefore, efficient, cost-effective, and sustainable methods with minimal environmental impact are still being researched for water purification and sterilization. There are over 7,000 public UV disinfection systems deployed globally, and small disinfection systems for household use are also available. UV light water treatment offers several advantages over traditional disinfection methods. UV radiation has antibacterial properties, leaves no residue or byproducts, has minimal adverse effects on the environment, and is compatible with current industrial operations. UV water treatment, unlike chemical water treatment, does not lead to the development of microorganisms resistant to antibiotics. Traditional UV sources have drawbacks such as being prone to breakage and requiring careful disposal due to the environmental impact of mercury lamps [36].

UV-LED water disinfection systems are used for a variety of purposes. These systems incorporate pathogen inactivation in water-environment engineering analysis, as well as microelectronics packaging techniques for UV-LED chips or module design. This issue necessitates technical experts who have a thorough understanding of the subject and can conduct analysis from a variety of angles. This essay is based on a thorough examination and assessment of various disinfection wavelengths, UV-LED energy consumption, and issues found throughout the water treatment process. For the sake of reference, certain positive conclusions are drawn. Moreover, while examining the inactivation effectiveness and energy consumption of UV-LEDs for pathogens at various wavelengths, choosing a wavelength and setting a system that corresponds with high-efficiency UV-LED water disinfection equipment are important considerations.

To prevent interference in the early stages of UV-LED water treatment, potential developments should include a pretreatment step in high-turbidity water. Radiant flux, exposure duration, and attenuation factors are all measured together, and attenuation factors including the divergence factor, water factor, sensor factor, and reflection factor are among the more complicated [4]. Moreover, these factors are the important aspects to consider when assessing the impacts of disinfection; additionally, as water treatment systems advance, attenuation factors will as well. As a result, we must analyze many parameters and establish regulations to increase the overall water treatment system's disinfection efficiency. The results of this study reveal that the disinfection effectiveness of the complete water treatment process is determined by two key components: regulating variables and settings. Wavelength, optical acuity, radiation fluxes, time of exposure, radiation range, and dose are all regulating factors obtained primarily from the UV-LED emission light source. The settings, on the other hand, include the material selection of the reactor device and operating model design, water sample pretreatment, and exterior auxiliary system [21].

14.6 SUMMARY

Hence, the applications of LED devices can be used for a variety of purposes, including phototherapy, UV curing, environmental sensing, lighting for plant development, and water purification, depending on their emission wavelength. Therefore, it is affirmed certainly that this technical invention of LED has enormous multipurpose potential. Moreover, there are many various application methods available at the moment, but it's vital to understand that many of them cannot be incorporated in the food system. Therefore, in order to achieve the best potential of LED efficiency especially in the field of food processing, improved study is required without compromising device stability, food quality, or compatibility with different plant systems.

KEYWORDS

- **disinfection**
- **horticulture**
- **light emitting diode**
- **non-thermal processing technique**
- **post-harvest quality**
- **wavelength**

REFERENCES

1. Alferez, F., Liao, H.-L., & Burns, J. K. (2012). Blue light alters infection by *Penicillium digitatum* in tangerines. *Postharvest Biology and Technology, 63*(1), 11–15.
2. Aurum, F. S., & Nguyen, L. T. (2019). Efficacy of photoactivated curcumin to decontaminate food surfaces under blue light-emitting diode. *Journal of Food Process Engineering, 42*(3), e12988.
3. Bhavya, M. L., & Umesh Hebbar, H. (2019). Efficacy of blue LED in microbial inactivation: Effect of photosensitization and process parameters. *International Journal of Food Microbiology, 290*, 296–304. https://doi.org/10.1016/J.IJFOODMICRO.2018.10.021.
4. Bolton, J. R., & Linden, K. G. (2003). Standardization of methods for fluence (UV dose) determination in bench-scale UV experiments. *Journal of Environmental Engineering, 129*(3), 209–215.
5. Bourget, C. M. (2008). An introduction to light-emitting diodes. *HortScience, 43*(7), 1944–1946.
6. Cope, K. R., & Bugbee, B. (2013). Spectral effects of three types of white Light-Emitting Diodes on plant growth and development: absolute versus relative amounts of blue light. *HortScience, 48*(4), 504–509.
7. D'Souza, C., Yuk, H.-G., Khoo, G. H., & Zhou, W. (2015). Application of Light-Emitting Diodes in food production, postharvest preservation, and microbiological food safety. *Comprehensive Reviews in Food Science and Food Safety, 14*(6), 719–740.
8. Dutta Gupta, S., & Jatothu, B. (2013). Fundamentals and applications of Light-Emitting Diodes (LEDs) in in vitro plant growth and morphogenesis. *Plant Biotechnology Reports, 7*(3), 211–220.
9. Ghate, V., Kumar, A., Zhou, W., & Yuk, H.-G. (2016). Irradiance and temperature influence the bactericidal effect of 460-nanometer Light-Emitting Diodes on *Salmonella* in orange juice. *Journal of Food Protection, 79*(4), 553–560.
10. Goins, G. D. (2001). Performance of salad-type plants grown under narrow-spectrum Light Emitting Diodes in a controlled environment. In Proceedings of Bioastronautics Investigators' Workshop.

11. Gunther IV, N. W., Phillips, J. G., & Sommers, C. (2016). The effects of 405-nm visible light on the survival of *Campylobacter* on chicken skin and stainless steel. *Foodborne Pathogens and Disease, 13*(5), 245–250.
12. Gupta, S. D., & Agarwal, A. (2017). Light emitting diodes for agriculture. In *LED Supplementary Lighting* (pp. 27–36). Springer Nature, Singapore.
13. Held, G. (2016). Introduction to light Emitting Diode Technology and Applications. Auerbach Publications, New York. https://doi.org/10.1201/9781420076639.
14. Hinds, L. M., O'Donnell, C. P., Akhter, M., & Tiwari, B. K. (2019). Principles and mechanisms of ultraviolet light-emitting diode technology for food industry applications. *Innovative Food Science & Emerging Technologies, 56*, 102153.
15. Hogewoning, S. W., Douwstra, P., Trouwborst, G., Van Ieperen, W., & Harbinson, J. (2010). An artificial solar spectrum substantially alters plant development compared with usual climate room irradiance spectra. *Journal of Experimental Botany, 61*(5), 1267–1276.
16. Hyun, J.-E., & Lee, S.-Y. (2020). Antibacterial effect and mechanisms of action of 460-470 nm light-emitting diode against *Listeria monocytogenes* and *Pseudomonas fluorescens* on the surface of packaged sliced cheese. *Food Microbiology, 86*, 103314.
17. Kneissl, M., Seong, T.-Y., Han, J., & Amano, H. (2019). The emergence and prospects of deep-ultraviolet light-emitting diode technologies. *Nature Photonics, 13*(4), 233–244.
18. Koutchma, T., & Popovi′ c, V. (2019). UV light-emitting diodes (LEDs) and food safety. In T. Koutchma (Ed.), *Ultraviolet LED Technology for Food Applications* (pp. 91–117). Academic Press. https://doi.org/10.1016/B978-0-12-817794-5.00005-4.
19. Kusuma, P., Pattison, P. M., & Bugbee, B. (2020). From physics to fixtures to food: Current and potential LED efficacy. *Horticulture Research, 7*, 56.
20. Li, Q., & Kubota, C. (2009). Effects of supplemental light quality on growth and phytochemicals of baby leaf lettuce. *Environmental and Experimental Botany, 67*(1), 59–64.
21. Li, X., Cai, M., Wang, L., Niu, F., Yang, D., & Zhang, G. (2019). Evaluation survey of microbial disinfection methods in UV-LED water treatment systems. *Science of the Total Environment, 659*, 1415–1427.
22. Ma, Y., Xu, A., & Cheng, Z.-M. M. (2021). Effects of light-emitting diode lights on plant growth, development and traits: a meta-analysis. *Horticultural Plant Journal, 7*(6), 552–564.
23. Mitterhofer, S., Korošak, Ž., Rojec, Ž., Jankovec, M., & Topič, M. (2020). Development and analysis of a modular LED array light source. *Photonics, 7*(4), 92.
24. Munir, Z., Banche, G., Cavallo, L., Mandras, N., Roana, J., Pertusio, R., Ficiarà, E., Cavalli, R., & Guiot, C. (2022). Exploitation of the antibacterial properties of photoactivated curcumin as a "green" tool for food preservation. *International Journal of Molecular Sciences, 23*(5), 2600.
25. Nair, G. B., & Dhoble, S. J. (2015). A perspective perception on the applications of light-emitting diodes. *Luminescence, 30*(8), 1167–1175.
26. Nair, G. B., & Dhoble, S. J. (2021). Current trends and innovations. In *The Fundamentals and Applications of Light-Emitting Diodes* (p. 253).
27. Nardelli, A., Deuschle, E., de Azevedo, L. D., Pessoa, J. L. N., & Ghisi, E. (2017). Assessment of light emitting diodes technology for general lighting: A critical review. *Renewable and Sustainable Energy Reviews, 75*, 368–379.

28. Nicole, M., Raymond, J., Penman, M., & McAllister, L. (2016). 12 tips for establishing a hospital-based student-led service. *MedEd Publish, 5,* 113. https://doi.org/10.15694/mep.2016.000113
29. Noodén, L. D., & Schneider, M. J. (2004). Light control of senescence. In *Plant Cell Death Processes* (pp. 375–383). Elsevier.
30. Noor, W., Sembok, W. A. N. Z. W. A. N., & Abdullah, W. A. N. Z. W. A. N. (n.d.). The effects of Light-Emitting Diodes (LED) irradiation treatment on the postharvest preservation and microbiological quality of strawberry (*Fragaria ananassa*) cv. Festival. *Journal of Undergraduate Research, 3*(4), 137–148.
31. Olle, M., & Viršile, A. (2013). The effects of light-emitting diode lighting on greenhouse plant growth and quality. *Agricultural and Food Science, 22*(2), 223–234.
32. Park, J. H., & Lee, H. S. (2017). Phototactic behavioral response of agricultural insects and stored-product insects to Light-Emitting Diodes (LEDs). *Applied Biological Chemistry, 60*(2), 137–144.
33. Pinho, P., Hytönen, T., Rantanen, M., Elomaa, P., & Halonen, L. (2013). Dynamic control of supplemental lighting intensity in a greenhouse environment. *Lighting Research & Technology, 45*(3), 295–304.
34. Pogson, B. J., & Morris, S. C. (2004). Postharvest senescence of vegetables and its regulation. In *Plant Cell Death Processes* (pp. 319–329). Elsevier.
35. Prasad, A., Du, L., Zubair, M., Subedi, S., Ullah, A., & Roopesh, M. S. (2020). Applications of Light-Emitting Diodes (LEDs) in food processing and water treatment. *Food Engineering Reviews, 12*(3), 268–289. https://doi.org/10.1007/S12393-020-09221-4.
36. Prasad, A., Gänzle, M., & Roopesh, M. S. (2021). Antimicrobial activity and drying potential of high-intensity blue light pulses (455 nm) emitted from LEDs. *Food Research International, 148*, 110601.
37. Samuolienė, G., Sirtautas, R., Brazaitytė, A., Viršilė, A., & Duchovskis, P. (2012). Supplementary red-LED lighting and the changes in phytochemical content of two baby leaf lettuce varieties during three seasons. *Journal of Food, Agriculture and Environment, 10*(3–4), 701–706.
38. Schwend, T., Beck, M., Prucker, D., Peisl, S., Mempel, H., & others. (2016). Test of a PAR sensor-based, dynamic regulation of LED lighting in greenhouse cultivation of *Helianthus annuus*. *European Journal of Horticultural Science, 81*, 152–156.
39. Singh, S. C. (2009). Basics of light-emitting diodes, characterizations and applications. In *Handbook of Light Emitting and Schottky Diode Research,* 1–36.
40. Sluszny, C., He, Y., & Yeung, E. S. (2005). Light-emitting diode-induced fluorescence detection of native proteins in capillary electrophoresis. *Electrophoresis, 26*(21), 4197–4203.
41. Song, K., Mohseni, M., & Taghipour, F. (2016). Application of ultraviolet Light-Emitting Diodes (UV-LEDs) for water disinfection: A review. *Water Research, 94*, 341–349.
42. Sorouraddin, M.-H., Rostami, A., & Saadati, M. (2011). A simple and portable multi-color light-emitting diode-based photocolorimetry for the analysis of mixtures of five common food dyes. *Food Chemistry, 127*(1), 308–313.
43. Srimagal, A., Ramesh, T., & Sahu, J. K. (2016). Effect of light-emitting diode treatment on inactivation of *Escherichia coli* in milk. *LWT-Food Science and Technology, 71*, 378–385.

44. Wang, F., Liu, X.-K., & Gao, F. (2019). Fundamentals of solar cells and light-emitting diodes. In *Advanced Nanomaterials for Solar Cells and Light Emitting Diodes* (pp. 1–35). Elsevier.
45. Wierer Jr, J. J., Tsao, J. Y., & Sizov, D. S. (2013). Comparison between blue lasers and Light-Emitting Diodes for future solid-state lighting. *Laser & Photonics Reviews, 7*(6), 963–993.
46. Zhao, M., Zhang, Q., & Xia, Z. (2020). Narrow-band emitters in LED backlights for liquid-crystal displays. *Materials Today, 40*, 246–265.
47. Zukauskas, A., Bliznikas, Z., Breivė, K., Novičkovas, A., Samuolienė, G., Urbonavičiūtė, A., Brazaitytė, A., Jankauskienė, J., & Duchovskis, P. (2009). Effect of supplementary pre-harvest LED lighting on the antioxidant properties of lettuce cultivars. *VI International Symposium on Light in Horticulture, 907*, 87–90.
48. Haughton, P. N., Grau, E. G., Lyng, J., Cronin, D., Fanning, S., & Whyte, P. (2012). Susceptibility of *Campylobacter* to high intensity near ultraviolet/visible 395±5nm light and its effectiveness for the decontamination of raw chicken and contact surfaces. *Int. J. Food Microbiol., 159*, 267–273. doi: 10.1016/j.ijfoodmicro.2012.09.006.

CHAPTER 15

LIGHT-EMITTING DIODES: SOURCES, EQUIPMENT, DESIGN CONSIDERATIONS, AND PARAMETERS AFFECTING ITS EFFICACY

SHWETA SURI and PINTU CHOUDHARY

ABSTRACT

Light-emitting diodes (LEDs) are non-thermal technology that offers high light output along with low heat generation, emit a limited wavelength range (typically less than ±10 nm of the dominant wavelength) and provide uniform light distribution in space if they are based on multiple individual LEDs. The spoilage-causing microorganisms commonly found in food and water could be removed with the light of an appropriate wavelength and continuous/pulsed mode of operation, which makes LEDs efficient. In contrast to traditional techniques, LEDs are a mercury-free, non-hazardous technique with a compact size which makes them suitable to integrate with the food processing sector. In addition, LEDs are a cost-effective high-performance approach that consumes less power, has a longer life (>10,000 hours), and a robust way of effective disinfection. Nowadays, LEDs are employed for disinfection of air, water, and surfaces as well as for curing. However, the effectiveness of LEDs in food processing varies based on the nature and type of food and its constituents, water activity, along with the surface morphology of the food. Additionally, the source of LED, type of equipment, and design significantly influence the efficiency of treatment to achieve desired results. This chapter focuses on different sources of LEDs, equipment design aspects, and various considerations that impact LED efficacy.

Nonthermal Light-Based Technologies in Food Processing.
C. K. Sunil, Preeti Birwal, Megh R. Goyal, and R. Mahendran (Eds.)

15.1 INTRODUCTION

In the food sector, the manufacture, processing, and delivery of safe food is a top priority. Increasing consumer awareness for safe, nutritious, and minimally processed foods along with intensifying marketplace competition have led food product processing industries to adopt new methods for preserving foods without compromising nutritional as well as sensory quality [4]. Some of the pathogenic microorganisms such as *Salmonella* species, *Staphylococcus aureus*, *Listeria monocytogenes*, etc., are listed as the main cause of food poisoning [32]. Therefore, their removal from food is a major concern.

Currently, two types of sterilization techniques are used in the food sector, namely thermal and non-thermal sterilization methods. For safety from microbial growth, thermal technology is considered the most effective way to eliminate pathogens, but it is not suitable for certain kinds of foods, such as ready-to-eat, fresh foods, etc. [12]. Although thermal sterilization techniques have a known history, a wide range of applications, and sterilization proficiencies, their high energy consumption and high temperatures could cause the loss of the inherent nutrition and taste of food. On the contrary, non-thermal sterilization techniques, such as the use of irradiation, antibiotics, high-voltage pulsed electric field treatment, hydrostatic pressure processing, cold plasma, and higher intensity ultrasound treatment, are continuously evolving [2, 36]. However, these non-thermal sterilization techniques also have some drawbacks, such as the excessive utilization of antibiotics, which has caused the development of resistance in bacteria, leading to the abandonment of antibiotic sterilization methods. The utilization of further non-thermal ways of sterilization, for example, irradiation and pulsed electric fields, requires highly expensive equipment. In addition, the excessive use of ultrasounds can generate free radicals or cause protein denaturation and the release of metal particles [2, 58]. Moreover, there are limitations concerning solid food applications. Also, in connection with high-pressure treatment, the food matrix requires more than 40% free water to achieve proper microbial inactivation [21].

The inactivation of microbes through photodynamicism is a new technology that is based on non-thermal photochemical and photophysical processes, requiring light and a photosensitizer in the presence of oxygen. Light-emitting diodes (LEDs) are made up of layers of semiconductor material that emit light when an electric current flows through them. LEDs that produce monochromatic light have been widely employed in the agricultural and food industries due to their various benefits over traditional light sources

[53]. Nowadays, LEDs operating in the visible range have shown great potential as new food preservation and sterilization techniques [22]. The cost of LEDs has recently dropped significantly due to technological advancements and their increased usage. The output power of LEDs is becoming increasingly efficient [12]. LED usage offers several advantages over UV lamps. The average lifespan of LEDs is significantly longer (>10,000 hours) than the UV lamps used in today's industry. Additionally, LEDs are harmless and non-hazardous as they contain no mercury and emit much less heat than UV lamps, making them more suitable for food sterilization and processing applications. LEDs consume less energy as they do not require warm-up time [1]. They provide high performance, robustness, low power consumption, and cost-efficiency.

LEDs operate on the electroluminescence principle which signifies that LED produces light rays with the application of a magnetic/electric field. Usually, in a magnetic/electric field, excited electrons enter a low-energy state, releasing light as well as energy as electromagnetic rays. LEDs are impurity-doped semiconductor materials that form boundaries or interfaces (called p-n junctions) between positive or p-type semiconductor matrix that has excess holes, and the negative or n-type that has extra electrons. In fact, the color and wavelength characteristics of the emitted light vary depending on two factors: semiconductors and the impurities used to form the LED [53]. As a result, LEDs can produce monochromatic light with a fine wavelength range that the eye perceives in different colors. Typically, the color of the emitted light depends on the bandgap energy of the semiconductor used. Gallium arsenide is employed for the emission of red or IR color, while yellow, green, orange, and red colors are emitted using indium gallium aluminum phosphide. Additionally, the blue color of light is observed with gallium nitride and silicon carbide [12].

LEDs are employed in numerous processes in the food sector, for example, for enhancing the nutritional and antioxidant activity of vegetables (such as stored tomatoes, Chinese cabbage, peas, etc.) (Figure 15.1). LED treatment on the overall nutritional characteristics of cabbage led to an increase in total phenol, total chlorophyll, and vitamin C content, and decreased reactive oxygen species [28]. Blue and red LEDs aid in building flavonoids and glycosides. Blue light plays a vital role in activating metabolic pathways for the synthesis of polyphenolic components, while red light promotes photosynthetic action as well as stomatal opening by inducing photophosphorylation [62].

In addition, LEDs can increase agricultural yields and the nutrient content of foods. Studies have shown that the application of blue LEDs is related to multiple metabolic pathways as well as the accumulation of polyphenolic components, ascorbic acid, carotenoids, and anthocyanins [47, 59]. Furthermore, one of the major roles of food processing methods is to lower the quality degradation of food. Post-harvest changes in foods such as fruit rot and diseases in standing crops are major issues for agricultural experts. LEDs are becoming more and more popular as a useful medium for sustainable farming and their use in the post-harvest management of diseases in foods [51]. In a study, blue LED (440–450 nm) was used for pre-treatment of green tomatoes to extend their shelf life before storing them in the dark [14]. The LED pre-treatment led to a delay in the softening of tomatoes. The tomatoes were completely ripened after 3 weeks of the storage period in the dark because of the increased concentration of lycopene. Also, several authors reported the antibacterial effects of LED treatment on food [18, 27, 33, 51]. Furthermore, LEDs have been effectively used for the disinfection of air, treatment of water, decontamination of surfaces, and curing. Longer wavelength LEDs, namely infrared and red (630–1,000 nm), are employed in a number of fields such as dyeing/coloring, phototherapy, coating curing, and ink curing [40, 52] (Table 15.1).

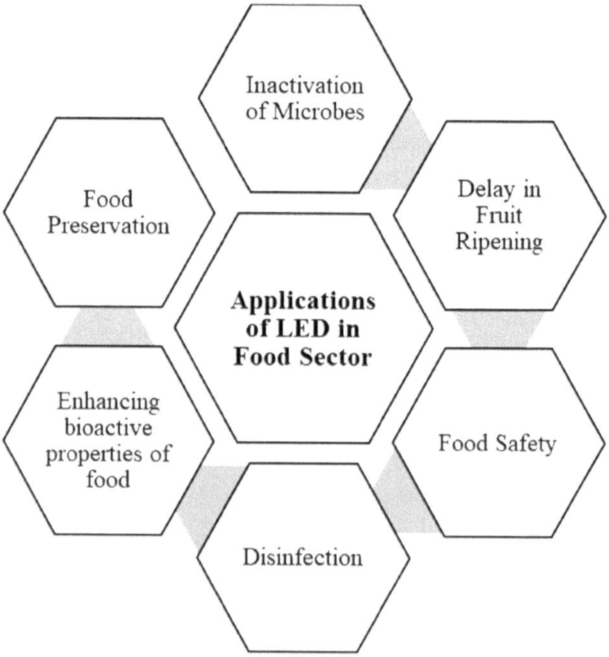

FIGURE 15.1 Illustration showing applications of LEDs in food sector.

Source: By Authors.

TABLE 15.1 Recent Advances in Utilization of LED

Sl. No.	Food Material	LED Treatment	Results	References
Increase in the Activity/Concentration of Biologically Active Compounds				
1.	Immature strawberries	Blue LED Wavelength: 470 nm Intensity: 40 Wm^{-2}	13% increase in total phenols.	[29]
2.	Green oak leaf	Blue: 460 nm Intensity: 133 ± 5 μmol s^{-1} m^{-2}	Highest chlorophyll content (1.31 mg/g) was observed on treatment with LED of red color along with florescent light.	[9]
3.	Pea sprouts	Blue LED Wavelength: 460 nm Yellow: 585 nm Red: 635 nm	1.46, 1.25 and 1.45 times increase in the total phenols upon treatment with blue, red, and white LEDs, respectively. 24.55, 21.01, and 24.29 times increase in the total flavonoids upon treatment with blue, red, and white LEDs, respectively.	[42]
4.	Chinese kale sprouts	Blue–White Intensity: 20 μmol s^{-1} m^{-2}	34.55 and 69.09% increase in the total phenols content of sprouts under treatment with blue as well as white LEDs.	[53]
5.	Mustard	Red LED Wavelength: 638 and 665 nm Intensity: 300 μmol m^{-2} s^{-1} Treatment time: 16 hours	An upsurge in total β-carotene level 0.028–0.073 mg/g.	[6]
Inactivation of Microorganisms/Reduction of Microbial Spoilage				
1.	Food matrix	Wavelength: 395 nm Treatment time: 1,115 seconds	1.37 Log CFU/ml decrease in *E. coli* K12.	[5]
2.	Fruits (apricot and plum) Vegetables (cauliflower)	Wavelength: 585 nm	0.7, 1.1, 1.3 log CFU/g drop in *Bacillus cereus* on the external covering of apricot, plum, and cauliflower, respectively, in comparison to controls.	[3]
3.	Freshly cut mango	Wavelength: 405 nm Temperature: 20?C Treatment time: 24 hours	LED inactivated the load of *Listeria monocytogenes*, *E. coli* O157: H7, and *Salmonella* (1.0–1.6 Log CFU/cm^2).	[56]

TABLE 15.1 *(Continued)*

Sl. No.	Food Material	LED Treatment	Results	References
4.	Citrus fruits	Wavelength: 272, 293 and 289 nm Treatment time: 9 days	Inhibition of infections caused by *Botrytis cinerea.*	[48]
5.	Shelled almonds	Wavelength: 405 nm	Microbial reduction up to 2.44 Log CFU/g in *E. coli* O157:H7, 0.96 Log CFU/g in *S. typhimurium*, 1.86 Log CFU/g in *E. coli* K12, and 0.7Log CFU/g in *S. enteritidis*, respectively.	[35]
6.	Apple juice	Blue LED Wavelength: 95, 405, 415 and 425 nm	6 to 7 Log decrease in *E. coli* O157:H7.	[31]
Role in Food Safety and Reducing Post-Harvest Losses				
1.	Citrus fruits	Blue LED: 40 μmolm^{-2} s^{-1} Treatment time: 5–7 days	Lowered soft rot, mycelium growth, and sporulation by fungus.	[38]
2.	–	LED Wavelength: 625–635 nm Hydrogen peroxide: 0.25%	Exposure to LED (after treatment with 0.25% hydrogen peroxide led to a reduction in *Porphyromonas gingivalis* biofilm.	[15]
3.	Broccoli	Red LED Wavelength: 660 nm	Delayed aging of vegetables; less yellowing and lower ethylene production.	[43]
4.	Lettuce	White LED Rate: 150 W/m^2 Temperature: 6?C	Improves lettuce quality for 1 week at 6?C LED treatment by reducing browning and curtailing loss in weight and respiration rates in lettuce.	[8]
5.	Broccoli	Red LED Fluence rate: 50 W/m^2	Prevents yellowing and chlorophyll deterioration and reduces weight loss in broccoli.	[26]
Applications in Phototherapy				
1.	–	LED Wavelength: 600–900 nm	Useful in light therapy or phototherapy. Helps in lowering pain and healing of wounds.	[23]
2.	–	Red LED Wavelength: 660 nm Infrared LED Wavelength: 840 nm	LED is useful for personal phototherapy. LED treatment exhibited no side effects for example skin burns during surgery when the phototherapy device touched the skin for a period of 10 minutes.	[39]

15.2 SOURCES OF LED

LEDs are semiconductor light sources that produce light once an electric current is passed. The electrons present in the semiconductor combine with the holes and lead to the emission of energy as photons. The light color (equivalent to the photon energy) is determined by the energy needed for the electron particles to move across the bandgap of the semiconductors. A white color of light is achieved with the help of luminescent phosphor layers on various semiconductors or semiconductor devices. The current state-of-the-art advancements have resulted in increased availability of LEDs in different wavelengths such as visible, ultraviolet (UV), and infrared wavelengths with low, medium, and high lumen outputs. Today, almost half a century after the rapid and sustainable growth of LEDs, traditional conventional lights such as incandescent and fluorescent lamps are gradually being replaced by LEDs [7].

LEDs are evaluated based on their light output, also called luminous efficacy. Luminous flux signifies the reaction of the eye to light of different wavelengths. The choice of color emission of LEDs is quite limited owing to the nature of the semiconductors used in manufacturing. Common LED colors available are blue, red, green, yellow, amber, and white (Figure 15.2). The impact of varying LED sources on the growth of lettuce and nitrogen

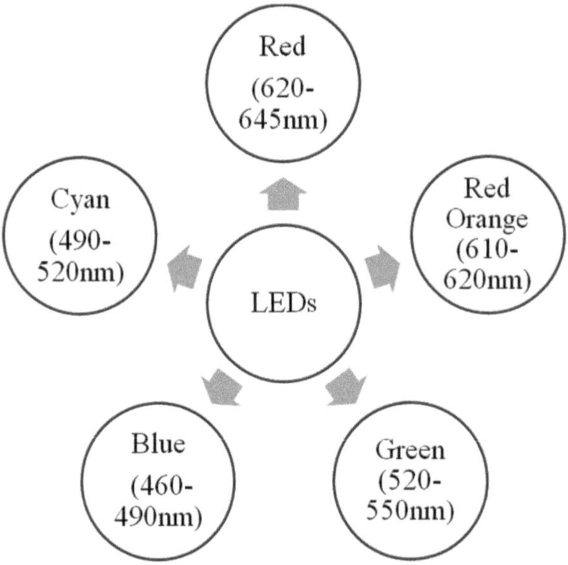

FIGURE 15.2 Different sources of LEDs.

Source: By Authors.

metabolism was investigated. Using LED as a light source, 8 different light treatment effects (red, white, blue, purple, green, yellow, red–blue in 9:1 ratio, and 4:1 ratio red–blue) for growth, overall quality, and nitrogen metabolism of lettuce were researched. The findings exhibited that in comparison with purple, white, red, blue, and the combination of red-blue light led to an upsurge in the biomass of the aerial portions of lettuce to different extents, whereas yellow and green light repressed the growth of lettuce [63, 64].

15.3 CONSIDERATIONS FOR DESIGNING OF LED EQUIPMENT

LEDs contain chips made up of semiconductor materials along with doped impurities which help in emitting light with a particular wavelength and color. P-type semiconductors could be designed by incorporating group II elements, for example, magnesium ions, into a substrate of group III elements to create an additional hole in the substrate. The n-type semiconductors are developed by incorporating group IV elements into the group III elements to increase the number of free electrons in the substrate. The additional holes are responsible for the increase in the number of free electrons in the substrate. Furthermore, the additional hole on the p-side as well as the free electrons present on the n-side merge at the p-n junction of the electrode to create a non-conductive depletion area [48]. The radiative combination of holes and free electrons is very essential in the LED arrangement. When the electric current passes through the LED, high-energy electrons from the conduction band (n-side) interact with the holes on the p-side, creating electron-hole duos and moving to the valence electron coupling of the p-type semiconductor where the energy is low in comparison to free electrons from the n band.

The bandgap can be direct or indirect. In a direct bandgap, the momentum is maintained constant; however, in an indirect bandgap, the momentum keeps changing and has a lower transition. Transition can be increased by adding isoelectric traps [13]. The energy difference carried by free electrons and electron-hole pairs is released as photons that carry electromagnetic radiation of a specific color and wavelength. The most efficient LEDs depend on semiconductors belonging to group III-V and are fabricated using straight gap alloys. The variation in the chemical composition of these alloys will influence the energy of the bandgap as well as the wavelength of the emitted light. By applying the energy conservation law, assuming that the thermal energy generated is very small compared to the photon energy formed, the amount of energy generated as light will be equivalent to the bandgap of diodes.

$$E_c - E_v \approx E_g \quad (1)$$

where; E_c stands for energy of electron in the conduction band; E_v stands for the energy of the holes in the valence bond; and E_g stands for energy in the bandgaps of diode and indicates total amount of energy produced all through the electroluminescence.

Conferring to the law of conservation of energy, when the thermal energy is very small as compared to E_g:

$$hv = E_c - E \approx E_g \quad (2)$$

where; hv represents energy of a photon released; 'h' represents Planck's constant and lastly; 'v' represents frequency of photon of light which is inverse of wavelength.

From Eqns. (1) and (2), we can see that the material with an advanced conduction band requires to produce light with a shorter wavelength and vice versa. The packaging of LED chips can influence the efficiency of the LED arrangement. For example, if the packaging material traps most of the light generated by an LED source due to total internal reflection, the total light perceived by the human eye is altered, leading to a change in the overall luminous efficacy. However, the complete luminous efficiency refers to the ability of light to convert electrical energy into visible power that can be perceived by the human eye under normal conditions. The current and voltage requirements for LEDs depend on the materials used for the semiconductor and the wavelength of the emitted light. Generally, the voltage ranges from 1.5 to 3 V, while the current ranges from 10 to 30 mA [20]. Maintaining the duty ratio (which is the proportion of time operated) and electric current helps control the light intensity and LED spectral output. The radiation (I) of the LED plays a significant role in the efficiency of the process. The power radiated per unit area of the sample matrix is known as irradiance. Radiometers can be used to measure the irradiance level of LEDs of different wavelengths at a specific distance from the source and can be monitored at regular intervals over the source's lifetime to track the source's power [52, 54]. The total irradiance multiplied by the exposure time equals the radiant energy (energy dose) of a material to light at a specific height and exposure period, as represented in the equation.

$$E = I \times t \quad (3)$$

where; 'E' stands for amount of energy LED light/area (mJ/cm^2); 'I' stand for irradiance of the LED light (mW/cm^2); and 't' stands for treatment time for the LED light (sec) [16].

Another factor is photon fluence, that is the ratio of dN and dA.

$$Ø = dN/dA \tag{4}$$

where; Ø represents fluence of photon (cm^2); dN represents incident photons on an imaginary sphere; however, the dA represents cross-sectional region of the imaginary sphere [50, 56].

Additionally, dN can be calculated by taking the ratio of the total energy of incident photons and the energy of a photon (hc/λ), in this formula, h represents Planck's constant (6.626×10^{-3} Js), c stands for the speed of light (3×10^8 m/s) while λ stands for the wavelength of light (m).

Various researchers designed the UV LED system for different applications. Merve Pelvan Akgün & Ünlütürk [1] designed a static UV-LED system for the inactivation of microorganisms and enzymes from apple juice (Figure 15.3). They attached four UV-LEDs with an 8.33 mm diameter that emits light with wavelengths of 254, 280, 365, and 405 nm. These LEDs can be used together or individually; the electronic circuit can be operated by four power supply channels. Individual UV-LEDs were powered by a 20-mA direct current supply. A similar system was previously fabricated by Chevremont et al. [10] for the inactivation of fecal bioindicators. They found that compared to individual wavelengths, coupling of wavelengths was more effective for the inactivation of bacteria.

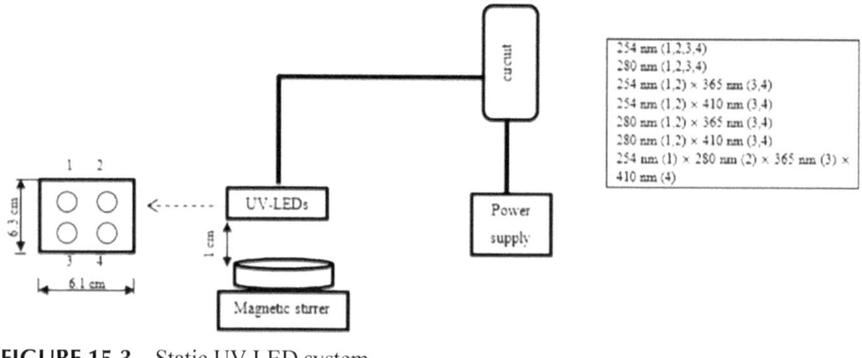

FIGURE 15.3 Static UV-LED system.

Source: Reprinted with permission from Ref. [1]. Copyright 2017. Elsevier.

15.4 FACTORS AFFECTING THE EFFICACY OF LED TREATMENT

15.4.1 TYPES OF FOOD MATRIX

The efficiency of the LED treatment for solid foods is usually based on the type as well as the food's nature, food components/ingredients, water activity (a_w), and surface morphology of food products. In addition, several factors such as treatment time, dose, light wavelength, illumination temperature, relative humidity, and microbial factors are equally important. The antimicrobial efficacy of LEDs is altered by different food and processing parameters such as product type, composition, treatment temperature, and environmental surroundings [52]. Different researchers worked on different food materials to check the efficacy of the LED treatment.

15.4.1.1 ANIMAL-BASED FOODS

An average 1.5 Log reduction in the activity of *Staphylococcus aureus* in chicken legs was detected upon treatment with 435 nm wavelength of LED along with curcumin [60, 61]. Total inactivation of *E. coli* O157:H7 was observed on the hot dog after exposure to a 405 nm LED with an irradiance value of 83.3 mW/cm^2 [18]. The application of a 460 nm LED along with the presence of riboflavin lessened the infection risk in smoked salmon with *Listeria monocytogenes* during storage [27]. Pacific oysters were analyzed for their shelf life upon treatment with LED coupled with curcumin at 51?C. The study showed an increase in shelf life from 8 to 12 days with no change in the textural properties. The treated oysters also exhibited low oxidation levels, indicating less nutrient degradation and metabolism by spoilage-causing microorganisms at the end of the shelf life [41].

15.4.1.2 PLANT-BASED FOODS

The impact of 3 different wavelengths (405, 415, and 450 nm) on *Botrytis cinerea*, commonly known as gray mold, was studied. Treatment with 405 nm not only led to the inhibition of spore growth of *Botrytis cinerea* on tomato leaves after 120 hours of light, but also led to the inhibition of spore germination within 18 hours of light [25]. Inactivation of *E. coli* in milk was investigated as an effect of blue LED treatment at varying wavelengths (405, 433, 460 nm), temperatures (5, 10, 15?C), and processing times (0–90 minutes). Inactivation of microorganisms was found after 60 minutes of irradiation at 405 nm, with *E. coli* O157:H7 decreasing by 5.27 log CFU/mL. Inactivation of microorganisms was greater at high temperatures as well as short wavelengths. A 2–5 logarithmic reduction was seen with the 460 nm LED [57]. LEDs that emit light at 405 nm caused a 1–1.2 log CFU/cm^2 decrease in freshly cut pieces of papaya inoculated with the bacteria *Salmonella*. Treatment of papaya at a temperature of 4?C for a 48-hour time interval with an overall dose of 1.7 kJ/cm^2 [32]. The antibacterial action of 460 nm wavelength LED treatment on fresh-cut pineapple infested with *S. enterica* cocktails at varying illuminance temperatures as well as irradiation. A reduction (1.72 Log CFU/g) in *E. coli* O157:H7, *S. typhimurium*, *E. coli* K12, and *S. enteritidis* was observed at an irradiance of 92 mW/cm^2 at an illumination temperature of 16?C [16].

The influence of 405 nm irradiation treatment on the blueberries was evaluated to destroy the Tulane virus on the external layer of blueberries. It revealed that photosensitizers or further chemical boosters are needed for the efficient inactivation of the virus by the photoactivation phenomenon [34]. A decrease of 2.43 Log and 3.18 Log was detected in *Escherichia coli* and *Staphylococcus aureus*, respectively, in freshly cut potato slices upon treatment with 30 μM curcumin coupled photodynamic inactivation at 420 nm LED treatment. In addition, a reduction in the polyphenol oxidase as well as peroxidase activity, while an increase in the activity of phenylalanine ammonolyzes, was observed with an increase in overall antioxidant capacity [63].

15.4.2 MICROBIAL LOAD

Microbial inactivation by using light exposure is an emerging area of study due to the persistent issue of microbial resistance [44, 46]. Currently, photodynamic inactivation of pathogenic microorganisms from food has been widely used for food decontamination. Many studies focused on utilizing either LED technology or LED treatment coupled with other treatments for lowering the microbial load from food. Antibacterial action of curcumin in LED (400–800 nm) to stop the growth of *E. coli* O157: H7 as well as *Listeria innocua* in the wash water of fresh produce was evaluated. The combined action of curcumin and blue LED illumination (470 nm) on foodborne pathogenic microorganisms such as *E. coli*, *S. aureus*, *Pseudomonas aeruginosa*, *Salmonella typhimurium*, and *Aeromonas hydrophila*, was investigated [49]. Effects of curcumin coupled with blue LED (455 nm) treatment on methicillin-resistant *Staphylococcus aureus* along with L929 fibroblasts were evaluated [55].

In a study, the population of *Campylobacter* species decreased from the initial microbial load (4 Log CFU/cm^2) to the undetectable load upon treatment with LED at 395 nm with a minimum intensity of 0.12 J cm^{-2} in chicken fillets [19]. In another study, *Listeria monocytogenes* and *E. coli* decreased by 26% and 13%, respectively, when exposed to light irradiance of 60 mWcm^{-2}. The rate of inactivation of *Listeria monocytogenes* and *E. coli* was 95% and 99%, respectively, once pretreatment was done with acid (pH 3) and with subsequent treatment using a 405 nm LED having an intensity of 36 J cm^{-2} [45].

15.4.3 PHYSICAL PROPERTIES OF FOOD

Physical properties of foods include characteristics such as color, texture, structure, water activity, relative humidity, rheology, interfacial features, and compositional properties. The effect of the physical features of food on the efficacy of LED was explored. Low water activity foods, for example, cereal grains, dried nuts, pet foods, etc., are more susceptible to microbial growth. These food-borne pathogens can endure extended periods of dormancy and become active when exposed to favorable environments. Investigations on the antimicrobial effects of LED actions in low-water activity foods are limited; however, recent research has shown positive results in microbial inhibition. Treatment of the shelled almonds with 405 nm LEDs led to reductions of up to 2.44 Log CFU/g in *E. coli* O157:H7, 0.96 Log CFU/g in *S. Typhimurium*, 1.86 Log CFU/g in *E. coli* K12, and 0.7 Log CFU/g in *S. Enteritidis*, respectively [36].

In the context of the color of liquid food commodities, studies have reported a variation in the color of liquid foodstuffs upon blue LED treatment [16, 17]. LEDs that emit blue light (range 400–480 nm) led to the destruction of pathogens found in liquid foodstuffs. LED illuminations in the blue range suppress microbial action primarily by photodynamic inactivation. The photons generated by the LED light are absorbed by the endogenous photosensitizers (porphyrins, cytochromes, flavins, etc.) and the bacterial NADH is sensitized after illumination [52]. Compared to water, other liquid foods comprise a complex system encompassing pigments, fibers, insoluble matter, etc. Turbidity as well as the color of liquid foods can have an impact on the antimicrobial efficacy of the treatment with LED.

In a study, the potential of UVA-LED was analyzed to check its disinfectant property in colorful solutions as well as orange juice injected with *E. coli* DH5α. Various coloring agents such as safflower flavonoids, carotenoids, and a blended color of grape and melon were arranged in concentrations ranging from 0.001 to 0.1% with inoculated *E. coli* DH5α besides treatment with UV-A LED light of 126 J/cm^2 was utilized to process the solutions [37]. However, this energy consumption is high, particularly for LEDs with wavelengths (365, 395, and 455 nm). The researchers employed a UVA LED with an intensity of 70 mW/cm^2 for 30 minutes [37]. It has usually been observed that high concentrations of colored solutions reduce antibacterial activity after LED treatment; however, the log decrease in cell numbers in diverse colored samples varies. The highest reduction in log of 1.75 log CFU/mL was attained in the 0.001% colored solution containing β-carotene, which is still much lower than the 2.5 log decrease in the phosphate-buffered saline control solution. Comparable findings were observed in orange juice, where the bacterial log reduction was lower than that of the control solution after the LED exposure. Colorants, as well as the different suspended elements found in liquid foodstuffs, can reflect light and scatter it, lowering the effectiveness of LED bacteria removal. Consequently, reactive oxygen species generated by UV-A rays ranging in the wavelength of 320–400 nm are fundamental to the action against bacteria. Thus, the antioxidative action of food colorings, for example, the carotenoids present in liquid food commodities, may be reduced, leading to oxidation and qualitative changes [52].

Red light (wavelength ranging from 610 to 760 nm) is produced through LEDs having an aluminum gallium arsenide semiconductor device. The bactericidal action of red LEDs is reported to speed up the healing of wounds in mice; besides, it is also utilized to sterilize equipment [55]. Investigations have reported that a combination of red LED (wavelength of 660 nm) and toluidine blue O lowered *Streptococcus olaris* in plaque in a dose-dependent way [24]. Introduction of LED (625–635 nm) subsequently after treatment with 0.25% hydrogen peroxide led to a reduction in *Porphyromonas gingivalis* biofilm (related to periodontitis) [15]. Besides, the red LED (660 nm) has a more prominent effect than blue as well as white LEDs in increasing certain growth rates plus the concentration of cells of the genus Chlorella [11].

A study was also conducted to examine the synergistic effect of LED with other food bioactives. Kim et al. [30] examined the impact of blue LED light with riboflavin to regulate *E. coli* O157:H7 along with *S. Typhimurium* in apple juice. They found that the combined effect of LED and riboflavin was more effective compared to individual treatment.

15.5 SUMMARY

LED is one of the novel non-thermal techniques for inactivation of microorganisms and enzymes. It emits uniform light distribution throughout the product with minimum heat generation and emits a limited wavelength range. There are various factors that influence the effectiveness of LED systems, such as the type of LED source (red, white, blue, purple, green, yellow), types of food, physicochemical properties of food, and microbial load. The application of combined light is more efficient compared to individual light sources. LED is an environmentally friendly and cost-effective technique. The low penetration depth and high intensity application are the main limitations of LED systems; optimization studies and improvements in design are needed to overcome these challenges.

KEYWORDS

- **cost-effective technique**
- **efficacy**
- **enzymes**
- **inactivation**
- **light-emitting diodes**
- **microorganisms**
- **non-thermal techniques**
- **wavelength**

REFERENCES

1. Akgün, M. P., & Ünlütürk, S. (2017a). Effects of ultraviolet Light-Emitting Diodes (LEDs) on microbial and enzyme inactivation of apple juice. *International Journal of Food Microbiology, 260*, 65–74.
2. Zhu, S., Song, Y., Pei, J., Xue, F., Cui, X., Xiong, X., & Li, C. (2021). The application of photodynamic inactivation to microorganisms in food. *Food Chemistry, 12*, 100150.
3. Aponiene, K., Paskeviciute, E., Reklaitis, I., & Luksiene, Z. (2015). Reduction of microbial contamination of fruits and vegetables by hypericin-based photosensitization: Comparison with other emerging antimicrobial. *Journal of Food Engineering, 144*, 29–35.

4. Bhavya, M. L., & Umesh Hebbar, H. (2019). Efficacy of blue LED in microbial inactivation: Effect of photosensitization and process parameters. *International Journal of Food Microbiology, 290*, 296–304. https://doi.org/10.1016/J.IJFOODMICRO.2018.10.021.
5. Birmpa, A., Vantarakis, A., Paparrodopoulos, S., Whyte, P., & Lyng, J. (2014). Efficacy of three light technologies for reducing microbial populations in liquid suspensions. *BioMed Research International, 673939*, 1–9.
6. Brazaityte, A., Sakalauskiene, S., Virsile, A., Jankauskiene, J., Samuoliene, G., Sirtautas, R., Vastakaite, V., Miliauskiene, J., Duchovskis, P., & Novickovas, A. Dabasinskas, L. (2014). The effect of short-term red lighting on Brassicaceae microgreens grown indoors. In *XXIX International Horticultural Con, Livelihoods and Landscapes* (Vol. 1123), 177–184. Acta Horticulturae. https://doi.org/10.17660/ActaHortic.2016.1123.25.
7. Bürmen, M., Pernuš, F., & Likar, B. (2008). LED light sources: a survey of quality-affecting factors and methods for their assessment. *Iopscience.iop.org, 19*(12), 122002.
8. Charles, F., Nilprapruck, P., Roux, D., & Sallanon, H. (2018). Visible light as a new tool to maintain fresh-cut lettuce post-harvest quality. *Postharvest Biology and Technology, 135*, 51–56.
9. Chen, X. L., Guo, W. Z., Xue, X. Z., Wang, L. C., & Qiao, X. J. (2014). Growth and quality responses of 'Green Oak Leaf' lettuce as affected by monochromic or mixed radiation provided by fluorescent lamp (FL) and light-emitting. *Scientia Horticulturae, 172*, 168–175.
10. Chevremont, A. C., Farnet, A. M., Sergent, M., Coulomb, B., & Boudenne, J. L. (2012). Multivariate optimization of fecal bioindicator inactivation by coupling UV-A and UV-C LEDs. *Desalination, 285*, 219–225. https://doi.org/10.1016/j.desal.2011.10.006.
11. Choi, B., Lim, J. H., Lee, J., & Lee, T. (2013). Optimum conditions for cultivation of Chlorella sp. FC-21 using light-emitting diodes. *Journal of Chemical Engineering, 30* (8), 1614–1619. https://doi.org/10.1007/s11814-013-0081-0.
12. D'Souza, C., Yuk, H. G., Khoo, G. H., & Zhou, W. (2015). Application of light-emitting diodes in food production, postharvest preservation, and microbiological food safety. *Comprehensive Reviews in Food Science and Food Safety, 14*(6), 719–740. https://doi.org/10.1111/1541-4337.12155.
13. DenBaars, S. P. (1993). Light-emitting diodes: materials growth and properties. In *Solid State Luminescence* (pp. 263–291). Springer.
14. Dhakal, R., & Baek, K. H. (2014). Short period irradiation of single blue wavelength light extends the storage period of mature green tomatoes. *Postharvest Biology and Technology, 90*, 73–77. https://doi.org/10.1016/J.POSTHARVBIO.2013.12.007.
15. Eick, S., Markauskaite, G., Nietzsche, S., Laugisch, O., Salvi, G. E., & Sculean, A. (2013). Effect of photoactivated disinfection with a light-emitting diode on bacterial species and biofilms associated with periodontitis and peri-implantitis. *Photodiagnosis and Photodynamic Therapy, 10*(2), 156–167. https://doi.org/10.1016/J.PDPDT.2012.12.001.
16. Ghate, V., Kumar, A., Kim, M. J., Bang, W. S., Zhou, W., & Yuk, H. G. (2017). Effect of 460 nm light-emitting diode illumination on survival of *Salmonella* spp. on fresh-cut pineapples at different irradiances and temperatures. *Journal of Food Engineering, 196*, 130–138.
17. Ghate, V., Kumar, A., Zhou, W., & Yuk, H. G. (2015). Effect of organic acids on the photodynamic inactivation of selected foodborne pathogens using 461 nm LEDs. *Food Control, 57*, 333–340.

18. Guffey, J. S., Payne, W. C., Motts, S. D., Towery, P., Hobson, T., Harrell, G., Meurer, L., & Lancaster, K. (2016). Inactivation of *Salmonella* on tainted foods: using blue light to disinfect cucumbers and processed meat products. *Food Science and Nutrition, 4*(6), 878–887. https://doi.org/10.1002/fsn3.354.
19. Haughton, P. N., Grau, E. G., Lyng, J., Cronin, D., Fanning, S., & Whyte, P. (2012). Susceptibility of *Campylobacter* to high-intensity near ultraviolet/visible 395±5 nm light and its effectiveness for the decontamination of raw chicken and. *International Journal of Food Microbiology, 159*(3), 267–273.
20. Held, G. (2016). *Introduction to Light Emitting Diode Technology and Applications*. Auerbach Publications, New York. https://doi.org/10.1201/9781420076639.
21. Hinds, L. M., O'Donnell, C. P., Akhter, M., & Tiwari, B. K. (2019). Principles and mechanisms of ultraviolet light-emitting diode technology for food industry applications. *Innovative Food Science & Emerging Technologies, 56*, 102153.
22. Hu, J., Lin, S., Tan, B. K., Hamzah, S. S., Lin, Y., Kong, Z., Zhang, Y., Zheng, B., & Zeng, S. (2018). Photodynamic inactivation of *Burkholderia cepacia* by curcumin in combination with EDTA. *Food Research International, 111*, 265–271.
23. Huang, Y. Y., Carroll, J., & Hamblin, M. R. (2011). Biphasic dose response in low level light therapy–An update. *Journals.sagepub.com, 9*(4), 602–618. https://doi.org/10.2203/dose-response.11-009.Hamblin.
24. Ichinose-Tsuno, A., Aoki, A., Takeuchi, Y., Kirikae, T., Shimbo, T., Lee, M. C. Il, Yoshino, F., Maruoka, Y., Itoh, T., Ishikawa, I., & Izumi, Y. (2014). Antimicrobial photodynamic therapy suppresses dental plaque formation in healthy adults: A randomized controlled clinical trial. *BMC Oral Health, 14*(1),152. https://doi.org/10.1186/1472–6831–14–152.
25. Imada, K., Tanaka, S., Ibaraki, Y., Yoshimura, K., & Ito, S. (2014). Antifungal effect of 405-nm light on Botrytis cinerea. *Letters in Applied Microbiology, 59*(6), 670–676. https://doi.org/10.1111/LAM.12330.
26. Jiang, A., Zuo, J., Zheng, Q., Guo, L., Gao, L., Zhao, S., Wang, Q., & Hu, W. (2019). Red LED irradiation maintains the postharvest quality of broccoli by elevating antioxidant enzyme activity and reducing the expression of senescence-related. *Scientia Horticulturae, 251*, 73–79.
27. Josewin, S. W., Ghate, V., Kim, M. J., & Yuk, H. G. (2018). Antibacterial effect of 460 nm light-emitting diode in combination with riboflavin against *Listeria monocytogenes* on smoked salmon. *Food Control, 84*, 354–361.
28. Kang, C. H., Yoon, E. K., Muthusamy, M., Kim, J. A., Jeong, M. J., & Lee, S. I. (2020). Blue LED light irradiation enhances L-ascorbic acid content while reducing reactive oxygen species accumulation in Chinese cabbage seedlings. *Scientia Horticulturae, 261*, 108924.
29. Kim, B. S., Lee, H. O., Kim, J. Y., Kwon, K. H., Cha, H. S., & Kim, J. H. (2011). An effect of light-emitting diode (LED) irradiation treatment on the amplification of functional components of immature strawberry. *Horticulture Environment and Biotechnology, 52*(1), 35–39. https://doi.org/10.1007/S13580-011-0189-2.
30. Kim, D. K., Shin, M., Kim, H. S., & Kang, D. H. (2022). Inactivation efficacy of combination treatment of blue Light-Emitting Diodes (LEDs) and riboflavin to control *E. coli* O157 and *S. typhimurium* in apple juice. *Innovative Food Science & Emerging Technologies, 78*, 103014. https://doi.org/ 10.1016/j.ifset.2022.103014.

31. Kim, D. K., & Kang, D. H. (2021). Efficacy of Light-Emitting Diodes emitting 395, 405, 415, and 425 nm blue light for bacterial inactivation and the microbicidal mechanism. *Food Research International, 141*, 110105.
32. Kim, M. J., Mikš-Krajnik, M., Kumar, A., & Yuk, H. G. (2016). Inactivation by 405±5 nm light-emitting diode on *Escherichia coli* O157: H7, *Salmonella typhimurium*, and *Shigella sonnei* under refrigerated condition might. *Food Control, 59*, 99–107.
33. Kim, S. S., Park, S. H., Kim, S. H., & Kang, D. H. (2019). Synergistic effect of ohmic heating and UV-C irradiation for inactivation of *Escherichia coli* O157: H7, *Salmonella typhimurium* and *Listeria monocytogenes* in. *Food Control, 102*, 69–75.
34. Kingsley, D. H., Perez-Perez, R. E., Boyd, G., Sites, J., & Niemira, B. A. (2018). Evaluation of 405-nm monochromatic light for inactivation of Tulane virus on blueberry surfaces. *Journal of Applied Microbiology, 124*(4), 1017–1022. https://doi.org/10.1111/JAM.13638.
35. Lacombe, A., Niemira, B. A., Sites, J., Boyd, G., Gurtler, J. B., Tyrell, B., & Fleck, M. (2016). Reduction of bacterial pathogens and potential surrogates on the surface of almonds using high-intensity 405-nanometer light. *Journal of Food Protection, 79*(11), 1840–1845.
36. Li, X., & Farid, M. (2016). A review on recent development in non-conventional food sterilization technologies. *Journal of Food Engineering, 182*, 33–45. https://doi.org/10.1016/J.JFOODENG.2016.02.026.
37. Lian, X., Tetsutani, K., Katayama, M., Nakano, M., Mawatari, K., Harada, N., Hamamoto, A., Yamato, M., Akutagawa, M., Kinouchi, Y., & Takahashi, A. (2010). A new colored beverage disinfection system using UV-A light-emitting diodes. *Biocontrol Science, 15*(1), 33–37.
38. Liao, H. L., Alferez, F., & Burns, J. K. (2013). Assessment of blue light treatments on citrus postharvest diseases. *Postharvest Biology and Technology, 81*, 81–88.
39. Lim, S. (2011). Phototherapy and the benefits of LEDs. *Journal of the Society for Information Display, 19*(12), 882. https://doi.org/10.1889/JSID19.12.882.
40. Lim, S., Lee, J., & Crystals, J. J.-M. C. (2015). Red/infrared led unit with moderate light intensity and low heat generation for personal phototherapy applications. *Taylor & Francis, 618*(1), 1–9. https://doi.org/10.1080/15421406.2015.1076297.
41. Liu, F., Li, Z., Cao, B., Wu, J., Wang, Y., Xue, Y., Xu, J., Xue, C., & Tang, Q. J. (2016). The effect of a novel photodynamic activation method mediated by curcumin on oyster shelf life and quality. *Food Research International, 87*, 204–210.
42. Liu, H., Chen, Y., Hu, T., Zhang, S., Zhang, Y., Zhao, T., Yu, H., & Kang, Y. (2016). The influence of Light-Emitting Diodes on the phenolic compounds and antioxidant activities in pea sprouts. *Journal of Functional Foods, 25*, 459–465.
43. Ma, G., Zhang, L., Setiawan, C. K., Yamawaki, K., Asai, T., Nishikawa, F., Maezawa, S., Sato, H., Kanemitsu, N., & Kato, M. (2014). Effect of red and blue LED light irradiation on ascorbate content and expression of genes related to ascorbate metabolism in postharvest broccoli. *Postharvest Biology and Technology, 94*, 97–103.
44. Maclean, M., MacGregor, S. J., Anderson, J. G., & Woolsey, G. (2009). Inactivation of bacterial pathogens following exposure to light from a 405-nanometer light-emitting diode array. *Applied and Environmental Microbiology, 75*(7), 1932–1937. https://doi.org/10.1128/AEM.01892-08.
45. McKenzie, K., Maclean, M., Timoshkin, I. V., MacGregor, S. J., & Anderson, J. G. (2014). Enhanced inactivation of *Escherichia coli* and *Listeria monocytogenes* by

exposure to 405 nm light under sub-lethal temperature, salt and acid stress. *International Journal of Food Microbiology, 170*, 91–98.
46. Mitchell, C. A., Both, A. J., Bourget, C. M., Burr, J. F., Kubota, C., Lopez, R. G., Morrow, R. C., & Runkle, E. S. (2012). LEDs: The future of greenhouse lighting! *Chronica Horticulturae, 52*(1), 6–12.
47. Nagar, A., & Pradeep, T. (2020). Clean Water through Nanotechnology: Needs, Gaps, and Fulfillment. *ACS Nano, 14*(6), 6420–6435. https://doi.org/10.1021/acsnano.9b01730.
48. Papoutsis, K., Mathioudakis, M. M., Hasperué, J. H., & Ziogas, V. (2019). Non-chemical treatments for preventing the postharvest fungal rotting of citrus caused by *Penicillium digitatum* (green mold) and *Penicillium italicum* (blue mold). *Trends in Food Science & Technology, 86*, 479–491.
49. Penha, C. B., Bonin, E., da Silva, A. F., Hioka, N., Zanqueta, É. B., Nakamura, T. U., de Abreu Filho, B. A., Campanerut-Sá, P. A. Z., & Mikcha, J. M. G. (2017). Photodynamic inactivation of foodborne and food spoilage bacteria by curcumin. *LWT – Food Science and Technology, 76*, 198–202.
50. Podgorsak, E. B. (2005). *Radiation Oncology Physics: A Handbook for Teachers and Students.* International Atomic Energy Agency, Vienna, Austria.
51. Poonia, A., Pandey, S., & Vasundhara. (2022). Application of Light-Emitting Diodes (LEDs) for food preservation, post-harvest losses and production of bioactive compounds: A review. *Food Production, Processing and Nutrition, 4*(1),1–10. https://doi.org/10.1186/S43014–022–00086 0.
52. Prasad, A., Du, L., Zubair, M., Subedi, S., Ullah, A., & Roopesh, M. S. (2020). Applications of Light-Emitting Diodes (LEDs) in food processing and water treatment. *Food Engineering Reviews, 12*(3), 268–289. https://doi.org/10.1007/S12393-020-09221-4.
53. Qian, H., Liu, T., Deng, M., Miao, H., Cai, C., Shen, W., & Wang, Q. (2016). Effects of light quality on main health-promoting compounds and antioxidant capacity of Chinese kale sprouts. *Food Chemistry, 196*, 232–1238.
54. Qu, X., Alvarez, P. J. J., & Li, Q. (2013). Applications of nanotechnology in water and wastewater treatment. *Water Research, 47*(12), 3931–3946.
55. Ribeiro, M. S., Núñez, S. C., Sabino, C. P., Yoshimura, T. M., Silva, C. R., Nogueira, G. E., Suzuki, H., & Garcez, A. S. (2015). Exploring light-based technology for wound healing and appliance disinfection. *Journal of the Brazilian Chemical Society, 26*, 2583–2589.
56. Santhirasegaram, V., Razali, Z., George, D. S., & Somasundram, C. (2015). Comparison of UV-C treatment and thermal pasteurization on quality of Chokanan mango (*Mangifera indica* L.) juice. *Food and Bioproducts Processing, 94*, 313–321.
57. Srimagal, A., Ramesh, T., & Sahu, J. K. (2016). Effect of light-emitting diode treatment on inactivation of Escherichia coli in milk. *LWT – Food Science and Technology, 71*, 378–385.
58. Su, J., & Cavaco-Paulo, A. (2021). Effect of ultrasound on protein functionality. *Ultrasonics Sonochemistry, 76*, 105653.
59. Taulavuori, E., Taulavuori, K., Holopainen, J. K., Julkunen-Tiitto, R., Acar, C., & Dincer, I. (2017). Targeted use of LEDs in improvement of production efficiency through phytochemical enrichment. *Journal of the Science of Food and Agriculture, 97*(15), 5059–5064. https://doi.org/10.1002/JSFA.8492.

60. Tortik, N., Spaeth, A., & Plaetzer, K. (2014). Photodynamic decontamination of foodstuff from *Staphylococcus aureus* based on novel formulations of curcumin. *Photochemical & Photobiological Sciences, 13*(10), 1402. https://doi.org/10.1039/c4pp00123k.
61. Wang, D., Zhou, F., Lai, D., Zhang, Y., Hu, J., & Lin, S. (2021). Curcumin-mediated sono/photodynamic treatment preserved the quality of shrimp surimi and influenced its microbial community changes during refrigerated storage. *Ultrasonics Sonochemistry, 7*, 105715.
62. Wang, S. Y., Chen, C. T., & Wang, C. Y. (2009). The influence of light and maturity on fruit quality and flavonoid content of red raspberries. *Food Chemistry, 112*(3), 676–684.
63. Yu, J., Zhang, F., Zhang, J., Han, Q., Song, L., & Meng, X. (2021). Effect of photodynamic treatments on quality and antioxidant properties of fresh-cut potatoes. *Food Chemistry, 362*, 130224.
64. Zhang, T., Shi, Y., Piao, F., & Sun, Z. (2018). Effects of different LED sources on the growth and nitrogen metabolism of lettuce. *Plant Cell, Tissue and Organ Culture, 134*(2), 231–240. https://doi.org/10.1007/S11240-018-1415-8.

CHAPTER 16

LIGHT-EMITTING DIODES: MECHANISM OF INACTIVATION OF MICROORGANISMS AND EFFECTS ON DIFFERENT MICROORGANISMS

YADIGAR SEYFI CANKAL and SEVCAN ÜNLÜTÜRK

ABSTRACT

UV lights are widely used for surface disinfection, sterilization of water, and various liquid foods. The usage of light-emitting diodes (LEDs) has become prominent as a light source for sterilization and decontamination of materials. LEDs have been widely investigated for water treatment. However, there are limited studies on the use of LEDs for food materials due to their low light transmittance capabilities. To develop the most effective treatment for sterilization and decontamination purposes in food processing, it is necessary to know what the different microbial inactivation mechanisms of LEDs are and what their limitations are for use. This chapter includes discussions on the microbial inactivation efficiency of light-emitting diodes (LEDs) of different wavelengths; and the inactivation mechanisms of various microorganisms (such as bacteria, viruses, molds, algae, and yeast).

16.1 INTRODUCTION

There has been increased interest in studies and applications for the use of Light-Emitting Diodes (LEDs) during the cold storage of foods, to improve the nutritional quality of agricultural products at the postharvest stage, and to inactivate pathogenic bacteria and molds in foods [3]. Additionally, the

Nonthermal Light-Based Technologies in Food Processing.
C. K. Sunil, Preeti Birwal, Megh R. Goyal, and R. Mahendran (Eds.)

increasing trend toward fresh-like products forced researchers to investigate alternative processing techniques [6, 50]. In particular, the findings of the studies demonstrating the inactivation efficiency of non-thermal and non-chemical light-based technology revealed that this technology has the potential to be used in food preservation. Therefore, the use of Light-emitting diodes (LEDs) has attracted the food industry in recent years.

The highest germicidal effect of light-based technologies is reported to be observed between 250 and 270 nm (UV-C region) [18, 52]. The antimicrobial effect of UV-C light is very well-known, and this technique is used for disinfection of fruit surfaces, packages, hospital equipment, water resources, etc. [7, 8, 41, 44]. Generally, low or medium-pressure mercury vapor lamps are used for UV-C irradiation, first, one emits predominantly monochromatic UV irradiation at 254 nm, and the second one emits polychromatic UV irradiation in the wavelengths ranging from 200 to 400 nm. However, these lamps contain mercury, which is known to have toxic effects on both the environment and the human body [40]. Therefore, it is necessary to use new UV sources of various sizes, which do not contain toxic substances and have a low energy consumption rate. In that regard, UV disinfection using LEDs (Light Emitting Diodes) seems a promising technology. LEDs are capable of emitting light at multiple individual wavelengths including UV-C (200–280 nm), UV-B (280–315 nm), UV-A (320–400 nm), and visible light spectrum (400–700 nm). Besides, it is possible to use the combination of different LEDs emitting light at different wavelengths.

Different microorganisms have different sensitivity to light, especially to UV-C rays. Cantwell & Hofmann [12] found that coliforms are sensitive to UV-C irradiation [12]. Unluturk et al. [53] reported that gram-negative bacteria (such as coliforms) are more sensitive to UV-C light than gram-positive bacteria due to their cell wall property. Gram-positive bacteria are indicated to have a thick layer of peptidoglycan which does not allow the UV light to penetrate inside the cell. However, gram-negative bacteria cannot resist UV-C light due to their thin peptidoglycan layer [53]. Additionally, it is stated that the inactivation of natural microflora is more difficult than inoculated microorganisms due to the differences in their growth phases and their resistance to environmental conditions [13, 14]. On the other hand, yeasts and molds are not significantly affected by UV-C radiation. It has been reported that yeasts are more resistant to UV-C light than bacteria due to their size, differences in genetic material, as well as cell wall composition, and thickness [19]. Additionally, some studies have revealed the antibacterial effect of LEDs of visible wavelengths on different foodborne pathogens under different lighting temperatures and pH conditions [15, 28].

This chapter explores the microbial inactivation efficiency of light-emitting diodes (LEDs) of different wavelengths and the inactivation mechanisms of various microorganisms such as bacteria, viruses, molds, and yeast. Pulsed irradiation is not considered in this section, as there is limited research on the pulsed irradiation of LEDs and the results are inconsistent.

16.2 INACTIVATION MECHANISM OF LED LIGHT

The photosensitivity of the microorganism depends on the wavelength. The highest germicidal effect of LEDs has been reported in the UV-C region of the electromagnetic spectrum, in particular between 260 and 265 nm, which is the peak absorption range for DNA in cells [22]. UV radiation in different regions (UV-A, UV-B, and UV-C) is absorbed by different biomolecules such as DNA, proteins, and lipids. The inactivation mechanism of UV-C irradiation is based on the absorption of UV photons by the genetic material and the formation of cyclobutene pyrimidine dimers (CPDs) and pyrimidine 6–4 primidone (6-4PP) photoproducts, which inhibit the transcription and replication of the cell (direct photoinactivation) [10, 22, 29, 43]. However, the main mechanism of action of UV light at 280 nm is protein damage because at this wavelength UV light is absorbed by aromatic amino acids [45].

The biological inactivation mechanism of UV-A is based on damaging lipids, proteins, and DNA via the photo-oxidation of O_2 (indirect photoinactivation). It increases the production of reactive oxygen species (ROS) (e.g., H_2O_2, O_2', and OH). Enhanced ROS production causes breakage of single-strand DNA indirectly, and oxidative damage to lipids in the membrane and membrane proteins [20, 46]. Although DNA damage caused by UV-C radiation can be repaired by the enzyme photolyase, it is not possible to repair the damage to bacterial membranes with UV-A radiation [42]. Exposure to UV-B causes not only direct DNA damage by inducing the formation of DNA lesions but also oxidative stress on the lipids and proteins of microorganisms [46].

The inactivation effect of visible light (400–700 nm), particularly blue light (450–495 nm), is based on photodynamic inactivation (PDI), known as the microbial inactivation mechanism resulting from light-induced ROS formation [37]. In the production of ROS, stimulation of endogenous or exogenous photosensitizing molecules is required. Hence, the addition of exogenous photosensitizers (nontoxic dyes) can be used to increase the germicidal efficacy of both UV-A and visible light, and this method is called antibacterial photosensitization-based treatment. For example, yeast cells are more resistant to UV light than bacteria due to their different cell wall structures and large cell sizes. Therefore, it requires multiple damages to vital organelles prior to cell death. Successful photoinactivation of fungi and yeasts with photosensitizers has been demonstrated [38]. On the other hand, viruses are more susceptible to photoinactivation since they have small sizes, receptors required for attachment to the host cell are located on the surface, and do not have a protective envelope. Antibacterial photosensitization-based treatment is a widely preferred practice in killing viruses [39].

16.3 EFFECT OF LED LIGHT ON DIFFERENT MICROORGANISMS

16.3.1 UV-C LED

UV-C light is capable of inactivating various pathogenic microorganisms [34]. Bacteria are very sensitive to UV irradiation because of their small size, short generation time, and inadequate UV-protective pigmentation. The inactivation of pathogenic and spoilage microorganisms using LEDs has been reported in different treatment media (Table 16.1). Table 16.1 summarizes the studies investigating the inactivation efficiency of LEDs in various liquid/solid food applications, excluding water disinfection.

A number of studies showed that UV-C LEDs could inactivate various foodborne pathogens, such as *E. coli* O157:H7, *L. monocytogenes*, and *S. typhimurium, Staphylococcus aureus* in solid media [26, 47], sliced cheese [48], apple juice [2], sliced deli meat and spinach [25], different food contact surfaces [24], coriander [23], wheat flour [49], and saline solution [17]. The inactivation of food spoilage yeasts such as *Pichia membranaefaciens* and *Saccharomyces pastorianus* in solid media and *Botrytis cinerea* (gray mold) on strawberry surfaces by UV-C LEDs has also been reported. In addition, Argyraki et al. [5] have demonstrated that 266 nm UV-C LED had an insignificant inactivation effect on *Pseudomonas aeruginosa* biofilms at different growth stages (24, 48, and 72 h grown) [5].

All these studies showed that the inactivation efficiency of UV-C LEDs depends on the treatment medium, UV wavelength, and the types of microorganisms. Foodborne bacterial pathogens are typically more UV sensitive than spoilage organisms, while biofilm-forming bacteria have been shown to be much more UV resistant (Table 16.1). Studies involving UV LEDs using

TABLE 16.1 Studies on Inactivation of Microorganisms Using UV LEDs

Electro-Magnetic Spectrum of LEDs	Wavelength	Pathogenic Microorganism	Treatment Material	Fluence (mJ/cm^2)	Log Reduction (log CFU/mL)	References
UV-C	275 ± 3	*Escherichia coli* O157:H7, *Salmonella enterica* serovar *Typhimurium* *Listeria monocytogenes*	Solid media	0.17–1.67	5–6	[47]
	254 280	*Escherichia coli* K12 (ATCC 25253)	Apple juice	707.2 ± 143.5 771.6 ± 133.9	4.4	[2]
	266	*Pseudomonas aeruginosa*	Biofilm	2,000	1 (mature biofilm)	[5]
	266 270 275 279	*Escherichia coli* O157:H7, *Salmonella enterica* serovar *Typhimurium, Listeria monocytogenes*	Sliced cheese	1–3	4–5	[48]
	259 268	*Escherichia coli* O157:H7	Saline	7	5.21 4.88	[17]
	259 268	*Listeria monocytogenes*	Saline	7	3.95 4.68	[17]
	266 271 276 279	*Pichia membranaefaciens*	Solid media	0.6	4.53 3.77 4.44 3.19	[26]
	266 271 276 279	*Saccharomyces pastorianus*	Solid media	0.6	1.62 1.43 1.2 0.78	[26]

TABLE 16.1 *(Continued)*

Electro-Magnetic Spectrum of LEDs	Wavelength	Pathogenic Microorganism	Treatment Material	Fluence (mJ/cm^2)	Log Reduction (log CFU/mL)	References
	272	*Botrytis cinerea*	Strawberry	20 mW/cm^2 (only UV irradiance provided)	No growth after 6 days	[11]
	280	*E. coli* O157: H7 *S. typhimurium* *L. monocytogenes*	Selective media Sliced deli meat Spinach	Up to 21.6	1.5–3	[25]
	280	*E. coli* O157: H7 *S. typhimurium* *L. monocytogenes*	Food contact surfaces: glass, PVC, stainless steel (SS), Teflon, and silicon.	0.5–3	0.5–1	[24]
	280	*Salmonella* *E. coli*	Coriander	UV irradiance: 80, 160, 240 $\mu W/cm^2$ Time: 5, 15, and 30 min	0.4–1.3	[23]
	275 (pulsed light)	*Salmonella enterica* spp.	Wheat flour	60.1 (J/cm^2)	1.07	[49]
UV-B	296	*Pseudomonas aeruginosa*	Biofilm	2,000	>3 (mature biofilm)	[5]
	289	*Escherichia coli* O157:H7	Saline	7	0.67	[17]
	289	*Listeria monocytogenes*	Saline	7	1.05	[17]
	289 293	*Botrytis cinerea*	Strawberry	20 mW/cm^2 (only UV irradiance provided)	No growth after 6 days	[11]
UV-A	365 (pulsed)	*Escherichia coli* K12 *Candida albicans*	Biofilm	0.28 mW/cm^2 (UV irradiance) and 60 min (1,008 mJ/cm^2).	>3	[32]

TABLE 16.1 *(Continued)*

Electro-Magnetic Spectrum of LEDs	Wavelength	Pathogenic Microorganism	Treatment Material	Fluence (mJ/cm^2)	Log Reduction (log CFU/mL)	References
	395 ± 5	*Campylobacter jejuni*	Skinless chicken fillet and contact surfaces.	60–1,800	2.62 and 2–4	[21]
	365	*Escherichia coli* strain DH5α	Cabbage	67.5	3.23	[1]
	365	*Escherichia coli* strain DH5α	Colored beverage	126 (J/cm^2)	1.58	[33]
	370	*Escherichia coli* O157:H7	Saline	7	0.06	[17]
	370	*Listeria monocytogenes*	Saline	7	0.18	[17]
	395	*Listeria innocua*	–	36 (J/cm^2)	2.74	[9]
	365 (pulsed light) 395 (pulsed light)	*Salmonella enterica* spp.	Wheat flour	1479.6 (J/cm^2) 2397.6 (J/cm^2)	2.42 3.67	[49]
Combined wavelengths	254/365 254/405 280/365 280/405 254/280/365/405	*Escherichia coli* K12 (ATCC 25253)	Clear apple juice	–	3.22 ± 0.16 3.77 ± 0.08 3.76 ± 0.07 3.58 ± 0.03 3.66 ± 0.15	[2]
	259/289	*Escherichia coli* ATCC 8739 *Listeria seelegri*	Saline	14	2.42 ± 0.31 4.27 ± 0.22	[17]
Visible light	461 521 642	*Salmonella typhimurium* *Escherichia coli* O157:H7, *Listeria monocytogenes* *Staphylococcus aureus*	Solid media	596.7 (J/cm^2)	4.6–5.2 at 461 nm 1–2 at 521 nm No bacterial effect at 642 nm	[16]
	400 nm + 5-aminolevulinic acid (ALA)	*Bacillus cereus*	Packaging material	24 (J/cm^2) at 400 nm	6.3	[35]

TABLE 16.1 *(Continued)*

Electro-Magnetic Spectrum of LEDs	Wavelength	Pathogenic Microorganism	Treatment Material	Fluence (mJ/cm^2)	Log Reduction (log CFU/mL)	References
	435 nm + Curcumin poly-vinylpyrrolidone and NovaSol R-Curcumin formulation	*Staphylococcus aureus*	Cucumber Peppers Chicken meat	812.16 (J/cm^2)	2.6 2.5 1.7	[51]
	585 nm + Hypericin	*Bacillus cereus* Mesophilic bacteria	Apricot Plums Cauliflower	Max 9.2 (J/cm^2)	4.4 0.6–0.72	[4]
	400 nm + Sodium chlorophyllin	*Listeria monocytogenes*	Strawberries	14.4 (J/cm^2)	1.8	[36]
	405	*Salmonella* spp.	Papaya	1.700	1.3	[27]
	405	*Escherichia coli* O157:H7 *Listeria monocytogenes* *Salmonella*	Mango	3,500	1.6	[28]
	405	*Vibrio parahemolyticus* *Lactobacillus plantarum* *Staphylococcus aureus*	Phosphate buffered saline (PBS)	6.0×102 (J/cm^2)	3.6	[31]
	455 (pulsed light)	*Salmonella enterica* spp.	Wheat flour	1756.8 (J/cm^2)	2.64	[49]

saline solution, solid growth media, or smooth surfaces such as food contact surfaces, sliced cheese, spinach, or sliced deli meat generally required lower UV fluence, ranging from 0.17 to 21.0 mJ/cm^2. However, UV-C treatment of opaque food materials such as apple juice, coriander, and wheat flour require the application of much higher UV fluence.

In addition, different microorganisms may show different spectral UV sensitivity when treated at different wavelengths. For example, Green et al. [17] found that *Escherichia coli* O157:H7 exhibited higher UV sensitivity compared to *Listeria monocytogenes* when a similar equivalent UV dose of 7 mJ/cm^2 was employed at 259 nm and 268 nm wavelengths. This was because the UV absorbance of their DNA was different at different wavelengths. While *Escherichia coli* O157:H7 exhibited the highest UV sensitivity at 259 nm, *L. monocytogenes* showed high UV sensitivity at 268 nm. Currently, UV-C LEDs are quite expensive and have low UV irradiance. Therefore, there is a need-to-know which wavelength of UVC LED to work with for effective inactivation. For this, it is necessary to know the action spectrum of food pathogens [30].

16.3.2 UV-B LED

There have been comparatively fewer studies involving UV LEDs in the UV-B range. Argyraki et al. [5] tested the inactivation efficiency of ultraviolet B (UVB) and ultraviolet C (UVC) irradiation on *Pseudomonas aeruginosa* biofilms at different growth stages [5]. They found that UVB irradiation was significantly more effective than UVC irradiation in inactivating *P. aeruginosa* biofilms grown for 24 h. However, the efficacy of the UVB treatment at the same dose and 296 nm drops drastically when performed on more mature biofilms (72 h), resulting in >3 log reduction. It was stated that mature biofilms have a thicker matrix preventing the penetration of light. Green et al. [17] studied the inactivation efficiency of UV-LEDs emitting light at 289 nm on *E. coli*, *Salmonella,* and *Listeria* in a saline solution [17]. When used alone, 289 nm did not produce very high inactivation efficiency with all tested bacteria. It was indicated that the UV sensitivity of all bacteria reduced at wavelengths beyond 268 nm or 275 nm. However, combining different LED wavelengths, which could lead to a possible synergistic effect for bacterial inactivation, yielded effective results. For example, UV treatment at 259/289 nm led to a hurdle effect and resulted in particularly higher log reduction values in *E. coli* ATCC 8739 and *L. seeligeri*. Britz et al. [11]

showed UV treatment at 289 nm and 293 nm extended the cold storage shelf life of strawberries up to 2-fold [11]. Mold growth, *Botrytis cinerea*, was not seen after UV treatment; strawberries were largely mold-free and undamaged even after 9 days. However, more studies are needed to fully understand what effect UVB light has on foods.

16.3.3 UV-A

The inactivation of food pathogens such as *Listeria, Escherichia coli, Salmonella*, and *Campylobacter* has been reported using UV-A LEDs. These studies have shown that the UV sensitivity of these microorganisms can vary depending on the wavelength and treatment medium. UV-A irradiation inactivates microorganisms through an indirect photoinactivation mechanism based on the formation of ROS, while UV-C radiation directly targets the DNA of microorganisms and inactivates them by a direct photoinactivation mechanism. The first mechanism takes a longer time than the second mechanism [30]. Nevertheless, Haughton et al. [21]; Aihara et al. [1]; Lian et al. [33]; Green et al. [17]; Birmpa et al. [9]; and Subedi et al. [49] demonstrated the inactivation capability of UV-A LEDs ranging from 365–395 nm wavelengths on different microorganisms present in different treatment environments [1, 9, 17, 21, 30, 33, 49]. For example, Li et al. [32] demonstrated that pulsed irradiation (100 Hz, 60 min) at 365 nm almost completely killed *Candida albicans* and *Escherichia coli* in biofilm [32]. Haughton et al. [21] investigated the effectiveness of 395 ± 5 nm LED application with a UV dose of 1,800 mJ/cm^2 for decontamination of raw chicken and contact surfaces [21]. Its results revealed reductions of 2.62 and 4 log CFU/g in *C. jejuni* on chicken surfaces and cutting board surfaces. Aihara et al. [1]; and Lian et al. [33] reported the inactivation of *Escherichia coli* strain DH5α on cabbage surface and colored beverages using UVA LED at 365 nm [1, 33]. Lian et al. [33] showed that the UV-A LED sterilization system could reduce the number of bacteria by 1.58 log in liquid beverages and colored solutions using 126 J/cm^2 of UV dose, although the disinfection efficiency was 3 logs in the clear liquid [33]. They suggested using UVA-LED with other methods to ensure the microbial safety of colored beverages. UVA light is able to penetrate deeper into turbid liquids and could possibly offer advantages for the UV treatment of beverages. Similarly, Aihara et al. [1] suggested that UVA-LED treatment may have great potential for surface sterilization of vegetables, based on a reduction of over 3.0 log in bacterial populations for

90 minutes (6.75×10^2 J/m^2) after UVA irradiation [1]. In addition, Green et al. [17] demonstrated that *Escherichia coli* O157:H7 and *Listeria monocytogenes* could be reduced by 0.06 and 0.18 log in saline solution, respectively, by applying UVA LED at 370 nm with UV dose of 7 mJ/cm^2 [17]. However, Subedi et al. [49] applied a much higher UV dose (up to 2397.6 J/cm^2) to achieve a 3.67 log reduction of *Salmonella enterica* spp. in wheat flour using high-intensity LED at 365 and 395 nm [49]. They pointed out that pulsed UVA LEDs could be used for both drying and decontamination purposes, due to a photothermal effect causing an increase in the surface temperature of wheat flour.

16.3.4 COMBINATION OF DIFFERENT UV-LEDS

In some studies, it has been shown that after simultaneous combined UV-LED treatment at different wavelengths, a synergistic inactivation effect was created due to the alternative inactivation mechanisms of each wavelength and a hurdle effect occurred. For example, Green et al. [17] reported 2.42 and 4.27 log reductions in *E. coli* 8739 and *Listeria seelegri*, respectively, after UV treatment with 259/289 nm [17]. In contrast, the theoretical additive effects of each wavelength resulted in 1.20 and 4.27 log reductions for *E. coli* 8739 and *Listeria seelegri,* respectively. Similarly, Akgün & Ünlütürk [2] studied the effects of combined ultraviolet light-emitting diodes (UV-LEDs) on the inactivation of *E. coli* K12 (ATCC 25253) and polyphenol oxidase (PPO) in cloudy and clear apple juice [2]. For clear apple juice, 40 min treatment time resulted in a maximum 3.9 log reduction when 280/365 nm wavelength combination was used. However, the same treatment time and combined wavelengths resulted in a 2.0 log reduction in cloudy apple juice. The difference was attributed to the fact that cloudy liquids had low absorption of UV due to the presence of color compounds, organic matter, and suspended solids, and therefore UV efficiency was lower than clear liquids. It was stated that UV-A and UV-C rays in combination showed a better inactivation effect on PPO than UV-C rays used separately. Additionally, the authors investigated the change in the color of juice samples after different UV treatments. They found that samples exposed to UV-LED irradiation at 280/365 nm had the lowest color difference (ΔE).

Therefore, the use of LEDs of different wavelengths can be used as a treatment strategy both to ensure microbial inactivation and to preserve the quality characteristics of food products.

16.3.5 VISIBLE LIGHT LEDS

A number of researchers have shown that visible wavelength LEDs are effective in reducing microbial loads in various foods [4, 16, 27, 28, 31, 35, 36, 49, 51]. Ghate et al. [16] studied the effect of visible wavelengths (461, 521, and 642 nm) and temperature on the inactivation, D values, and sublethal injury of *Escherichia coli* O157:H7, *Salmonella typhimurium, Listeria monocytogenes*, and *Staphylococcus aureus* [16]. For all bacterial strains, the log reduction was in the range of 4.6–5.2 log CFU/ml at 10 and 15C after illumination with 461 nm LED, while illumination with 521 nm LED resulted in only 1.0–2.0 log reductions after 7.5 hours. On the other hand, they reported that no antibacterial effect was observed using the 642 nm LED treatment. In this study, photodynamic inactivation by 461 and 521 nm LEDs was found to be greater at temperatures of 10 and 15?C than at 20?C. There was no significant difference between the D values of the bacterial strains except *L. monocytogenes* at 10 and 15?C, and their sensitivity to LED lighting was found to be the same. Regardless of temperature, sublethal injury was observed in all bacterial strains during illumination with 461 and 521 nm LEDs, and the percentage of injured cells increased as treatment time increased. The results showed that the antibacterial effect of the LEDs was highly dependent on the wavelength and the temperature.

5-aminolevulinic acid (ALA)–based photosensitization treatment was found to be effective against *Bacillus cereus* spores on packaging surfaces when treated with LED light at 400 nm and a total energy dose of 24 J/cm^2 [35]. The log reduction after illumination was found to be 6.3 logs. Photosensitization, with its antimicrobial activity, could serve for the development of completely safe, non-thermal surface decontamination and food preservation techniques in the future. Tortik et al. [51] focused on photodynamic inactivation (PDI), employing curcumin bound to polyvinylpyrrolidone (PVP-C) and NovaSol®-curcumin as photosensitizers for the decontamination of cucumber, pepper, and chicken meat from *Staphylococcus aureus* using visible light at 435 nm [51]. Photodynamic inactivation of *S. aureus* averaged 2.6 log (99.8%) reductions for cucumber, 2.5 log (99.7%) for pepper, and 1.7 log (98%) for chicken meat. Moreover, no visible change in the external appearance of the foodstuff was observed after photodynamic decontamination.

Aponiene et al. [4] investigated the removal of natural microflora on fruits and vegetables by hypericin (Hyp)-based photosensitization at 585 nm [4]. Results indicated that Hyp-based photosensitization effectively reduced the population of *Bacillus in vitro* by 4.4 log CFU/mL. However, inactivation of mesophilic bacteria on the food surfaces reached approximately 0.6–0.72 log. Apricot, plum, and cauliflower treated with hypericin and irradiated with LED light were found to have similar antioxidant activity and color compared to control samples. During treatment, the temperature on the surface of the products changed from 20 to 25?C. The study revealed that Hyp-based photosensitization could be an effective non-thermal method as an alternative to chemical decontamination processes. However, the possible toxicity of treated fruits and vegetables must be evaluated in the future.

Luksiene & Paskeviciute [36] studied the possibility of controlling microbial contamination of strawberries by sodium chlorophyllin (Na-Chl)-based photosensitization. They inoculated the strawberries with *Listeria monocytogenes* and illuminated them for 30 minutes using visible light at 400 nm. The number of *Listeria monocytogenes* was reduced by 1.8 log, and the shelf life of treated strawberries was extended by 2 days. While the total antioxidant activity of the treated strawberries increased, no effect was detected on the amount of phenol, anthocyanin, or surface color. It was revealed that Na-Chl-based photosensitization could be a useful tool in a non-thermal food decontamination process. However, in the future, more detailed studies on the quality and organoleptic properties of processed fruits were suggested.

Kim et al. [26] found that blue light LEDs at 405 nm wavelengths could reduce the *Salmonella* count on fresh-cut papaya by up to 1.3 logs using a UV dose of 1.7 kJ/cm^2, whereas *Escherichia coli* O157:H7 and *Listeria monocytogenes* levels on mango surface could be reduced by 1.6 logs using a dose of 3.5 kJ/cm^2 [27, 28].

Kumar et al. [31] have examined the antibacterial efficacy of 405, 460, and 520 nm light-emitting diodes on *Lactobacillus plantarum, Staphylococcus aureus*, and *Vibrio parahemolyticus* in phosphate-buffered saline (PBS) at 4, 10, and 25?C, respectively [31]. As a result, the researchers found that 520 nm did not have any antibacterial effect, whereas 405 nm LED illumination significantly reduced the populations of all three bacteria, while 460 nm LED produced photodynamic inactivation only in *Lactobacillus plantarum* and *Vibrio parahemolyticus* populations. This study demonstrated that photodynamic inactivation using 405 nm LEDs has the potential to be used as an antimicrobial control method for food preservation and safety.

Similarly, Subedi et al. [49] investigated the inactivation of Salmonella and quality changes in wheat flour after pulsed light-emitting diode (LED) treatments at 455 nm and different environmental relative humidity (RH) conditions [49]. The exposure to LED radiant energy of 1756.8 J/cm^2 at the wavelength of 455 nm resulted in a *Salmonella* reduction of 2.64 logs CFU/g at 25?C and an RH of 75%. Although the authors observed significant color changes in wheat flour, as well as significant drying, oxidation of gluten proteins, and changes in the secondary structure of gluten after LED treatment at 455 nm, they indicated that LED treatment has the potential for simultaneous drying and microbial decontamination of foods.

16.4 SUMMARY

LEDs in the range of 200 and 700 nm wavelength are generally used for the inactivation of pathogens and spoilage microorganisms. The effectiveness of the treatment depends on the application wavelength, type of microorganisms (bacteria, molds, yeasts, and viruses), and type of food or food-contact material. The inactivation mechanisms of LEDs are based on direct or indirect photoinactivation and photodynamic inactivation mechanisms. UVC-LED wavelength is known as the most germicidal range and causes DNA damage (direct photoinactivation). On the other hand, UVA-LED treatment results in damaging proteins, lipids, and DNA photooxidation via ROS production (indirect photoinactivation). UVB-LEDs provide both DNA damage and oxidative stress of proteins and lipids. In addition to the UV range, the inactivation mechanism of visible light is photodynamic inactivation via light-induced ROS formation in the presence of photosensitizing molecules that are generally used to increase the efficacy of treatment. In addition to the individual applications of LED treatment at each wavelength, a synergistic effect can be obtained by combining different wavelengths with each other. Besides, for an effective inactivation of the UVC LED, the spectrum of action of food pathogens must be known. The effect of UVB light on foods is still not understood; more studies are needed for this. Pulsed UVA LEDs can also be used for both drying and decontamination purposes. In order to increase the inactivation efficiency of light in the visible region and to use it in decontamination, different wavelengths should be used with photosensitizers. In addition, by using moderate heat treatment, the inactivation efficiency of LEDs can be increased on food surfaces and liquid food environments.

KEYWORDS

- **beverages**
- **food quality**
- **food safety**
- **fruit juice**
- **light-emitting diodes**
- **liquid food products**
- **UV-LEDs**

REFERENCES

1. Aihara, M., Lian, X., Shimohata, T., Uebanso, T., Mawatari, K., Harada, Y., & Takahashi, A. (2014). Vegetable surface sterilization system using UVA light-emitting diodes. *The Journal of Medical Investigation: JMI, 61*(3–4), 285–290. https://doi.org/10.2152/jmi.61.285.
2. Akgün, M. P., & Ünlütürk, S. (2017). Effects of ultraviolet light emitting diodes (LEDs) on microbial and enzyme inactivation of apple juice. *International Journal of Food Microbiology, 260*, 65–74. https://doi.org/10.1016/j.ijfoodmicro.2017.08.007.
3. Anbazhagi, M., Shilpa, V., & Munthikote, S. (2018). Role of light emitting diode in food industry. *CPQ Nutrition, 1*(2), 1–9.
4. Aponiene, K., Paskeviciute, E., Reklaitis, I., & Luksiene, Z. (2015). Reduction of microbial contamination of fruits and vegetables by hypericin-based photosensitization: Comparison with other emerging antimicrobial treatments. *Journal of Food Engineering, 144*, 29–35. https://doi.org/10.1016/j.jfoodeng.2014.07.012.
5. Argyraki, A., Markvart, M., Bjørndal, L., Bjarnsholt, T., & Petersen, P. M. (2017). Inactivation of Pseudomonas aeruginosa biofilm after ultraviolet light-emitting diode treatment: A comparative study between ultraviolet C and ultraviolet B. *Journal of Biomedical Optics, 22*(6), 65004. https://doi.org/10.1117/1.JBO.22.6.065004.
6. Basaran-Akgul, N., Churey, J. J., Basaran, P., & Worobo, R. W. (2009). Inactivation of different strains of *Escherichia coli* O157 in various apple ciders treated with dimethyl dicarbonate (DMDC) and sulfur dioxide (SO_2) as an alternative method. *Food Microbiology, 26*(1), 8–15. https://doi.org/10.1016/j.fm.2008.07.011.
7. Begum, M., Hocking, A. D., & Miskelly, D. (2009). Inactivation of food spoilage fungi by ultra violet (UVC) irradiation. *International Journal of Food Microbiology, 129*(1), 74–77. https://doi.org/10.1016/j.ijfoodmicro.2008.11.020.
8. Bintsis, T., Litopoulou-Tzanetaki, E., & Robinson, R. K. (2000). Existing and potential applications of ultraviolet light in the food industry – A critical review. *Journal of the Science of Food and Agriculture, 80*(6), 637–645. https://doi.org/10.1002/(SICI)1097-0010(20000501)80:6<637::AID-JSFA603>3.0.CO;2-1.

9. Birmpa, A., Vantarakis, A., Paparrodopoulos, S., Whyte, P., & Lyng, J. (2014). Efficacy of three light technologies for reducing microbial populations in liquid suspensions. *BioMed Research International, 2014*, 673939. https://doi.org/10.1155/2014/673939.
10. Bolton, J. R., Stefan, M. I., Shaw, P.-S., & Lykke, K. R. (2011). Determination of the quantum yields of the potassium ferrioxalate and potassium iodide–iodate actinometers and a method for the calibration of radiometer detectors. *Journal of Photochemistry and Photobiology A: Chemistry, 222*(1), 166–169. https://doi.org/10.1016/j.jphotochem.2011.05.017.
11. Britz, S., Gaska, I., Shturm, I., Bilenko, Y., Shatalov, M., & Gaska, R. (2013). Deep ultraviolet (DUV) Light-Emitting Diodes (LEDs) to maintain freshness and phytochemical composition during postharvest storage. In *CLEO: 2013* (pp. 1–2). https://doi.org/10.1364/CLEO_AT.2013.ATh3N.3.
12. Cantwell, R. E., & Hofmann, R. (2008). Inactivation of indigenous coliform bacteria in unfiltered surface water by ultraviolet light. *Water Research, 42*(10–11), 2729–2735. https://doi.org/10.1016/j.watres.2008.02.002.
13. El-Hag, A. H., Dadarwal, R., Gonzalez, O. R., Jayaram, S. H., & Griffiths, M. W. (2010). Survivability of inoculated versus naturally grown bacteria in apple juice under pulsed electric fields. *IEEE Transactions on Industry Applications, 46*(1), 9–15. https://doi.org/10.1109/TIA.2009.2036513.
14. El-Hag, A. H., Jayaram, S. H., & Griffiths, M. W. (2006). Inactivation of naturally grown microorganisms in orange juice using pulsed electric fields. *IEEE Transactions on Plasma Science, 34*(4), 1412–1415. https://doi.org/10.1109/TPS.2006.878382.
15. Ghate, V., Leong, A. L., Kumar, A., Bang, W. S., Zhou, W., & Yuk, H.-G. (2015). Enhancing the antibacterial effect of 461 and 521 nm light emitting diodes on selected foodborne pathogens in trypticase soy broth by acidic and alkaline pH conditions. *Food Microbiology, 48*, 49–57. https://doi.org/10.1016/j.fm.2014.10.014.
16. Ghate, V. S., Ng, K. S., Zhou, W., Yang, H., Khoo, G. H., Yoon, W.-B., & Yuk, H.-G. (2013). Antibacterial effect of light emitting diodes of visible wavelengths on selected foodborne pathogens at different illumination temperatures. *International Journal of Food Microbiology, 166*(3), 399–406. https://doi.org/10.1016/j.ijfoodmicro.2013.07.018.
17. Green, A., Popović, V., Pierscianowski, J., Biancaniello, M., Warriner, K., & Koutchma, T. (2018). Inactivation of *Escherichia coli, Listeria* and *Salmonella* by single and multiple wavelength ultraviolet-light emitting diodes. *Innovative Food Science & Emerging Technologies, 47*, 353–361. https://doi.org/10.1016/j.ifset.2018.03.019.
18. Guerrero-Beltrán, J. A., & Barbosa-Cánovas, G. V. (2005). Reduction of *Saccharomyces cerevisiae*, *Escherichia coli* and *Listeria innocua* in apple juice by ultraviolet light. *Journal of Food Process Engineering, 28*(5), 437–452. https://doi.org/10.1111/j.1745-4530.2005.00040.x.
19. Hakguder Taze, B., Unluturk, S., Buzrul, S., & Alpas, H. (2015). The impact of UV-C irradiation on spoilage microorganisms and color of orange juice. *Journal of Food Science and Technology, 52*(2), 1000–1007. https://doi.org/10.1007/s13197-013-1095-7.
20. Hamamoto, A., Mori, M., Takahashi, A., Nakano, M., Wakikawa, N., Akutagawa, M., & Kinouchi, Y. (2007). New water disinfection system using UVA light-emitting diodes. *Journal of Applied Microbiology, 103*(6), 2291–2298. https://doi.org/10.1111/j.1365-2672.2007.03464.x.
21. Haughton, P. N., Grau, E. G., Lyng, J., Cronin, D., Fanning, S., & Whyte, P. (2012). Susceptibility of Campylobacter to high intensity near ultraviolet/visible 395±5nm

light and its effectiveness for the decontamination of raw chicken and contact surfaces. *International Journal of Food Microbiology, 159*(3), 267–273. https://doi.org/10.1016/j.ijfoodmicro.2012.09.006.
22. Hinds, L. M., O'Donnell, C. P., Akhter, M., & Tiwari, B. K. (2019). Principles and mechanisms of ultraviolet light emitting diode technology for food industry applications. *Innovative Food Science & Emerging Technologies, 56*, 102153. https://doi.org/10.1016/j.ifset.2019.04.006.
23. Jiang, Y., Ai, C., Liao, X., Liu, D., & Ding, T. (2020). Effect of slightly acidic electrolyzed water (SAEW) and ultraviolet light illumination pretreatment on microflora inactivation of coriander. *LWT, 132*, 109898. https://doi.org/10.1016/j.lwt.2020.109898.
24. Kim, D., & Kang, D.-H. (2020). Effect of surface characteristics on the bactericidal efficacy of UVC LEDs. *Food Control, 108*, 106869. https://doi.org/10.1016/j.foodcont.2019.106869.
25. Kim, D. K., & Kang, D. H. (2020). Inactivation efficacy of a sixteen UVC LED module to control foodborne pathogens on selective media and sliced deli meat and spinach surfaces. *LWT, 130*, 109422. https://doi.org/10.1016/j.lwt.2020.109422.
26. Kim, D.-K., Kim, S.-J., & Kang, D.-H. (2017). Bactericidal effect of 266 to 279 nm wavelength UVC-LEDs for inactivation of Gram-positive and Gram-negative foodborne pathogenic bacteria and yeasts. *Food Research International (Ottawa, Ont.), 97*, 280–287. https://doi.org/10.1016/j.foodres.2017.04.009.
27. Kim, M. J., Bang, W. S., & Yuk, H. G. (2017). 405 ± 5 nm light emitting diode illumination causes photodynamic inactivation of *Salmonella* spp. on fresh-cut papaya without deterioration. *Food Microbiology, 62*, 124–132. https://doi.org/10.1016/j.fm.2016.10.002.
28. Kim, M. J., Tang, C. H., Bang, W. S., & Yuk, H. G. (2017). Antibacterial effect of 405±5nm light emitting diode illumination against *Escherichia coli* O157, *Listeria monocytogenes*, and *Salmonella* on the surface of fresh-cut mango and its influence on fruit quality. *International Journal of Food Microbiology, 244*, 82–89. https://doi.org/10.1016/j.ijfoodmicro.2016.12.023.
29. Koutchma, T. (2009). Advances in ultraviolet light technology for non-thermal processing of liquid foods. *Food and Bioprocess Technology, 2*(2), 138–155. https://doi.org/10.1007/s11947-008-0178-3.
30. Koutchma, T., & Popović, V. (2019). Chapter 5 – UV Light-Emitting Diodes (LEDs) and food safety. In T. B. T. U. L. E. D. T., & F. A. Koutchma (Eds.), (pp. 91–117). Academic Press. https://doi.org/10.1016/B978-0-12-817794-5.00005-4.
31. Kumar, A., Ghate, V., Kim, M. J., Zhou, W., Khoo, G. H., & Yuk, H. G. (2016). Antibacterial efficacy of 405, 460 and 520 nm light emitting diodes on *Lactobacillus plantarum*, *Staphylococcus aureus* and *Vibrio parahaemolyticus*. *Journal of Applied Microbiology, 120*(1), 49–56. https://doi.org/10.1111/jam.12975.
32. Li, J., Hirota, K., Yumoto, H., Matsuo, T., Miyake, Y., & Ichikawa, T. (2010). Enhanced germicidal effects of pulsed UV-LED irradiation on biofilms. *Journal of Applied Microbiology, 109*(6), 2183–2190. https://doi.org/10.1111/j.1365-2672.2010.04850.x.
33. Lian, X., Tetsutani, K., Katayama, M., Nakano, M., Mawatari, K., Harada, N., & Takahashi, A. (2010). A new colored beverage disinfection system using UV-A light-emitting diodes. *Biocontrol Science, 15*(1), 33–37. https://doi.org/10.4265/bio.15.33.
34. Lui, G. Y., Roser, D., Corkish, R., Ashbolt, N., Jagals, P., & Stuetz, R. (2014). Photovoltaic powered ultraviolet and visible Light-Emitting Diodes for sustainable

point-of-use disinfection of drinking waters. *Science of The Total Environment, 493*, 185–196. https://doi.org/10.1016/j.scitotenv.2014.05.104.
35. Luksiene, Z., Buchovec, I., & Paskeviciute, E. (2009). Inactivation of food pathogen *Bacillus cereus* by photosensitization in vitro and on the surface of packaging material. *Journal of Applied Microbiology, 107*(6), 2037–2046. https://doi.org/10.1111/j.1365-2672.2009.04383.x.
36. Luksiene, Z., & Paskeviciute, E. (2011). Novel approach to the microbial decontamination of strawberries: Chlorophyllin-based photosensitization. *Journal of Applied Microbiology, 110*(5), 1274–1283. https://doi.org/10.1111/j.1365-2672.2011.04986.x.
37. Luksiene, Z., & Brovko, L. (2013). Antibacterial photosensitization-based treatment for food safety. *Food Engineering Reviews, 5*(4), 185–199. https://doi.org/10.1007/s12393-013-9070-7.
38. Lyon, J. P., Moreira, L. M., de Moraes, P. C. G., dos Santos, F. V., & de Resende, M. A. (2011). Photodynamic therapy for pathogenic fungi. *Mycoses, 54*(5), e265–e271. https://doi.org/10.1111/j.1439-0507.2010.01966.x.
39. Mohr, H., Bachmann, B., Klein-Struckmeier, A., & Lambrecht, B. (1997). Virus inactivation of blood products by phenothiazine dyes and light. *Photochemistry and Photobiology, 65*(3), 441–445. https://doi.org/10.1111/j.1751-1097.1997.tb08586.x.
40. Mori, M., Hamamoto, A., Takahashi, A., Nakano, M., Wakikawa, N., Tachibana, S., Ikehara, T., Nakaya, Y., Akutagawa, M., & Kinouchi, Y. (2007). Development of a new water sterilization device with a 365 nm UV-LED. *Med. Biol. Eng. Comput. 45*, 1237–1241. https://doi.org/ 10.1007/s11517-007-0263-1.
41. Nigro, F., Ippolito, A., & Lima, G. (1998). Use of UV-C light to reduce Botrytis storage rot of table grapes. *Postharvest Biology and Technology, 13*(3), 171–181. https://doi.org/10.1016/S0925-5214(98)00009-X.
42. Oguma, K., Katayama, H., & Ohgaki, S. (2002). Photoreactivation of *Escherichia coli* after low- or medium-pressure UV disinfection determined by an endonuclease sensitive site assay. *Applied and Environmental Microbiology, 68*(12), 6029–6035. https://doi.org/10.1128/AEM.68.12.6029-6035.2002.
43. Oguma, K., Kita, R., Sakai, H., Murakami, M., & Takizawa, S. (2013). Application of UV light emitting diodes to batch and flow-through water disinfection systems. *Desalination, 328*, 24–30. https://doi.org/10.1016/j.desal.2013.08.014.
44. Pan, J., Vicente, A. R., Martínez, G. A., Chaves, A. R., & Civello, P. M. (2004). Combined use of UV-C irradiation and heat treatment to improve postharvest life of strawberry fruit. *Journal of the Science of Food and Agriculture, 84*(14), 1831–1838. https://doi.org/10.1002/jsfa.1894.
45. Popović, V., Koutchma, T., & Pagan, J. (2021). Emerging applications of ultraviolet Light-Emitting Diodes for foods and beverages. In K. Knoerzer & K. Muthukumarappan (Eds.), *Innovative Food Processing Technologies* (pp. 335–344). Oxford: Elsevier. https://doi.org/10.1016/B978-0-08-100596-5.22667-X.
46. Santos, A. L., Oliveira, V., Baptista, I., Henriques, I., Gomes, N. C. M., Almeida, A., & Cunha, Â. (2013). Wavelength dependence of biological damage induced by UV radiation on bacteria. *Archives of Microbiology, 195*(1), 63–74. https://doi.org/10.1007/s00203-012-0847-5.
47. Shin, J.-Y., Kim, S.-J., Kim, D.-K., & Kang, D.-H. (2016). Fundamental characteristics of deep-UV Light-Emitting Diodes and their application to control foodborne pathogens.

Applied and Environmental Microbiology, 82(1), 2–10. https://doi.org/10.1128/AEM.01186-15.
48. Soo-Ji, K., Do-Kyun, K., Dong-Hyun, K., & Lee, S. J. (2016). Using UVC Light-Emitting Diodes at wavelengths of 266 to 279 nanometers to inactivate foodborne pathogens and pasteurize sliced cheese. *Applied and Environmental Microbiology, 82*(1), 11–17. https://doi.org/10.1128/AEM.02092-15.
49. Subedi, S., Du, L., Prasad, A., Yadav, B., & Roopesh, M. S. (2020). Inactivation of *Salmonella* and quality changes in wheat flour after pulsed light-emitting diode (LED) treatments. *Food and Bioproducts Processing, 121*, 166–177. https://doi.org/10.1016/j.fbp.2020.02.004.
50. Tahiri, I., Makhlouf, J., Paquin, P., & Fliss, I. (2006). Inactivation of food spoilage bacteria and *Escherichia coli* O157 in phosphate buffer and orange juice using dynamic high pressure. *Food Research International, 39*(1), 98–105. https://doi.org/10.1016/j.foodres.2005.06.005.
51. Tortik, N., Spaeth, A., & Plaetzer, K. (2014). Photodynamic decontamination of foodstuff from *Staphylococcus aureus* based on novel formulations of curcumin. *Photochemical & Photobiological Sciences, 13*(10), 1402–1409. https://doi.org/10.1039/c4pp00123k.
52. Tran, M. T. T., & Farid, M. (2004). Ultraviolet treatment of orange juice. *Innovative Food Science & Emerging Technologies, 5*(4), 495–502. https://doi.org/10.1016/j.ifset.2004.08.002.
53. Unluturk, S., Atılgan, M. R., Baysal, A. H., & Unluturk, M. S. (2010). Modeling inactivation kinetics of liquid egg white exposed to UV-C irradiation. *International Journal of Food Microbiology, 142*(3), 341–347.

CHAPTER 17

APPLICATIONS OF LIGHT-EMITTING DIODES IN LIQUID FOODS

YADIGAR SEYFI CANKAL and SEVCAN ÜNLÜTÜRK

ABSTRACT

Recently, there has been increased interest in investigating alternative non-thermal processes to maintain the quality of liquid food products while preserving their nutritional value. Ultraviolet (UV) irradiation technology is one such method and has been widely explored for liquid food products. Low or medium-pressure mercury vapor lamps are used in this technology. Since these lamps contain mercury known to have toxic effects on the human body and the environment [18], the use of light-emitting diodes (LEDs) has become prominent as an ultraviolet light source. LEDs are created using non-toxic semiconductor material, which allows emitting in different wavelengths. Recently, Light Emitting Diode (LED) technology has been emerging as a non-thermal and non-chemical treatment for surface disinfection and preservation of solid and liquid food materials. This chapter emphasizes the application and effectiveness of LEDs of different wavelengths in liquid food processing and storage.

17.1 INTRODUCTION

Food safety is an important issue for the food industry. Food safety is defined as the methods that ensure the protection of food products from foodborne pathogens and microorganisms that cause spoilage during the transportation, processing, and storage of food products [37]. Liquid foods are suitable media for pathogens and spoilage microorganisms due to the high-water activity

Nonthermal Light-Based Technologies in Food Processing.
C. K. Sunil, Preeti Birwal, Megh R. Goyal, and R. Mahendran (Eds.)

(a_w) and carbohydrate content. In addition to the microbial load induced by microorganism growth, these microorganisms cause negative effects on the quality of liquid foods and reduce the shelf life of the products. In order to ensure food safety for liquid products and increase their shelf life, several food preservation techniques have been applied to liquid food products from the past to the present. Some authorities such as the FDA (Food and Drug Administration) approve/inspect these techniques. According to the FDA, the most well-known and preferred preservation technique is thermal treatment, such as pasteurization, sterilization, etc. The FDA specifies a 5-log reduction in pathogen counts as a microbial safety standard for juice products under HACCP guidelines. Although thermal treatment provides effective microbial inactivation, it causes some defects in liquid foods such as losses in nutrients (ascorbic acid, riboflavin), rheological and sensory properties of liquid foods [11, 28, 38]. Therefore, non-thermal methods (such as pulsed electric field (PEF), high pressure processing (HPP), radiation, and irradiation) can be applied as alternative treatments to prevent harmful effects of thermal techniques.

Light-emitting diodes (LEDs) made of a semiconductor and producing monochromatic wavelengths are new sources of irradiation. Treatment with LEDs is one of the preferred non-thermal techniques due to its non-chemically invasive nature [22]. Widely applied LED wavelengths for anti-microbial activity in the food industry can be viewed under four categories: 200–280 nm (UVC), 280–320 nm (UVB), 320–400 nm (UVA), and 400–470 nm (blue light). Recently, it has been noticed that LEDs are generally used for air disinfection, water treatment, and surface decontamination. However, longer wavelengths such as infrared and red (630–1,000 nm) are generally preferred for curing, drying, ink curing, and phototherapy applications [24].

LEDs are mostly used for water disinfection purposes rather than liquid beverages such as fruit juice, wine, beer, and milk. Extensive research has been carried out in the treatment and purification of water and air. Thus, studies of LED applications in the field of different liquid food products are restricted. According to these studies, LED treatment is generally used alone or in combination with a different preservation method for microbial inactivation [1–3, 8, 11, 13, 20, 21, 33, 38], improving the quality of the liquid product, and enhancing the storage conditions of the products [5, 6, 14, 17, 23, 26, 29, 32, 34–36]. Additionally, LEDs are also used in different analytical devices or in the improvement of existing analytic methods. For instance, Su et al. [30] used capillary electrophoresis-blue light emitting diodes (CE-LED) to develop a simple, inexpensive, and reliable method for routine riboflavin analysis in beer [30]. LED-assisted processes or devices in different applications are covered in various studies [4, 7, 10, 12, 15, 20, 25, 27, 31].

This chapter explores the potential application of LED technology in the processing of liquid products such as apple juice, orange juice, milk, water, and other beverages (alcoholic and non-alcoholic).

17.2 APPLICATIONS OF LED TECHNOLOGY IN LIQUID FOODS

Recent studies on LED technology to treat liquid food products are listed in Table 17.1.

TABLE 17.1 Recent Studies on LED Technology to Treat Liquid Food Products

Product Type	Aim of the Application	References
Apple juice	Microbial inactivation	[1, 11, 13, 38]
Alcoholic beverages	Quality improvement in storage conditions	[17]
Orange juice	Microbial inactivation	[3, 8]
Other beverages	Microbial inactivation	[2, 33]
	Quality improvement in storage conditions	[14]
Milk	Microbial inactivation	[9, 16]
	Quality improvement in storage conditions	[5, 6, 23, 26, 29, 32, 34, 35]
Water	Microbial inactivation	[20, 21]
	System configuration to optimize the inactivation efficiency	[36]

17.2.1 MILK

Milk is rich in lipids, proteins, vitamins (such as vitamin A and riboflavin), and minerals (such as iron, calcium, phosphorus, potassium, etc.). It is vulnerable to microbial spoilage and quality losses due to its rich content. Gram-positive bacteria such as *Pseudomonas* sp. are the main groups responsible for quality losses. This group of bacteria is responsible for sensory defects, such as fruity, barny, and shrimp flavors after processing. Additionally, gram-positive spore-forming bacteria can also be found in the natural flora of raw milk and can survive processing conditions. They can then multiply during refrigerated storage and lead to microbial spoilage.

Dos Anjos et al. [9] studied the inactivation efficiency of LED treatment on UHT bovine milk products. They investigated the effects of blue light treatment at 413 nm on microbial inactivation and the nutritional quality of milk [9]. *S. aureus*, *E. coli*, *Mycobacterium fortuitum*, *Salmonella Typhimurium*, and *Pseudomonas aeruginosa* cultures were used in this study. Microbial growth kinetics were investigated based on the response of these cultures to the blue light used at a wavelength of 413 nm at an irradiation level of 100 mW/cm^2. The nutritional components (sugars, proteins, and lipids) were determined using near-infrared (NIR) and Fourier transform infrared spectroscopy (FTIR). The results showed that at 600 J/cm^2 (corresponding to less than 100 minutes of exposure), a reduction of more than 5 logs was achieved in all bacteria. It was observed that there was no statistically significant difference in protein, sugar, and lipid contents between irradiated and unirradiated samples. Conversely, the riboflavin content of milk slightly decreased after irradiation, but it was suggested that this could be adjusted with a small supplementation.

Although LED treatment provides adequate microbial inactivation in milk, it can induce the oxidation of some photosensitive components such as riboflavin, porphyrin, chlorophyll, etc., during treatment or storage conditions. For example, Martin et al. [16] studied the storage of milk under LED light. The effect of LED lighting on milk quality was examined by sensory evaluation, consumer testing, microbiological testing, and analysis of some nutritional components of milk (casein as a percentage of true protein, CS/TP, color assessment, mineral analysis including % calcium, % phosphorus, % magnesium, % potassium, % sodium, zinc, copper, iron, manganese, and molybdenum in ppm unit) [16]. Skim and 2% milk samples were used in this study and coded as fresh milk (FR) and near-code milk (NC). All samples were exposed to LED light with an average intensity of 1,200 lx for 4 hours. FR samples were evaluated immediately after exposure to light, while NC samples were stored for ~13 days between exposure and testing. Total bacteria count (TBC), psychotropic bacteria count (PBC), total aerobic plate count (APC), and total gram-negative bacteria count were performed to evaluate the microbial quality of samples. TBC in all samples was 4.64-log CFU/mL on average. APC was below the pasteurized milk ordinance of 4.31-log cfu/mL for all FR samples and the majority of NC samples. The CS/TP ratio had a correlation with APC and gram-negative bacteria count. Moreover, panelists detected the creamy aroma, buttery taste, and sweet aftertaste in milk samples. Product age was associated with a plastic and putrid aroma. Additionally, consumers rated the light-exposed FR samples lower than the non-exposed ones. Overall, LED lighting exposure resulted in a significantly lower liking of fluid milk samples. The data showed that consumers' preference for skim milk and 2% milk was more affected by exposure to light than by storage time.

In other studies, light-induced oxidation was examined. Light-induced oxidation occurs in two different molecules contained in milk. Milk may undergo light oxidation due to the degradation of fats to aldehydes, ketones, or alcohols and/or sulfur-containing amino acids. The newly formed components then cause aroma degradation and off-flavor. Additionally, riboflavin and vitamin A, other components contained in milk, may also degrade under light. Riboflavin splits into free radicals at wavelengths between 200 and 500 nm, while tyrosine and tryptophan, which are types of amino acids, break down into indoles.

Brothersen et al. [5] investigated the effect of fluorescent light (FL) and white light (LED) on the oxidation process of milk [5]. For this purpose, pasteurized and homogenized milk samples in HDPE (high-density polyethylene) containers were stored for 24 hours in a light-free environment, illuminated with 4,000 lx white LED light and 2,000 lx fluorescent light (FL). At the end of the storage period, the vitamin A content decreased by 30% after 12 hours under FL, and then a slight decrease continued, but no statistically significant difference was observed between FL and LED treated milk samples after 24 hours. Volatile compounds in milk generally showed an increasing trend under both lighting conditions. Furthermore, the riboflavin content of milk samples treated with FL and LED remained nearly the same. It was confirmed that LED treatment provided better quality milk properties than FL treatment.

Srimagal et al. [29] investigated the effect of blue light LED treatment at a 405–460 nm wavelength and a treatment time of 0–90 min at 5–15?C on microbial inactivation of *E. coli* ATCC 25922 in skim milk. The shelf-life of the LED-treated skim milk was also evaluated [29]. LED-treated skim milk was aseptically packaged with low-density polyethylene (LDPE) and stored at 37 and 4?C. The results indicated that LED application at 405 nm and 5?C for 60 minutes resulted in a maximum reduction of 4.69 log, whereas a 75-minute application at 460 nm resulted in a maximum reduction of 4.11 log. It was demonstrated that maximum log reductions were achieved after 60 minutes (5.27 log) at 405 nm, 75 minutes (5.04 log) at 433 nm, and 90 minutes (4.64 log) at 460 nm. Hence, the higher temperature and the shorter wavelength ensured the maximum inactivation. Furthermore, the maximum color change was observed at a shorter wavelength and higher temperatures.

The overall color also varied with an increase in time and a decrease in wavelength. Finally, the shelf-life of sterilized skim milk was determined as 19 h at 37?C and 9 days at 4?C.

In order to prevent the oxidation effect of light on milk, researchers have investigated various combined process applications that can be used with LED treatment, such as using different packaging materials (e.g., HDPE, PET, glass), adding antimicrobial substances (e.g., natamycin) and photoprotective substances (e.g., mycosporine-like amino acids (MAA)), and supplementing with vitamins (e.g., riboflavin, vitamin A and D). For example, Schiano et al. [26] examined the effect of vitamin fortification on LED light oxidation in skim milk [26]. For this purpose, fluorescent (FL) and 2,000 lx LED light at 4?C was applied to milk samples for 0, 2, 4, 8, or 24 h. It was concluded that there was no correlation between vitamin supplementation and the light-induced oxidation of milk, hence the off-flavor of milk. On the other hand, packaging was important to prevent milk products from oxidation during storage.

Potts et al. [23], unlike Brotherson et al. [5], examined the effect of PET (polyethylene terephthalate) packaging on the acceptability of milk in addition to HDPE [23]. It was found that the milk in HDPE and PET packages treated with LED light had identifiable differences and both were acceptable to the panelists. However, it was confirmed that fluorescent light caused a decrease in the acceptability of milk in both HDPE and PET packaging.

Wang et al. [35] studied the protection efficiency of PET packaging comprising different amounts of light protective additives (LPA) used for storage of UHT 2% milk fat milk in a refrigerated LED light retail case for 26 weeks [35]. In this study, the synergistic efficacy of TiO_2 and carbon black (CB) pigments in PET packaging in preserving milk nutrients and flavor was also evaluated. Furthermore, some quality parameters of milk such as riboflavin content, Thiobarbituric acid reactive substance (TBARS) assay, and volatiles produced in milk during storage were determined. Five types of PET packaging were evaluated: light-exposed control (LE, containing no LPA), light-protected control (LP, clear PET bottle covered with foil), moderate-PET (M-PET, 3.7 wt.% TiO_2 added into PET), high-PET (H-PET, 6.9 wt.% TiO_2 added into PET), and superior-PET (S-PET, 6.8 wt.% TiO_2 and 23 ppm CB added into PET). White LED light with an average intensity of 3,823 lx was used for up to 26 weeks in a 4?C refrigerator condition. It was determined that the riboflavin amount of UHT milk before being exposed to LED light was 1.39 mg/L and this amount was completely preserved for up to 26 weeks by packaging with PET containing a high level of LPA. In comparison, PET with intermediate and low levels of LPA was less successful in maintaining riboflavin levels. The lipid oxidation level in milk samples was determined with TBARS assay. The results indicated that S-PET provided full protection up to 13 weeks, H-PET provided full protection up to 4 weeks and partial protection up to 13 weeks, and M-PET provided limited protection up to 13 weeks of storage. The volatiles produced in milk were commonly aldehydes such as hexanal, heptanal, and octanal, which cause an unacceptable flavor in milk. It was confirmed that LP and S-PET prevented the formation of heptanal and hexanal for the first week and octanal formation up to 2 weeks of storage. On the other hand, H-PET slowed down the formation of octanal after week 13 and limited the formation of hexanal for 26 weeks of storage. M-PET and LE had the same level of protection against volatiles. It was concluded that light-induced oxidation could be prevented for up to 26 weeks in UHT 2% milkfat by choosing the appropriate combination of LPA and PET for an LED lighted retail case.

Wang et al. [34] conducted an extensive study to examine the effect of LED light at low and high intensities, and fluorescent light on milk fat oxidation level [34]. In addition, the effects of different packaging materials such as glass, HDPE, and PET with and without light protective additives (LPA) such as titanium dioxide (TiO_2) on milk fat oxidation, the oxygen barrier properties of these materials, riboflavin and vitamin A degradations were also determined. Oxygen permeation and light transmission levels of packaging materials were the main reasons for oxidation in packaged milk. The paperboard package prevented light transmission very effectively, but its oxygen permeation was greater than PET (54 times) and HDPE (19 times) bottles. Experiments were conducted under two conditions: closed-door retail case (Case 1) and walk-in/closed-door retail case (Case 2) with 18 W LED lights. The average light intensity in Case 1 was 1,068 lx whereas it was 4,094 lx in Case 2. Experiment was done at three different time intervals (0, 4, and 24 h) for each case and each packaging type. Oxidation level was determined by TBARS (thiobarbituric acid reactive substance) assay, which are the secondary oxidation products including malondialdehyde and other secondary oxidation aldehyde products. Concentration of dissolved oxygen (DO) was determined as 11.3 mg/L in raw milk. It decreased to 10.2 mg/L after 4 h exposure to low intensity LED light and reduced to 9.1 mg/L, 9.2 mg/L and 9.3 mg/L after 24 h exposure to low intensity LED in glass, HDPE and PET bottles, respectively. Additionally, riboflavin retentions at low-intensity LED light for long treatment times (24 h) reduced by 81.6% in glass, 85.1% in HDPE bottles, and 83.6% in PET bottles compared to fresh milk. Riboflavin degradation at high-intensity LED light was faster than low-intensity LED treatment for all bottles. Moreover, there was no significant difference in vitamin A levels of irradiated milks compared to fresh milk except for HDPE bottles exposed to low-intensity LED light for 24 h. According to TBARS analysis, it was confirmed that high-intensity LED treatment for long exposure time (>24 h) caused a rapid increase in oxidation rate.

Chang & Dando [6] also examined the effects of fluorescent light and LED light on milk oxidation [6]. Fat-free milks were exposed to 2,000 lx of LED and FL with varying exposure times from 0.5 to 48 h. Sensory thresholds of exposures, consumer acceptance, and the effectiveness of LED treatment supplemented with antioxidants on milk oxidation were compared. The results showed that sensory thresholds of LED and fluorescent were indifferent, but packaging with antioxidants (tocopherols and ascorbic acid) was not discriminated from LED treatment by the panelists. Panelists classified the aroma of FL treated milk samples as cardboard aroma whereas the aroma of LED treated milk samples was described as cardboard, old oil, and marginally plastic. It was confirmed that light-protective packaging could produce similar results to LED-treated milk samples, but the appearance of LED-treated milk was not appreciated.

Antimicrobial agents such as natamycin, which is one of the permitted food preservatives, protect a wide variety of products against yeasts and molds. For example, it has been employed in dairy-based food products to control spoilage by fungi, especially in cheese. Additionally, photoprotective agents such as mycosporine-like amino acids (MAA) can be introduced into milk to prevent the negative effects of light exposure during storage. Teixeira et al. [32] determined the effects of MAA and natamycin in cow's milk stored for up to 8 days at 4?C in glass bottles placed in retail commercial display cabinets illuminated by fluorescent and high-intensity LEDs [32]. According to their results, LED light caused less nutrient loss due to light-induced oxidation than fluorescent light, but more natamycin degradation than fluorescent light. Moreover, the addition of MAA to milk samples decreased natamycin degradation. However, it has been suggested that further study is needed to consider the interaction between MAA and milk.

17.2.2 WATER

Water treatment is usually achieved by chemical disinfectants, such as chlorine, sodium hypochlorite, and ozone. These treatments can cause negative effects on human health due to the formation of unwanted by-products [36]. UV irradiation processes have emerged as an alternative to these treatments. Water treatment topics are generally focused on the inactivation of microorganisms, system optimization, and investigation of the inactivation mechanism of irradiation treatment.

Recently, Park et al. [21] studied the effect of 278 nm wavelength UVC-LED treatment on microbial inactivation of foodborne gram-positive and gram-negative bacteria such as *Escherichia coli*, *Staphylococcus aureus*, *Bacillus cereus*, *Salmonella Typhimurium*, and *Listeria monocytogenes* in drinking water using different exposure times (A, 0–60 min), stirring speeds (B, 0–100 rpm), and water volume (C, 400–1,200 mL) [21]. The UVC dose was determined as 3.55 mJ/cm^2 considering the water volume of 800 mL and an irradiation time of 0.6 s. The number of all strains was reduced by 5 logs after 60 min of exposure time. Similarly, increasing stirring speed decreased the cell viability, and a 3-log reduction was obtained at 50 rpm. Finally, the change in water volume resulted in a reduction from 2 to 4 logs, and there was no significant change in cell viability above a water volume of more than 800 mL. In conclusion, optimized conditions were defined as 60 min of exposure time, 100 rpm stirring speed, and a water volume of 400 mL. It was suggested that UVC-LED treatment with proper circulation could be used as a disinfection method for drinking water.

Oliveira et al. [19] investigated the effects of 255 nm and 265 nm UVC-LEDs on DNA damage of *Aspergillus niger*, *Aspergillus fumigatus*, and *Aspergillus terreus* microorganisms in filtered water [19]. The irradiance (UV intensity) of UVC-LEDs was measured as 54 μW/cm^2 at 255 nm and 250 μW/cm^2 at 265 nm wavelengths. The exposure times were determined as 30 and 60 min at both wavelengths. It was found that the inactivation of *A. terreus*, *A. niger*, and *A. fumigatus* was inadequate at 255 nm UVC-LED application. Whereas UVC-LED treatment at 265 nm provided 3-log, 2-log, and 4-log reductions in *A. fumigatus*, *A. niger*, and *A. terreus*, respectively. DNA damage occurred through the formation of cyclobutane pyrimidine dimers (CPDs). The resulting dimers of *A. fumigatus* were similar at both wavelengths. On the other hand, while no dimer formation was observed at 255 nm in *A. niger*, dimer formation was observed at 265 nm for both exposure times. In *A. terreus*, dimer formations were similar for two wavelengths and exposure times.

Wang & Lin [36] focused on different aspects of UVC-LED application [36]. The objective of their study was to investigate the effect of UVC-LEDs at 278 nm wavelength on the disinfection of water using a flow reactor with a spiral-channel configuration having different spiral pitch, flow rate, and radiative power. *E. coli* was selected as the target microorganism. Around 24 LEDs, each with a radiant power of 10 mW and a rated current of 100 mA, were used in the study. A spiral channel was employed to provide sufficient exposure time for inactivation and to induce the Dean vortices. Spiral pitches of the spiral channel were 10, 20, and 30 mm. The flow rate of water was between 40 and 120 mL/min. According to the study results, the optimal inactivation was obtained at a flow rate of 60 mL/min for all configurations. The inactivation rate of bacteria was determined as 5.1, 4.5, and 3.4-log for spiral pitches of 10, 20, and 30 mm, respectively. Hence, the highest performance was obtained for the 10 mm pitch. In addition, exposure time and Dean number decreased with increasing spiral pitches, and the optimal Dean number was achieved at a flow rate of 60–70 mL/min. Lastly, at a fixed spiral pitch (10 mm), the radiative power had no significant impact on the log reduction. The logarithmic reductions of 5.1, 4.6, and 4.2 log were determined at 10, 8, and 6 mW radiant power, respectively.

17.2.3 APPLE JUICE

Fruit juices are preferred as an alternative to fruits due to their easy consumption. Fruit juices generally contain phenolic compounds such as phenolic acids (gallic acid, caffeic acid, etc.) and flavonoids (catechin, quercetin, etc.) which have beneficial health impacts in the human diet. Despite their high acidity, microbial spoilage is an important problem in fruit juices. Juices are rich in carbohydrates and complex nitrogen sources, which provide a suitable growth environment for yeasts, molds, and some acid-resistant bacteria [38]. In order to prevent spoilage, thermal treatments such as pasteurization are widely applied techniques. In addition to providing microbial inactivation, heat treatment also provides inactivation of some enzymes responsible for quality losses such as color, taste, and aroma [1].

Fruit juices are often subjected to microbial contamination, including those associated with *E. coli* O157:H7 and *Salmonella Typhimurium*. Inactivation of these pathogens is essential in the pasteurization of fruit juices. For example, Kim et al. [11] demonstrated the effect of a combination treatment of blue light-emitting diodes (BL) and riboflavin (Rb) on the inactivation of *Escherichia coli* O157:H7 and *Salmonella typhimurium* in apple juice [11]. They also investigated the bacterial inactivation mechanism under BL+Rb treatment. BL was determined to be an ineffective process for the inactivation of microorganisms when applied alone, due to the long exposure time requirements resulting in poor product quality. Therefore, a combination of BL with different photosensitizers (curcumin, chitosan, caffeic acid, riboflavin, folic acid, etc.) was found to be necessary to obtain high-quality liquid products within short exposure times. In addition, it was demonstrated that an efficient treatment could be achieved using four blue LED chips at 405–460 nm to inactivate the indicated microorganisms. Reactive oxygen species (ROS) probe was also used to measure ROS level after the BL, Rb, and BL+Rb treatments. It was stated that ROS could be generated by electron transfer to oxygen due to the presence of riboflavin under UVA/UVB irradiation. Thus, a direct correlation between ROS concentration and RB concentration after BL treatment was noted. According to the experimental result, no microbial reduction in *E. coli* O157:H7 was observed in BL (30 mJ/cm^2) and Rb treatments. However, 1-log and 2.5-log reductions were found in *S. Typhimurium* with BL and Rb treatments, respectively. On the other hand, 6.4-log and 4.5-log microbial reductions were obtained in *E. coli* O157:H7 and *S. typhimurium*, respectively, under the BL+Rb treatment condition at 30 mJ/cm^2. Thus, BL+Rb treatment at 30 mJ/cm^2 showed a synergistic effect. Furthermore, it was concluded that *S. typhimurium* was more sensitive than *E. coli* O157:H7 to BL and Rb treatments. A higher ROS concentration was measured during *E. coli* O157:H7 inactivation. Thus, the concentration level of ROS was highly correlated with the microbial inactivation due to the fact that ROS was responsible for weakening the enzymes and reducing the cell membrane robustness of pathogens.

Akgün & Ünlütürk [1] investigated the effect of UV-LEDs on microbial and enzyme (polyphenol oxidase, PPO) inactivation in clear (AJ) and cloudy apple juices (CAJ) [1]. *E. coli* K12 ATCC 25253, a surrogate of *E. coli* O157:H7, was chosen as a target microorganism in this study. Two UVC (254 nm), two UVA (365 nm and 405 nm) wavelengths, and their combinations (254/365 nm, 254/410 nm, 280/365 nm, 280/410 nm, and 254/280/365/410 nm) were applied to AJ and CAJ using a batch reactor equipped with four LEDs. They also measured pH, turbidity, total soluble solids, color, and ascorbic acid content of AJ and CAJ before and after the treatments. The minimum exposure time for CAJ was determined to be 35 minutes at 254 nm. Since negative color change was observed in juice samples lasting longer than 40 minutes, the maximum exposure time was specified as 40 minutes. The highest inactivation of *E. coli* K12 in CAJ was obtained at both 280 nm and 280/365 nm wavelengths, while the highest microbial reduction in AJ was found using 4 lamps at 254 nm during the 40-minute exposure time. In addition to this, complete degradation of ascorbic acid was observed after UV-LED and thermal treatment. Furthermore, the L* parameter of color remained almost constant, while the lowest change was noticed when using the combination treatment of 280/365 nm wavelengths. The a* and b* values were highly affected by UV-LED treatment. The highest change was observed in the combination treatment of 280/365 nm wavelengths. Finally, PPO activity was reduced in UV-LED treatment reaching the level of 67.42% and 65.62% reduction achieved when using 280/365 nm and 280/410 nm combination treatments, respectively.

Similarly, Lee et al. [13] studied the effects of 365, 385, and 405 nm LEDs used individually or combined with 10 μM folic acid (LEDF) on the inactivation of foodborne pathogens (*E. coli* O157:H7 ATCC 35150 and methicillin-resistant *S. aureus* ATCC 33591). Inactivation kinetics were modeled by the Weibull model to quantitatively compare the efficacy [13]. The highest inactivation of pathogens was obtained when using LEDF treatment at wavelengths of 365 and 385 nm. The synergistic effect was found to occur when treatments were combined rather than used individually. It was also confirmed that ROS generation via folic acid after treatment with LED was responsible for the bactericidal effect.

Molds and yeasts might also cause spoilage in apple juices. Xiang et al. [38] investigated the effect of UVC-LED irradiation on the inactivation of *Zygosaccharomyces rouxii.* Additionally, changes in quality parameters (pH, electrical conductivity, total soluble solids, titratable acidity, reducing sugar, total phenolics, antioxidant capacity, and color) in apple juice subjected to UV-LED treatment were also determined [38]. UVC-LED treatment at 275 nm with different dose levels (200, 400, 800, and 1,200 mJ/cm^2) provided using 64 UVC-LEDs resulted in 2.79, 3.42, 4.86, and 5.46-log reduction of *Z. rouxii*, respectively. It was found that pH, titratable acidity, reducing sugar, and total soluble content of apple juice were not affected by the treatment at 200–1,200 mJ/cm^2, whereas the total phenolic content of juices was prominently decreased with increasing UV dose levels. Additionally, the antioxidant capacity of apple juices was diminished after the treatment due to the degradation of phenolic compounds. Furthermore, the color of the apple juice changed in response to the applied UV dose. The authors found slightly noticeable differences in the color of apple juice, but they stated that the color change mechanism is still uncertain. In juice pasteurization, it was concluded that UVC-LED technology could be an alternative to thermal methods, but optimization may be necessary to avoid the negative effects of UVC-LED treatment.

17.2.4 ORANGE JUICE

The increase in the use of light-emitting diodes instead of conventional UV lamps in light-based food processing has accelerated the studies on their use in the decontamination of different fruit juices such as orange juice. For example, Bhavya & Hebbar [3] studied the sono-photodynamic inactivation of *E. coli* and *Staphylococcus aureus* in orange juice [3]. Orange juice samples were processed under two processing conditions: one with a combined treatment of blue light-emitting diodes (BL) (462 nm, 70 J/cm^2) and curcumin (PS, photosensitizer) (BL+PS), and the other with ultrasound (US, 20 kHz, 5 min) combined with BL+PS using the barrier concept treatment (BL+PS+US). In addition, the effect of treatment parameters such as PS concentration (50 and 100 μM), US amplitude (10 and 30 μm), vibration mode (5 on–10 off), and energy density of different juice volumes (605.5 and 302.8 J/cm^3) of BL treatment on microorganisms was investigated. The researchers also determined the color, cloud value, browning index, total phenolic content, total flavonoid content, hesperidin content, and antioxidant capacity before and after treatments. According to the results of this study, PS + BL treatment provided a 1-log reduction in *E. coli* at both concentration levels, while a significant difference was found between the two concentration levels in *S. aureus* inactivation. The maximum microbial reduction of *S. aureus* was obtained using 100 μM PS+BL treatment. Individual treatments of PS and BL showed no effect on both bacteria. Moreover, gram-positive bacteria were found to be more sensitive to the treatments than gram-negative bacteria. The effect of PS alone and US+PS treatment on *E. coli* and *S. aureus* inactivation was also examined. The inactivation pattern of both bacteria by US alone was significantly different due to the structural differences among the bacteria. US+PS treatment was determined ineffective in the inactivation of *E. coli*. Additionally, the inactivation efficiency of US+PS on *S. aureus* was found to be low compared to US alone. However, PS+US+BL treatment provided a 4.26-log reduction in *E. coli* and a 2.35-log reduction in *S. aureus*. It was confirmed that PS+US+BL treatment had a high inactivation efficiency in *E. coli* but had a lower effect on *S. aureus* compared to PS+BL treatment. Furthermore, the color of the juice showed a slight change when using US, PS+BL, and PS+US+BL treatments due to the occurrence of browning reactions. US treatment had a dominant effect on color change and cloud values of the orange juice. It was determined that individual US and BL treatments caused no significant change in total phenolic content (TPC), but a significant change was found when using combined treatments. No significant change in total flavonoid content (TFC), as well as hesperidin content, was observed when different treatments were used.

Prado-Silva et al. [8] investigated the effectiveness of antimicrobial Photodynamic Treatment (aPDT) on the inactivation of the spores of Alicyclobacillus acidoterrestris in orange peel, orange juice, and *in vitro* conditions [8]. 1-methylpyridinium-4-yl porphyrin tetraiodide (Tetra-Py+-Me) and new methylene blue (NMB) were chosen as photosensitizers (PSs). White LED, with a wavelength range between 400 and 700 nm, was used at a light irradiance of 140 or 65 mW/cm^2. Potassium iodide (KI) was also used to increase aPDT efficiency. The experiments were conducted under two conditions: absence and presence of KI. Furthermore, the total phenolic content and antioxidant activity of the orange juice after the aPDT were determined. It was confirmed that there was no reduction in the viability of spores in the presence of both PSs and the absence of KI. On the other hand, inactivation of *A. acidoterrestris* spores was observed in the presence of both PSs and KI (PS+KI), and the logarithmic reduction was 5-log after 10 hours of light exposure. Moreover, TPC and antioxidant capacity levels decreased compared to the control after treatments.

17.2.5 ALCOHOLIC BEVERAGES

17.2.5.1 WINE

The effect of light radiation on wine quality has been known since the 1970s. Photo-degradation of riboflavin, cysteine, and methionine, and the formation of volatile sulfur compounds (VSCs) are undesirable consequences of light on wine quality. The lighting in the wineries causes some undesirable flavors in wines such as rotten egg, cabbage, or cooked onion due to the formation of VSCs. However, little has been reported regarding the effect of currently used LEDs on the quality of wine.

Recently, Mislata et al. [17] evaluated the use of new LED lights to avoid bottled wine spoilage by determining the photo-degradation of riboflavin, cysteine, and methionine, the formation of VSCs, and changes in color during exposure to light [17]. The researchers used three different types of lamps: an ultraviolet lamp like the current one used in wineries (LA), an LED that minimized the 400–450 nm radiation (LB), and another LED that eliminated the radiation emitted into the 400–450 nm region (LC). The exposure time was determined as 10 days (240 h), and samples were collected at different analysis times (3, 6, 24, 48, 120, and 240 h). Riboflavin, methionine, cysteine contents, and VSCs were determined with ultra-performance liquid chromatography. In addition, oenological parameters such as color, titratable acidity, pH, free and total sulfur content, alcoholic degree, and volatile acidity were also measured without exposure to light (time 0 h) and after the last exposure time (240 h) with each of the lamps (LA, LB, and LC). After 10 days of storage, the color of wines was stable compared to 0 hours of exposure. There were some losses of riboflavin, methionine, and cysteine contents in rosé and white wines. In white wines, riboflavin content decreased by 44% after 240 hours of exposure to LA, whereas LB light treatment caused 38% losses at the end of the 240 hours. On the other hand, in rosé wines, LA treatment resulted in an 85% loss in riboflavin content after 10 days of storage, but riboflavin content declined by 45% with LB light treatment. In general, LA treatment caused more degradation of the studied components than LB treatment, and rosé wines were more affected than white wines by the types of light treatment. These degradation ratios were used to determine the formation of VSCs. It was confirmed that LA light increased VSC formation mainly in rosé wines compared to LB light. Contrary to these lights, LC light-treated wines other than rosé wines were not affected much. As a result, it was revealed that LC light did not cause high VSC formation during storage, and it was concluded that this light has the potential for use in storage in wineries.

17.2.6 OTHER BEVERAGES

17.2.6.1 GREEN TEA

Epigallocatechin gallate (EGCG) is one of the phenolic compounds in green tea and the most abundant catechin due to its antioxidant potential and quality indicator of green tea. Although EGCG is a quality indicator of green tea, it is sensitive to light, oxygen, temperature, brewing methods, storage conditions, and packaging materials, and can undergo epimerization and oxidation under unsuitable conditions.

Natamycin is a natural food additive and generally recognized as safe (GRAS). It is used to suppress the growth of molds and yeasts. Like EGCG, natamycin is sensitive to UV light. However, it has been reported that mycosporine-like-amino acids (MAA) added to acidified beverages might protect EGCG and natamycin stability under LED or fluorescent illumination.

Teixeira et al. [32] examined the photoprotective effect of mycosporine-like-amino acids (MAA) in acidified green tea containing natamycin as an additive stored under different lighting conditions [33]. The acidified green tea samples were stored in the presence of 3,500 K LED lamps in a closed-door commercial display case and fluorescent lamps in a walk-in retail case for up to 8 days at 4C. They examined the effect of different pH (5.4 – natural pH of green tea and 3.5 – acidified condition) on natamycin and EGCG photostability, efficacy of MAA extracts on natamycin and EGCG photostability, and efficacy of MAA extracts and EGCG on natamycin and EGCG photostability. The results showed that the natural pH of green tea caused less natamycin photodegradation in the absence of light. On the contrary, natamycin concentration slightly decreased to 72.94% in the presence of high-intensity LED. Maintaining the pH at 3.5 caused natamycin to degrade intensely and drop to concentration levels of 82.14%, 66.79%, and 67.77% in the absence of light, fluorescent, and LED, respectively. Light and reduction of EGCG content in pH 5.4 mainly caused natamycin photodegradation. To prevent this situation, MAA has been added to acidified beverages. The results showed that MAA had an adverse effect when interacting with EGCG and led to a sharp decrease in EGCG content, which directly affected natamycin retention in LED light and high MAA levels. In conclusion, maintaining EGCG level was more effective than adding MAA to prevent photodegradation of natamycin in acidified green tea.

17.2.6.2 MIXED BEVERAGES

Baykuş et al. [2] investigated the effect of UV-LEDs on microbial inactivation (*E. coli* K12), quality, physicochemical (color, browning index), and phytochemical (total phenolic content) parameters of a mixed beverage [2]. The mixed beverage consists of carob, carrot, ginger, grape, and lemon juices in decreasing amounts, respectively. Beverage samples were treated with UV-LEDs that emit light at 280 nm and 365 nm and 280/365 nm at different exposure times at room temperature. The maximum inactivation (> 4-log reduction) of *E. coli* K12 was obtained at 280/365 nm UV light for 40 min. Also, the number of mesophilic bacteria was reduced to 2.59 log from 5.69 log initial load, while the number of yeast and mold was reduced from 3.28 log to 0.17 log initial load. There was no significant color change after the irradiation of the samples. Total phenolic content and antioxidant capacity increased 1.75-fold and 4.60-fold compared to heat treatment (70?C–120 s), respectively. It was indicated that UV-LED protected the beverage better than heat treatment and may be an alternative to heat treatment for liquid food pasteurization.

Conversely, Lian et al. [14] examined the effects of UVA-LED treatment at 365 nm with the irradiance of 70 mW/cm^2 to inactivate *E. coli* DH5 α in a colored beverage and commercial orange juice [14]. Carotenoids (β-carotene, paprika color, marigold), flavonoids (sun yellow, cabbage color, grape juice color, sun red, and purple red pigment), and mixed food colorants (green colorant, melon color-L, purple colorant, and grape color) were used to prepare colored solutions at different concentrations (1%, 0.1%, 0.01%, and 0.001%). It was indicated that some colors (β-carotene, sun red, and sun yellow) inhibited the bactericidal effect. Flavonoid and carotenoid coloring agents had no effect on the bactericidal effect of UVA, but there was a positive correlation between the log survival ratio and hydrophilic food colors. Furthermore, according to the results of UVA-LED treatment applied to two different orange juices (juice A and B), 0.35-log and 1.58-log reduction were achieved in juice A and juice B, respectively. The researchers confirmed that UVA-LED application can inhibit bacteria in a colored solution of artificial food colors and liquids containing other compounds.

17.3 SUMMARY

Light-emitting diodes (LEDs) made of a semiconductor and producing monochromatic wavelengths are new sources of irradiation. Studies on the use of LEDs in the treatment of liquid food products, usually UV-LEDs, are mainly aimed at microbial inactivation, extending storage life, and determining product quality characteristics. It turns out that significantly higher UV-A doses are required in food safety applications compared to doses in the UV-C range. For this, hurdle technology can be considered as a method of improving the microbial inactivation efficiency of UV-LEDs. In addition, it has been revealed that the quality of the liquid product stored under LED light can be improved with the help of different packaging materials, antioxidants, and photo-protective agents, but more research is required to confirm the results obtained in previous studies.

KEYWORDS

- **antioxidants**
- **beverages**
- **food quality**
- **food safety**
- **light emitting diodes**
- **liquid food products**
- **photo-protective agents**
- **UV-LEDs**

REFERENCES

1. Akgün, M. P., & Ünlütürk, S. (2017). Effects of ultraviolet light emitting diodes (LEDs) on microbial and enzyme inactivation of apple juice. *International Journal of Food Microbiology, 260*, 65–74. https://doi.org/10.1016/j.ijfoodmicro.2017.08.007.
2. Baykuş, G., Akgün, M. P., & Unluturk, S. (2021). Effects of ultraviolet-light emitting diodes (UV-LEDs) on microbial inactivation and quality attributes of mixed beverage made from blend of carrot, carob, ginger, grape and lemon juice. *Innovative Food Science and Emerging Technologies, 67,* 102572. https://doi.org/10.1016/j.ifset.2020.102572.
3. Bhavya, M. L., & Hebbar, H. U. (2019). Sono-photodynamic inactivation of *Escherichia coli* and *Staphylococcus aureus* in orange juice. *Ultrasonics Sonochemistry, 57*, 108–115. https://doi.org/10.1016/j.ultsonch.2019.05.002.
4. Brasil, M. A. S., Sasaki, M. K., & Kamogawa, M. Y. (2021). New approach for direct analysis of opaque solutions with high content of suspended solids: Determination of calcium and phosphate in raw. *Talanta, 226*, 122095. https://doi.org/10.1016/j.talanta.2021.122095.
5. Brotherson, C., McMahon, D. J., Legako, J., & Martini, S. (2016). Comparison of milk oxidation by exposure to LED and fluorescent light. *Journal of Dairy Science, 99*(4), 2537–2544. https://doi.org/10.3168/jds.2015-9849.
6. Chang, A. C., & Dando, R. (2018). Exposure to Light-Emitting Diodes may be more damaging to the sensory properties of fat-free milk than exposure to fluorescent light. *Journal of Dairy Science, 101*(1), 154–163. https://doi.org/10.3168/jds.2017-13519.
7. Chang, K. H., Chen, R. L. C., Hsieh, B. C., Chen, P. C., Hsiao, H. Y., Nieh, C. H., & Cheng, T. J. (2010). A hand-held electronic tongue based on fluorometry for taste assessment of tea. *Biosensors and Bioelectronics, 26*(4), 1507–1513. https://doi.org/10.1016/j.bios.2010.07.100.
8. do Prado-Silva, L., Gomes, A. T. P. C., Mesquita, M. Q., Neri-Numa, I. A., Pastore, G. M., Neves, M. G. P. M. S., & Sant'Ana, A. S. (2020). Antimicrobial photodynamic treatment

as an alternative approach for *Alicyclobacillus acidoterrestris* inactivation. *International Journal of Food Microbiology, 333*. https://doi.org/10.1016/j.ijfoodmicro.2020.108803.
9. Dos Anjos, C., Sellera, F. P., de Freitas, L. M., Gargano, R. G., Telles, E. O., Freitas, R. O., & Sabino, C. P. (2020). Inactivation of milk-borne pathogens by blue light exposure. *Journal of Dairy Science, 103*(2), 1261–1268. https://doi.org/10.3168/jds.2019-16758.
10. Gros, N. (2011). Microdiffusion-based UV-LED spectrometric setup for determining low levels of ethanol in fruit juice. *Talanta, 87*(1), 174–179. https://doi.org/10.1016/j.talanta.2011.09.058.
11. Kim, D. K., Shin, M., Kim, H. S., & Kang, D. H. (2022). Inactivation efficacy of combination treatment of blue Light-Emitting Diodes (LEDs) and riboflavin to control *E. coli* O157 and *S. typhimurium* in apple juice. *Innovative Food Science & Emerging Technologies, 78*, 103014. https://doi.org/10.1016/j.ifset.2022.103014.
12. Kimoto, A., & Kanazawa, A. (2009). A new plane type sensor for concentration measurement in liquid. In *Procedia Chemistry* (Vol. 1, pp. 140–143). https://doi.org/10.1016/j.proche.2009.07.035.
13. Lee, J. I., Kim, S. S., & Kang, D. H. (2022). The combined effect of folic acid and 365–405 nm light emitting diode for inactivation of foodborne pathogens and its bactericidal mechanisms. *International Journal of Food Microbiology, 373*, 109704. https://doi.org/10.1016/j.ijfoodmicro.2022.109704.
14. Lian, X., Tetsutani, K., Katayama, M., Nakano, M., Nawatari, K., Harada, N., & Takahashi, A. (2010). A new colored beverage disinfection system using UV-A light emitting diodes. *Biocontrol Science, 15*(1), 33–37.
15. Lima, M. J. A., & Reis, B. F. (2017). Fully automated photometric titration procedure employing a multicommuted flow analysis setup for acidity determination in fruit juice, vinegar, and wine. *Microchemical Journal, 135*, 207–212. https://doi.org/10.1016/j.microc.2017.09.016.
16. Martin, N., Carey, N., Murphy, S., Kent, D., Bang, J., Stubbs, T., & Dando, R. (2016). Exposure of fluid milk to LED light negatively affects consumer perception and alters underlying sensory properties. *Journal of Dairy Science, 99*(6), 4309–4324. https://doi.org/10.3168/jds.2015-9603.
17. Mislata, A. M., Puxeu, M., Nadal, M., de Lamo, S., Mestres, M., & Ferrer-Gallego, R. (2022). Influence of different types of LEDs lights on the formation of volatile sulfur compounds in white and rosé wines. *Food Chemistry, 371,* 131144. https://doi.org/10.1016/j.foodchem.2021.131144.
18. Mori, M., Hamamoto, A., Takahashi, A., & Nakano, M. (2007). Development of a new water sterilization device with a 365 nm UV-LED. *Medical & Biological Engineering & Computing, 45*, 1237–1241. https://doi.org/10.1007/s11517-007-0263-1.
19. Oliveira, B. R., Marques, A. P., Asif, M., Maria, M. T., & Pereira, V. J. (2021). Light Emitting Diodes effect on Aspergillus species in filtered surface water: DNA damage, proteome response and potential reactivation. *Environmental Pollution, 287,* 117553. https://doi.org/10.1016/j.envpol.2021.117553.
20. Pandit, C., Alajangi, H. K., Singh, J., Khajuria, A., Sharma, A., Hassan, M. S., & Kaur, I. P. (2022). Development of magnetic nanoparticle assisted aptamer-quantum dot-based biosensor for the detection of Escherichia coli in water samples. *Science of the Total Environment, 831*, 154857. https://doi.org/10.1016/j.scitotenv.2022.154857.
21. Park, S. K., Jo, D. M., Kang, M. G., Khan, F., Hong, S. D., Kim, C. Y., & Ryu, U. C. (2021). Bactericidal effect of ultraviolet C light-emitting diodes: Optimization of efficacy

toward foodborne pathogens in water. *Journal of Photochemistry and Photobiology B: Biology, 222,* 112277. https://doi.org/10.1016/j.jphotobiol.2021.112277.

22. Popović, V., Koutchma, T., & Pagan, J. (2020). Emerging applications of ultraviolet Light-Emitting Diodes for foods and beverages. In *Innovative Food Processing Technologies: A Comprehensive Review* (pp. 335–344). Elsevier. https://doi.org/10.1016/b978-0-08-100596-5.22667-x.
23. Potts, H. L., Amin, K. N., & Duncan, S. E. (2017). Retail lighting and packaging influence consumer acceptance of fluid milk. *Journal of Dairy Science, 100*(1), 146–156. https://doi.org/10.3168/jds.2016-11673.
24. Prasad, A., Du, L., Zubair, M., Subedi, S., Ullah, A., & Roopesh, M. S. (2020). Applications of Light-Emitting Diodes (LEDs) in food processing and water treatment. *Food Engineering Reviews, 12,* 268–289. https://doi.org/10.1007/s12393–020–09221–4.
25. Saetear, P., Khamtau, K., Ratanawimarnwong, N., Sereenonchai, K., & Nacapricha, D. (2013). Sequential injection system for simultaneous determination of sucrose and phosphate in cola drinks using paired emitter-detector diode sensor. *Talanta, 115,* 361–366. https://doi.org/10.1016/j.talanta.2013.05.051.
26. Schiano, A. N., Jo, Y., Barbano, D. M., & Drake, M. A. (2019). Does vitamin fortification affect light oxidation in fluid skim milk? *Journal of Dairy Science, 102*(6), 4877–4890. https://doi.org/10.3168/jds.2018-15594.
27. Singla, M. L., Singh, M., Walia, M. S., Singla, S., & Mahapatra, P. K. (2006). An optoelectronic instrument for the determination of sulphite in beverages. *Food Chemistry, 97*(4), 737–741. https://doi.org/10.1016/j.foodchem.2005.06.009.
28. Skåra, T., & Rosnes, J. T. (2016). Emerging methods and principles in food contact surface decontamination/prevention. In *Innovation and Future Trends in Food Manufacturing and Supply Chain Technologies* (pp. 151–172). Elsevier Inc. https://doi.org/10.1016/B978-1-78242-447-5.00006-X.
29. Srimagal, A., Ramesh, T., & Sahu, J. K. (2016). Effect of light emitting diode treatment on inactivation of *Escherichia coli* in milk. *LWT – Food Science and Technology, 71,* 378–385. https://doi.org/10.1016/j.lwt.2016.04.028.
30. Su, A. K., Chang, Y. S., & Lin, C. H. (2004). Analysis of riboflavin in beer by capillary electrophoresis/blue Light Emitting Diode (LED)-induced fluorescence detection combined with a dynamic pH junction technique. *Talanta, 64*(4), 970–974. https://doi.org/10.1016/j.talanta.2004.04.011.
31. Teerasong, S., Chan-Eam, S., Sereenonchai, K., Amornthammarong, N., Ratanawimarnwong, N., & Nacapricha, D. (2010). A reagent-free SIA module for monitoring of sugar, color and dissolved CO_2 content in soft drinks. *Analytica Chimica Acta, 668*(1), 47–53. https://doi.org/10.1016/j.aca.2010.01.021.
32. Teixeira, G. H. A., & O'Keefe, S. F. (2019). Short communication: Mycosporine-like amino acids protect natamycin against photodegradation in milk exposed to fluorescent or light-emitting diode light. *Journal of Dairy Science, 102*(6), 4972–4977. https://doi.org/10.3168/jds.2018-15658.
33. Teixeira, G. H. de A., Ma, S., Stewart, A. C., Wang, H., & O'Keefe, S. F. (2019). Natamycin photostability in acidified green tea beverage is dependent on mycosporine-like amino acids and epigallocatechin gallate interaction. *LWT, 116.* https://doi.org/10.1016/j.lwt.2019.108572.
34. Wang, A., Dadmun, C. H., Hand, R. M., O'Keefe, S. F., Phillips, J. B., Anders, K. A., & Duncan, S. E. (2018). Efficacy of light-protective additive packaging in protecting

milk freshness in a retail dairy case with LED lighting at different light intensities. *Food Research International, 114*, 1–9. https://doi.org/10.1016/j.foodres.2018.07.053.
35. Wang, A., Stancik, C. M., Yin, Y., Wu, J., & Duncan, S. E. (2022). Performance of cost effective PET packaging with light protective additives to limit photo-oxidation in UHT milk under refrigerated LED-lighted storage condition. *Food Packaging and Shelf Life, 31*, 100773. https://doi.org/10.1016/j.fpsl.2021.100773.
36. Wang, C. P., & Lin, W. C. (2021). Combination of ultraviolet-C Light-Emitting Diodes and a spiral-channel configuration in a water disinfection reactor. *Journal of Water Process Engineering*, *42*(11), 102160. https://doi.org/10.1016/j.jwpe.2021.102160.
37. WHO. (2022). *Food Safety*. https://www.who.int/health-topics/food-safety.
38. Xiang, Q., Fan, L., Zhang, R., Ma, Y., Liu, S., & Bai, Y. (2020). Effect of UVC Light Emitting Diodes on apple juice: Inactivation of *Zygosaccharomyces rouxii* and determination of quality. *Food Control, 111,* 107082. https://doi.org/10.1016/j.foodcont.2019.107082.

CHAPTER 18

APPLICATIONS OF LIGHT-EMITTING DIODES IN SOLID AND SEMI-SOLID FOODS

ASHOK KUMAR CHAKKA, KUMARA KRISHNANAYAKA, and SURENDRA BABU AYENAMPUDI

ABSTRACT

In recent years, Light-emitting diodes (LED) technology has been gaining attention from researchers in agriculture, food preservation, and food safety. LED possesses the potential to be an alternative light source, effective surface decontaminant, and the ability to inactivate microorganisms. It is widely used in water disinfection (bacteria, viruses, and fungi). LED utilizes light energy between 200 and 780 nm to produce unique properties such as high-efficiency photon and luminous intensity, higher monochromatic light and orientation emission, stable peak wavelength, low radiant heat emission, and long-life expectancy. LED application in the food industry is more suitable. LEDs find various applications in food processing, such as disinfection of food products and water. The antimicrobial activity of LEDs in liquid, solid, and semi-solid food systems is ascribed to photodynamic inactivation owing to photosensitization of chemical compounds leading to DNA lysis in the microbial cell. On the other hand, microbial inactivation depends on the wavelength of light used. It also provides numerous advantages in food processing, such as non-toxic, monochromatic light production, robustness, and high durability compared to conventional light treatments. Therefore, LEDs are suitable for nonthermal operations, enhancing nutrients, reducing infection during post-harvesting, and the inactivation of pathogens in food. This chapter introduces the principles of LED, antimicrobial efficacy of LEDs, mechanism of inactivation, and limitations.

Nonthermal Light-Based Technologies in Food Processing.
C. K. Sunil, Preeti Birwal, Megh R. Goyal, and R. Mahendran (Eds.)

18.1 INTRODUCTION

18.1.1 LIGHT EMITTING DIODES (LEDS)

Light-emitting diodes are commonly known as LEDs. An LED is a solid-state photonic semiconductor device. The symbol of LEDs is shown in Figure 18.1. Technically, a standard p-n junction diode operates under suitable forward bias conditions. Thus, LEDs are compound semiconductors that emit an incoherent narrow spectrum of spontaneous radiation under suitable electrically forward bias conditions. This compound semiconductor comprises III-V group, II-IV group, and IV-IV group elements in the periodic table. Moreover, compound semiconducting materials with a direct energy band gap are suitable for fabricating LEDs due to momentum conservation in optical transition (*see* Figure 18.2). LEDs emit light in the near-ultraviolet-visible and near-infrared regions of the electromagnetic spectrum.

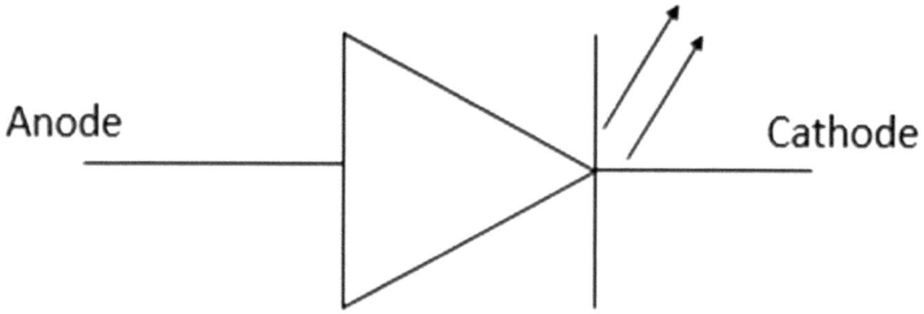

FIGURE 18.1 Symbol of light emitting diode.

Primarily, the basic structure of the LEDs is a p-n junction formed by some specific semiconductor materials. A chip of semiconducting material is doped or impregnated with optically active isoelectronic impurities to form a p-n heterojunction. Here, the excess positive charge (holes) and the excess negative charge (electrons) are present in the p-side and n-side, respectively (*see* Figure 18.3). The p-n heterojunction is a structure that, under certain conditions, can spatially separate these charge carriers (electrons and holes). During the electrically forward biasing of the device, the current flows through it, causing the electrons and holes to meet in the same region of space. This process is known as carrier injection.

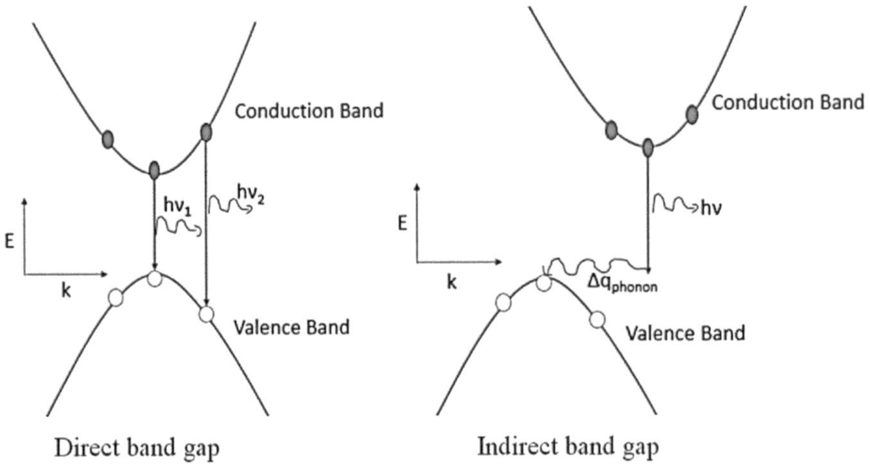

FIGURE 18.2 Parabolic electron and hole dispersion relation showing the electron-hole recombination and photon emission.

Source: Adapted from Ref.[24]

When the electrons and holes in the same region of space are more in number when compared to their number in equilibrium, the carrier recombination is enhanced, and the energy is emitted. When the free electrons in the conduction band recombine with the holes in the valence band, the free electrons occupy a bond in the atomic structure by filling the empty state in the bond. In this process, the electrons are released as electromagnetic radiation in the heterojunction semiconductor. Therefore, the energy is emitted during the carrier recombination process. This energy corresponds to the energy band gap of that semiconductor. The energy band is the energy difference between the bottom of the conduction band and the top of the semiconductor's valence band. The probability of spontaneous recombination of electron-hole pairs increases by increasing the active region doping concentration (or carrier concentration) in direct band gap materials. The electron-hole recombination rate 'r' is,

$$\Delta E = 1.8kT \text{ } or \text{ } \Delta\lambda = {1.8kT\lambda^2}/{hc} \tag{1}$$

where; 'n' and 'p' are the carrier concentration of electrons and holes in atoms/cm^3, respectively; and the constant 'B' depends on semiconducting materials.

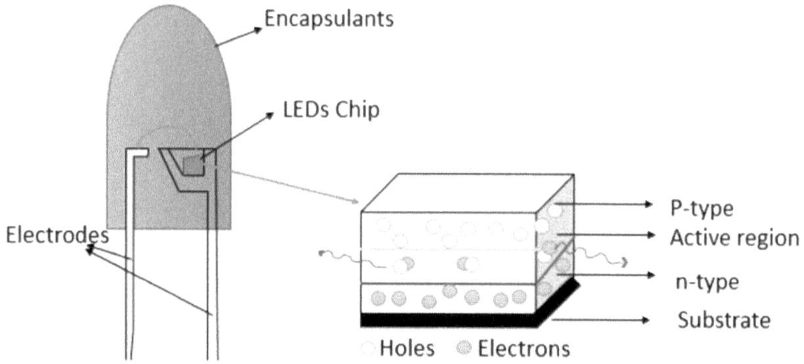

FIGURE 18.3 Schematic diagram of LED structure having LED chip and encapsulants and p–n junction [39].

Source: Adapted from Ref. [39]

18.2 SOURCES OF LED

Light emitting diodes illuminate various colors of light based on their semiconductor material. A list of commercially available LED sources is listed in Table 18.1.

According to Schubbert [39], the basic optical properties of LEDs include:

1. **Emission Spectrum:** The spontaneous recombination of electron-hole pairs results in the emission of photons (Figure 18.1). The spectral width of the LED's emission is narrower with respect to the spectral width of the single color. The human eye perceives the LED emission as monochromatic. The full width at half maximum of the emission spectrum is given by $\Delta E = 1.8kT \text{ or } \Delta\lambda = 1.8kT\lambda^2 / hc$
2. **Encapsulants:** Transparent materials with an extensive refractive index have been used as encapsulants for enhancing the light-extraction efficiency in the LED (Figure 18.3). Usually, encapsulants are in dome-shaped geometry; this increases the angle of total internal reflection through the top surface of the semiconductor:

$$\frac{\eta_{encap}}{\eta_{air}} = \frac{1 - \cos\varphi c, encap}{1 - \cos\varphi c, air}$$

where; $\varphi c, encap$ and $\varphi c, air$ are the critical angles for total internal reflection at the semiconductor-encapsulants and semiconductor-air interface, respectively. η_{encap} and η_{air} are the refractive indices of the encapsulants and air, respectively.

TABLE 18.1 List of Commercially Available LEDs

Semiconductor Material	Wavelength (nm)	Color	Biasing Voltage (V)
Aluminum nitride (AlN)	< 400	Ultraviolet	3.1–4.4
Aluminum gallium nitride (AlGaN)			
Aluminum gallium indium nitride (AlGaInN)			
Indium gallium nitride (InGaN)	400–450	Violet	2.8–4.0
Indium gallium nitride (InGaN)	450–500	Blue	2.5–3.7
Silicon carbide (SiC)			
Gallium phosphide (GaP)	500–570	Green	1.9–4.0
Aluminum gallium indium phosphide (AlGaInP)			
Aluminum gallium phosphide (AlGaP)			
Gallium arsenide phosphide (AlGaP)	570–590	Yellow	2.1–2.2
Aluminum gallium indium phosphide (AlGaInP)			
Gallium phosphide (GaP)			
Gallium arsenide phosphide (GaAsP)	590–610	Orange/ amber	2.0–2.1
Aluminum gallium indium phosphide (AlGaInP)			
Gallium phosphide (GaP)			
Aluminum gallium arsenide (AlGaAs)	610–760	Red	2.0–2.1
Gallium arsenide phosphide (GaAsP)			
Aluminum gallium indium phosphide (AlGaInP)			
Gallium phosphide (GaP)			
Gallium arsenide (GaAs)	>760	Infrared	< 1.9
Aluminum gallium arsenide (AlGaAs)			

Sources: Nair & Dhoble [35]; Dutta Gupta & Jatothu [12].

3. **Temperature Dependence of Emission Intensity:** The temperature of the LEDs affects their performance. As a result, the emission intensity decreases with increasing temperature. Non-radiative recombination, surface recombination, and carrier loss over heterostructure barriers will contribute to non-radiative recombination at higher temperatures.

Nair & Dhoble [35] reported various advantages of LEDs as mentioned below:

- LEDs are available in small sizes;

- LEDs are cost-effective devices and require low power for their operation;
- LEDs can have a long life that can work for 25,000 to 1,00,000 hours. So, the durability of LEDs is high;
- LEDs have high energy efficiency;
- Low maintenance cost and fast response.

18.3 LED TECHNOLOGY FOR DECONTAMINATING

Light-emitting diodes (LEDs) are semiconductor devices that can illuminate monochromatic light, which finds wide application in agro-food processing industries. They can destroy foodborne pathogenic microorganisms by treating them with a specific wavelength, intensity, and dose of light. LED processing is considered to be chemical-free, harmless, and easy to operate in various food systems [31]. The antimicrobial efficiency of different wavelengths of light ranging between 200 and 280 nm (UV-C), 280–320 nm (UV-B), 320–400 nm (UV-A), 400–470 nm (Blue light) has been extensively studied in the literature. Most of the studies in the literature have focused on UV-C LEDs for the disinfection of food samples [13, 18, 37]. Nevertheless, LEDs capable of illuminating light with wavelengths such as 365 nm, 395 nm, and 455 nm have gained a lot of attention in the field of food preservation.

18.4 LED TREATMENT OF SOLID FOOD SYSTEMS

The antimicrobial efficiency of LED light on solid food systems depends on various parameters of food products such as food type and its nature, moisture content, and surface characteristics. Furthermore, LED treatment conditions including treatment time and temperature, dosage, wavelength of light, and humidity also play a vital role in its efficacy [36]. Currently, fresh-cut produce (fruits and vegetables) is an immediate source of nutrients and has gained more popularity and demand in the market. Although the products are stored under refrigerator conditions, they are still prone to microbial contamination where the growth seems to be limited. In such conditions, LED treatments have been reported to be effective against a wide range of microorganisms in various food products. The effect of surface sterilization of fresh vegetables (cabbage and lettuce) using UVA-LED on the microbial load was tested by Aihara et al. [1]. LED irradiation of fresh produce with a wavelength of 365 nm for 90 minutes showed a 3.23 log CFU/g reduction of *E. coli* cells. Kim et al. [21] demonstrated the antimicrobial efficiency of

LED emitting light with a wavelength of 405 nm on Salmonella-inoculated fresh-cut papaya. The dosage applied in the study ranged between 0.9 and 1.7 kJ/cm^2 for a period of 48 h at 4?C (Table 18.2). LED treatment shown to reduce *Salmonella* population in the sample to 0.30–1.30 log CFU/cm^2. Similar results were reported by the same authors [22] in another study by treating freshly-cut mangoes with LED of 405 nm with a dose of 1.7–3.5 kJ/cm^2 for 36–48 hours. The treatment resulted in a reduction of bacterial count (*Escherichia coli* O157:H7, *Listeria monocytogenes, Salmonella* spp.) in the range of 1.00–1.60 log CFU/cm^2. The sterilization efficiency of UVC-LED on coriander leaves was studied by Jiang et al. [18]. UVC-LED irradiation showed a 0.4 to 1.3 log CFU/g reduction in *Salmonella* and *E. coli* depending on the time of exposure and intensity. Salmonella was found to be more sensitive to LED treatment. A similar experiment was performed on fresh-cut carrots by subjecting them to LED processing with a wavelength of 390 nm for a varied period of time resulting in 0.47–1.80 CFU/g log reductions in TBC (Total bacterial count) while a maximum reduction in fungal count (1.27 log CFU/g) was observed in the samples after LED treatment. The reports of the study confirmed that LED treatment conditions can vary the antimicrobial efficiency [33].

Poultry meat is one of the richest sources of protein for the global population. Chicken and its products are widely consumed owing to their low fat and high protein content [9]. Nevertheless, their quality gets affected in terms of texture, flavor, color, and the formation of slimy surfaces during storage [8] when contaminated with bacteria such as *Enterobacteriaceae, Pseudomonas* spp., and *Lactic acid bacteria* [5]. While the microorganisms associated with the deterioration of raw meat include *Escherichia coli, Campylobacter, Salmonella Typhimurium, Listeria monocytogenes,* and *Yersinia*, causing serious illness to consumers [38]. Consequently, appropriate techniques like LED treatments can be applied for the decontamination of meat and meat products. There are limited studies demonstrating the antimicrobial efficacy of LED on meat surfaces. In a study by Steele et al. [44], the decontamination efficacy of LED lighting for 24 hours on pork loin chops, beef longissimus lumborum steaks, ground beef, ground turkey, and beef semimembranosus steaks were investigated. The aerobic plate count (APC) of beef longissimus lumborum steaks, ground beef, and ground turkey increased with treatment time (2.90–5.40 log CFU/g). While *Enterobacteriaceae* count in ground turkey displayed a decreased trend (3.44 log CFU/g). Sommers et al. [42] reported that LED treatment at 405 nm reduced the foodborne pathogens (*E. coli, Staphylococcus spp., Listeria monocytogenes*) to 0.40 log CFU/g in chicken skin and 1.01 log CFU/g in chicken purge. In another study, Calle et al. [4] showed a greater reduction (>3 log CFU/cm^2) in *Salmonella* count inoculated on chicken breast after LED illumination for 900s. Similarly, in a recent study, McSharry et al. [32] observed a significant reduction in *Listeria monocytogenes, Salmonella Typhimurium*, Total viable counts (mesophiles and psychrophiles), when treating the beef broth with LED with different wavelengths over 60 min.

Generally, seafood is more prone to microbial spoilage than meat because the former has low pH and moisture. Application of LED technology has also been studied on the contamination of seafood. Shirai et al. [40] noticed a decreased viable cell count after LED treatment of boiled young sardine from 3.67 log CFU/g to 3.40 log CFU/g. In a study conducted by Josewin et al. [19], it was revealed that LED illumination at a wavelength of 460 nm exerted a bactericidal effect against *Listeria monocytogenes* on smoked salmon. The exposure to LED had resulted in a reduction of *L. monocytogenes* to 0.70–1.20 log CFU/g. While exposure of Ready-to-eat (RTE) fresh salmon to LED light at a lower wavelength (405 nm) showed a lower reduction (0.30 log CFU/cm^2) of *Listeria monocytogenes* and *Salmonella* spp. by 0.50 log CFU/cm^2 [29]. Thus, the literature survey confirms that LED technology has the potential for applications in the decontamination of solid foods.

18.5 LED TREATMENT OF SEMI-SOLID FOOD SYSTEMS

Studies of antimicrobial efficacy of LED light on semi-solid foods were found to be scanty in the literature (Table 18.2). Semi-solid foods such as soft cheese, tinned fruits, meat stews, etc., will deteriorate as quickly as liquid food systems, while solid foods first get contaminated from the outer surface and progress to the inner surface [34]. Ready-to-eat foods such as cheese and other minimally processed foods are developed for immediate consumption without much further processing; nevertheless, the microbial safety of such food products is questionable. *Listeria monocytogenes* is a potential pathogenic microorganism in cheese that can often cause serious illness or even be life-threatening [10]. LED irradiation with a wavelength of 266 nm and a dose of 3 mJ/cm^2 reduced the *L. monocytogenes, E. coli,* and *S. Typhimurium* counts to 3.52, 4.88, and 4.72 log CFU/g, respectively, in Camembert cheese. Furthermore, the application of LED at higher wavelengths (266–279 nm) in sliced Camembert cheese resulted in 4 to 5 log reductions in *E. coli*, while

3 to 4 log reductions in *Listeria* spp. [23]. In a recent study, Ricciardi et al. [37] reported that LED application had completely destroyed the microbial (*P. fluorescens*) load in cheese and was found to be below the detection level. The application of UVC-LED light with 275 nm presented a 5.66, 5.14, 5.10 log CFU/g reduction in *S. Typhimurium, L. monocytogenes*, and *E. coli*, respectively, in raw tuna fillets [13]. Similarly, in another study, UV-LED irradiation on chicken breast fillets resulted in a decline of total viable count from 1.17 to 1.67 log CFU/g, while a reduction as low as 2 log CFU/g was noticed for *Pseudomonas* and *lactic acid* bacteria [43]. Further studies need to be carried out to investigate the effect and mechanism of action of LED for microbial inactivation in semi-solid foods. Additionally, LED treatments combined with other nonthermal processing techniques can be explored in the future to enhance antimicrobial efficiency and produce microbiologically safe food.

18.6 EFFECTS OF LED ON FOOD QUALITY

Light Emitting Diodes (LEDs) and Ultraviolet Light Emitting Diodes (UVLEDs) contain a semiconductor of light. The LEDs and the UVLEDs have become of market importance due to their solid-state cold light source. This property of these semiconductors might also be important in food preservation. The LEDs have been observed to cease chemical reactions and increase the defense activity in fruits and vegetables. Meanwhile, the UVLED can inactivate food pathogens like *Escherichia coli*, *Pseudomonas fluorescens,* and *Salmonella spp*. The inactivity, however, can be brought upon depending on the target organism and the food product [14].

The impact of LED irradiation on the postharvest features of bananas gave promising, positive results. Mature green bananas treated with blue light (464–475 nm), green light (515–525 nm), and red light (617–627 nm) for 8 days showed rapid ripening as compared with the control. The ripening was the highest in blue light, followed by red and green. Further, it was also observed that the fruit's ascorbic acids, total phenols and sugar content increased due to LED exposure. Such findings pave new ways to induce chemical-free ripening in the fruits [17]. The effects of LED have also been tested on cabbage. The cabbage stored at 4–5?C under blue, red, green, and white LED on the 15th day of storage showed increased chlorophyll content with green, followed by white, red, and blue. The vitamin C content was the highest with blue, followed by white, green, and red. On the 18th day of

TABLE 18.2 *In Vitro* Antimicrobial Efficiency of LED Irradiation on Solid and Semi-Solid Food Systems

Type of Food System	Food Sample	LED Treatment Conditions	Microorganisms Inactivated	Findings Reported	References
Semi solid	Sliced camembert cheese	LED type: UV-LED Wavelength (nm): 266, 270, 275, and 279 Intensity (W/cm^2): 4 Treatment time: NA Dose: 1 to 3 mJ/cm^2	*E. coli* O157:H7 *S. typhimurium* *L. monocytogeneses*	LED irradiation with a dose of 3 mJ/cm^2 showed 4–5 log CFU/g reduction in *E. coli*, *S. typhimurium*, and *L. monocytogeneses*.	[23]
–	Ricotta cheese	LED type: UVC-LED Wavelengths (nm): 254 Intensity: NA Treatment time (sec): 30 Dose (mJ/cm^2): 6.54	*P. fluorescens*	LED application completely destroyed the microbial load in cheese and was below the detection level.	[37]
–	Raw tuna fillets	LED type: UVC-LED Wavelengths (nm): 275 Intensity: 0.5 mW/cm^2 Treatment time (sec): 0, 10, 20 Dose (mJ/cm^2): 0, 5, 10	*S. Typhimurium*, *L. monocytogeneses* and *E. coli* O157:H7	UVC-LED treatment presented 5.66, 5.14, 5.10 log CFU/g reduction in *S. typhimurium*, *L. monocytogeneses* and *E. coli*, respectively.	[13]
–	Chicken breast fillets	LED type: UV-LED Wavelengths (nm): 255, 280, 300, and 365 Intensity (W/cm^2): 0.003, 0.043, 0.049, 0.068. Treatment time (min): 2, 4, 6, 8, 10 Dose (mJ/cm^2): NA	Total viable counts (TVC) of mesophilic, psychrophilic bacteria, *entero-bacteriaceae*, *Pseudomonas*, and *Lactic acid bacteria* count.	UV-LED illustration resulted in TVC from 1.17 and 1.67 log CFU/g while a reduction of as low as 2 log CFU/g was noticed for *Pseudomonas*, and *Lactic acid bacteria*.	[43]

TABLE 18.2 *(Continued)*

Type of Food System	Food Sample	LED Treatment Conditions	Microorganisms Inactivated	Findings Reported	References
Solid	Cabbage, lettuce	LED type: UVA-LED Wavelength (nm): 365 Intensity (W/cm^2): NA Treatment time (min): 0–90 Dose (mJ/cm^2)	*E. coli* DH5α	LED irradiation for 90 min showed 3.23 log CFU/g reduction.	[1]
–	Apricots, plums, and cauliflowers	LED type: LED Wavelengths (nm): 585 Intensity (mW/cm^2): 3.84 Treatment time (min): 30 Dose (J/cm^2): 6.80	*Bacillus cereus* Total mesophilic count	LED treatment demonstrated a decrease in the *Bacillus* count to 4.40 log CFU/g while the mesophilic count ranged between 0.60 and 0.72 log CFU/g in all fruits and vegetables after LED processing.	[2]
–	Shelled almonds	LED type: UVC-LED Wavelengths (nm): 405 Intensity (μW/cm^2): NA Treatment time (min): 0, 1, 2, 4, 6, 8, 10 Dose (mJ/cm^2): NA	*E. coli* O157:H7 *E. coli* K12 *S. enteritidis* *S. typhimurium*	LED treatment resulted in 2.44 and 1.44 log CFU/g. 1.85 and 1.63 log CFU/g. 0.70 and 0.55 log CFU/g. 0.54 and 0.97 log CFU/g. Reduction for higher and lower inoculum levels.	[26]
–	Fresh-cut pineapples	LED type: Blue LED Wavelengths (nm): 460 Intensity (mW/cm^2): 92, 147.70, 254.70 Treatment time (h): 24, 13.91, 8.66 Dose (kJ/cm^2): 15–27	*Salmonella enterica*	LED illumination of pineapple cuts showed an antibacterial effect ranging between 0.61 and 1.72 log CFU/g. The maximum antimicrobial efficiency was achieved at intensity of 92 mW/cm^2	[15]

TABLE 18.2 *(Continued)*

Type of Food System	Food Sample	LED Treatment Conditions	Microorganisms Inactivated	Findings Reported	References
–	Fresh-cut papaya	LED type: LED Wavelengths (nm): 405 Intensity (mW/cm^2): 10 Treatment time (h): 24–48 Dose (kJ/cm^2): 0.9–1.7	*S. agona* *S. newport* *S. saintpaul* *S. typhimurium*	*Salmonella* population in the sample reduced to 0.30–1.30 log CFU/cm^2.	[21]
–	Fresh-cut mango	LED type: LED Wavelengths (nm): 405 Intensity (W/cm^2): 20 Treatment time (h): 24–48 Dose (kJ/cm^2): 1.70–3.50.	*Escherichia coli* O157:H7 *Listeria monocytogeneses* *Salmonella* spp.	LED illumination had showed a potential antimicrobial effect by decreasing the bacterial count in the range of 1.00–1.60 log CFU/cm^2.	[22]
–	Spinach leaves	LED type: Blue LED Wavelengths (nm): 405 Intensity (mW/cm^2): 0.14 Treatment time (days): 14 Dose (J/cm^2): NA	Total standard plate count (TSPC)	The results of the study indicated that there is a reduction in TSPC of about 1.50 log CFU/g for 14 days of LED illumination.	[3]
–	White mushrooms	LED type: UVC-LED Wavelengths (nm): 280 Intensity (μW/cm^2): NA Treatment time (min): NA Dose (mJ/cm^2): 2, 5.	*S. typhimurium*, *L. monocytogeneses* and *E. coli* O157:H7	UVC-LED treatment presented 2 to 3 log CFU/g reduction in *S. typhimurium* and *E. coli* while 1 log CFU/g in *L. monocytogeneses* was observed.	[20]

TABLE 18.2 *(Continued)*

Type of Food System	Food Sample	LED Treatment Conditions	Microorganisms Inactivated	Findings Reported	References
–	Fresh-cut Hami melon	LED type: LED + curcumin treatment. Wavelengths (nm): 460 Intensity (mW/cm^2): NA Treatment time (min): 5, 30, 60, 90 Dose (mJ/cm^2): NA	Total aerobic microorganism	Illumination of Hami melon treated with 50 μmol/L curcumin using blue LED decreased the total aerobic microbial count to 1.80 log CFU/g.	[30]
–	Wheat flour	LED type: UVA-LED Wavelengths (nm): 395 Intensity (W/cm^2): 0.45 Treatment time (min): 0, 10, 30, 60 Dose (J/cm^2): NA	*Salmonella* spp.	The study showed that UVA-LED application in wheat flour can reduce the *Salmonella* to 2.91 log CFU/g.	[11]
–	Coriander	LED type: UVC-LED Wavelengths (nm): 280 Intensity (μW/cm^2): 80, 160, 240 Treatment time (min): 5, 15, 30 Dose (mJ/cm^2): NA	*Salmonella* and *E. coli*	UVC-LED reduced *Salmonella* and *E. coli* by 0.4 to 1.3 log CFU/g, with sensitivity of Salmonella to LED treatment observed.	[18]
–	Fresh cut-carrots	LED type: LED Wavelengths (nm): 390 Intensity (W/cm^2): 25 Treatment time (min): NA 10, 20, 30, 60, 120, 180 Dose (J/cm^2): 0–30	Total bacterial count Total fungal count	The results of the study indicated a log reduction ranged between 0.47 and 1.80 CFU/g while a maximum reduction in fungal count (1.27 log CFU/g) was observed in the samples after LED treatment.	[33]

TABLE 18.2 *(Continued)*

Type of Food System	Food Sample	LED Treatment Conditions	Microorganisms Inactivated	Findings Reported	References
–	Black peppercorns	LED type: UVC-LED Wavelengths (nm): 280, 300, 365 Intensity (mW/cm^2): NA Treatment time (min): 5, 10, 20 Dose (mJ/cm^2): NA	*Bacillus subtilis* DSM618	The LED treatment on black peppercorns for 20 min showed a reduction in *B. subtilis* count from initial 7.18 log CFU/g to 6.20 log CFU/g at 280 nm.	[16]
–	Fresh-cut cherry tomatoes	LED type: LED Wavelengths (nm): 634, 520, 442 Intensity (mW/cm^2): 3 Treatment time (days): 7 Dose (mJ/cm^2): NA	Aerobic count, yeast, and mold count.	The total microbial counts of the fresh-cut cherry tomatoes treated with red, green, and blue light was found to be lower than 7 log CFU/g at the end of the treatment time. While the yeast and mold count also decreased after illustration with red light.	[25]
–	Carrot, celery, paprika, cabbage	LED type: UVC-LED Wavelengths (nm): 275 Intensity ($\mu W/cm^2$): 23 Treatment time (min): 1, 3 Dose (mJ/cm^2): 4.14.	*E. coli* O157:H7 *S. aureus*	LED irradiation of vegetables showed a decrease in *E. coli* count of 0.30–0.82 log CFU/g and *S. aureus* 0.34–1.37 log CFU/g.	[27]
–	Pork loin chops, beef longissimus lumborum steaks, ground beef ground turkey, beef semimembranosus steaks.	LED type: LED Wavelengths (nm): NA Intensity (lm): 1712 Treatment time (h): 24 Dose (mJ/cm^2): NA.	Aerobic plate count (APC) *Enterobacteriaceae*	APC of Beef longissimus lumborum steaks, ground beef ground turkey increased with treatment time (2.90–5.40 log CFU/g). *Enterobacteriaceae* count in ground turkey displayed a decreased trend (3.44 log CFU/g).	[44]

TABLE 18.2 *(Continued)*

Type of Food System	Food Sample	LED Treatment Conditions	Microorganisms Inactivated	Findings Reported	References
–	Chicken skin and chicken purge	LED type: LED Wavelengths (nm): 405 Intensity (mW/cm^2): 150, 300 Treatment time (min): 10 Dose (J/cm^2): 0, 45, 90, 135, or 180	*E. coli*, *Staphylococcus* spp., *Listeria monocytogenes*	LED treatment at 405 nm reduced the foodborne pathogens to a level of 0.40 log CFU/g in chicken skin and 1.01 log CFU/g in chicken purge.	[42]
–	Smoked salmon	LED type: LED Wavelengths (nm): 460 Intensity (mW/cm^2): 15, 31, 58 Treatment time (min): NA Dose (kJ/cm^2): 2.40	*Listeria monocytogenes*	The smoked salmon exposed to LED had demonstrated a reduction of *L. monocytogenes* to 0.70–1.20 log CFU/g.	[19]
–	Ready-to-eat (RTE) fresh salmon.	LED type: LED Wavelengths (nm): 405 Intensity (mW/cm^2): 16 Treatment time (h): 8 Dose (J/cm^2): 460.80	*Listeria monocytogenes* *Salmonella* spp.	LED illumination of RTE salmon samples had demonstrated a significant reduction in *L. monocytogenes* and *Salmonella* spp. by 0.30 log CFU/cm^2 and 0.50 log CFU/cm^2, respectively.	[29]
–	Skinless chicken breast	LED type: UVC-LED Wavelengths (nm): 250–280 Intensity (mW/cm^2): 2, 4 Treatment time (sec): 60, 180, 300, 600, 900 Dose (mJ/cm^2): 0–3.60.	*Salmonella enterica*	A greater reduction (>3 log CFU/cm^2) in *Salmonella* count was observed in chicken breast after LED illustration for 900s.	[4]

TABLE 18.2 *(Continued)*

Type of Food System	Food Sample	LED Treatment Conditions	Microorganisms Inactivated	Findings Reported	References
–	Beef broth and diced beef	LED type: UV-LED Wavelengths (nm): 255, 265, 285, and their combinations. Intensity (mW/cm^2): NA Treatment time (min): 2, 30, 60 Dose (mJ/cm^2): 0.60–301.30	*Listeria monocytogenes*, *Salmonella typhimurium*, total viable counts (mesophiles and psychrophiles), total *Enterobacteriaceae* counts.	UV-LED treatment presented a significant reduction in all the bacterial counts when treating the beef broth beyond 60 min. While a reduction in *L. monocytogenes*, total viable counts of mesophiles and total, and *Enterobacteriaceae* count was observe during illustration at 285 nm and all combined wavelength after 60 min.	[32]
–	Boiled young sardine	LED type: Violet-blue LED Wavelengths (nm): 405, 412, 421, 455 Intensity (mW/cm^2): 12.48 Treatment time (min): 5, 10, 15, 30 Dose (J/cm^2): 7.5, 15, 22, 45	Viable cell count	The viable cell count after LED treatment of boiled young sardine was found to be decreased from 3.67 log CFU/g to 3.40 log CFU/g.	[40]

*NA: Not available in the study.

storage, the phenolic content increased with the blue, followed by white, red, and green. The non-illuminated control had the lowest levels of all the above compounds [28]. In another study, physiological changes in broccoli sprouts during postharvest were conducted. The sprouts were subjected to blue light, Far-Red LED, and darkness, and fluorescent light illumination were considered controls. Physical and microbiological changes were determined at 5?C for 15 days. It was observed that the Far-Red LED facilitated increased hypocotyl elongation compared to the controls. The blue LED, however, reduced the antioxidant activity by 50% [6]. The effect of LED illumination has also been performed on strawberries to test the inactivation of *Rhizopus stolonifera* and *Botrytis cinerea.* The samples were illuminated for 12 days and were observed to have an antifungal effect of 1.9 log CFU/g and 3.4 log CFU/g on *Botrytis cinerea* and *Rhizopus stolonifera,* respectively. Results also showed a significant increase in the strawberry samples' total phenolic content and anthocyanin content. The illumination, however, did not affect the strawberries' mass, color, and texture. It was obtained from this study that the 405 nm LED was effective against mold spoilage, aiding in preserving the strawberries' quality during refrigeration [7].

18.7 COMMERCIAL LIMITATIONS AND CONSUMER ACCEPTANCE OF LED

The application of LED technology in the food processing industry has several challenges to overcome before developing from a niche to general applications. In the commercialization of LED technology, technical, infrastructure, and market barriers must be addressed. A disconnect between LED designers and end users of food industries is one of the infrastructure barriers. However, displacing the older lighting and post-harvesting technologies is also likely to be challenging. Furthermore, food industries are diversified with their products; this has resulted in the development of specific types of LEDs per the industry requirement. Moreover, comparing different manufacturers' LED effectiveness on food quality and safety is difficult because the LED industry has yet to establish common standards for food applications [41].

The potential of energy efficiency, nonthermal nature, ensuring food safety, and quality of LED technology has been a driver in the development of this emerging technology in the food processing sector. This technology pushes processors to accelerate practical usage in the industry. However, any technology will be sustainable in the industry only when consumers are satisfied with its potential usage. Few studies have rigorously assessed the LED effect on the organoleptic properties and retention of different functional properties of various foods, which are the primary key to consumer acceptability [6, 14, 18, 28]. More importantly, photovoltaic-powered LEDs provide nutritious and microbiologically safe food that is acceptable to consumers. From the perspective of food processors as consumers, this technology offers postharvest applications, maintains sanitation in food handling areas, and treats waste generated in the industry. An in-depth quantitative and instrumental analysis of food quality parameters would be required to make this technology more acceptable for consumers. Hence, clear, comprehensive, and conclusive information about this emerging technology should be provided to consumers through concerned government and non-government bodies. Further, it is also essential that academicians, researchers, scientists, and engineers educate society about this technology by various means. LED technology brings numerous advantages to the food supply chain; furthermore, the development will benefit food processors and consumers.

18.8 SUMMARY

Various studies confirm that LED technology is novel and can be adopted for different food systems (solid and semi-solid), from agricultural products to animal and seafood products, to enhance the shelf life by destroying pathogenic microorganisms. LEDs light has numerous advantages over conventional light sources in terms of high purity, cost-effectiveness, less power consumption, narrow light range, and compact size. The penetration ability, moisture content, size of the food sample, and sample transparency will decide the LED treatment's antimicrobial efficacy. The benefits of LED technology can be enhanced by combining different wavelengths of light because light with varying wavelengths can significantly reduce microbial counts. Hence, further research needs to be carried out to understand the response of microorganisms to different wavelengths, treatment times, and doses to eliminate them entirely for delivering safe food to consumers. Future trends include a combination of LED treatments with other nonthermal food processing techniques such as pulsed light, ultrasound, pulsed electric field, irradiation, etc., to improve the antimicrobial efficacy of the treatment.

KEYWORDS

- **antimicrobial efficacy**
- **disinfection**
- **LED**
- **light energy**
- **pathogenic microbes**
- **shelf-life**
- **solid food**
- **ultrasound**

REFERENCES

1. Aihara, M., Lian, X., Shimohata, T., Uebanso, T., Mawatari, K., Harada, Y., & Takahashi, A. (2014). Vegetable surface sterilization system using UVA light-emitting diodes. *The Journal of Medical Investigation, 61*(3.4), 285–290.
2. Aponiene, K., Paskeviciute, E., Reklaitis, I., & Luksiene, Z. (2015). Reduction of microbial contamination of fruits and vegetables by hypericin-based photosensitization: Comparison with other emerging antimicrobial treatments. *Journal of Food Engineering, 144*, 29–35.
3. Boyles, C. (2018). Visible LED light exposure of spinach during storage: effect on Vitamin C and microbial load. Project Paper. Cornell University, Ithaca, New York. www.ecommons.cornell.edu.
4. Calle, A., Fernandez, M., Montoya, B., Schmidt, M., & Thompson, J. (2021). UV-C LED irradiation reduces *Salmonella* on chicken and food contact surfaces. *Foods, 10*(7), 1459.
5. Casaburi, A., Piombino, P., Nychas, G. J., Villani, F., & Ercolini, D. (2015). Bacterial populations and the volatilome associated to meat spoilage. *Food Microbiology, 45*, 83–102.
6. Castillejo, N., Martínez-Zamora, L., Gómez, P. A., Pennisi, G., Crepaldi, A., Fernández, J. A., & Artés-Hernández, F. (2021). Postharvest LED lighting: Effect of red, blue and far red on quality of minimally processed broccoli sprouts. *Journal of the Science of Food and Agriculture, 101*(1), 44–53.
7. Chong, L., Ghate, V., Zhou, W., & Yuk, H. G. (2022). Developing an LED preservation technology to minimize strawberry quality deterioration during distribution. *Food Chemistry, 366*, 130566.
8. Dave, D., & Ghaly, A. E. (2011). Meat spoilage mechanisms and preservation techniques: a critical review. *American Journal of Agricultural and Biological Sciences, 6*(4), 486–510.

9. de Almeida, A. M. (2017). Poultry and rabbit meat proteomics. In *Proteomics in Food Science* (pp. 215–223). Academic Press.
10. Donnely, C. W. (2004). Growth and survival of microbial pathogens in cheese. In Fox, P. F., McSweeney, P. L., Cogan, T. M., & Guinee, T. P. (Eds.), *Cheese: Chemistry, Physics and Microbiology, Volume 1, General Aspects* (3rd Ed., pp. 541–559). Elsevier Academic Press.
11. Du, L., Prasad, A. J., Gänzle, M., & Roopesh, M. S. (2020). Inactivation of *Salmonella* spp. in wheat flour by 395 nm pulsed Light Emitting Diode (LED) treatment and the related functional and structural changes of gluten. *Food Research International, 127*, 108716.
12. Dutta Gupta, S., & Jatothu, B. (2013). Fundamentals and applications of Light-Emitting Diodes (LEDs) in in vitro plant growth and morphogenesis. *Plant Biotechnology Reports, 7*(3), 211–220.
13. Fan, L., Liu, X., Dong, X., Dong, S., Xiang, Q., & Bai, Y. (2021). Effects of UVC Light-Emitting Diodes on microbial safety and quality attributes of raw tuna fillets. *LWT, 139*, 110553.
14. Finardi, S., Hoffmann, T. G., Schmitz, F. R. W., Bertoli, S. L., Khayrullin, M., Neverova, O., & De Souza, C. K. (2021). Comprehensive study of Light-Emitting Diodes (LEDs) and ultraviolet-LED lights application in food quality and safety. *Journal of Pure and Applied Microbiology, 15*(3), 1125–1135.
15. Ghate, V., Kumar, A., Kim, M. J., Bang, W. S., Zhou, W., & Yuk, H. G. (2017). Effect of 460 nm light emitting diode illumination on survival of *Salmonella* spp. on fresh-cut pineapples at different irradiances and temperatures. *Journal of Food Engineering, 196*, 130–138.
16. Hinds, L. M., Charoux, C. M., Akhter, M., O'Donnell, C. P., & Tiwari, B. K. (2020). Effectiveness of a novel UV light emitting diode-based technology for the microbial inactivation of *Bacillus subtilis* in model food systems. *Food Control, 114*, 106910.
17. Huang, J. Y., Xu, F., & Zhou, W. (2018). Effect of LED irradiation on the ripening and nutritional quality of postharvest banana fruit. *Journal of the Science of Food and Agriculture, 98*(14), 5486–5493.
18. Jiang, Y., Ai, C., Liao, X., Liu, D., & Ding, T. (2020). Effect of Slightly Acidic Electrolyzed Water (SAEW) and ultraviolet light illumination pretreatment on microflora inactivation of coriander. *LWT, 132*, 109898.
19. Josewin, S. W., Ghate, V., Kim, M. J., & Yuk, H. G. (2018). Antibacterial effect of 460 nm light-emitting diode in combination with riboflavin against *Listeria monocytogenes* on smoked salmon. *Food Control, 84*, 354–361.
20. Kim, D. K., & Kang, D. H. (2018). Elevated inactivation efficacy of a pulsed UVC light-emitting diode system for foodborne pathogens on selective media and food surfaces. *Applied and Environmental Microbiology, 84*(20), e01340-18.
21. Kim, M. J., Bang, W. S., & Yuk, H. G. (2017a). 405±5 nm light emitting diode illumination causes photodynamic inactivation of *Salmonella* spp. on fresh-cut papaya without deterioration. *Food Microbiology, 62*, 124–132.
22. Kim, M. J., Tang, C. H., Bang, W. S., & Yuk, H. G. (2017b). Antibacterial effect of 405±5 nm light emitting diode illumination against *Escherichia coli* O157: H7, *Listeria monocytogenes*, and *Salmonella* on the surface of fresh-cut mango and its influence on fruit quality. *International Journal of Food Microbiology, 244*, 82–89.

23. Kim, S. J., Kim, D. K., & Kang, D. H. (2016). Using UVC Light-Emitting Diodes at wavelengths of 266 to 279 nanometers to inactivate foodborne pathogens and pasteurize sliced cheese. *Applied and Environmental Microbiology, 82*(1), 11–17.
24. Kittel, C. (2004). *Introduction to Solid State Physics* (8th ed.). John Willey and Sons Inc, USA.
25. Kong, D., Zhao, W., Ma, Y., Liang, H., & Zhao, X. (2021). Effects of light-emitting diode illumination on the quality of fresh-cut cherry tomatoes during refrigerated storage. *International Journal of Food Science & Technology, 56*(4), 2041–2052.
26. Lacombe, A., Niemira, B. A., Sites, J., Boyd, G., Gurtler, J. B., Tyrell, B., & Fleck, M. (2016). Reduction of bacterial pathogens and potential surrogates on the surface of almonds using high-intensity 405-nanometer light. *Journal of Food Protection, 79*(11), 1840–1845.
27. Lee, J. Y., Yang, S. Y., & Yoon, K. S. (2021). Control measures of pathogenic microorganisms and shelf-life extension of fresh-cut vegetables. *Foods, 10*(3), 655.
28. Lee, Y. J., Ha, J. Y., Oh, J. E., & Cho, M. S. (2014). The effect of LED irradiation on the quality of cabbage stored at a low temperature. *Food Science and Biotechnology, 23*(4), 1087–1093.
29. Li, X., Kim, M. J., & Yuk, H. G. (2018). Influence of 405 nm light-emitting diode illumination on the inactivation of *Listeria monocytogenes* and *Salmonella* spp. on ready-to-eat fresh salmon surface at chilling storage for 8 h and their susceptibility to simulated gastric fluid. *Food Control, 88*, 61–68.
30. Lin, Y., Hu, J., Li, S., Hamzah, S. S., Jiang, H., Zhou, A., & Lin, S. (2019). Curcumin-based photodynamic sterilization for preservation of fresh-cut *Hami melon. Molecules, 24*(13), 2374.
31. Matioli, E., Brinkley, S., Kelchner, K. M., Hu, Y. L., Nakamura, S., DenBaars, S., & Weisbuch, C. (2012). High-brightness polarized light-emitting diodes. *Light: Science & Applications, 1*(8), e22–e22.
32. McSharry, S., Koolman, L., Whyte, P., & Bolton, D. (2022). Inactivation of *Listeria monocytogenes* and *Salmonella typhimurium* in beef broth and on diced beef using an ultraviolet light emitting diode (UV-LED) system. *LWT, 158*, 113150.
33. Mishra, A., Chakraborty, S. K., Tripathi, M. K., & Kotwaliwale, N. (2020). Effect of 390 nm light on surface microbes of cut carrots. *Current Journal of Applied Science and Technology, 39*(44), 52–57.
34. Modi, H. A. (2009). An introduction to microbial spoilage of foods. In Modi, H. A. (Ed.), *Microbial Spoilage of Foods* (pp. 43–66). Aavishkar Publishers.
35. Nair, G. B., & Dhoble, S. J. (2020). *The Fundamentals and Applications of Light-Emitting Diodes.* Woodhead Publishing.
36. Prasad, A., Du, L., Zubair, M., Subedi, S., Ullah, A., & Roopesh, M. S. (2020). Applications of Light-Emitting Diodes (LEDs) in food processing and water treatment. *Food Engineering Reviews, 12*(3), 268–289.
37. Ricciardi, E. F., Pedros-Garrido, S., Papoutsis, K., Lyng, J. G., Conte, A., & Del Nobile, M. A. (2020). Novel technologies for preserving ricotta cheese: Effects of ultraviolet and near-ultraviolet–visible light. *Foods, 9*(5), 580.
38. Rouger, A., Tresse, O., & Zagorec, M. (2017). Bacterial contaminants of poultry meat: Sources, species, and dynamics. *Microorganisms, 5*(3), 50.
39. Schubert, E. F. (2018). *Light-Emitting Diodes*. Cambridge University Press, New York.

40. Shirai, A., Yasutomo, Y. K., & Kanno, Y. (2022). Effects of violet-blue light-emitting diode on controlling bacterial contamination in boiled young sardine. *Biocontrol Science, 27*(1), 9–19.
41. Slocum, A. (2005). A technology assessment of Light Emitting Diode (LED) solid-state lighting for general illumination (No. 2168–2018–8077). Working paper series. U.S Environmental Protection Agency. Washington, DC.
42. Sommers, C., Sheen, S., Scullen, O. J., & Mackay, W. (2017). Inactivation of *Staphylococcus saprophyticus* in chicken meat and purge using thermal processing, high pressure processing, gamma radiation, and ultraviolet light (254 nm). *Food Control, 75*, 78–82.
43. Soro, A. B., Whyte, P., Bolton, D. J., & Tiwari, B. K. (2021). Application of a LED-UV based light technology for decontamination of chicken breast fillets: Impact on microbiota and quality attributes. *LWT, 145*, 111297.
44. Steele, K. S., Weber, M. J., Boyle, E. A. E., Hunt, M. C., Lobaton-Sulabo, A. S., Cundith, C., & Roenbaugh, T. L. (2016). Shelf life of fresh meat products under LED or fluorescent lighting. *Meat Science, 117*, 75–84.

INDEX

D

E

F

I

J

K

L

M

P

U

V